Ecological Modeling for Mitigating Environmental and Climate Shocks

Achieving the UNSDGs

Ecological Modeling for Mitigating Environmental and Climate Shocks

Achieving the UNSDGs

Hock Lye Koh

Sunway University, Malaysia

Su Yean Teh

Universiti Sains Malaysia, Malaysia

World Scientific

NEW JERSEY · LONDON · SINGAPORE · BEIJING · SHANGHAI · HONG KONG · TAIPEI · CHENNAI · TOKYO

Published by

World Scientific Publishing Co. Pte. Ltd.

5 Toh Tuck Link, Singapore 596224

USA office: 27 Warren Street, Suite 401-402, Hackensack, NJ 07601

UK office: 57 Shelton Street, Covent Garden, London WC2H 9HE

Library of Congress Cataloging-in-Publication Data
Names: Koh, Hock Lye, 1948–   author. | Teh, Su Yean, author.
Title: Ecological modeling for mitigating environmental and climate shocks :
    achieving the UNSDGs / Hock Lye Koh, Sunway University, Malaysia;
    Su Yean Teh, Universiti Sains Malaysia, Malaysia.
Description: New Jersey : World Scientific, [2022] | Includes bibliographical references and index.
Identifiers: LCCN 2021009773 | ISBN 9789811236334 (hardcover) |
    ISBN 9789811236341 (ebook for institutions) | ISBN 9789811236358 (ebook for individuals)
Subjects: LCSH: Sustainable Development Goals. | Ecosystem management. |
    Ecosystem management--Mathematical models. | Restoration ecology. |
    Water security. | Climate change mitigation. | Sustainable development.
Classification: LCC QH75.A3 K64 2022 | DDC 333.95--dc23
LC record available at https://lccn.loc.gov/2021009773

British Library Cataloguing-in-Publication Data
A catalogue record for this book is available from the British Library.

For any available supplementary material, please visit
https://www.worldscientific.com/worldscibooks/10.1142/12258#t=suppl

Desk Editors: Aanand Jayaraman/Amanda Yun

Typeset by Stallion Press
Email: enquiries@stallionpress.com

Preface

Lakes, wetlands and coastal regions provide essential services critical to the survival of humans, wildlife and the larger ecosystems. They are constantly threatened by anthropogenic activities, environmental degradation and climate shocks. Half of all lakes in the world are eutrophic, characterized by turbid water with abundance of nutrients and algae. Lake eutrophication undermines global water and food security, potentially leading to social unrest.

Marine resources, particularly mangroves and corals, are vulnerable to coastal developments, including coastal reclamation and human settlements that discharge large quantities of wastes into the seas. Climate change impacts, such as increased salt intrusion and sea level rise, may provide the additional impact to induce regime shifts detrimental to these delicate ecosystems. Warming climate has increased the frequency, duration and intensity of catastrophic coastal disturbances, leading to profound uncertainty in the sustainability of coastal infrastructures and resources essential for human populations.

Numerical simulations coupled with onsite monitoring can improve the state-of-the-art knowledge and technical skills regarding ecosystem vulnerability and resilience to environmental degradation, climate change and anthropogenic disturbances. Such scientific knowledge and technical skills provide valuable insights for mitigating the predicted adverse impacts and for developing a sustainable development strategy, incorporating climate and environmental adaptations.

Reference texts and books on sustainable development viewed from the perspectives of providing solutions via model simulation are rare. This book provides a unique approach to solving sustainable development issues related to human population growth and climate change impact.

This book was written to complement Mission 4.7 by inspiring education for sustainable development programs and action toward achieving the United Nations Sustainable Development Goals (UNSDGs). Chaired by Tan Sri Dr. Jeffrey Cheah, Mission 4.7 is a joint project among UNESCO, the Ban Ki-moon Center for Global Citizens, Columbia University and Sunway University. Mission 4.7 brings together national, regional and global leaders from government, academia, civil society and business to accelerate the implementation of Education for Sustainable Development around the world.

About the Authors

Hock Lye Koh is currently a Senior Fellow and Professor at the Jeffrey Sachs Center on Sustainable Development in Sunway University. He received his B.Sc. from the University of Malaya (UM) in 1970 and M.A. and Ph.D. in mathematics in 1971 and 1976, respectively, from the University of Wisconsin, Madison, USA. He was the recipient of the Oppenheim Prize of UM, Fulbright Scholarship USA and DAAD Fellowship. He served as an Associate Member of the International Centre for Theoretical Physics (ICTP) from 1986 to 1992. Specializing in environmental/ecological system modeling, integrated river management modeling, and tsunami and epidemiological modeling, he has many journal publications in diverse fields. Koh is the editor of two international journals, namely, the *International Journal of Environmental Science and Development* and the *International Journal of Sustainability in Higher Education.*

Su Yean Teh received her B.Sc., M.Sc. and Ph.D. in Mathematical Modeling in 2004, 2005 and 2008, respectively, all from Universiti Sains Malaysia. Her research interests revolve around mathematical modeling with particular focus on computational simulation of real-life problems to provide insights and to suggest possible solutions. In 2006, she was awarded the UNESCO/Keizo Obuchi Research Fellowship to undertake research at the University of Miami, Florida, USA. In 2017, she was awarded the prestigious L'Oréal-UNESCO Malaysia Fellowship for Women in Science. She currently serves as an Associate Editor of *Hydrogeology Journal* and has published numerous articles in various journals.

Contents

Chapter 1

Introduction

1.1 Water Security

Water is a fundamental requirement for the planet's environmental, economic, cultural and social systems. Being a vital element for all healthy ecosystems (UN WATER, 2018), water is essential for planetary health. Freshwater is central to the maintenance of global health and is a critical component of climate systems, human society and adaptation (UNDESA, 2015). However, freshwater is under tremendous threat since its availability and quality are under stress around the world (Rodell *et al.*, 2018). This water stress can directly undermine the United Nations sustainable development goals (UNSDGs) Freshwater constitutes only 35% of the total water in the world, and less than 1% of freshwater is readily usable by humans (WWF, 2010). A major portion of this 1% is shared among 145 nations along common transboundary water basins. Scarce transboundary water has the potential to cause social turbulence and lead to conflict within and among countries. Transboundary water crises and conflicts have many dimensions and are complex and tough to manage in a changing meta-coupled world (Liu, 2017). If the current unsustainable use of water resources continues, there is a growing potential of crises and conflicts around the world, especially in developing countries, over the scarcity of water resources and over its management and consumption practices (Sivakumar, 2011). For example, there have already been conflicts between India and Bangladesh, between Ethiopia and Egypt, and between China and India, over transboundary water resources (Postel, 2014). The crisis in Syria is a good example of how water can be a factor that triggers conflicts (Gleick, 2014). The world freshwater supplies are already being degraded and used unsustainably. Yet, the global demand for freshwater has been increasing consistently at a rate of about 1% per year over the past few decades and will keep increasing in the future (WWAP/UN-Water, 2018). An additional 2.3 billion people will experience severe water stress in many parts of the world by 2050, particularly in North and South Africa as well as in South and Central Asia (Kitamori *et al.*, 2012). The world could face a 40% global water deficit by 2030 in a business-as-usual scenario (WWAP, 2016). This enormous global water deficit will undermine the UNSDGs, as freshwater is vital to all aspects of sustainable development (UNDESA, 2015). Water is the main theme that links the diverse dimensions of the UNSDGs. Water is the primary driver of this book that attempts to demonstrate the intimate interconnectivity between water and the UNSDGs.

1.2   Concepts of Sustainable Development

The concepts embodied in the UNSDGs have a long and interesting history. Sustainable development (or Sustainability) is basically a fair and equitable distribution of wealth and responsibility between the present and future generations with the goal to ensure the perpetual continuity of the human–nature ecosystem. It is intended for guiding human activities and social behaviors toward a secure future, in which the renewability of natural resources is the central keystone. Essentially, sustainability calls for a balance between the three pillars of sustainability (Environment, Society, Economics), sometimes referred to as 3P (People, Planet, Profit) to ensure the security of the future generations. A historical perspective on the evolution of the sustainability concept is beneficial for laying a firm foundation for understanding sustainability and its inherent complexity. The seminal concept of sustainability was first introduced in Germany more than two centuries ago for managing forest resources to achieve maximum sustainable yield (MSY). Little success was achieved by MSY because of the utter simplicity in the MSY management despite the complexity of forest resource sustainability issues.

1.2.1 *Sustainability in a historical perspective*

The seminal concept of sustainability was first introduced in Germany for managing forest resources more than two centuries ago by a German forestry scientist Georg Ludwig Hartig, an Honorary Professor at the University of Berlin. In around 1804–1805, Hartig conceptualized "sustainability" as the guiding principles and practices for maximum utilization of the resources of the forests in such a manner to enable future generations to enjoy the same benefits (Vehkamäki, 2005). In Germany in the early 1800s, rapid exploitation of natural resources, especially forest resources such as wood, was needed for mining, shipyards, construction, manufacturing and household consumption. The excessive exploitation of forests in the manner practiced then, however, resulted in extensive deforestation of large areas and in widespread shortages of wood and forest products (Vehkamäki, 2005). The sustainability of maximum forest exploitation practiced in Germany in the early 1800s was thus in doubt. To overcome this shortfall, sustainable forest management principles and practices were subsequently introduced to promote silviculture for continuous and sustainable utilization of timber (Johnston *et al.*, 1967). These basic concepts, methods and criteria developed for sustainable exploitation of forest resources in the 1800s were subsequently adapted for managing wildlife and fish sustainability (Clark, 1976), based on the concept of MSY, with limited success.

1.2.1.1 *Flaw in maximum sustainable yield*

The fundamental flaw in MSY is that it assumed that surplus fish stocks always existed in nature that could be harvested in perpetuity without harming the fish population. Because of this fundamental flaw, the MSY concept was subsequently replaced by the concept of optimum sustainable yield (OSY). This fundamental shift in sustainability principle from MSY to OSY suggested that sustainability is a constantly evolving and shifting concept prone to imperfect science, to high uncertainty, to miscalculations and to flawed assumptions. The compelling lesson is that good sustainability principles and practices should be based upon holistic models that integrate all important variables and factors into a complete system that incorporates complex, uncertain and multidimensional phenomena. Forest management in the early 1800s in Germany ignored exogenous dimensions consisting of unpredictable variables such as biotic and abiotic environment, timber market fluctuations, technological changes and social mobility. It neglected the three important dimensions consisting of the environment, the economy and the society that had critical impact on a successful forest management strategy. The exclusion of these three interconnected dimensions from the forest management model then had severe implications on forest management. The failure to integrate environment, economics and society into a holistic system led to severely flawed management of forest resources. This flawed one-dimensional forest management model was based solely upon the biological characteristics of the forest stands, notably plant growth rates. The model was totally detached and disintegrated from other important exogenous variables or dimensions that are critical to sustainable forest management. The management model was not holistic, in that it did not integrate the three important dimensions of the environment, the economy and the society into an interconnected and complex system.

1.2.1.2   *Holistic management*

"Holistic" generally refers to the belief that the parts of something are intimately interconnected and explicable only by reference to the whole. Holistic medicine is therefore the treatment of the whole person, considering mental, social and other factors, rather than just the symptoms of a disease. Holistic management of forests must therefore consider the forests as an "integrated holistic system" comprising many interconnected elements such as timber market forces, technological innovations, social impacts, biotic and abiotic environment, weather conditions and tree growth. The simplistic one-dimensional model used in the early 1800s for managing forestry reserves, however, permitted the simplistic calculation of the maximum utilization of the forest stands based purely on either area or volume allocations. Two centuries later, the modern concepts of sustainability, as enshrined in the Brundtland report, were primarily shaped by the seminal concept of Hartig, its failure and its subsequent evolutions. In principle, both the Hartig and the Brundtland principles endorse the fair and equitable distribution of wealth and responsibility between the current and future generations. Both implicitly acknowledge the limited carrying capacity of the planet to support unlimited demands of humans on the natural resources for economic profits. With the goal of creating human wealth, the exploitation of natural resources invariably would lead to some degree of damage to nature and its resources. Managing the intricate balance between human wealth and nature health is the essence of sustainability and the focus of this book.

1.2.2 *1972 Clean Water Act of USA*

Sustainability can also be traced back, arguably, to campaigns for the protection of the natural environment, human society and the planetary ecosystems in the early 1970s. Official campaigns for the protection of the environment and the ecosystems were first initiated by the enactment of the 1972 Clean Water Act of USA and the 1974 Environmental Quality Act of Malaysia. The enactments of these acts were the direct and indirect consequence of the severe mercury contamination that occurred in the marine water around Tokyo, commonly known as Minamata disease. Pollution of wetlands and coastal estuaries by persistent organic pollutants (POPs) such as mercury and PCB will harm wildlife that thrives there and might eventually lead to grave consequences to humans. Mercury contamination in the Minamata Bay around the 1960s that ultimately led to many human casualties served as a painful wakeup call regarding the need to control release of POPs and other pollutants into the marine environments including mangroves and wetlands so as to protect wildlife and humans. The concepts and practices of the modern sustainability (or sustainable development) agenda minimally existed then. The "sustainability" concepts and practices in the 1970s then were primarily confined to the protection of the environments. Environmental education then was dominated by the cognitive goals of acquiring knowledge, but largely ignored the affective goals of cultivating desirable social attitudes and personal behaviors that were favorable to the preservation of nature and its ecosystems.

1.2.3 *The Brundtland Report 1987*

Embracing but going beyond the environment and the ecosystems, the Brundtland Report (WCED, 1987) initiated a transformational process for developing integrated sustainability concepts and best practices. The goals were to holistically embrace the three pillars of sustainability consisting of the environmental, societal and economic aspirations. The ultimate objectives are to promote equitable distribution of benefits and responsibility between the current and future generations. This ongoing transformation entails an educational paradigm shift that requires integrating social-cultural and economic considerations with ethical and esthetical values in ESD. The Brundtland report defines sustainable development as development that meets the needs of the present without compromising the ability of future generations to meet their own needs. Being non-specific and ambiguous in many ways, the 1987 Brundtland Report carries an undeniable implication that sustainable development must deal with complex and uncertain planetary systems and issues. These complex and uncertain planetary systems confronting the current and future generations are viewed and interpreted by a diverse spectrum of societies, beset with vastly different values, beliefs and cultures.

1.3   Contested Concept of Sustainability

It is therefore inevitable that vagueness in the definition of sustainability has rendered sustainability a contested concept, with multiple interpretations and different ways of implementation. It is not surprising that the progress of sustainability has been characterized as sluggish, uneven and negligible (Leal Filho *et al.*, 2015), three decades after the launch of the Brundtland report. This dismal assessment is amply reflected in recent studies that indicated that undergraduate economics education in the UK did not engage with sustainability (Plumridge, 2010). Broad consensus has emerged to show that Economic 101 courses across North American universities are highly standardized because the course textbook is typically selected from a handful of leading contenders (Colander, 2000). These leading standardized Economic 101 textbooks devoted a mere 3.2 percent of their total printed area to environmental externalities, public goods and other sustainability-related topics (Green, 2012). This disconnect and contest within the environment-social-economic nexus is alarming, given that much of the current environmental woes were the consequences of economic drivers that ignored social-environmental dimensions (Dietz *et al.*, 2007). Noting that environment, society and economy form the three-pillar model of sustainability, Hueting (2009) suggested that improved prospects for sustainability would require new economic goals and novel integrative economic models that address social-environmental issues.

The complexity of sustainability was not fully understood nor fully recognized. Further, three additional pillars (cultural, political and spiritual) were suggested for addition to the traditional three-pillar model for sustainable development (Burford *et al.*, 2013). The suggestion of an expended and complex six-pillar model implies further fragmentation and deepened disconnect and contest between the expended six pillars. This fragmentation, contest and disconnect could result in an uncertain road to the UNSDGs. What is broadly agreed upon is the desired outcome of a sustainable future in which the natural resources are renewable. However, the actions and pathways by which specific outcomes are to be achieved are often contested (Carew and Mitchell, 2008). Given the phobia of climate change, some advocated decarbonization as the ideal path to a sustainable future, with limited success anticipated. Others opted for innovative engineering and social solutions as mitigation and adaptation measures (Boyle, 1999). Other contestants put forward a strong case in favor of community consultations in all major decision processes in the sustainability discourse (Clift, 1998). This call for active community engagement is relevant and is consistent with the Maslow Hierarchy of Human Needs (MHHN), in which the emotional need for self-esteem and community-esteem is ranked the second highest tier of human needs. Ignoring this high-level human need would invariably undermine sustainability.

1.3.1 *Sustainability curriculum and pedagogy*

While confusing to some, this contested nature of sustainability is indeed healthy as ambiguity in the Brundtland report encourages deep reflections, promotes social innovations, enables cultural-economic dialogs and enhances active community engagement. These desirable attributes should be highlighted and incorporated into ESD by innovations in Sustainability Curriculum and Pedagogy (SCP). These attributes are crucial in reaching a common goal for maintaining a delicate balance between the three pillars of sustainability. It is generally accepted that sustainability may be divided into three subordinate concepts: environmental, social and economic sustainability. However, there are two competing representations of this subordination, which have deep implications for ESD and Sustainable Development Goals (SDGs). One representation conceptualizes the three sub-systems as concentric and nested circles, in which the environmental system represents the most important and hence the outermost and the largest circle. The second smaller social system is nested within the larger environmental system, with the third and smallest economic system being nested within the social system (Mitchell, 2000). This representation suggests that the economic system is subordinate to both the social and environmental systems, with vast implications for ESD and the associated SCP. In contrast, Clift (1995) and many others chose to interpret the active interaction among the three systems as an intersecting Venn diagram, in which the small core zone of the triple common intersection represents concurrent environmental, social and economic sustainability. These two representations may have implications on the design of SCP in ESD. For example, how a course like Economic 101 should be taught or learned will differ greatly between the two contesting interpretations, with vastly different SCP designs and implementations. This contest of subordination leads to a search for best practices in SCP designs founded on core values of educational institutions and the relationship with society and the environment. This contest also has implications on the content, context and presentation of this book, in achieving the UNSDGs.

1.4   The 17 UNSDGs

The UNSDGs consist of the following seventeen SDGs. These are not stand-alone or independent goals, but rather many of these goals are interconnected and interdependent. For example, SDG 6 (Clean water and sanitation) is highly interdependent on SDG 14 (Healthy life under water) and on SDG 15 (Vibrant life on land). All three SDGs are vulnerable under global climate change (GCC) and sea level rise (SLR) scenarios, and the mitigation and adaptation of these vulnerabilities calls for appropriate and timely climate action SDG 13. SDGs 1 to 4 are obviously interconnected and interdependent.

Quality education SDG 4 has deep implications on the eradication of poverty SDG 1 and on the attainment of good health and wellbeing SDG 3, in addition to ensuring zero hunger SDG2. The aims of this book are directed to the provision of suitable SCP for achieving the goals of ESD, with a focus on the five SDGs (4, 6, 13, 14, 15) mentioned. The central and paramount chain in these interconnected and interdependent SDGs is *water* of high quality in sufficient quantity provided at the right time and at the right place in a sustainable manner.

SDG 1: No Poverty
SDG 2: Zero Hunger
SDG 3: Good Health and Well-being
SDG 4: Quality Education
SDG 5: Gender Equality
SDG 6: Clean Water and Sanitation
SDG 7: Affordable and Clean Energy
SDG 8: Decent Work and Economic Growth
SDG 9: Industry, Innovation, and Infrastructure
SDG 10: Reduced Inequality
SDG 11: Sustainable Cities and Communities
SDG 12: Responsible Consumption and Production
SDG 13: Climate Action
SDG 14: Life Below Water
SDG 15: Life on Land
SDG 16: Peace and Justice Strong Institutions
SDG 17: Partnerships to achieve the SDG

1.5 Water is Life

Water is a critical component of the biosphere, linking land, water bodies, estuaries, oceans and atmosphere to form a complex and complete system, whose intimate interrelationships are just beginning to emerge under GCC initiatives. Failures in the water sector will have severe consequences on the eight SDGs mentioned earlier. Water sustains both life below water (SDG 14) and life on land (SDG 15), supports socioeconomic development (SDG 8) and maintains sustainable cities (SDG 11). Water resource management entails the competence to (1) understand the entire water cycle, (2) assess the impacts of human activities on the water cycle, (3) evaluate the viability of associated resources and ecosystems and (4) develop the strategy to make environmentally responsible decisions and actions based upon sound knowledge and desirable attitudes.

1.5.1   *Integrated water resources management*

Integrated water resources management (IWRM) may, thereby, be defined as the process that promotes the coordinated development and management of water, land and related natural–human resources, in order to maximize the economic and social welfare in an equitable manner without compromising the sustainability of vital ecosystems (Snellen and Schrevel, 2004) and without compromising the well-being of the future generations. The three key roles of SCP in ESD may be summarized in three keywords, denoted by the acronym SSS: skills, social-cultural-coherence and strategy. SCP in ESD must seek to enable the following: (1) the transmission of *skills* and knowledge crucially needed for understanding the intimate relationships between the interconnected and interdependent components in the ecosystems, (2) the cultivation of *social*-cultural coherence through the transformation of attitudes and behavior that lead to a sustainable environment and (3) the *strategic* capability to negotiate and implement IWRM programs. Hence, IWRM seeks to promote the conservation of water resources and aquatic ecosystems by (a) enhancing *skills* and knowledge of water-related problems and their solutions, (b) inducing *social*, attitudinal and behavioral changes toward the goals of IWRM and (c) empowering the *strategic* capacity to deliver the goals of IWRM.

Hence, IWRM education must develop the competence to recognize and conceptualize the intricate connections within this complex aquatic ecosystem. Most students recognize water as an important natural resource. However, they might not have the competence to comprehensively assess the impacts of anthropogenic activities on water resources and the associated aquatic ecosystems. This integrated system thinking involves the capabilities to do the following: (1) distinguish individual elements of the aquatic ecosystem and the intimate relation between the elements, (2) understand the behavior of the entire aquatic ecosystem by analyzing the complex interdependencies of its elements and interconnected relations and (3) derive qualitative and quantitative strategies for sustainable management for the future (Covitt *et al.*, 2009). Global Climate change (GCC) and population growth will exert ever-increasing pressure on the balance between demand and supply of water to society, particularly in highly urbanized areas in emerging economies without adequate resources. Two approaches are generally used to address this mismatch between water demand and supply. The first approach is based upon science and technology to increase water supply, while the second relies on behavioral innovation to reduce water consumption (Clark and Finley, 2008). The technology-based development of new water resources will be constrained by GCC and population growth and should not be relied upon exclusively as a solution in this mismatch, particularly in urbanized future cities. Concerted and innovative efforts must be devoted to changing consumer behavior to reduce water consumption demand. SCP and ESD need to innovate to address this water mismatch and seek long-term sustainable solutions to the water crisis looming over the horizon. A key innovation toward meeting the critical water demand is rainwater harvesting.

1.5.2 *Rainwater harvesting*

Chronic water stress around the world has led to a fundamental shift in the approach to the provision of urban water services in many countries. In future urban areas, rainwater should be considered a valuable resource rather than a risk. The harvesting of urban rainwater to supply non-potable water demands is emerging as a viable option to augment increasingly stressed urban water supply systems. The feasibility of utilizing non-traditional water sources, such as rainwater, for the supply of non-potable urban water demands has been actively investigated. Key research issues are as follows: (a) how reliable is this supply source, (b) how much rainwater can be harvested and (c) how large a storage is required. The three most significant parameters that have an impact on the estimation of volumetric reliability are length and quality of rainfall record, interannual variability of seasonal demand and storage type. In sizing storages, two types of storages are considered: (a) within-year storages that go through the full–empty cycle several times a year, typical of a small household rainwater tank and (b) long-term over-year storages that go through this cycle over a much longer time. Urban rainwater stores are often designed for within-year storages, due to storage capacity often being limited by space constraints. Due to the within-year nature of urban rainwater stores, the within-year inflow variability has the potential to influence storage. The wish to overcome the variability between the within-year and over-year storage designs has aroused the interest in incorporating household rainwater tanks into the larger urban water system throughout the world to improve reliability (Mitchell *et al.*, 2008).

Evolving trends are emerging in which rainwater harvesting (RWH) is being promoted through ambitious regulations and incentives. Local regulations and partial subsidies appear as an effective means of fostering the installation of RWH where citizens acknowledge the value of rainwater. For example, the Metropolitan Area of Barcelona (MAB) has started to promote the use of RWH through specific regulations and incentives and by influencing users' practices and perceptions. Surveys regarding consumers' perceptions, reactions and satisfaction toward RWH systems suggest that both regulations and subsidies are good strategies to advocate and expand RWH technologies in residential areas. However, an integrated and multidimensional learning environment needs to be promoted to ensure a proper use of RWH systems with minimal risk. Skepticism regarding the use of RWH technologies still prevails today, particularly in low precipitation areas. Despite low precipitation and high variability of precipitation, toilet flushing demand of a single-family house can be practically met with a relatively small tank. Rooftop rainwater can also meet more than 60% of the landscape irrigation demand in both single- and multi-family buildings. The main drawback is the long payback period that RWH systems present today (Domènech and Saurí, 2011). RWH systems should be based on local skills, materials and equipment. Harvested rainwater can then be used for rain-fed agriculture or water supply for households. Unfortunately, rainwater might be polluted by bacteria and hazardous chemicals requiring treatment before usage. Slow sand filtration and solar technology are methods to reduce the pollution. Membrane technology would also be a potential disinfection technique for a safe drinking water supply.

1.6   Focus of This Book

Education in modern technically advanced countries has empowered humans with cutting-edge technologies to dominate nature rather than to work with nature. Equipped with the strong beliefs that nature has an infinite carrying capacity to provide services demanded by humans, modern educated persons rarely worry about the limited carrying capacity of human–nature ecosystems. Armed with seemingly unlimited capabilities of modern technologies that are perceived as capable of driving unlimited economic growth, highly educated people bear much of the responsibility for our current environmental crises (Weisser, 2017). Reversing our current environmental crisis begins with ESD, a major focus of this book.

1.6.1 *Education for sustainable development*

Higher education institutes (HEIs) bear heavy responsibility for correcting the wrongs committed in the past and for preventing further wrongs in the future. This responsibility can be partially discharged by providing effective ESD conducted via good SCP, which is a focus of this book. To effectively contribute to the UNSDGs (2016–2030), HEIs need to cultivate core sustainability competences among students and educators critical to ESD. Competence broadly refers to an ability, aptitude and attribute to do something successfully or efficiently, like managing an automobile plant, which can be acquired by self-learning or through formal education. SCP and ESD aim to train professionals with the competences such as technical *skills*, the correct *spiritual*-cultural attitude, and the *strategic* capability to transform society toward sustainability-led lifestyles. Skill generally has a vocational context and has a reference to specific tasks like renovating a house. Generic skills, knowledge and attributes considered desirable to graduates from HEIs have been a subject of much debate (Barrie, 2006; Green *et al.*, 2009), leading to a profusion of definitions of skills and competences. Strategic competence empowers an individual to manage and adapt to complex, constantly changing and uncertain situations, like managing the economy of a nation or the finance of a large transnational corporation. The choice and implementation of the sustainability strategy may be grounded in a sound spiritual or cultural belief in the spirituality of the human–nature nexus. HEIs can play an important role in equipping graduate professionals with "skills", "competences" and "capability" to manage and adapt to complex and uncertain sustainability challenges in the future, while fulfilling the needs of the present generation (Blewitt, 2004). The concept of sustainability as enshrined in the Brundtland Report carries the implication that sustainability should be modeled and practiced in a multi-dimensional, multi-stakeholder and integrative framework in a holistic way. To achieve these SDGs, ESD through good SCP should embrace competence-based learning outcomes to equip students and educators with three core competences, skills, spirituality and strategy, critically needed to successfully manage complex, uncertain and unpredictable sustainability challenges (Koh and Teh, 2019). The SCP in ESD should adopt problem-based, solution-oriented and sustainability-focused projects as case studies, using learner-centered approaches, consistent with the approaches used in this book. This book provides examples and practices that help students develop research skills and competences essential to successfully perform sustainability-led research leading to in-depth findings for actions. Science, Technology, Mathematics and Engineering (collectively, STEM) remain important elements in action-motivated ESD to provide deep learning to drive effective sustainability actions.

1.6.1.1   *STEM*

As commented by Vare and Scott (2007), sustainable development, if it is going to happen, is going to be a learning and evolving process. Hence, ESD should be learning-led, flexible, reflective and adaptive, constantly scoping for both incremental and transformational changes, a process that requires creativity, as there are no ready-made recipes (Wals and Jickling, 2002). The focus of most HEIs, however, appears to have been based on integrating contents relevant to sustainability into the existing curriculum of different subject areas or on developing stand-alone specialist courses on sustainable development (Tilbury, 2011). What appears missing is a commitment to reorienting the learning experiences of students and educators so that they fully understand their professional motivations, responsibilities, competences and capabilities. Also, what appears to be missing is the integration of sustainability principles and practices into core university activities (COPERNICUS Alliance, 2012), which requires policy shifts and institutional incentives. The Paris climate change agreement and other sustainability commitments toward low-carbon economies have created new job opportunities in the cross-disciplinary field of "green jobs". However, most HEIs are behind the learning curve in filling up these new vacancies because their graduates are deficient in job-specific skills in Science, Technology, Engineering and Math (STEM) (Bird and Lawton, 2009). Hence, acquiring competences in STEM would enhance opportunity in green jobs, a principle that is fully endorsed in this book. However, the students must also direct their attention toward addressing the inherent complexity of sustainability issues, beyond the confines of STEM, by adopting a holistic approach to sustainable development.

1.6.2 *Holistic approach in sustainability*

ESD should equip students with the capability to think holistically about the complexities and uncertainties of the future and to act in a responsible manner (Sterling, 2011; Tilbury, 2011). The contested concept of sustainability implicitly considers the human–nature relationship as a complex, convoluted, interconnected and vulnerable ecosystem that connects current and future generations, in ways that are not yet clearly understood. Hence, sustainability must embrace uncertainty and complexity as a guiding principle. The obvious implication is that the actions of the current generations will affect the well-being of the future generations. Nature and its resources may be viewed through three perspectives: (1) nature as Object of scientific study, (2) nature as Resource to be utilized and (3) nature as Spirituality to be revered. The three pillars of sustainability, or the triple bottom line (environment, society and economics), highlight this systemic, dynamic and inherently vulnerable connectivity between humans and nature mediated by environmental constraints, economic forces and social factors. Despite its vagueness, the 1987 Brundtland definition of sustainability has become a central focus around which other more situational and local definitions revolve. In this book, the Brundtland definition of sustainability is interpreted as supportive of the triple bottom line definition and is used holistically to construct the framework of SCP for the three core competences as the key learning outcomes in ESD.

The priority in ESD should be focused on future-oriented systemic thinking, on learning for transformational change and on stakeholder engagement for action (Cotton and Winter, 2010; Tilbury, 1995).

For this to happen, HEI educators should develop the capability of shaping students' learning experiences and of providing the professional development that graduates need to address the sustainability challenge (UE4SD, 2014). ESD should equip staff and students with a unique set of core competences that enable critical future-oriented systemic thinking that is holistic, and with capabilities that empower visionary change leading to transforming learning systems (UNECE, 2012). These core competences would enable critical issues to be addressed, including climate change, water resources, renewable energy, biodiversity, reduction of poverty and economic instability (UNESCO, 2014). Closing the gap between sustainability research and policies is essential (Future Earth, 2013). The three core competences proposed in this book would play a critical role in resolving problems on natural resource depletion, water shortages, mangrove wetland management for carbon control and pollution-related issues at the local and regional level. The challenges ahead are indeed formidable.

1.7   Organization of the Book

This book is organized into eleven chapters as follows.

1.7.1   *Chapter 1: Introduction*

The (UNSDGs, 2015–2030) were adopted in January 2016 (UN, 2015) as a template for sustainable development programs globally. The UNSDGs were established to succeed the United Nations Millennium Development Goals (UNMDGs, 2000–2015). The UNSDGs, officially known as "Transforming our world: the 2030 Agenda for Sustainable Development", consist of a set of 17 goals, 169 constituent targets and 230 indicators. Each of these SDGs targets is designed to meet the criteria for SMART (specific, measurable, ambitious, realistic and time-bound) for them to be useful and for which the achievement can be objectively evaluated (Maxwell *et al.*, 2015). To promote SDGs, UNESCO sponsors the program of ESD, covering a broad range of goals, from poverty reduction to sustainable energy. Recognizing that education is an indispensable prerequisite for achieving SDGs, the United Nations initiated the Decade of Education for Sustainable Development (DESD) 2005–2014. DESD highlights areas of action for education and promotes integrating key SDGs and issues into effective teaching and learning via SCP. These key SDGs include clean water and sanitation (SDG 6), climate change action (SDG 13), poverty elimination (SDG 1) and hunger reduction (SDG 2). ESD contributes to sustainable development by promoting personal, societal, cultural and political transformations via specific cognitive, socio-emotional and behavioral outcomes that empower individuals and societies to deal with the daunting challenges facing SDGs. ESD aims to integrate relevant skills and competences in STEM with other knowledge domains, and develop desirable personal attitudes and social-cultural beliefs to achieve a holistic approach to SDGs. The fundamental principles of the SDGs can be traced to the Brundtland report (WCED, 1987) that proclaimed that sustainable development is "Development that meets the needs of the present without compromising the ability of future generations to meet their own needs". This noble goal is to ensure a future in which our children and grandchildren will look back with deep appreciation for our foresight and actions. This is also the goal of this book, which is designed to help achieve SDGs via effective SCP for ESD.

1.7.2 *Chapter 2: Lake eutrophication and rehabilitation*

The functions and structures of lakes and wetlands are significantly influenced by the depth and temperature. A shallow tropical lake with year-round high temperature responds to nutrient loadings in a manner significantly different from that in a deep temperate lake that is subject to seasonal variations in temperature. High tropical temperatures year-round promote algae growth in the presence of nutrients such as phosphorus (P) and nitrogen (N) to cause eutrophication. Depth allows the lake water to stratify vertically due to temperature variation over the depth. Vertical stratification can buffer the lake water from internal nutrient input out of the lake sediments to reduce the degree of eutrophication. The input of external loadings consisting of P and N into a lake also provides the nutrients that stimulate algae growth and drive the processes of photosynthesis and subsequent decay of algae. Over time, persistently high nutrient inputs from external loadings and internal recycling will lead to a lake condition known as eutrophication. The lake water changes from a desirable clear-water state known as oligotrophic to an undesirable turbid-water state known as eutrophic, dominated by abundant growth of algae. More than 50% of lakes worldwide, including 60% in Malaysia, are contaminated by eutrophication due to excessive nutrient inputs derived from agriculture, industry and domestic waste discharges. Decaying organic matter from algae death ultimately settles onto the bottom sediments to form a pool of nutrient source. The interaction between accumulated sediments and the water column governs the eutrophication of lakes. This chapter presents mathematical and numerical analysis of eutrophication in a shallow tropical lake known as Tasik Harapan, located in the main campus of the Universiti Sains Malaysia. Tasik Harapan is characterized as highly eutrophic, caused by high nutrients particularly phosphorus, with wild fluctuation of dissolved oxygen over the diurnal cycle. Nutrients accumulated in the sediments over the past decades are the primary sources of internal nutrient input. Several methods of lake rehabilitation have not been successful, including the use of mechanical aerators. Water hyacinth is inappropriate as it is very invasive and constant removal of the invasive plants is required to prevent the lake from being overrun. Being an irreversible lake, Tasik Harapan requires dredging and frequent flushing to remove nutrients as the only way to rehabilitate the lake. Model simulations indicate that dredging the sediments to remove nutrients and to increase the depth of the lake appears to be a viable solution to rehabilitate the lake. The added depth and storage volume allow a higher rate of flushing to remove nutrients at a higher rate.

1.7.3   *Chapter 3: Regime shift in eutrophication*

Ecosystems are constantly undergoing shifts between multiple locally stable and unstable states. For example, a lake may shift from a desirable oligotrophic state with clear water to an undesirable eutrophic state with highly turbid water. Such regime shifts or bifurcations are fascinating ecological phenomena, involving multiple causes and many variables that change at different spatial-temporal scales, potentially altering the direction of feedbacks. Regime shifts in lakes may impair valuable ecosystem services provided by nature, for example, clean water supply and ecotourism attractions. Once shifted to a new undesirable eutrophic state, recovery to the original desired oligotrophic states may be difficult and costly. Induced by increasing human disturbance and climate change, significant regime shifts in the ecosystem structures and functions in many lakes have been observed in recent decades worldwide. Knowledge regarding the characterization of regime shifts in lakes, including early warning signals prior to regime shifts, is a critical issue in lake conservation and management. However, identification and prediction of regime shifts in lakes is a formidable task. Lakes may undergo regime shifts between two alternative steady states, the desirable oligotrophic state and the undesirable eutrophic state. The process of regime shift in lakes is governed by interaction between external phosphorus (P) input and internal exchange between water and sediment P. A desirable oligotrophic clear-water state is characterized by abundant macrophytes and low algae growth and low chlorophyll concentrations. On the contrary, an undesirable eutrophic state with turbid-water condition is dominated by high algae growth and high chlorophyll concentrations but sparse macrophytes. A regime shift from a clear-water oligotrophic state to a turbid eutrophic condition may occur in response to a combination of increased exogenous nutrient loadings and a strong feedback involving P release from the sediments. Many large shallow lakes in China such as Dianchi, Chaohu and Taihu, have already shifted from a clear-water oligotrophic state to a highly turbid eutrophic condition, seriously impairing the valuable ecosystem services provided such as freshwater supply and ecotourism.

This chapter performs bifurcation analysis for two distinctively different lakes: (a) a shallow (depth of 1 m) tropical lake Tasik Harapan in the (*Universiti Sains Malaysia*) USM main campus and (b) a deep (depth of 89 m) temperate Lake Fuxian in Kunming, China. Because of their fundamental differences in temperature and depth, the two lakes display distinctively dissimilar lake responses regarding regime shift thresholds and restoration methods. The deep temperate Lake Fuxian is identified as a reversible lake with a higher regime shift threshold of $l_p = 0.0765\,\mu\text{g/L/d}$. Conversely, the shallow tropical Tasik Harapan is an irreversible lake with a lower regime shift threshold of $l_p = 0.01595\,\mu\text{g/L/d}$. A lower value of regime shift threshold for Tasik Harapan implies that eutrophication would occur more easily at a lower threshold value of $l_p = 0.01595\,\mu\text{g/L/d}$. With a higher threshold value of $l_p = 0.0765\,\mu\text{g/L/d}$, the reversible Lake Fuxian is more resilient to eutrophication. Further, eutrophication can be reversed by controlling l_p alone. However, for the highly eutrophic and irreversible Tasik Harapan, bifurcation model analysis suggests that a combination of reduction in l_p with other restoration methods such as flushing and sediment dredging is required to restore Tasik Harapan. The differences in regime shift behavior for Lake Fuxian and TH are attributable to two critical factors, i.e., temperature and depth of the lake.

1.7.4 *Chapter 4: Role of coastal aquifers in water and food security*

Coastal aquifers are important resources for the supply of fresh groundwater. However, GCC and SLR may bring about certain consequences to the coastal environments and ecosystems in many coastal zones in the world. GCC might alter patterns of precipitation, causing significant changes to the hydrological regimes that feed freshwater to the coastal zones. In low-lying coastal areas, the predicted local SLR might increase groundwater salinity, and induce associated regime shift in coastal vegetation to halophytic species at the expense of freshwater species. The predicted GCC and its associated local SLR may therefore impact abiotic and biotic processes and induce fundamental regime shifts in some coastal ecosystem's functions and structures.

Low-lying atoll islanders in the Pacific and Indian Oceans depend heavily on fresh groundwater for their daily domestic use and agriculture. These atolls are highly vulnerable to the loss in fresh groundwater supply that may result from SLR particularly if the island is small in size. Mitigating this threat of diminishing fresh groundwater supply calls for appropriate climate action and adaptation (SDG 13) at both the global and local levels. As local sea level rises under GCC impact, more coastal areas in the atolls will be inundated with saline seawater. This increase in seawater inundation over an atoll results in a reduction in the surface area through which fresh rainfall can infiltrate vertically into the ground, a process known as freshwater recharge. Hence, SLR may reduce freshwater recharge into the ground and cause a reduction in fresh groundwater stored in the aquifer, known as freshwater lens. The consequence is the reduction in freshwater lens thickness and volume. Further, following SLR, more highly saline sea water will seep horizontally into the ground, a process known as saltwater intrusion, which will result in increase in salinity in the soil and in the groundwater. The combined impact of seawater inundation and saltwater intrusion may lead to the reduction of fresh groundwater lens thickness and volume, as well as to the salinization of groundwater. Salinization of groundwater will in turn have severe impacts on water security (SDG 6) and food security (SDG 2) in coastal communities.

This chapter discusses salinization of coastal fresh groundwater caused by several factors such as GCC and SLR. The simulation model known as SUTRA is used to evaluate the role of SLR on the process of soil salinization. For flat and low-elevation coastal zones subject to the threats of SLR, a climate adaptation strategy is needed to enhance water and food security. A viable alternative source of freshwater is rainwater harvesting. As noted, sustainable water and food security along low-elevation atolls is highly vulnerable to soil salinization induced by GCC impacts and SLR. Therefore, sustainable food security is placed at the top of the international agenda because of several compelling factors that may derail the attainment of sustainable food security. A combination of factors can indeed pose severe threats to sustainable food security. These threats include population growth, declining freshwater resources, soil salinization, land degradation, inadequate agricultural infrastructures, plant disease, poor soils, and unfavorable climate. The world demand for cereals is predicted to increase from 1.84 billion tons in 1997 to 2.50 billion tons in 2020. To meet this ever-increasing demand for cereals, we need to resort to sustainable hydrogeology, to achieve climate-resilient agriculture. Resilient agriculture must produce more crops per drop of water from the same area of land used, while reducing the environmental impact and negative externalities. Water is the main limiting factor for crop production; hence, producing more crop per drop of water used is critical. Given the physical limitations of natural resources required for food production, primarily water and land, understanding the tight linkage between food security, water security and land use is paramount. New strategies and effective management options are required in order to address water use, land management and food productivity of agricultural systems. Sustainable use of water and other natural resources at various scales (farm, district, nation, and region) must be maintained to ensure proper utilization and conservation of both surface water and groundwater.

1.7.5   *Chapter 5: Comprehensive everglades restoration plan*

The Florida Everglades has been under threat for decades because of diversion of water to coastal regions. Recognizing the urgent need for restoration of the Everglades ecosystems, the US Congress authorized the CERP in 2000 with a budget of USD 10 billion for implementation over 30 years. Chapter 5 summarizes collaborative research findings on the Everglades that contribute to the goals of CERP. Mangrove forests provide essential ecosystem services to humans and wildlife. They mitigate the adverse impacts of climate change by way of carbon sequestration. Mangroves also help to mitigate the impacts of coastal disturbances such as coastal erosion, tsunamis and extreme storms. Despites their critical importance, mangroves do not catch the attention of policy debate because the stakeholders are diverse and diffuse, and therefore the discourse suffers a lack of focus and devotion. This undesirable situation can be improved by partnership and collaboration (SDG 17) between stakeholders for mangrove conservation by a coordinated integration of social, ecological and economic dimensions of conservation efforts. The ecosystem services provided by mangroves are highlighted to draw attention to the need to quantify the economic valuation provided by mangroves. This economic quantification is necessary for mangrove inclusion in policy decisions beyond mere advocacy. For example, the role of mangroves in protecting coastal areas from coastal erosion, tsunamis and other extreme storm surges such as hurricanes and typhoons should be properly quantified. This quantification will facilitate policy debates leading to rational integrated coastal zone development plans involving mangroves. Chapter 5 deals with a regime shift from hammocks to mangroves in the Florida Everglades under the influence of sea level rise and altered precipitation. Chapter 6 provides an overview on the resilience and vulnerability of mangroves to natural processes and anthropogenic disturbances, including SLR and extreme storms. A primary objective of these two chapters is to promote public and institutional awareness and education regarding the urgent need to preserve and further enhance these ecosystem services. Improving mangrove replanting, restoration and resilience techniques worldwide is critical in mangrove conservation. Further research is needed to address several knowledge gaps. The ecological functions and services provided by mangroves are undoubtedly valuable. Conservation and protection of coastal wetlands and mangrove forests must be properly managed and integrated into coastal zone management planning, taking into consideration the impact of GCC and SLR.

1.7.5.1 *Role of GCC and SLR on vegetation shift*

Model simulations performed by Sternberg *et al.* (2007) indicate that (1) hammock and mangrove species aggregate into clusters rather than being dispersed randomly along a topographic gradient, (2) mangroves grow in a large range of salinities, whereas hammocks are severely restricted by their salinity intolerance and (3) micro-topography has a large effect on the mangrove and hammock distribution. Changes in the average precipitation or alteration in salinization of the underlying saline water table will affect the distribution of hammocks and mangroves. Areas occupied by hammock species have vadose pore water salinities of 5 ppt or less throughout the year. In contrast, the surrounding areas dominated by mangroves have vadose pore water salinity fluctuating between 10 ppt during the wet season and 20 ppt during the dry season. How GCC and SLR may alter the vadose pore water salinity and exert their impact on the vegetation regime shift between mangroves and hammocks in southern Florida is the main focus of Chapter 5.

1.7.5.2 *Early detection of regime shift*

Coastal vegetation community structures may be altered by salinization of soil induced by GCC, associated local SLR and reduced precipitation. In many important coastal habitats such as the Florida Everglades, a combination of soil salinization, reduced precipitation and increased storm surges may induce a regime shift from salinity-intolerant glycophytic vegetation to salinity-tolerant halophytic species. Early detection of regime shift signals and identification of thresholds in coastal vegetation are crucial in coordinated and integrated conservation management. It has been shown that the $\delta^{18}O$ value of water in the xylem of trees can be used as a reliable surrogate for salinity in the rooting zone of plants. Coupling measured $\delta^{18}O$ values in the tree xylem with simulated $\delta^{18}O$ values in trees and salinity in the vadose zone can be used to investigate competition outcome between glycophytic and halophytic trees. MANTRA-O18 numerical model simulations suggest that the impacts of soil salin-ization on reducing the resilience of salinity-intolerant trees can be detected 25 years ahead of actual occurrence of the regime shift (Teh *et al.*, 2019). This early detection provides critical lead time and valu-able information and insights useful for planning adaptation strategy to mitigate against the adverse impacts of GCC and local SLR.

1.7.6   *Chapter 6: Mangrove current status and future trends*

Mangroves are vulnerable and resilient to a variety of coastal dis-turbances, both natural and anthropogenic. Their vulnerability and resilience depend on the intensity, duration and frequency of the dis-turbances. Despite their ecological and economic values, around half of total global mangrove coverage has been lost since pre-industrial times (Giri *et al.*, 2011). Mangrove forests are currently disappear-ing at an alarming annual rate of 1 to 2% globally. Found mainly in the intertidal zones of coastal tropical and subtropical regions of the world, mangroves have evolved to be able to tolerate a wide range of salinities, from 1 part per thousand in low-salinity fresh-water habitats to hypersaline seas with salinity exceeding 100 parts per thousand. Although they can thrive under lower salinity condi-tions, their abundance among inter-tidal plant communities is lim-ited by competition with other species that are better adapted to the low-salinity terrestrial environment. As mangroves are physiolog-ically a tropics-adapted group, their poleward expansion is curtailed by minimum temperature requirements and is severely limited by frost frequency and severity (Twilley, 1998; Saintilan *et al.*, 2014). However, under GCC scenarios, mangroves might exploit the warm-ing climate to migrate poleward, and expand their habitats. Over the past few decades, mangrove coverage worldwide was reported to have progressively been reduced by destructive natural processes and damaging anthropogenic activities. Initially at 18.1 million ha as reported by Spalding *et al.* (1997), mangrove coverage has been reduced to 13.7 million ha according to Giri *et al.* (2011), and then to 8.3 million ha according to Hamilton and Casey (2016). Mangrove resilience has enabled their distribution to cover over 118 countries, as described in detail by Tomlinson (2016). Mangrove forests provide critical ecosystem services worldwide to both human populations and the ecosystems they share. However, anthropogenic activities have led to more than 50% losses of mangrove habitats in some parts of the world. Recognizing the importance of mangroves and being cognizant of the threats to their persistence, actions have been taken by national governments, coordinated through international agreements for the protection of mangroves. It is important to thoroughly explore the status of mangrove forests and seek the necessary efforts to protect them. These efforts entail examining the threats to their persistence and the potential solutions for their effective conservation. Case studies from disparate regions of the world will be presented in this chapter to illustrate that the integration of human livelihood needs with mangrove conservation goals can present viable solutions that could lead to long-term sustainability of mangrove forests throughout the world. Identifying and implementing suitable approaches to the conservation and restoration of mangrove habitats is a daunting task. We document efforts for the conservation and restoration of mangroves across the globe with the hope of accelerating the steep learning curve and motivating action crucial to protection efforts locally and globally. We provide insights gained from research and community efforts in many countries around the world on various successes and failures in mangrove conservation and restoration. These insights can provide a platform to identify successful solutions and to avoid potential failures. Such learning and action may secure the future of mangrove forests and the ecosystem services they provide worldwide.

Many mangroves have been severely degraded by excessive human exploitation and rapid urbanization. One of the most well-known wetland restoration projects is the 30-year federally funded project in the US state of Florida, known as the CERP. The source of the most serious threats to mangroves is increasing human population density. Human populations and urban areas are concentrated on coastlines, displacing native vegetation such as mangroves. Conversion of mangrove habitat to agriculture and aquaculture is a major factor in mangrove loss. Other direct effects are mining and the overexploitation of mangrove timber, including clear-cutting. Indirect effects include factors such as changes in freshwater or tidal flow, pollution from oil exploration, and runoff from solid waste. Disease and pest impacts may be furthered by some of these human-related activities (Kathiresan, 2002). Hurricanes and cyclones cause periodic heavy damage (Doyle *et al.*, 1995). Although these storms are natural, cumulative impacts can lead to habitat change (Smith *et al.*, 2009). Tsunamis and their associated earthquakes have also caused long-term change (Roy and Krishnan, 2005). Tsunami and hurricanes are infrequent disturbances with high intensity and short duration. Mangrove deforestation and SLR, on the contrary, are chronic events that evolve gradually over a longer period. A relentless government–community campaign is needed to cultivate awareness and education regarding the vulnerability of mangroves and their invaluable ecological services. Both soft and hard approaches are needed to reduce the adverse impacts of coastal disturbances on humans and mangroves.

1.7.7   *Chapter 7: Tsunami: Sendai disaster risk reduction*

Catastrophic natural disasters such as earthquakes, tsunamis, hurricanes, typhoons and pandemics have the tendency to cause tremendous damage and suffering to affected local communities. They can cause disruptions of employment, property damage and loss of lives, and create poverty and food insecurity. These disruptions can negatively impact several SDG goals such as reduction of poverty SDG 1, elimination of hunger SDG 2 (hunger) and meaningful employment SDG 8. Over the last decade, natural disasters such as earthquakes, tsunamis and storm surges have inflicted severe damage and loss to over 1.5 billion people across the world. The damages have been estimated to have exceeded US$1.5 trillion (Hemingway and Gunawan, 2018). To mitigate the risk of such natural disasters, it is critical to develop Disaster Risk Reduction and Resilience (DRRR) capability among affected communities. For this purpose, the United Nations has formulated a comprehensive and integrated disaster risk reduction framework to coordinate efforts among nations. The framework also provides working guidelines for nations and local communities to mitigate the impacts of disasters. The first framework is the Hyogo Framework for Action 2005–2015 (HFA), with the Title: Building the Resilience of Nations and Communities to Disasters (UN, 2005). This framework has been subsequently updated by the Sendai Framework for Disaster Risk Reduction: 2015–2030 (SFDRR) (UNISDR, 2015). The HFA and SFDRR call for an integrated and holistic approach to DRRR, by incorporating multisectoral and transdisciplinary international collaboration and partnerships (ICPs) across borders. DRRR entails the capability to understand hazard, to identify vulnerability, to anticipate risk, to mitigate the impact and to bounce back rapidly through survival and adaptation. Catastrophic disasters such as massive tsunami inundations, large hurricanes and severe storm surges would invariably result in local poverty, create disruptions in food supply and cause unemployment among communities affected. Hence, to be effective in addressing issues affecting local communities, DRRR should formulate policy and programs that contribute to the UNSDGs, such as poverty elimination (SDG 1), hunger reduction (SDG 2) and meaningful employment (SDG 8) for local communities before and after the occurrence of disasters. As it is impossible to address DRRR across all possible disasters, it is inevitable that we focus on one disaster at a time. Hence, disaster in this chapter refers primarily to large tsunami inundations that can inflict substantial loss of life and damage to properties.

A tsunami is a series of waves travelling at high speeds, generated by an abrupt displacement of large volumes of water in the ocean. Tsunamis may be caused by a submarine earthquake or landslide, or by a volcanic eruption or nuclear explosion in the deep ocean, or by meteorite impact in the ocean. Characterized as a catastrophic disaster of epic proportions, the 26 December 2004 Indian Ocean Tsunami, hereafter referred to as the Andaman Cross-boundary Tsunami (ACT), served as a rude wake-up call. This disastrous ACT killed about 250,000 people worldwide, including 68 deaths in Malaysia. Since then, a series of eight destructive tsunamis has occurred. The catastrophic Fukushima Triple Disaster on 11 March 2011 killed more than 20,000 people and caused major radiation leaks that have lingered until now. The next mega tsunami is long overdue, according to recent research (Sieh *et al.*, 2015). Chapter 8 focuses on addressing tsunami DRRR programs in Malaysia. The ACT has demonstrated the vulnerability of Malaysia to tsunami risk originating from the Andaman Ocean (Koh *et al.*, 2009). Malaysia, in coordination with countries in the region, has developed DRRR programs after the ACT. DRRR research and programs must identify beaches that harbor dangerous tsunami hazards, vulnerabilities and risks. The community living in tsunami hazard zones must be warned against such hazards and risks and make advance preparations before the occurrence.

1.7.8   *Chapter 8: Integrated coastal zone management*

Situated at the confluence between terrestrial and ocean systems, coastal zones are complex systems that are vulnerable due to the combined interactions of these systems. Coastal zones are highly resourceful in providing valuable ecosystem services to humans and wildlife and are immensely critical in supporting vibrant economic development in many countries. Their health and resilience are essential for the continuing prosperity of a significant proportion of human populations. However, they are vulnerable to various anthropogenic hazards such as coastal reclamation and sewage discharge, environmental hazards such as storm surges, and future climatic hazards such as sea level rise. Together, these coastal hazards may disrupt the delicate balance between economic development and ecosystem resilience. Research and management of these complex and vulnerable coastal zone systems subject to these multiple hazards is a daunting challenge, and is underpinned by knowledge and technical skills in natural sciences, engineering and social sciences. Hence, effective coastal zone management for sustainable utilization of coastal resources is premised on integrating the strengths of both sciences and engineering, and on integrating the engagement of multiple stakeholders such as communities, scientists, non-government organizations and government agencies. In this context, an Integrated Coastal Zone Management (ICZM) approach would empower the sustainable utilization of coastal resources. Global climate challenges and sea level rise implications would compound the complexity of ICZM by the requirement of integrating climate action planning (SDG 13) into ICZM best practices, particularly for wetlands because of their heightened sensitivity to climate factors.

An essential component of coastal zones, coastal wetlands play an important role in regulating climate, improving water quality and maintaining biodiversity. Physical hazards and ecosystem risks associated with or resulting from reclamation activities at the coastal wetlands remain a highly controversial topic worldwide. These hazards and risks might translate into permanent damage to coastal ecosystems and into increased exposure to social-economic risks. Reclamation projects in coastal zones can induce a series of negative impacts on the marine ecological environment, both short term and long term. In particular, coastal reclamation of salt marshes, tidal flats, mangroves and wetlands represents a major risk to the sustainability of the coastal environment and marine ecosystems. Coastal reclamations in wetlands reduce critical wetland areas and cause coastal fragmentation leading to loss of biodiversity. Following reclamation in environmentally sensitive wetlands, destruction of habitats for fish and feeding grounds for shorebirds would lead to decline in bird species and fishery resources. The reclamation around Tekong Island, by the Singapore government, at the confluence between the Johor Straits and the Johor River constitutes such a case. For this Tekong reclamation, it is essential to evaluate the impact, plan subsequent ecological restoration and quantify damage compensation. This highly controversial reclamation project was referred to the International Tribunal for the Laws of the Sea for arbitration, which is highlighted in this chapter to lay the foundation for subsequent deliberations.

Challenges confronting coastal reclamation abound worldwide, particularly in China and other parts of Asia. The experiences and lessons learned from coastal reclamations in several Asian coastal cities constitute the discourses on the merits and demerits of coastal reclamations. These discourses lead to doubts regarding the future of coastal reclamations in Asia and the challenges to be encountered. Of particular concern is the land subsidence in recent coastal reclamations. Land subsidence is widely regarded as a major characteristic of recently reclaimed coastal land and is a major challenge for coastal reclamations in Asia. To mitigate the various adverse impacts of reclamation, China has attached great importance to strengthening reclamation management and to protecting the marine environment. Much newly reclaimed land remains idle, or vegetated as coastal wetlands, with Shanghai and Singapore leading the pact contributing substantial proportions of idle land (58% and 29% respectively) in their newly reclaimed land. Many lessons can be learned from the insights and experience gained from countries such as Singapore, Korea and China that have been active in coastal reclamation in the past few decades.

1.7.9 Chapter 9: Oil and gas offshore disposal of drilling wastes

Offshore oil and gas exploration in deep water involves costly activities such as well drilling, well completion and installation of oil flow pipes as well as subsea facilities. Strained by current low prices of crude oil, cost rationalization is essential for economic viability of oil and gas exploration offshore. Well drilling and pipe laying generate drill cuttings that may be disposed either onshore or offshore. Onshore disposal is not cost-effective and operationally complicated and should be the option of last resort. More cost-effective offshore disposal of drill cuttings should therefore be utilized if it can be proven to be environmentally sustainable to the marine environment and ecosystems. A critical concern in cuttings offshore disposal is the suspended solids that might pose adverse impacts on marine ecosystems, particularly corals. Model simulation results indicate that this offshore disposal of drill cuttings will not harm corals living in offshore Malaysia at the modest rates of well drilling in Malaysia. The drill cutting impact simulation performed indicated that a precautionary rule of thumb for impact on corals at Kimanis off the coast of Sabah, with a depth of 100 m, is that the impact would be negligible to the corals if the distance between the drill site and coral sites is at least 100 m for drill cuttings of volume 500 m³ or a mass of 1.2 million kg or less. The maximum thickness of sediment deposited on the corals is 3.3 mm. The current threshold level for sediment thickness adopted for the Norwegian Continental Shelf in environmental risk assessment models by the offshore industry is 6.3 mm thickness (Smit *et al.*, 2008). The thresholds that must not be exceeded for sediment thickness and sedimentation rates to ensure coral safety are not quantified in most guidelines. Generally, corals are susceptible to high levels of sediment cover (more than 6.3 mm) over extended periods (weeks). However, many species are able to overcome exposure to low and medium levels of sediment cover over several days. Cuttings that settle onto the coral would lead to oxygen deficiency or depletion on the coral surface, depending on the thickness deposited and the duration of exposure, as well as on the sea current flow conditions that help to flush off sediment from the coral beds.

1.7.10   *Chapter 10: Governance for water security*

Governance for water security in an era of climate action, amid an environment of ever-growing population density in cities, is the focus of this chapter. Climate change, sea level rise and population growth will exert more pressure on the balance between demand and supply of water resources to densely populated cities. Transboundary conflicts over water issues should be resolved amicably through international laws and partnerships. The United Nations Convention on the Law of the Sea (UNCLOS) and the Coral Triangle Initiative (CTI) are two excellent examples. The dispute between Singapore and Malaysia over coastal reclamations by Singapore around the Tekong Island was amicably resolved through arbitration by the International Tribunal for the Law of the Sea (ITLOS), created by the mandate of UNCLOS. The CTI partnership was created by the six member countries (known as CT6), comprising Indonesia, Malaysia, the Philippines, Papua New Guinea, Solomon Islands, and Timor-Leste. The CTI aims to resolve daunting challenges confronting the six countries regarding sustainable exploitation of the coastal and marine resources. These critical resources are shared by 350 million, most of who live around vulnerable coastal zones fringing the coral triangle (CT). Severe land subsidence has occurred across many Asian cities, due to overexploitation of groundwater to meet the demand in cities, one of which is Jakarta. As a consequence, the government of Indonesia has announced a plan to relocate the capital Jakarta to the island of Borneo. The new future Jakarta capital city must be rebuilt to be sustainable (SDG 11), in the sense of environmental, social and economic viability over the long run. The new city must be resilient to internal and external risks (SDG 9), resilient to climate impacts (SDG 13), resilient to the hazards of poverty (SDG 1), resilient to health concerns (SDG 3), resilient to water shortage (SDG 6) and resilient to food insecurity (SDG 2). Good governance is essential to ensure water security and food security. Protection and conservation of vital coastal resources such as coastal wetlands, mangroves, seagrasses and coral reefs is critical. The preservation of the integrity of these ecosystems and their services is crucial to the coastal communities' survival. In the context of climate change, the task ahead is challenging, given the vast uncertainty and complexity in understanding the drivers, their impacts and responses. For countries that share borders with numerous neighboring countries, such as the CT6, partnership for the goals (SDG 17) is essential to reap the benefits of the synergy of collaboration and to attain peace and inclusiveness (SDG 16). Good governance over transboundary water is crucial to preserving life under water (SDG 14) and to sustaining decent work and economic growth (SDG 8).

1.7.11   *Chapter 11: Concluding remarks*

The 1987 Brundtland Report (WCED, 1987) initiated a transformational process for developing integrated sustainability concepts and best practices to embrace the three pillars of sustainability (environment, society and economy). The goal was to promote an equitable distribution of benefits and responsibility between the current and future generations. This ongoing transformation stimulates an educational paradigm shift toward integrating socio-cultural and economic dimensions with ethical and esthetical values in ESD. ESD plays two important roles in achieving the UNSDGs. First, ESD bridges the gaps and disconnect among the three sustainability pillars. Second, ESD trains future professionals and citizens to manage and adapt to complex sustainability challenges. Sustainability-focused universities should have the capacity and the mandate to develop curricula and pedagogy that are problem-driven and solution-oriented. These universities should devise curricula and pedagogy that are effective in developing holistic, integrated and use-inspired knowledge.

References

Barrie, S.C. (2006). Understanding what we mean by the generic attributes of graduates. *Higher Education*, 51(2), 215–241.

Bird, J. and Lawton, K. (2009). The Future's green: Jobs and the UK low carbon transition. Available at: https://www.ippr.org/research/publications/the-futures-green-jobs-and-in-the-uk-low-carbon-transition (accessed 10 January 2019).

Blewitt, J. (2004). Introduction. In: Blewitt, J. and Cullingford, C. (Eds.), The Sustainability Curriculum: The Challenge for Higher Education, Earthscan, Oxon, pp. 1–1.

Boyle, C. (1999). Education, sustainability and cleaner production. *Journal of Cleaner Production*, 7, 83–87, doi: 10.1016/S0959-6526(98)00045-6.

Burford, G., Hoover, E. and Velasco, I. *et al.* (2013). Bringing the "Missing Pillar" into sustainable development goals: Towards intersubjective values-based indicators. *Sustainability*, 5, 3035–3059.

Carew, A.L. and Mitchell, C.A. (2008). Teaching sustainability as a contested concept: Capitalizing on variation in engineering educators' conceptions of environmental, social and economic sustainability. *Journal of Cleaner Production*, 16(1), 105–115, doi: 10.1016/j.jclepro.2006.11.004.

Clark, C. (1976). Mathematical bioeconomics. The optimal management of renewable resources. John Wiley & Sons, New York, London, Sydney, Toronto.

Clark, W.A. and Finley, J.C. (2008). Household water conservation challenges in Blagoevgrad, Bulgaria: A descriptive study. *Water International*, 33(2), 175–188, doi: 10.1080/02508060802023264.

Clift, R. (1995). The challenge for manufacturing. In J. McQuaid J (Ed.), Engineering for sustainable development. London: The Royal Academy of Engineering.

Clift, R. (1998). Engineering for the environment: The new model engineer and her role. *Transactions of the Institution for Chemical Engineering*, 76(B), 151–160, doi: 10.1205/095758298529443.

Colander, D. (2000). *The lost art of economics*. Edward Elgar, Aldershot, England.

COPERNICUS Alliance (2012). The Rio_20 treaty on higher education. Available at: www.copernicusalliance.org/about-ca/rio-20-treaty (accessed 20 December 2016).

Cotton, D.R.E. and Winter, J. (2010). It's not just bits of paper and light bulbs: A review of sustainability pedagogies and their potential use in higher education. In: Jones, P., Selby, D. and Sterling, S. (Eds.), *Sustainability Education: Perspectives and Practice Across Higher Education*, Earthscan, London, pp. 39–54.

Covitt, B., Gunckel, K.L. and Anderson, C.W. (2009). Students' developing understanding of water in environmental systems. *The Journal of Environmental Education*, 40(3), 37–51.

Dietz, T., Rosa, E.A. and York, R. (2007). Driving the human ecological footprint. *Frontiers in Ecology and the Environment*, 5(1), 13–18.

Domènech, L. and Saurí, D. (2011). A comparative appraisal of the use of rainwater harvesting in single and multi-family buildings of the Metropolitan Area of Barcelona (Spain): Social experience, drinking water savings and economic costs. *Journal of Cleaner Production*, 19, 598–608.

Doyle, T.W., Smith III, T.J. and. Robblee, M.B. (1995). Wind damage effects of Hurricane Andrew on mangrove communities along the southwest coast of Florida, USA. *Journal of Coastal Research*, 21, 159–168.

Future Earth (2013). Future earth initial design: Report of the transition team. Available at: www.icsu.org/future-earth/media-centre/relevant_publications/futureearth-initial-design-report (accessed 17 January 2014).

Giri, C., Ochieng, E., Tieszen, L.L. *et al.* (2011). Status and distribution of mangrove forests of the world using earth observation satellite data.

Global Ecology and Biogeography, 20, 154–159, doi: 10.1111/j.1466-8238.2010.00584.x.

Gleick, P.H. (2014). Water, drought, climate change, and conflict in Syria. *Weather, Climate, and Society*, 6(3), 331–340.

Green, T.L. (2012). Introductory economics textbooks: What do they teach about sustainability? *International Journal of Pluralism and Economics Education*, 4(3), 189–223.

Green, W., Hammer, S. and Star, C. (2009). Facing up to the challenge: Why is it so hard to develop graduate attributes? *Higher Education Research & Development*, 28(1), 17–29.

Hamilton, S.E. and Casey, D. (2016). Creation of a high spatio-temporal resolution global database of continuous mangrove forest cover for the 21st century. *Global Ecology and Biogeography*, 25, 729–738, doi: 10.1111/geb.12449.

Hemingway, R. and Gunawan, O. (2018). The natural hazards partnership: A public-sector collaboration across the UK for natural hazard disaster risk reduction. *International Journal of Disaster Risk Reduction*, 27, 499–511, doi: 10.1016/j.ijdrr.2017.11.014.

Hueting, R. (2009). Why environmental sustainability can most probably not be attained with growing production. *Journal of Cleaner Production*, 18(6), 525–530.

Johnston, D., Grayson, A. and Bradley, R. (1967). *Forest Planning, Faber and Faber Limited*, London.

Kathiresan, K. (2002). Why are mangroves degrading? *Current Science*, 83, 1246–1249.

Kitamori, K., Manders, T., Dellink, R. and Tabeau, A.A. (2012). OECD environmental outlook to 2050: The consequences of inaction. Available at: https://www.oecd.org/g20/topics/energy-environment-green-growth/oecdenvironmentaloutlookto2050theconsequencesofinaction.htm.

Koh, H.L. and Teh, S.Y. (2019). Disaster risk reduction and resilience through partnership and collaboration. In: Leal Filho, W., Azul, A., Brandli, L., Özuyar, P., Wall, T. (Eds.) *Partnerships for the Goals. Encyclopedia of the UN Sustainable Development Goals.* Springer, Cham, doi: 10.1007/978-3-319-71067-9_49-1.

Koh, H.L., Teh, S.Y., Liu, P.L.-F., Md Ismail, A.I. and Lee, H.L. (2009). Simulation of Andaman 2004 tsunami for assessing impact on Malaysia. *Journal of Asian Earth Sciences*, 36, 74–83, doi: 10.1016/j.jseaes.2008.09.008.

Leal Filho, W., Manolas, E. and Pace, P. (2015). The future we want. *International Journal of Sustainability in Higher Education*, 16(1), 112–129.

Liu, J. (2017). Integration across a meta-coupled world. *Ecology and Society*, 22(4), 29.

Maxwell, S.L., Milner-Gulland, E.J., Jones, J.P.G. *et al.* (2015). Being smart about SMART environmental targets. *Science*, 347(6226), 1075–1076.

Mitchell, C. (2000). Integrating sustainability in chemical engineering practice and education: Concentricity and its consequences. *Transactions of the Institution for Chemical Engineering*, 78(B), 237–242, doi: 10.1205/095758200530754.

Mitchell, V.G., Siriwardene, N., Duncan, H. and Rahilly, M. (2008). Investigating the Impact of Temporal and Spatial Lumping on Rainwater Tank System Modelling. International Conference on Water Resources and Environment Research, Modbury, SA, Engineers Australia, Causal Productions.

Plumridge, A. (2010). Costing the earth: The economics of sustainability in the curriculum. In: Jones, P., Selby, D. and Sterling, S. (Eds.), *Sustainability Education: Perspectives and Practice Across Higher Education*, Earthscan, London, pp. 273–293.

Postel, S. (2014). The last oasis: Facing water scarcity. Routledge, London.

Rodell, M., Famiglietti, J.S., Wiese, D.N., Reager, J.T., Beaudoing, H.K., Landerer, F.W. and Lo, M.H. (2018). Emerging trends in global freshwater availability. *Nature*, 557, 651–659.

Roy, S.D. and Krishnan, P. (2005). Mangrove stands of Andamans vis-à-vis tsunami. *Current Science*, 89, 1800–1804.

Saintilan, N., Wilson, N., Rogers, K., Rajkaran, A. and Krauss, K. W. (2014). Mangrove expansion and salt marsh decline at mangrove poleward limits. *Global Change Biology*, 20(1), 147–157.

Sieh, K., Daly, P., Mckinnon, E.E., Pilarczyk, J.E., Chiang, H.-W., Horton, B., Rubin, C.M., Shen, C.-C., Ismail, N., Vane, C.H. and Feener, R.M. (2015). Penultimate predecessors of the 2004 Indian Ocean tsunami in Aceh, Sumatra: Stratigraphic, archaeological, and historical evidence. *Journal of Geophysical Research*, 120, 308–325, doi: 10.1002/2014JB011538.

Sivakumar, B. (2011). Water crisis: From conflict to cooperation-an overview. *Hydrological Sciences Journal*, 56(4), 531–552.

Smit, M.G.D., Holthaus, K.I.E., Trannum, H.C., Neff, J.M., Kjeilen-Eilertsen, G., Jak, R.G., Singsaas, I., Huijbregts, M.A.J. and Hendriks, A.J. (2008). Species sensitivity distributions for suspended clays, sediment burial, and grain size change in the marine environment. *Environmental Toxicology and Chemistry*, 27, 1006–1012.

Smith, T.J. III, Anderson, G.H., Balentine, K., Tiling, G., Ward, G.A. and Whelan, K.R.T. (2009). Cumulative impacts of hurricanes on

Florida mangrove ecosystems: Sediment deposition, storm surges, and vegetation. *Wetlands*, 29, 24–34.

Snellen, W.B. and Schrevel, A. (2004). IWRM: For sustainable use of water 50 years of international experience with the concept of integrated water. Wageningen, the Netherlands: Ministry of Agriculture, Nature and Food Quality.

Spalding, M., Blasco, F. and Field, C. (1997). World mangrove atlas. *The International Society for Mangrove Ecosystems*, Okinawa, Japan.

Sterling, S. (2011). Transformative learning and sustainability: Sketching the conceptual ground. *Learning and Teaching in Higher Education*, 5, 17–33.

Sternberg, L., Teh, S.Y., Ewe, S., Miralles-Wilhelm, F. and DeAngelis, D. (2007). Competition between Hardwood Hammocks and Mangroves. *Ecosystems*, 10(4), 648–660.

Teh, S.Y., Koh, H.L., DeAngelis, D.L., Voss, C.I. and Sternberg, L. (2019). Modeling $\delta^{18}O$ as an early indicator of regime shift arising from salinity stress in coastal vegetation. *Hydrogeology Journal*, 27(4), 1257–1276, doi: 10.1007/s10040-019-01930-3.

Tilbury, D. (1995). Environmental education for sustainability: Defining the new focus of environmental education in the 1990s. *Environmental Education Research*, 1(2), 195–212.

Tilbury, D. (2011). Education for sustainable development: An expert review of processes and learning, UNESCO, Paris, France.

Tomlinson, P.B. (2016). The botany of Mangroves. Cambridge, UK: Cambridge University Press, doi: 10.1017/CBO9781139946575.

Twilley, R.R. (1998). Mangrove wetlands. In: M.G. Messina and W.H. Conner (Eds.), *Southern Forested Wetlands: Ecology and Management*. Lewis Publishers, Boca Raton, Florida, USA, pp. 445–473.

UE4SD (2014). Mapping opportunities for professional development of university educators in Education for Sustainable Development: A state of the art report across 33 UE4SD partner countries. Authors: Mader, M., Tilbury, D., Dlouhá, J., Benayas, J., Michelsen, G., Mader, C., Burandt, S., Ryan, A., Barton, A., Dlouh., J. and Alba, D. University of Gloucestershire, Cheltenham, p. 53.

UN (United Nations) (2005). Hyogo framework for action 2005–2015: Building the Resilience of Nations and Communities to Disasters, 22 January 2005, A/CONF.206/6. Available at: https://www.refworld.org/docid/42b98a704.html (accessed 14 December 2020).

UN (2015). Sendai framework for disaster risk reduction 2015–2030, available at: https://www.undrr.org/publication/sendai-framework-disaster-risk-reduction-2015-2030 (accessed 10 June 2020).

UN WATER (2018). Sustainable Development Goal 6: Synthesis Report on Water and Sanitation. Available at: https://sustainabledevelopment .un.org/content/documents/19901SDG6_SR2018_web_3.pdf.

UNDESA (United Nations Department of Economic and Social Affairs) (2015). Water and sustainable development. http://www.un.org/wa terforlifedecade/water_and_sustainable_development.shtml.

UNECE (2012). Learning for the future: Competences in education for sustainable development. UNECE, Geneva. Available at: www.unece.org/fileadmin/DAM/env/esd/ESD_Publications/Co mpetences_Publication.pdf (accessed 27 July 2016).

UNESCO (2014). UNESCO roadmap for implementing the global action programme on education for sustainable development. Available at: http://unesdoc.unesco.org/images/0023/002305/230514e.pdf (accessed March 2016).

Vare, P. and Scott, W. (2007). Learning for a change: Exploring the relationship between education and sustainable development. *Journal of Education for Sustainable Development*, 1(2), 191–198.

Vehkamäki, S. (2005). The concept of sustainability in modern times. In Sustainable use of renewable natural resources — from principles to practices, University of Helsinki Department of Forest Ecology Publications, p. 34.

Wals, A.E.J. and Jickling, B. (2002). Sustainability in higher education: From doublethink and newspeak to critical thinking and meaningful learning. *International Journal of Sustainability in Higher Education*, 3(3), 221–232.

WCED (World Commission on Environment and Development) (1987). Our common future – The Bruntland Report. Oxford University Press, Oxford.

Weisser, C.R. (2017). Defining sustainability in higher education: A rhetorical analysis. *International Journal of Sustainability in Higher Education*, 18(7), 1076–1089.

WWAP (United Nations World Water Assessment Programme) (2016). The United Nations world water development report 2016: Water and Jobs. Paris, UNESCO.

WWAP/UN-Water (2018). The United Nations world water development report 2018: Nature-Based Solutions for Water. Paris, UNESCO.

WWF (World Wildlife Fund) (2010). Global Water Scarcity: Risks and challenges for business. Available at: http://awsassets.panda.org/do wnloads/lloyds_global_water_scarcity.pdf.

Chapter 2

Lake Eutrophication and Rehabilitation

2.1   Introduction

Aquatic eutrophication occurs when a water body becomes enriched with nutrients and minerals that stimulate excessive growth of algae and aquatic plants. In the absence of human inputs of nutrients, natural eutrophication is a slow process that takes place over a long period. However, anthropogenic activities introduce excessive nutrients into the aquatic environment that vastly speed up eutrophication, a process commonly referred to as cultural eutrophication. Eutrophication was recognized as a serious water pollution problem in European and North American lakes and reservoirs in the mid-20th century. Finland initiated phosphorus removal measures in the mid-1970s, targeting rivers and lakes polluted by industrial and municipal discharges to combat eutrophication. These efforts have achieved a 90% removal efficiency. The main source of eutrophication is dysfunctional sewage treatment plants that produce untreated, or at best partially treated, sewage effluent that discharges into rivers and dams. Eutrophication has become more widespread worldwide. Surveys showed that 54% of lakes in Asia are eutrophic; in Europe, 53%; in North America, 48%; in South America, 41%; and in Africa, 28%. In South Africa, a study using remote sensing has shown that more than 60% of the dams surveyed were eutrophic. Cultural eutrophication induced by human activities can accelerate the rate of eutrophication by increasing the rate at which nutrients enter aquatic ecosystems. Runoff from agriculture and urban development, as well as pollutants from septic systems and sewers, increases the flow of both inorganic nutrients and organic substances into the aquatic ecosystems. The oversupply of nutrients into aquatic environment, particularly nitrogen and phosphorus, is derived from the discharge of nitrate or phosphate containing substances, such as detergents, fertilizers, industrial and domestic runoffs, and sewage into an aquatic system. With the phasing out of phosphate-containing detergents in the 1970s, industrial and domestic runoff and agricultural wastes have emerged as the dominant contributors to eutrophication. These nutrients promote overgrowth of aquatic plants and algae in aquatic ecosystems. After such organisms die, bacterial degradation of their biomass results in oxygen consumption, thereby creating the state of oxygen depletion known as hypoxia. This oxygen depletion of the water body may cause fish kills and induce a range of other ill effects such as reduced biodiversity and impaired ecosystem services. Lake eutrophication may cause three particularly troubling ecological impacts: decreased biodiversity, changes in species composition and toxicity effects. Eutrophication can occur in many aquatic and terrestrial ecosystems that can lead to disturbing ecological impacts.

2.2   Eutrophication in Aquatic and Terrestrial Ecosystems

2.2.1   Causes and effects of eutrophication

Clearing of land and building of towns and cities are typical examples of how anthropogenic activities accelerate land runoffs that supply excessive nutrients such as phosphates and nitrate to nearby lakes and rivers, and eventually to coastal estuaries and bays. Extra nutrients are also supplied by sewage treatment plants and untreated sewage, as well as by fertilizers from golf courses, and terrestrial and aquatic farms. When excess nutrients in fertilizers are applied to the soil, some nutrients leach into the soil where they can remain for years. As phosphate adheres tightly to soil, it is mainly transported by soil erosion, particularly after heavy rains. Once translocated to lakes, the release of phosphate from the sediment layer into the water column is slow. The excess nutrients accumulate over time in the water column and in the sediments. These nutrients can cause an algal bloom when the conditions are suitable, such as high temperatures and strong sunlight during the summer. Most of these nutrients sink to the bottom sediments and remain there over a long period. Hence, the reversal of the effects of eutrophication is difficult and time consuming. Enhanced growth of phytoplankton and algal blooms disrupts normal functioning of the aquatic ecosystem, and causes a host of problems such as depletion of dissolved oxygen crucially needed for fish and shellfish to survive. The water becomes cloudy, typically colored with various shades of green, yellow, brown or red. Eutrophication decreases the ecosystem value of rivers and lakes, and impairs aesthetic enjoyment. Health problems can occur when eutrophic conditions interfere with drinking water treatment. Some algal blooms can be toxic to plants, animals and humans. The toxic compounds produced by these toxic algae can make their way up the food chain, posing a threat to livestock and humans. An example of algal toxins working their way into humans is the case of shellfish poisoning. Biotoxins created during harmful algal blooms are taken up by shellfish such as mussels and oysters, leading to these human foods acquiring the toxicity and poisoning humans. Other marine animals can be vectors for such toxins, as in the case of ciguatera, a predator fish that accumulates the toxin and then poisons humans. Research indicates that intercepting pollution between the source and the water body is a successful means of prevention. These buffer zones, typically wetlands, are interfaces between a body of water and the polluting land source and have been created near waterways in an attempt to filter out pollutants. Sediments and nutrients are deposited or retained in these buffer zones and removed from the water body. Creating buffer zones near farms and roads is another possible way to prevent nutrients from reaching the water body such as a lake.

2.2.2 *Eutrophication in lakes*

The primary limiting factor for eutrophication in freshwater bodies is phosphate, although some works of literature suggested that nitrogen is the primary limiting nutrient for the proliferation of algal biomass in freshwater bodies. The concentration of algae and the trophic state of lakes correspond well to the phosphorus levels in freshwater bodies. Many studies conducted worldwide have shown a close positive relationship between the rate of addition of phosphorus and the rate of algal growth. Phosphorus is a crucial nutrient for plants to live and grow, and is generally considered as the limiting factor for plant growth in many freshwater ecosystems such as lakes. The availability of abundant phosphorus promotes excessive plant growth and subsequent decay, in a process that favors simple algae and plankton growth over other more complicated higher plants, causing a severe reduction in water quality, biodiversity and ecosystem services. The dense algal bloom blocks sunlight from reaching the lower layers of the water body, causing the aquatic plants beneath the algal bloom to die because they cannot receive sunlight to photosynthesize. Eventually, the algal bloom itself dies and sinks to the bottom of the lake, to be decomposed by bacteria, consuming oxygen for respiration. The decomposition causes the water to become depleted of oxygen, suffocating fish and shellfish to death.

Nutrients may be concentrated in a bottom anoxic zone and may only be available during autumn turnover or in conditions of turbulent flow. The dead algae and the organic loads settle at the bottom and undergo anaerobic digestion, releasing greenhouse gases like methane and carbon dioxide in the process. Some of the methane gas is consumed by the anaerobic methane oxidation bacteria, serving as food sources to the zooplankton. With adequate dissolved oxygen at all depths, the aerobic methane oxidation bacteria can consume the methane thereby releasing carbon dioxide, helping the production of algae. In this manner, a self-sustaining biological process can take place to generate primary food source for the phytoplankton and zooplankton, if adequate dissolved oxygen in the water bodies is available. Algae generate dissolved oxygen by releasing oxygen from photosynthesis during the daytime and consume oxygen by its respiration during the night. Adequate dissolved oxygen in water bodies is crucial for fishery production and elimination of greenhouse gas emissions, especially during the absence of sunlight in eutrophic water bodies. The carbon dioxide released by the algae during the absence of sunlight is stored in the water by reducing the water alkalinity and pH for its use during sunshine.

2.2.3   *Eutrophication in coastal waters*

Eutrophication is a common phenomenon in coastal waters. In contrast to freshwater systems where phosphorus is usually the limiting nutrient, nitrogen is more commonly the key limiting nutrient of marine waters. Thus, nitrogen levels have greater importance for the understanding of eutrophication problems in sea water. As the interface between freshwater and saltwater, many estuaries worldwide exhibit symptoms of eutrophication, whose growth can be both phosphorus and nitrogen limited. Eutrophication in estuaries often results in bottom water hypoxia or anoxia, leading to fish kills, habitat degradation and impairments of ecosystem services. In addition to external nutrient sources, marine environments also receive nutrients from upwelling of deeper water. By conveying deep, nutrient-rich waters to the surface, upwelling in coastal systems promotes enhanced productivity where the upwelled nutrients can be assimilated by algae. Examples of anthropogenic sources of nitrogen-rich pollution to coastal waters include sea cage fish farming and discharges of sewage from nearby cities. Fish farming wastes, runoff from land and industrial waste discharges are important nutrient sources in the estuary and coastal area. It has been suggested that accumulating reactive nitrogen in the environment may prove as serious as putting carbon dioxide in the atmosphere. One proposed solution to eutrophication in estuaries is to restore shellfish populations, such as oysters and mussels. Oyster and mussel reefs remove nitrogen from the water column and filter out suspended solids, subsequently reducing the likelihood or the extent of harmful algal blooms or anoxic conditions. Shellfish filter feeding activity is considered beneficial to water quality by controlling phytoplankton density and sequestering nutrients. These nutrients can be removed from the system by shellfish harvest, by burial in the sediments or by loss through denitrification. Foundational work toward the idea of improving marine water quality through shellfish cultivation has been conducted in Sweden and in the East, West and Gulf coasts of the United States of America. Seaweed such as kelp absorbs phosphorus and nitrogen and is thus useful in removing nutrients from overfertilized parts of the sea.

2.2.4 *Eutrophication in terrestrial ecosystems*

Terrestrial ecosystems are subject to similar adverse impacts from soil eutrophication caused by high nutrient contents. Increased nitrates in soil are frequently undesirable for some species of plants. Many terrestrial plant species, such as most orchid species in Europe, are endangered as a result of soil eutrophication. Meadows, forests and bogs are characterized by low nutrient content, where slowly growing species adapted to low nitrogen levels. Hence, they can be overtaken by faster-growing and more competitive species when the soil conditions are right, such as soil having high nitrogen contents. Tall grasses that can take advantage of higher nitrogen levels may overgrow the meadows so that the natural meadow species may be lost. Fens that are rich in species diversity can be overtaken by reed or reed grass species when soil is overly rich in nutrients. Forest undergrowth enriched by high-nutrient runoff from nearby fertilized fields can be turned into a nettle and bramble thicket. Because terrestrial plants have high nitrogen requirements, additions of nitrogen compounds will stimulate plant growth. Nitrogen is not readily available in soil directly to higher plants. Terrestrial ecosystems rely on microbial nitrogen fixation to convert nitrogen gas in the atmosphere into other forms such as nitrates. However, there is a limit to how much nitrogen can be utilized by plants. Ecosystems receiving more nitrogen than the plants require are called nitrogen saturated. Nitrogen-saturated terrestrial ecosystems then can contribute both inorganic and organic nitrogen to freshwater, river, coastal and marine eutrophication, where nitrogen may be a limiting nutrient. This is also the case with increased levels of phosphorus in terrestrial ecosystems. However, because phosphorus is generally much less soluble than nitrogen, it is leached from the soil at a much slower rate than nitrogen. Consequently, phosphorus is much more important as a limiting nutrient in fresh aquatic systems of eutrophication.

2.3   Control of Eutrophication

As eutrophication poses serious problems to natural ecosystems and humans as well, reducing eutrophication should be a key concern in policy for a sustainable future. Natural runoff, in the absence of human settlements, can cause algal blooms in the wild. Hence, control of nutrient concentrations should not aim to reduce nutrient levels below the normal levels in the wild. In general, clean-up measures have not been very successful as many targeted point-source nutrient reduction efforts did not show a sustained decrease in runoff nutrient level, because the recalcitrant non-point sources are not properly managed. The literature suggests, however, that when both these point sources and non-point sources are controlled, eutrophication decreases in a sustained manner. Laws regulating the discharge and treatment of sewage have led to dramatic nutrient reductions to the surrounding ecosystems. However, an effective policy and enforcement regulating agricultural use of fertilizer and treatment of animal waste are lacking. In Japan, as in many other countries, the amount of nitrogen produced by livestock is adequate to serve the fertilizer needs for the agricultural industry. Livestock waste treatment needs to be integrated with agricultural fertilizer use to reduce livestock waste discharge and to minimize fertilizer consumption, which will vastly reduce nutrients released into the ecosystems. The policy concerning the prevention and reduction of eutrophication can be broken down into four sectors: Technologies, public participation, economic instruments and cooperation.

Non-point sources of pollution are the primary contributors to eutrophication and must be minimized through best agricultural practices and technology. Reducing the amount of pollutants that reach a watershed can be achieved through the protection of its forest cover, thereby reducing the amount of soil erosion leeching into a watershed. Sustainable agricultural practices will minimize land degradation and reduce the amount of soil runoff and fertilizers reaching a watershed. Waste disposal technology constitutes another important contribution to eutrophication prevention. Untreated domestic sewage from urban cities in underdeveloped nations is a major non-point-source nutrient loading into water bodies. It is essential to provide adequate sewage treatment facilities to highly urbanized areas in underdeveloped nations, where treatment of domestic wastewater is scarce. The technology to reuse wastewater safely and efficiently, both from domestic and industrial sources, should be a primary concern for policy regarding eutrophication.

The public can play a major role in the effective prevention of eutrophication. For a policy to have any effect, the public must be aware of their contribution to the problem and understand the ways by which they can reduce eutrophication. Programs instituted to promote public participation in the recycling and elimination of wastes, as well as in education on the issue of rational water use, are necessary to protect water quality within urbanized areas and the adjacent water bodies. Economic instruments, that include property rights, water markets, fiscal and financial instruments, and charge systems and liability systems, are gradually becoming a substantive component of the management tool set used for pollution control and water allocation decisions. Incentives for those who practice clean, renewable water management technologies are an effective means of encouraging pollution prevention. By internalizing the costs associated with the negative effects on the environment, governments can encourage cleaner water management. Because a body of water can influence a range of people reaching far beyond that of the watershed, cooperation among organizations of different scales is necessary to prevent the intrusion of contaminants into a watershed that can lead to eutrophication. Agencies ranging from nation and state governments to those of water resource management and non-governmental organizations, and the local communities, are collectively responsible for preventing eutrophication of water bodies.

2.4   Eutrophication of Tasik Harapan

Tasik Harapan (TH) was constructed in the 1990s to serve as a flood retention pond in the Universiti Sains Malaysia (USM) Minden campus in Penang. It has a surface area of about 1.5 acres with a depth of between 1 and 1.5 m. It receives sullage water originating from within the campus as well as from domestic effluents originating from the surrounding human settlements. This polluted sullage water contains nutrients, particularly phosphorus and nitrogen. Over the 30 years since its construction, these nutrients have accumulated in the sediments of the pond, contributing to a situation of eutrophication with wide variation in dissolved oxygen over the 24-hour diurnal cycle.

2.4.1   *Eutrophic status of TH*

The chlorophyll *a* concentration in TH occasionally reached 300 μg/L. Dissolved oxygen (DO) at the surface water varied between 4 and 18 mg/L over the diurnal cycle, indicating the effect of strong photosynthesis induced by high nutrients and strong tropical sunlight and high temperatures. Being open and shallow, the atmospheric reaeration rate is high, of the order of about 3 per day. The photosynthesis rates and respiration rates are estimated to be 30 mg/L/d

Figure 2.1:   Eutrophic Tasik Harapan in USM.

oxygen and 3 mg/L/d oxygen, respectively (Teh *et al.*, 2008). The low (near anaerobic) DO level at the sediment layers, coupled with a pH of about 7, promotes the release of phosphorus into the water column as bio-available phosphorus to stimulate algal growth. The lake water regularly looks greenish due to a high level of algae, as can be seen in Figure 2.1. This chapter presents simulation approaches to describe the DO diurnal patterns in TH with the purpose of providing useful insights and guidelines for an integrated lake rehabilitation program for TH.

Chlorophyll *a* concentration in TH has been measured to reach a level of 330 μg/L (Haslim *et al.*, 2006), indicating a high level of eutrophication. Acceptable levels of chl *a* in clean water are below 2.0 μg/L, while a level of 9.0 μg/L is an indication of eutrophic water (Chapra and Tarapchak, 1976). For perspective, it is noted that in the Pearl River Estuary, chl *a* levels were reported to vary in the range of 2 to 8 μg/L in July 1999 (Chen *et al.*, 2004), while in Lake Kinneret Israel, the chl *a* levels fluctuated between

Figure 2.2:   Tasik Harapan with a running aerator.

7 and 70 μg/L during the period from November 2001 to April 2002 (Yacobi and Schlichter, 2004). For Pasad Jolasid Reservoir in Thailand, a high chl a concentration ranging from 20 to 80 μg/L was observed between April and October 2002 (Charumas, 2004). Various attempts have been made to rehabilitate TH to a level appropriate for recreational activities such as canoeing and angling. A previous attempt in this direction employed the so-called Bio Reco Effective Aerator (Figure 2.2), with little documented success (Nielsen and Pillay, 2006). The lack of success of the aerator to rehabilitate TH does not come as a surprise. DO in the surface water of TH has been observed to vary between 4 and 18 mg/L over the diurnal cycle, indicating an excessive level of DO in the water column. Hence, the aerator is of little use in such a situation of high DO. Further, the aerator will not remove nutrients, the source of eutrophication. It should also be noted that aerator impact is limited to a radius of some 3 m around the aerator (see Figure 2.2).

Table 2.1: Mean water quality parameters for Tasik Harapan (Hashim *et al.*, 2006).

Parameter	Mean ± s.d
DO (mg/L)	5.20 ± 1.900
BOD (mg/L)	8.00 ± 2.000
Chlorophyll *a* (mg/L)	0.33 ± 0.064

Table 2.2: Mean water quality parameters used for Tasik Harapan.

Parameter	Mean values
DO (mg/L)	5.20, 6.00
DO (saturation) (mg/L)	7.50
BOD (mg/L)	8.00
Chlorophyll *a* (mg/L)	0.33

2.4.2 *Simple photosynthesis model for TH*

We develop a simple analytical photosynthesis model to demonstrate the ineffectiveness of the aerator in rehabilitating TH. This simple model also serves as an introduction to an enhanced photosynthesis model in later sections. For this purpose, we use relevant parameter values as indicated in Tables 2.1 and 2.2. This analytical model simulates the DO levels in TH subject to the oxidation of biochemical oxygen demand (BOD) and the generation of DO due to photosynthesis in the top 1 m of water in TH.

The analytical model that we developed for TH consists of the following two ordinary differential equations following the approach adopted by Thomann and Mueller (1987) and Chapra (1997). The definition and unit of the symbols used in Eqs. (2.1) and (2.2) are listed in Table 2.3.

$$\frac{d\ell}{dt} = -\alpha\ell + \gamma \tag{2.1}$$

$$\frac{dc}{dt} = -\alpha\ell + \beta(c_s - c) + P_{net}\sin(\sigma t - 7\pi/12) \tag{2.2}$$

Table 2.3:   Definition of symbols used in Eqs. (2.1) and (2.2).

Symbol	Unit	Definition
t	day	Time
ℓ	mg/L	BOD level
c	mg/L	DO level
c_s	mg/L	DO saturation level
β	day^{-1}	Reaeration rate
γ	mg/L/day	BOD loading
P_{net}	mg/L/d	Maximum net photosynthesis rate
σ	2πper day	Frequency for a 1-day cycle
$P_{net}\sin(\sigma t - 7\pi/12)$	mg/L/day	diurnal rate of O_2 generation by n *et al* gal photosynthesis

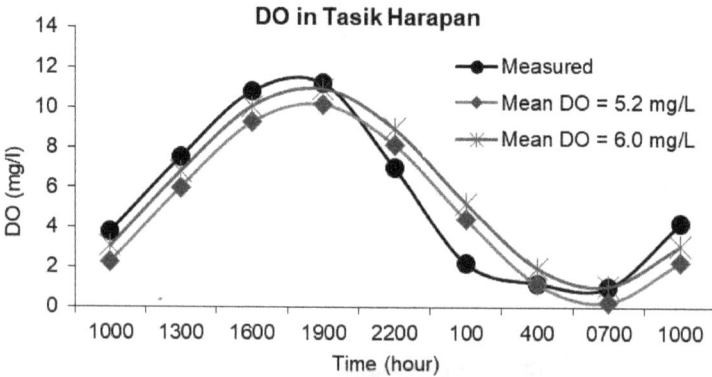

Figure 2.3:   Measured and simulated DO levels in Tasik Harapan.

In Eq. (2.1), the BOD loading into TH from the surrounding catchment is represented by the term $+\gamma$; while the oxidation of BOD is represented by the term $-\alpha\ell$. Similarly, in Eq. (2.2), the oxidation of BOD is represented by the term $-\alpha\ell$, while atmospheric reaeration is represented by the term $+\beta(c_s - c)$. The photosynthesis during daytime and respiration at night are represented by the term $P_{net}\sin(\sigma t - 7\pi/12)$. The term $(-7\pi/12)$ refers to the time of sunrise at 7:00 am in the morning and σ is the frequency of a 1-day cycle. Figure 2.3 demonstrates the analytical model for photosynthesis fitted to the data available from Hashim *et al.* (2006), indicating reasonable agreement between the simulated model result and the measured mean data (Teh *et al.*, 2006). We have used two mean DO levels of 5.2 and 6.0 mg/L. Dissolved oxygen levels vary between 1 mg/L and 12 mg/L over the diurnal cycle with minimum DO attained around 6:00 am to 7:00 am and maximum DO achieved around 5:00 pm to 6:00 pm in the middle layer of TH. It should be noted here that this simple analytical model for photosynthesis applied to the top 1 m of TH assumes that the rate of photosynthesis equals that of respiration. As we will show soon in the later section, the rate of photosynthesis (around 30 mg/L/d oxygen) far exceeds that of respiration (around 3 mg/L/d oxygen) by one order of magnitude. The difference between photosynthesis and respiration is compensated by a high reaeration rate β of 3 to 7 per day. This high reaeration rate allows oxygen generated by photosynthesis during the daytime to escape into the atmosphere, hence reducing the level of DO measured. On the contrary, this high reaeration rate also allows the accumulated DO to escape into the atmosphere at night, producing the apparent effect of respiration. The net effect is that the measured DO in TH indicated an apparent balance between photosynthesis and respiration. To verify this observation, we therefore proceeded to measure the rates of photosynthesis, respiration and atmospheric reaeration in TH. Further, these three rate constants are critical parameters crucially important to describe or model algal dynamics in TH, which is important for any effort of lake rehabilitation. Moreover, these rate constants are important for lake ecology and limnology in general.

2.4.3 *Mechanical aerator is not effective*

We demonstrate next the ineffectiveness of the aerator for TH rehabilitation. The aerator used in the field test conducted in TH had a power of 180 W. For an aerator with 180 W, the O_2 delivery rate Φ_1 is estimated to be 0.864 mg/L/d O_2, according to the conversion provided in Mara (1998). We intend to simulate with four distributed aerators running simultaneously with a combined power rating of 720 W, with an equivalent O_2 delivery rate of $\Phi_4 = 3.456$ mg/L/d. Compared with P_{net} (around 30 mg/L/d oxygen), Φ is too small to be effective in delivering sufficient O_2 to counteract the depletion of O_2 due to algal respiration at night. To demonstrate this, we simulated the analytical photosynthesis model of Eqs. (2.1) and (2.2) with and without Φ_1 for three scenarios: the aerator is (1) not switched

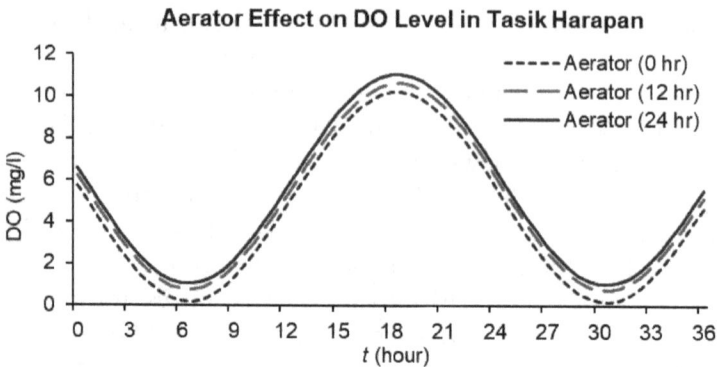

Figure 2.4:   Aerator effect on DO levels in Tasik Harapan.

on, (2) switched on for the entire 24 hours non-stop or (3) switched on only at night for only 12 hours. It can clearly be seen in Figure 2.4 that the mechanical aerator plays a very minor role in increasing the level of DO in TH.

In the above analytical model, we have assumed that the oxygen delivered by the aerator is uniformly distributed over the entire area of TH, for convenience of analysis and presentation. It may be argued that the effect may be different if the uniformity of distribution of oxygen supplied is violated. For this purpose, we simulate the DO spatial-temporal variations in TH by more refined two-dimensional partial differential equations of the advection–diffusion type Eqs. (2.3) and (2.4), solved by the finite element method (Teh *et al.*, 2006). The definition and unit of the symbols used in Eqs. (2.3) and (2.4) are listed in Table 2.4.

$$\overset{\circ}{\ell} + u\ell_x + v\ell_y - D_x\ell_{xx} - D_y\ell_{yy} + \alpha\ell = \gamma \tag{2.3}$$

$$\overset{\circ}{c} + uc_x + vc_y - D_xc_{xx} - D_yc_{yy}$$
$$+ \alpha\ell - \beta(c_s - c) - P_{net}\sin(\sigma t - 7\pi/12) = \Phi \tag{2.4}$$

The simulated results for DO at 6:00 am in the morning in TH with one aerator Φ_1 (top) and four distributed aerators Φ_4 (bottom) are shown in Figure 2.5. As can be seen, the aerator does increase the local oxygen level by 2 to 3 mg/L in the immediate vicinity (radius of several meters) of the aerator. Elsewhere, the aerator has little or no impact on DO. Therefore, an aerator with the intended power rating

Table 2.4: Definition of symbols used in Eqs. (2.1) and (2.2).

Symbol	Unit	Definition
t	s	Time
ℓ	kg/m^3	BOD level
c	kg/m^3	DO level
c_s	kg/m^3	DO saturation level
β	s^{-1}	Reaeration rate
γ	kg/m^3/s	BOD loading
$P_{net}\sin(\sigma t - 7\pi/12)$	kg/m^3/s	diurnal rate of O$_2$ generation by n *et al* gal photosynthesis
u	m/s	Advection in the x direction
v	m/s	Advection in the y direction
D_x	m^2/s	Diffusion in the x direction
D_y	m^2/s	Diffusion in the y direction
Φ	kg/m^3/s	Mechanical aeration loading

has little impact on DO in TH. What is more important to observe is that the aerator will play no useful role in the rehabilitation of TH since the lake is characterized by excessive DO in the surface water, not by the lack of DO.

2.4.4 *Photosynthesis, respiration and reaeration rates for TH*

In any lake study, a good estimate of the rates of photosynthesis, respiration and reaeration will contribute toward an improved understanding of the ecosystem's performance, since photosynthesis is the foundation upon which that ecosystem is built. With this in mind, we proceeded to provide good estimates of these three critically important process rates for TH. Technical details on the field study and model simulations are further discussed in Section 2.6. Tables 2.5 to 2.7 provide a tabulation of the rates for respiration (R_e mg/L/d), reaeration (β d^{-1}) and photosynthesis (P_{net} mg/L/d) for TH, estimated during a field study conducted in the months of November and December 2004. It should be noted that the respiration rates given in Table 2.5 for 26 November 2006 are inaccurate due to experimental error. The respiration rates for TH vary between 3.4 and 5.6 mg/L/d. The atmospheric reaeration rates vary between 3.3 and 7.0 per day.

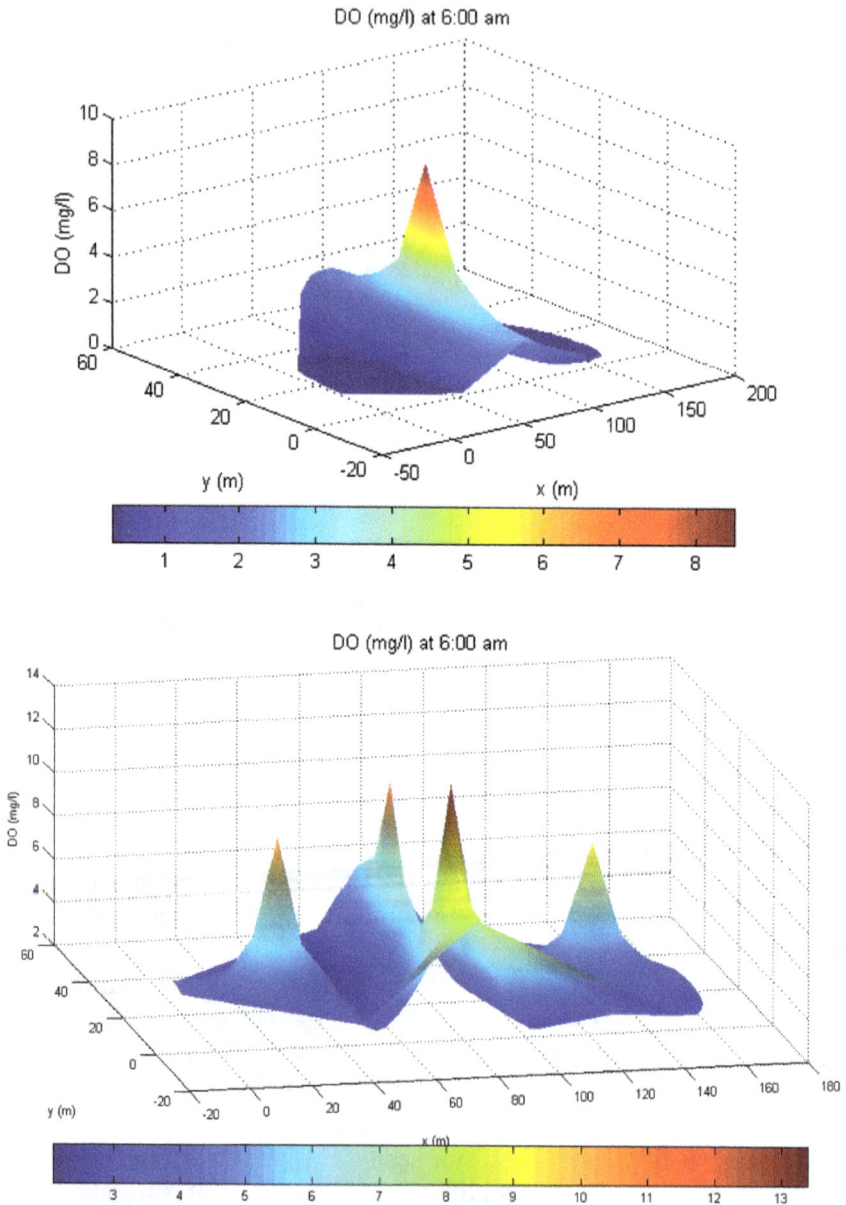

Figure 2.5:   DO contours at 6:00 am with one (top) and four (bottom) mechanical aerators.

The photosynthesis rates vary between 25.5 and 43.3 mg/L/d. Noting that TH has a high reaeration rate, we decided to investigate the role of reaeration in the estimation of photosynthesis rates. For this purpose, we submerged two sealed BOD bottles filled with water from TH in the lake and monitored the DO levels continuously. For Bottle A, we estimated photosynthesis rates P_{net} to vary between 37.7 and 50.7 mg/L/d, while in Bottle B, they vary between 37.7 and 51.6 mg/L/d. The photosynthesis rate estimated for both bottles appears to be higher than that of TH for a good reason. The strong reaeration in TH permits DO to escape into the atmosphere, with the result that the reduced DO is reflected in reduced photosynthesis rates. Removing atmospheric reaeration by performing photosynthesis in a small stationary tank will provide higher estimates of photosynthesis. In a later section, we will describe an experiment conducted between December 2010 and January 2011 to monitor DO levels in TH and in four small stationary tanks containing water taken from TH to demonstrate the impact of strong reaeration in TH, as compared to weak aeration in the small tanks. We note that the values of reaeration (between 3.3 and 7.0 per day) estimated in this study correspond to the higher range of reaeration rates estimated by a rating curve (Figure 2.6) provided by USEPA (1985), which is based upon depth and velocity of water. On the contrary, the reaeration rates estimated in this study make use of time series of DO measured in TH. The measured time of DO peak (ϕ hours after 12:00 pm) is related to reaeration rate β as shown in Figure 2.7, given by the formula $\phi = \tan^{-1}(\sigma/\beta)$, where σ and β were defined earlier.

2.4.5   *Mudball and EM are not effective*

An attempt had been made beginning in August 2010 to use mudball and effective microorganisms (EMs) to improve water quality in TH, with undocumented success (Asha *et al.*, 2010). We therefore attempted to verify the effectiveness of the EM solution method by comparing DO level in TH for the period November–December 2004 and the period December 2010–January 2011. Figure 2.8 shows the three sampling locations in TH. Figure 2.9 shows the surface DO at Site 1 in TH for the period 4–5 December 2004, replotted from data available in Teh *et al.* (2008), in which additional details may

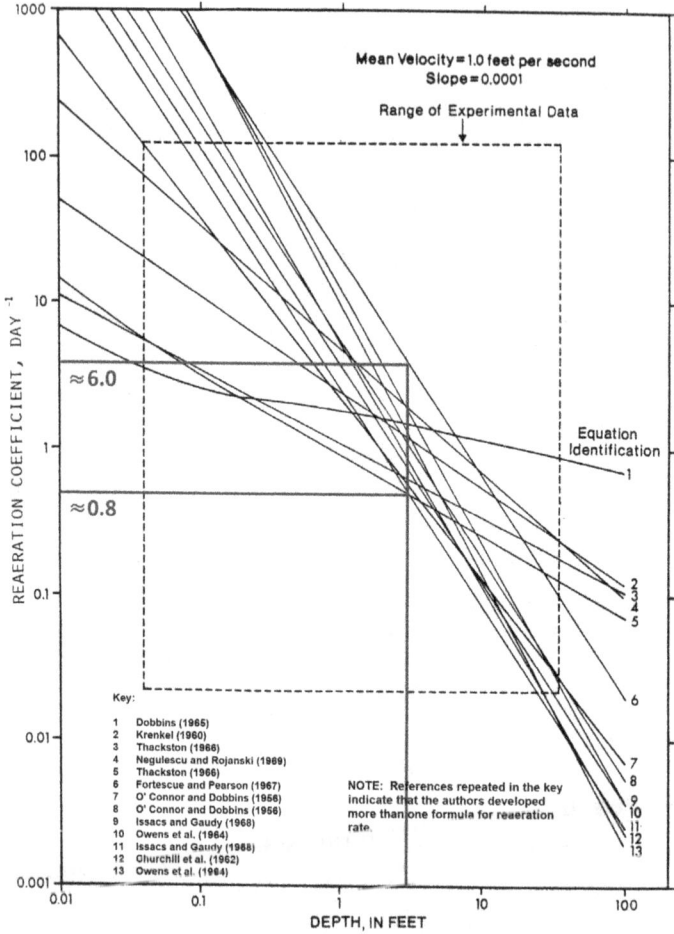

Figure 2.6: Estimation of β for streams vs depth in feet (USEPA, 1985).

be found. Over the diurnal cycle, surface DO varies from a minimum of around 3.5 mg/L in the early part of the morning (6:00 am) to a maximum of 15 mg/L in the late afternoon (5:00 pm) on 4–5 December 2004. On the contrary, for the period four months after continuous EM release (December 2010 to January 2011), the surface DO varies between a low of about 4 mg/L in the early morning and a high of 17 mg/L in the late afternoon (Figures 2.10 and 2.11). The diurnal patterns of DO before and after the continuous

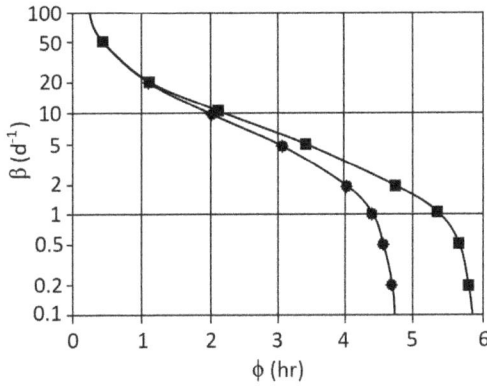

Figure 2.7: Estimation of β using (i) Delta method, (ii) time of DO peak, $\phi = \tan^{-1}(\sigma/\beta)$.

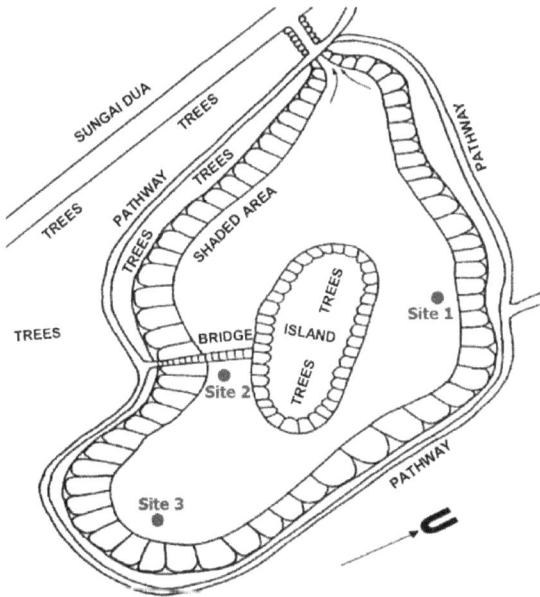

Figure 2.8:   Three sampling sites at Tasik Harapan.

release of EM demonstrate close similarity. Indeed, the post-EM-release DO indicates increased DO levels due to increased photosynthesis. Since photosynthesis is an indicator of eutrophication, this carries the implication that the eutrophic state in TH has not been

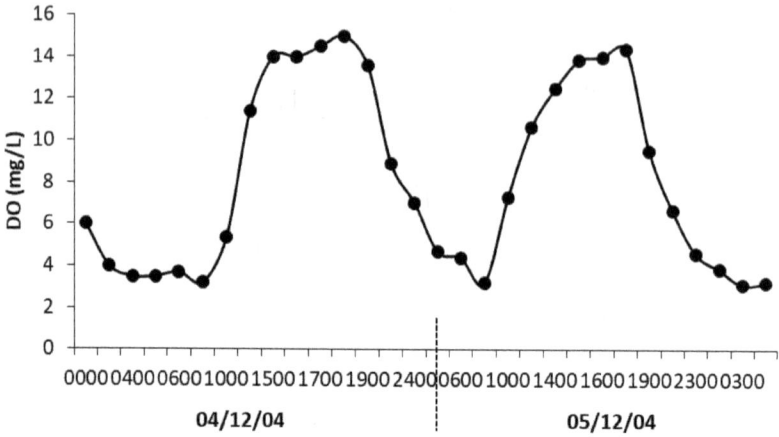

Figure 2.9:   Surface DO measured at Tasik Harapan at Site 1 between 4 and 5 December 2004.

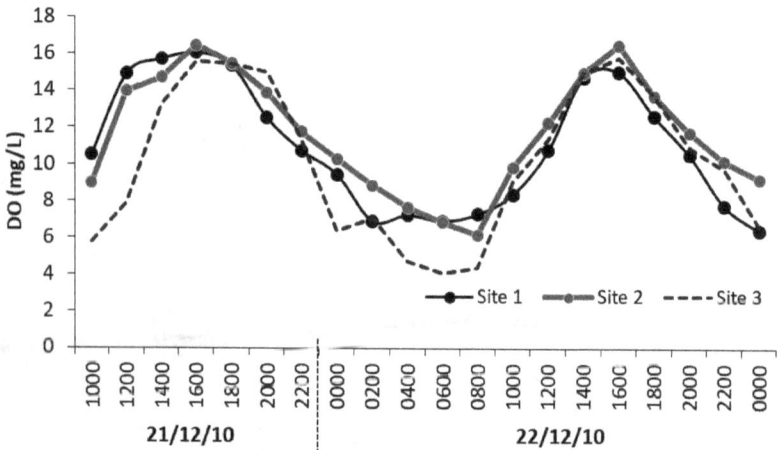

Figure 2.10:   Surface DO measured at Tasik Harapan between 21 and 22 December 2010.

reduced by four months of release of EM. Indeed, based upon this preliminary data, it may be surmised that the continuous release of EM in TH may have even caused a slight increase in photosynthesis in TH. If we allow ourselves the liberty to use photosynthesis as an indication of eutrophication, then we may arrive at the tentative findings that EM solution application in TH might have even contributed slightly to increase eutrophication.

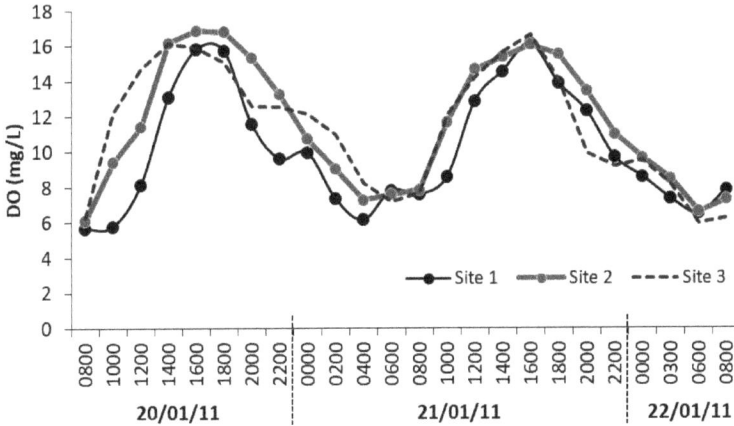

Figure 2.11: Surface DO measured at Tasik Harapan between 20 and 22 January 2011.

2.5 Strong Atmospheric Reaeration in TH

As we have mentioned earlier, the reaeration rate in TH was estimated to vary between 3.3 and 7.0 per day in a previous study (Teh *et al.*, 2008), which is in the higher range of the rates estimated by the rating curve (Figure 2.6) provided by USEPA (1985). To illustrate the impact of high reaeration on DO in TH, we conducted a DO experiment in four tanks of approximately 100 liters each, with a diameter of 0.5 m and a depth of 0.5 m. Tanks of this small size were chosen to minimize atmospheric reaeration. Surface water from TH was collected and immediately put in the tanks for continuous DO measurement every two hours. Figure 2.12 shows the diurnal patterns of DO in Tank 1 indicating a minimum of 9 mg/L and a maximum of 17 mg/L. The minimum DO level in Tank 1 of 9 mg/L (attained during early morning at about 6:00 to 7:00 am before sunrise) is 5 mg/L higher than the low DO measured in TH (of 4 mg/L). Since the field experiment was conducted under similar conditions for the tanks and TH, this difference in DO needs to be explained. The difference is due to the reduced level of DO released from the tank as compared to TH. We then repeated this experiment over four tanks, making sure that the tanks were not being shaken in order to minimize movement-induced reaeration. Figure 2.13 shows the diurnal patterns of DO in the four tanks, indicating a minimum

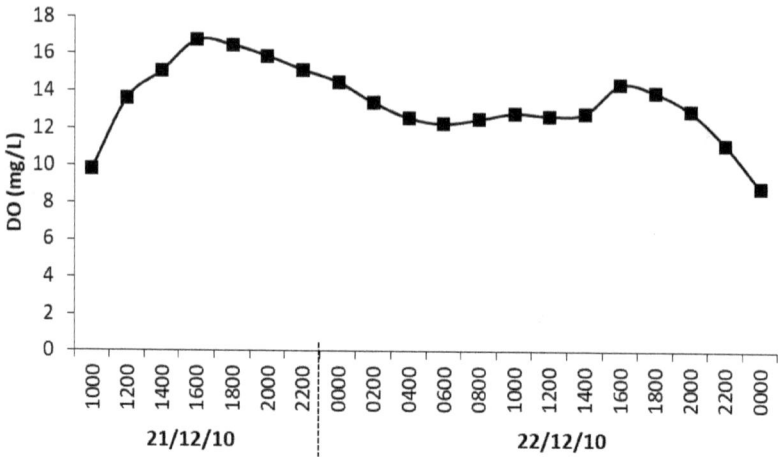

Figure 2.12:   Surface DO measured in Tank 1 between 21 and 22 December 2010.

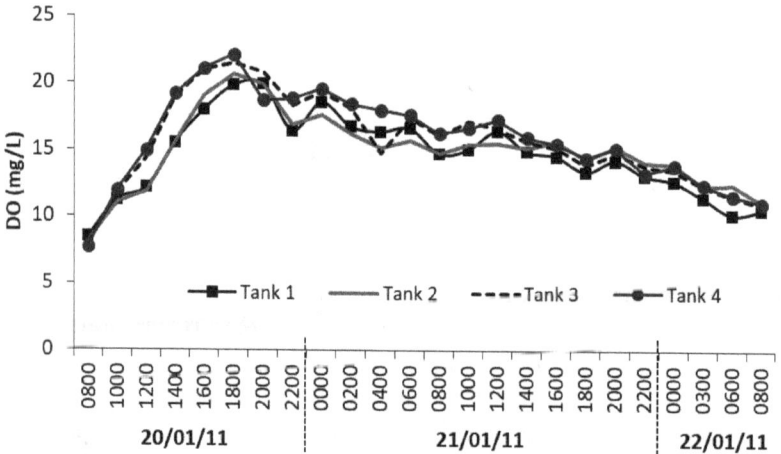

Figure 2.13:   Surface DO measured in Tanks 1–4 between 20 and 22 January 2011.

value of 12 mg/L and a maximum value of 22 mg/L. These heightened levels of DO in the four tanks as compared to those in TH further demonstrate that TH is indeed subjected to high atmospheric reaeration. Therefore, the main issue with TH is not the lack of oxygen but the abundance of it, due to high photosynthesis induced by high nutrients (phosphorus and nitrogen) and high solar radiation.

2.6   Refined Photosynthesis Model for TH

In any lake research or management, a good estimate of the rate constants for three key processes, namely, photosynthesis, respiration and atmospheric reaeration, will contribute toward an improved understanding of the ecosystem's performance, since photosynthesis is the foundation upon which that lake ecosystem is built. With this in mind, we proceed to estimate rate constants for these three critically important processes for TH. The rate constants for respiration (R_e mg/L/d), reaeration (β d^{-1}) and photosynthesis (P_{net} mg/L/d) for TH are estimated from field study conducted in the months of November and December 2004.

2.6.1   *DO field study*

Designed to supplement field data collected by Mansor *et al.* (2004), additional field study was conducted on five days between 26 November and 5 December 2004 for the purpose of estimating the rate constants of respiration R_e, reaeration β and primary production P_h. The field data for DO for these 5 days are shown in Figure 2.14, (a) to (e). For each day, DO levels in mg/L were measured under eight different conditions. Two clear bottles were immersed at the water surface to simulate photosynthesis without reaeration since the bottles were airtight (A and B in the square panel in Figure 2.14). Two more light bottles were placed at a depth of 2 m (C and D in the square panel in Figure 2.14) to assess the extent of light attenuation at this depth of 2 m. Two dark bottles were similarly immersed to simulate respiration (E and F in Figure 2.14). BOD bottles were used for this purpose. DO levels in all bottles were measured at one-hour intervals for 24 hours. DO levels in the surface water were also measured, one directly in situ and the other from water taken from the surface (G and H in Figure 2.14). DO levels were also taken in the lake at a depth of 2 m and 3 m (I and J in Figure 2.14). The results indicate vertical stratification of the lake, despite its shallow depth.

2.6.2   *Respiration rates R_e*

These sets of DO data are now used to estimate process rate constants R_e, β and P_{net} that will be used to develop an algal

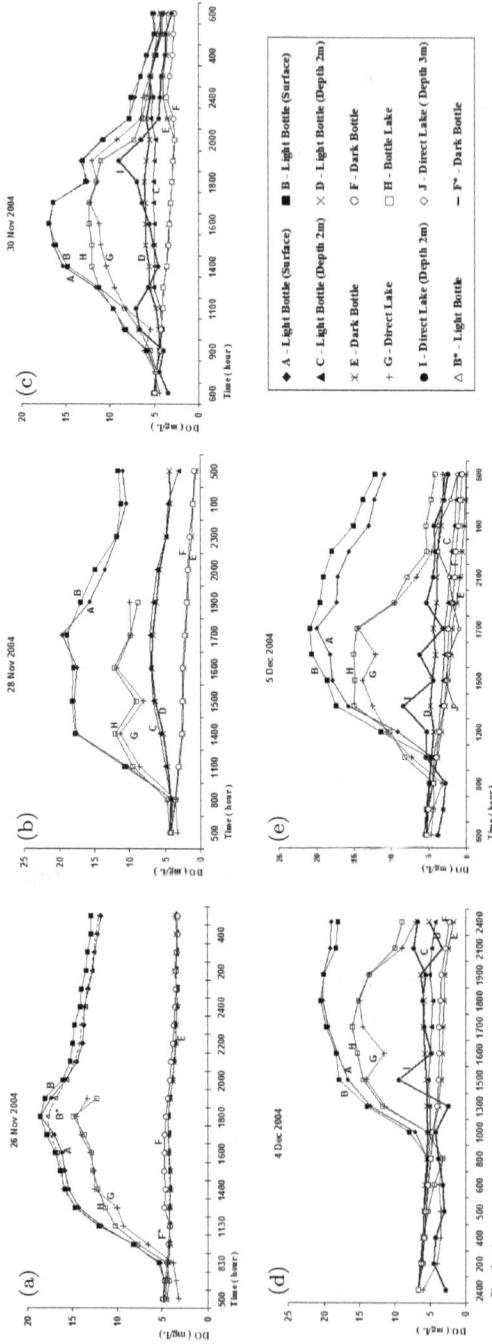

Figure 2.14: DO level curves on (a) 26 November, (b) 28 November, (c) 30 November, (d) 4 December and (e) 5 December 2004. The labels A to J in the square parel refer to the eight different conditions of photosynthesis.

photosynthesis model for lake management. The DO data from the dark bottles are fitted to a linear regression as shown in Figure 2.15, which indicate a good fit with high coefficient of determination R^2 values exceeding 0.9 except for 26 November. From these regression lines, respiration rates R_e are then estimated in the range of 3.4 mg/L/d to 5.6 mg/L/d as shown in Table 2.5. Respiration rates in the dark bottles on 26 November (1.6 and 1.4 mg/L/day) are much lower than those for the other days. On further analysis of the dark bottles used on 26 November, it was found that some sunlight had penetrated the dark bottles, as the caps of the bottles were not adequately sealed. On the subsequent days, the caps were tightly wrapped in black tapes to stop sunlight penetration. On December 4 and 5, additional black tapes were adhered to the dark bottles and their caps to further reduce any potential penetration of sunlight. This may account for the slight increase in the respiration rates measured on December 4 and 5 as indicated in Table 2.5.

2.6.3 *Photosynthesis models*

A simple photosynthesis model for TH given as Eqs. (2.1) and (2.2) is proposed, in which the net photosynthesis (after subtracting respiration) is represented by a sine function $P_{net} \sin(\sigma t - 7\pi/12)$. Other models to simulate algae and photosynthesis are also available elsewhere (Salas and Thomann, 1978; Park and Lee, 2002). A previous study has estimated that the campus contributes about 23 kg/d of BOD to the lake (Teh *et al.*, 2006). Further, $\sigma = 2\pi d^{-1}$ is the frequency for a 1-day solar cycle. The value of $7\pi/12$ radian signifies sunrise at 7:00 am and sunset at 7:00 pm with a photoperiod of 12 hours, which is indicative of a normal sunlight period for USM. It is observed that there is a persistent phase shift between the time of solar peak (1:00 noon) and time of DO peak (varying between 4:00 and 6:00 pm, Figure 2.14), denoted by ϕ radian (with 2π radians = 24 hours), which can be used to estimate β. From Eq. (2.2), we may derive Eq. 2.5 below, after some mathematical manipulation (Teh *et al.*, 2006):

$$\phi = \tan^{-1}(\sigma/\beta) \tag{2.5}$$

We therefore propose that Eq. (2.5) be used to estimate the reaeration rate β. There are two other methods available for

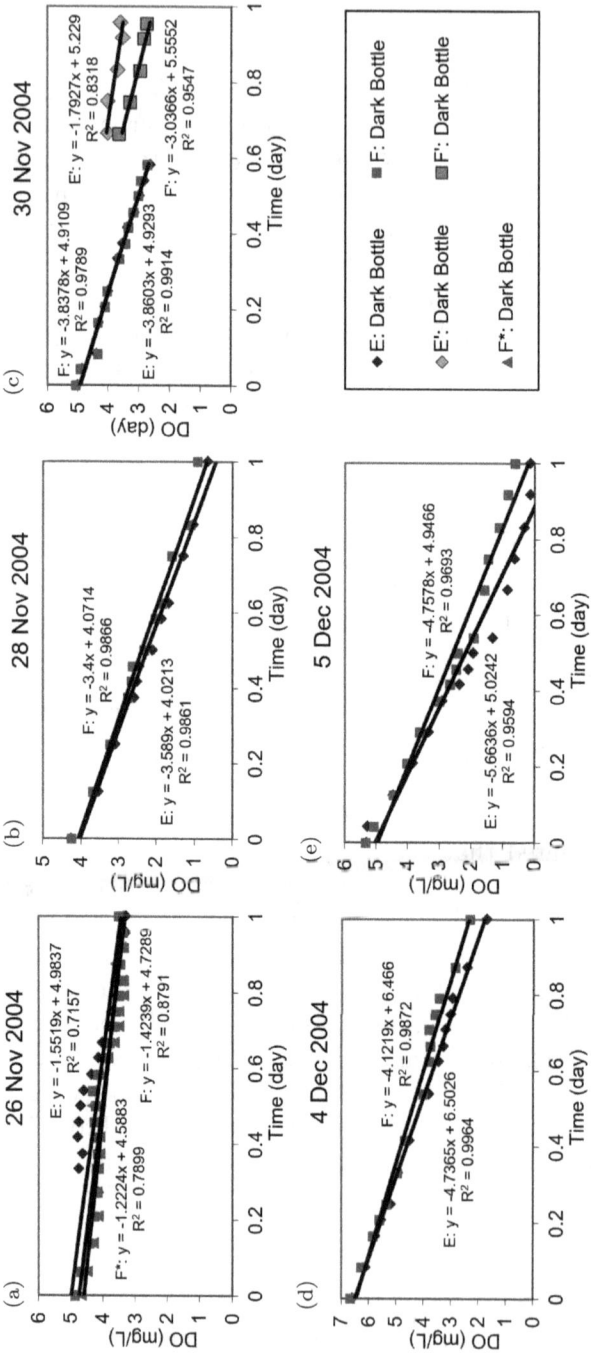

Figure 2.15: Linear regression of the dark bottles' DO levels on (a) 26 November, (b) 28 November, (c) 30 November, (d) 4 December and (e) 5 December 2004.

Table 2.5: Respiration rates R_e (mg/L/d) estimated using linear regression.

Day	Bottle E		Bottle F	
	R_e	R^2	R_e	R^2
1 (26/11/04)	1.6	0.716	1.4	0.879
2 (28/11/04)	3.6	0.986	3.4	0.987
3 (30/11/04)	3.9	0.991	3.8	0.979
4 (04/12/04)	4.7	0.996	4.1	0.987
5 (05/12/04)	5.6	0.959	4.8	0.969

estimating reaeration rate, known as the Delta method based upon Fourier Series (Chapra and Di Toro, 1993) and the Extreme Value Method (EVM) (Wang *et al.*, 2003). The EVM is based upon the maximum and minimum values of DO in the diurnal DO cycle, where DO saturation is a function of temperature. The DO saturation level used in the EVM is obtained from Benson and Krause (1984) and APHA (1985) according to the temperature when the minimum and maximum DO levels in the lake are measured. According to EVM, values of β may then be determined by Eqs. (2.6) and (2.7):

$$R_e = \beta(c_s - c_{\min}) \tag{2.6}$$

$$P_h(t_{\max C}) = R_e - \beta(c_s - c_{\max}) \tag{2.7}$$

Equation (2.6) is used for estimating β by using the nighttime minimum DO $c_{\min}$, while Eq. (2.7) is applicable for the determination of β using maximum DO $c_{\max}$ attained during the daytime. However, only Eq. (2.6) is used in this paper. Equation (2.7) is not used because $t_{\max C}$ (time of DO maximum) cannot be estimated accurately, which would in turn yield inaccurate estimation of β.

2.6.4 Reaeration rates β

We now proceed to estimate reaeration rate β by several methods including those shown in Figures 2.6 and 2.7. A rate constant such as β has been essential in many studies on DO, hence its accurate estimation is important. Figure 2.6 is a graphical presentation of reaeration rates as a function of water depths compiled from several sources (Bennett and Rathbun, 1972; USEPA, 1985). For a water depth of about 4 to 5 feet for Lake Harapan, the estimated reaeration rates fall in the range of 0.8 d^{-1} to 6.0 d^{-1}. Lake Harapan has no significant flow, implying that the reaeration rate should be at the lower values in this range. Hence, a reasonable estimate for reaeration for Lake Harapan without flow would be 1.0 d^{-1} to 1.6 d^{-1}, for a mean depth of about 4 to 5 feet. Figure 2.7 provides the reaeration rates estimated by the Delta method for a 12-hour photoperiod (graph i) and by Eq. (2.5) (graph ii). The reaeration rate β is a function of the phase shift ϕ between the time of solar peak (1:00 noon) and the time of DO maximum (4:00 to 6:00 pm) in both methods, but the estimation techniques are slightly different as shown in Figures 2.6 and 2.7. As shown in Figure 2.14 (curves G and H), however, the phase shift is difficult to determine accurately as the DO curve is nearly flat around the maximum. This is mainly the consequence of a mass balance between respiration, reaeration and photosynthesis, which are in dynamic equilibrium around the maximum. Therefore, the phase shift ϕ cannot be determined as a single point value, but rather it is estimated as a range of values. The phase shift ϕ for Lake Harapan is estimated to be in the range of 3.5 hours to 4.5 hours. Thus, from Figures 2.6 and 2.7, the range of β is estimated to be 0.8 d^{-1} to 4.0 d^{-1} by the Delta method (graph i). Based upon Eq. 2.5, β is estimated to be 2.6 d^{-1} to 4.8 d^{-1} (graph ii).

EVM is now used to provide an alternative estimation for β. At daytime, maximum DO is attained when there is equilibrium between photosynthesis, respiration and reaeration. At nighttime, the minimum value of DO indicates a balance between respiration and reaeration. Therefore, by using the EVM proposed by Wang *et al.* (2003), β at daytime and nighttime can be estimated by using Eq. (2.7) and (2.6), respectively. The reaeration rate estimated by Eq. (2.6) which is applicable at the minimum DO value at nighttime in the EVM method falls in the range of 1.1 d^{-1} to 1.6 d^{-1}. The evaluation of β estimated by Eq. (2.7) which is applicable at the maximum DO attained during daytime by the EVM is unreliable due to the difficulty in estimating the time of DO maximum, as can be seen from Figure 2.14 (curves G and H). Therefore, β is not estimated in this chapter using Eq. (2.7). Instead, a new method to estimate β is used in the following subsection.

2.6.5 *Proposed new method*

We therefore propose a new method to estimate β by using the difference between DO curves in the light bottles (A and B) and in the lake (G and H). Photosynthesis occurs both in the light bottles and in the lake. As the bottles were tightly closed, reaeration does not occur in the bottles, as it would in the lake. This accounts for the difference in the DO levels in the bottles (A and B) and in the lake (G and H) as shown in Figure 2.14. Hence, we propose another approach to estimate β based upon this observation that would avoid the uncertainty in determining the time of DO maximum. We propose Eq. (2.8) for the estimation of β by *integrating* DO levels in the light bottles c_{bottle} and in the lake c_{lake} in order to overcome the deficiency in EVM and the Delta methods that are sensitive to accurate *point estimation* of time of DO maximum:

$$\beta \int_{t_1}^{t_2} [c_s - c(t)]dt = c_{bottle}(t_2) - c_{lake}(t_2) \qquad (2.8)$$

Here, time t is measured in day, t_1 is the time of sunrise (7/24 day) and t_2 is the end time of integration, in day. The results of this integration method, as opposed to point estimation, are less sensitive to the choice of t_2. The integration method Eq. (2.8) utilizes the full range of DO curves (G and H) in contrast to the other methods, which use only extreme point values. Near the extreme values, the maximum DO tends to oscillate due to the dynamic interaction between the variability of three parameters that contribute to DO levels, namely, β, R_e and P_h. Hence, estimations of extreme point values are not accurate or reliable, particularly for time of maximum DO. This characteristic renders EVM not reliable, as the time of maximum DO is not well observed, as can be seen in the DO data for Lake Harapan. However, this deficiency does not affect the integration method, which utilizes the full range of DO values, not the extreme point value only. The values of β estimated from Eq. (2.8) for the five monitoring days are in the range of 3.3 d^{-1} to 7.0 d^{-1} as shown in Table 2.6.

It should be noted that this range of β as shown in Table 2.6 reflects the reaeration near the lake surface (0.25 m), while the β values (of about 1.0 d^{-1} to 1.6 d^{-1}) estimated earlier by Figure 2.14

Table 2.6:   Reaeration rates β d^{-1} estimated using Eq. (2.8).

			Date		
Bottle	26 Nov	28 Nov	30 Nov	4 Dec	5 Dec
A	4.3	6.2	4.4	3.3	3.7
B	5.2	7.0	4.4	3.3	4.3

are more indicative of averaged reaeration over the entire water depth of some 1.0 m to 1.5 m.

2.6.6   *Photosynthesis rates P_h*

We will now estimate the maximum net photosynthesis by the means of curve regression and Eq. 2.9:

$$a_m = \frac{P_{net}}{\sqrt{(\beta^2 + \sigma^2)}} \text{ mg/L} \tag{2.9}$$

The amplitude a_m of the diurnal DO curve in the lake is obtained by fitting the lake DO curves in Figure 2.14. Eq. (2.9), derived from the solutions to Eqs. (2.1) and (2.2), is then used to estimate the maximum net photosynthesis rate P_{net} (mg/L/d), with values of β and σ having been estimated previously. The regression produces correlation coefficients C of more than 0.8 as shown in Table 2.5. Similarly, by fitting the DO curve of the light bottles (A and B), the dynamics of the DO in the bottles are expressed in the form of an equation where it involves the mean DO value and the amplitude of the diurnal DO sine curve. The photosynthetic rate mg/L/day in the light bottle can then be derived by differentiating the DO equation. Hence, the primary production rate P_m mg/L/day in the lake and in the bottles is then obtained by adding back the respiration rates R_e to P_{net}. P_{net} and P_m in the lake are estimated by Eq. (2.9) by using β of 1.0 d^{-1} in TH and $\beta = 0$ in the bottles. The estimated values of P_{net} and P_m in the lake are shown in Table 2.7. The values of P_{net} in the bottles vary between 37.7 mg/L/d and 51.6 mg/L/d, whereas P_{net} in the lake varies from 25.5 mg/L/d to 43.3 mg/L/d. The difference between the net photosynthetic rates P_{net} in the bottles and in the

Table 2.7: Maximum net photosynthesis rate P_{net} mg/L/d and primary production rate P_m mg/L/d in lake and bottles estimated using curve regression.

	Bottle A			Bottle B			Tasik Harapan		
Day	P_{net}	C	P_m	P_{net}	C	P_m	P_{net}	C	P_m
1	43.7	0.849	48.2	43.4	0.827	47.9	33.5	0.973	38.0
2	47.1	0.873	51.6	44.0	0.892	48.5	25.5	0.920	30.0
3	37.7	0.944	42.2	37.7	0.947	42.2	27.0	0.951	31.5
4	48.1	0.921	52.6	50.1	0.936	54.6	43.3	0.951	47.8
5	50.7	0.926	55.2	51.6	0.927	56.1	34.4	0.945	38.9

lake is due to the presence of reaeration in the lake, resulting in loss of DO, but not in the bottle since the bottle is sealed.

2.6.7 *Photosynthesis model for TH*

In summary, respiration rates R_e in Lake Harapan are estimated to be in the range of 3.4 to 5.6 mg/L/d. As the coefficients of determination R^2 exceed 0.90 for each of the monitoring days, it may be reasonably concluded that these rates are reliable. The reaeration rate β between 1.0 d^{-1} and 1.6 d^{-1} (for the entire water column depth of 1.5 m) estimated by the graphical method of Figure 2.6 appears to be reliable since this range of values does fall within the range of β determined by other methods such as the Delta method and EVM. It should be noted that β values are higher for the surface layer (0.25 m) as compared to those applicable for the entire water column (1.5 m). The maximum net photosynthesis rates P_{net} (after subtracting respiration rate R_e) are within the range of 25.5 mg/L/d to 43.3 mg/L/d. Hence, these parameter values are used to develop a model for algal photosynthesis to assist the understanding of algal dynamics and to assess efficient management options to rehabilitate Lake Harapan. These high values of photosynthesis indicate an urgent need to rehabilitate TH. Noting that TH is a stagnant lake with no outflow, an obvious option of rehabilitation is to create water outflow from TH to flush out nutrients. This can be accomplished if there is sufficient rainwater storage within the campus. Hence, it is essential to understand the role of hydrology on the USM campus in providing adequate rainwater for rehabilitation of TH.

2.6.8   *Role of hydrology*

Tasik Harapan is classified as highly eutrophic with a high chl a concentration and wild diurnal fluctuations in DO levels, impairing ecosystem services such as recreation and causing potential health risks. The two main reasons for this eutrophic state are (a) high nutrient input flowing into TH from settlements outside the campus and (b) accumulation of nutrients in the bottom sediments over the past decades as TH has virtually no outflows. Regular flushing of lake water would help to remove nutrients from the lake provided there is sufficient water storage. To promote flushing, rainwater should be harvested and stored in TH as the water source. Hence, analysis of the USM campus's hydrology is essential to provide the water input for flushing approaches. Further, TH should be dredged to remove accumulated nutrients in the sediments and to deepen the lake to increase storage volume. The hydrology in USM is mainly driven by campus surface runoff after a heavy rain event.

2.7   Surface Runoff

When rainfall volume exceeds the soil infiltration capacity, excess rainwater known as runoff is created on the land surface. This surface runoff either remains onsite as initial abstraction or flows to lower elevations, starting the process of a local flood. A quick discharge of high runoff into the river and drainage systems causes an overflow, and results in floods in the watershed. Conventional drainage by rapid removal of surface runoff and immediate transport to downstream areas is not a good flood management practice. Back-to-nature flood control measures using retention ponds, undisturbed forests, wetlands and lakes are the best forms of flood defense by retaining rainwater onsite and releasing it slowly. Slow release of stormwater allows the river and urban drainage systems ample time to gradually drain the excess runoff, thereby reducing flood volumes and peak flows, both onsite and downstream. The surface runoff volumes depend on several factors, including rainfall intensity and duration, soil types and conditions as well as land-use patterns. Rainfall of high intensity and long duration creates high surface runoff. Soil with higher infiltration capacity or higher permeability allows higher infiltration of rainwater into the underground and reduces runoff volume and minimizes floods. High runoff can result from a combination of factors such as low soil infiltration rate, high groundwater table and urbanization that reduces permeability. A simple simulation model known as WinTR-55 is used to examine storm surface runoff in USM.

2.7.1 *WinTR-55 for flood simulation*

To calculate storm runoff characteristics, a flood simulation model known as WinTR-55 was developed by the US Department of Agriculture. WinTR-55 (Urban Hydrology for Small Watersheds) was first developed in 1975 as a procedure to calculate the storm runoff characteristics such as runoff volume, peak rate of discharge, hydrographs and storage volumes required for stormwater management structures (SCS, 1975). After several revisions, the model has become a standard tool for analyzing storm runoff characteristics caused by urbanization for design purposes. The latest revision adapted for windows applications, which is known as WinTR-55 (USDA NRCS, 2009), is used to simulate flooding on the USM campus. The most important empirical parameter used in WinTR-55 is the runoff curve number (CN).

2.7.1.1 *Runoff curve number CN*

The runoff CN is an empirical parameter used in hydrology for predicting direct runoff from rainfall excess. The CN method was developed by the USDA Natural Resources Conservation Service, formerly called the Soil Conservation Service or SCS. The CN number is popularly known as the "SCS runoff curve number" in the literature. The runoff curve number CN was developed from an empirical analysis of runoff from small catchments and hillslope plots monitored by the USDA. It is widely used and is an efficient method for determining the approximate amount of direct runoff from a rainfall event in a particular area. The runoff curve number is dependent on the area's hydrologic soil group, land-use pattern and hydrologic conditions. The lower the curve number CN, the more permeable the soil is and the lower the runoff volume. Runoff cannot begin until the initial abstraction has been met. CN has a range from 30 to 100; lower CN numbers indicate low runoff potential, while larger numbers indicate increasing runoff potential. In addition to CN, there is another important parameter known as the time of concentration Tc that characterizes how fast the runoff flows over the land and through flow channels. Larger Tc indicates smaller peak flow and longer time to peak flow. Smaller Tc indicates larger peak flow and faster time to peak flow. Smaller Tc is indicative of urbanization.

2.7.1.2   *Tc and Tt*

The time of concentration (Tc) is the time required for a drainage area to contribute to runoff at the point of interest. This time is calculated as the time for runoff to flow from the most hydraulically remote point of a chosen drainage area to the point under investigation. The time of concentration Tc is equal to the summation of the travel times (Tt) for each flow component along the hydraulic path. A hydraulic path may consist of several flow components. The Tt for a flow component along the hydraulic path is the length of time it would take a drop of water to flow across that component of land. The hydraulic path has three flow components known as (i) sheet flow, (ii) shallow concentrated flow and (iii) open channel flow. The runoff curve numbers CN and time of concentration Tc are critical parameters essential for flood simulation of USM runoff by WinTR-55.

2.8   Flood Simulation for USM

To model urban flood, the hydrological and topographic characteristics in the watershed must be ascertained. For USM, the hydrological and topographic characteristics on campus are analyzed by dividing the USM campus watershed into eight subareas as shown in Figure 2.16, contributing to a total drainage area of 222 acres. Catchment areas that are not hydrologically connected to TH, such as the football field, are ignored. Runoff simulation in USM is performed by WinTR-55.

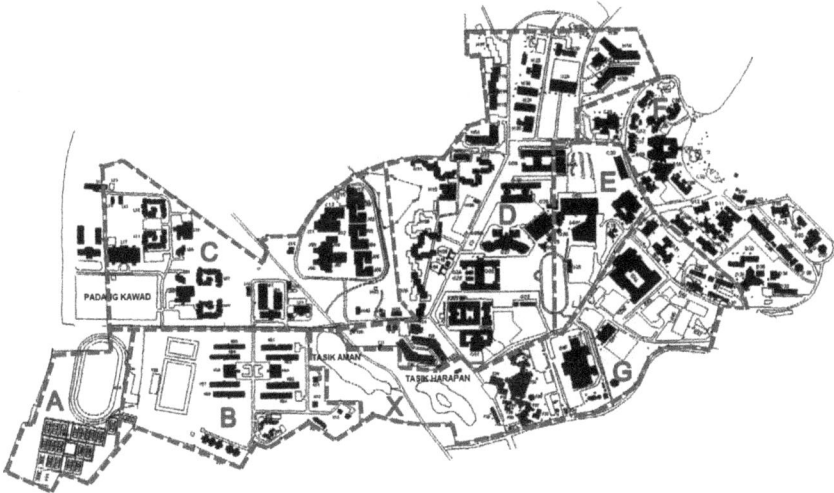

Figure 2.16: Subareas of developed USM campus.

2.8.1 *Hydrological and topographic characteristics*

By analyzing the maps of the USM campus obtained from the Development Department of USM and site survey, parameter values indicative of the topography and hydrology of each subarea can be estimated. These estimated land-use and hydrological characteristics for USM are listed in Tables 2.8 and 2.9, respectively. These land-use and hydrologic characteristics are used for WinTR-55 simulation of the flood to obtain runoff characteristics such as runoff volume, peak rate of discharge, hydrographs and retention volume required for the present developed campus. Type III rainfall appropriate for USM is used, with rainfall of two return periods, with a total rain depth over 24 hours of (a) 6 inches (150 mm) and (b) 4 inches (100 mm), corresponding to return periods of 4 years and 2 years, respectively. This WinTR-55 simulation would provide useful information on the flood situation in the campus and the size of the retention pond required. For flood mitigation, USM constructed a retention pond known as TH in 1990 to attenuate runoff generated in the campus. Tasik Harapan has a surface area of 6080 m^2 (1.5 acres) and a depth of 1.0 to 1.5 m. With an average annual rainfall of about 2500 mm, USM receives between 75 mm and 400 mm of rainfall per month.

Table 2.8:   Topography characteristics of developed USM campus.

Sub area	Area (acre)	Land-use characteristics	Downstream subarea	CN	Slope (ft/ft)
A	11	Open space; fair condition; grass cover 50–70%; group B, Impervious areas	B	91	0.052
B	33	Open space; poor condition; grass cover <50%; group B, Urban districts; commercial and business; group C, Residential districts – 1 ac; group B, Impervious areas	X	93	0.025
C	45	Open space; poor condition; grass cover <50%; group B, Urban districts; commercial and business; group C, Impervious areas	X	92	0.063
D	56	Open space; poor condition; grass cover <50%; group B, Urban districts; commercial and business; group C, Impervious areas	X	92	0.048
E	22	Open space; poor condition; grass cover <50%; group B, Urban districts; commercial and business; group C, Impervious areas	D	91	0.089
F	11	Urban districts; commercial and business; group C, Impervious areas	E	95	0.089
G	22	Open space; poor condition; grass cover <50%; group B, Urban districts; commercial and business; group C, Impervious areas	X	94	0.175
X	22	Open space; fair condition; grass cover 50–70%; group B, Impervious areas, Streets and roads; paved; curbs and storm sewers	Outlet	81	0.125

Table 2.9:   Hydrological characteristics of developed USM campus.

| | Length (ft) | | | Rainfall depth (inch) | | | |
| | | | | 6 | | 4 | |
Sub area	Sheet flow	Conc flow	Channel flow	Tc (hr)	Tt (hr)	Tc (hr)	Tt (hr)
A	300	300	300	0.05	0.02	0.05	0.02
B	300	890	840	0.15	0.12	0.16	0.12
C	300	500	980	0.07	0.04	0.07	0.04
D	300	780	1600	0.39	0.09	0.46	0.09
E	300	480	600	0.05	0.03	0.06	0.03
F	300	180	480	0.04	0.02	0.04	0.02
G	300	280	560	0.03	0.02	0.04	0.02
X	300	400	760	0.05	0.03	0.05	0.03

Table 2.10:   Tc and Tt for each subarea with different CN before development.

| | Rainfall = 6 inches | | | | | | Rainfall = 4 inches | | | | | |
| | CN = 71 | | CN = 75 | | CN = 79 | | CN = 71 | | CN = 75 | | CN = 79 | |
Sub area	Tc (hr)	Tt (hr)	Tc (hr)	Tt (hr)	Tc (hr)	Tt (hr)	Tc (hr)	Tt (hr)	Tc (hr)	Tt (hr)	Tc (hr)	Tt (hr)
A	0.32	0.03	0.25	0.03	0.12	0.03	0.38	0.03	0.30	0.03	0.13	0.03
B	0.53	0.14	0.44	0.14	0.25	0.14	0.61	0.14	0.50	0.14	0.28	0.14
C	0.33	0.07	0.27	0.07	0.14	0.07	0.39	0.07	0.32	0.07	0.16	0.07
D	0.42	0.13	0.35	0.13	0.21	0.13	0.49	0.13	0.40	0.13	0.23	0.13
E	0.28	0.05	0.22	0.05	0.11	0.05	0.33	0.05	0.26	0.05	0.13	0.05
F	0.26	0.02	0.20	0.02	0.09	0.02	0.31	0.02	0.24	0.02	0.11	0.02
G	0.20	0.02	0.16	0.02	0.07	0.02	0.24	0.02	0.19	0.02	0.08	0.02
X	0.24	0.04	0.19	0.04	0.10	0.04	0.29	0.04	0.23	0.04	0.11	0.04

Frequent floods occur due to the heavy rain during the inter-monsoon seasons between April and May and between October and November. For comparison, simulation of runoff characteristics on the USM campus after development and before development are performed under three curve numbers where CN = 71, 75 and 79 as shown in Table 2.10.

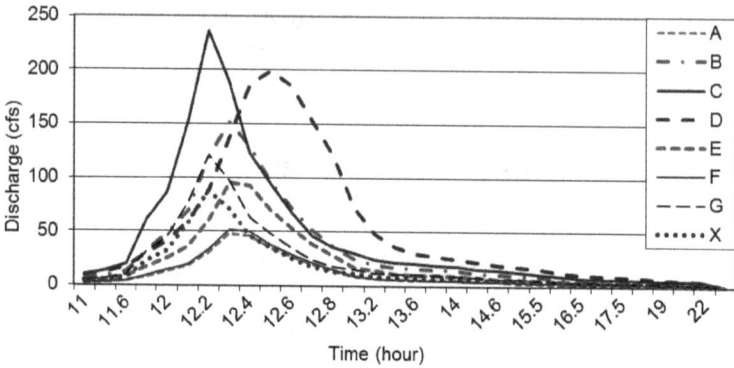

Figure 2.17:   Hydrograph for subareas on the present USM campus with 6 inches rainfall.

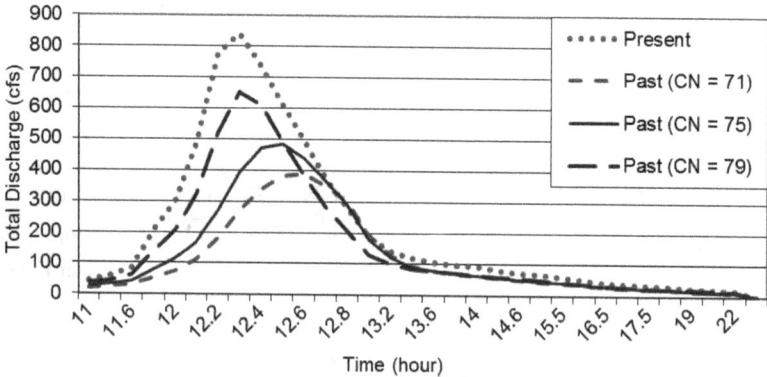

Figure 2.18:   Hydrograph for the past and present USM campus with 6 inches rainfall.

2.8.2   *Simulation results*

2.8.2.1   *Rain of 4-year return period*

WinTR-55 simulations are performed for the present USM with rainfall type III under two rainfall volumes over a period of 24 hours: (a) 6 inches (150 mm) and (b) 4 inches (100 mm), corresponding to return periods of 4 years and 2 years, respectively. The topographic and hydrological characteristics are given in Tables 2.8 and 2.9, respectively. Figure 2.17 shows the hydrograph for each subarea on the present USM campus subject to rain of a 4-year return period with 6 inches of rainfall over a period of 24 hours. Figure 2.18 shows

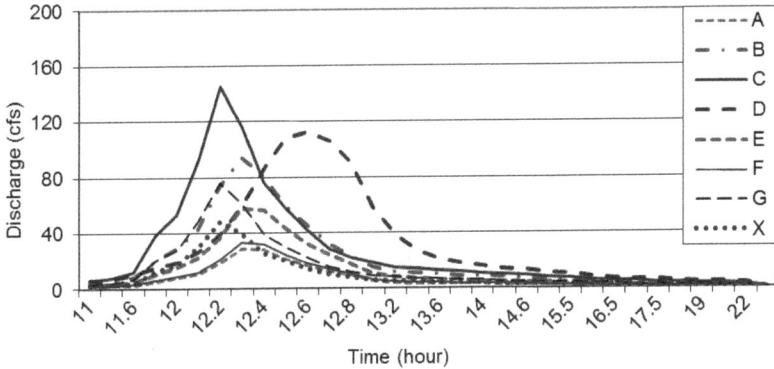

Figure 2.19: Hydrograph for subareas on the present USM campus with 4 inches rainfall.

the hydrograph for the past (with CN = 71, 75, 79) and present USM campus with 6 inches of rainfall over a period of 24 hours. For rain of a 4-year return period with 6-inch rainfall over 24 hours, WinTR-55 simulations indicated a runoff volume of 27 acre-ft or a runoff depth of 1.5 inches, which is beyond the storage capacity of TH, resulting in flooding on the campus. Hence, TH should be deepened to 4 m to accommodate additional storage of the runoff after heavy rain. Further, the rainwater can be harvested from campus rooftops and stored in TH to improve TH flushing and to reduce eutrophication.

2.8.2.2 *Rain of 2-year return period*

Figure 2.19 shows the hydrograph for each subarea on the present USM campus subject to rain of a 2-year return period with 4 inches of rainfall over a period of 24 hours. Figure 2.20 shows the hydrograph for the past (with CN = 71, 75, 79) and present USM campus with 4 inches of rainfall over a period of 24 hours. For rain of a 2-year return period with 4-inch rainfall over 24 hours, WinTR-55 simulations indicated a runoff volume of 20 acre-ft or a runoff depth of 1.0 inches, which is slightly beyond the storage capacity of TH, resulting in flooding on the campus. To accommodate this runoff volume, TH should be dredged to a depth of 4 m. Further, rainwater can be harvested from campus rooftops and stored in TH to improve TH flushing and to control eutrophication.

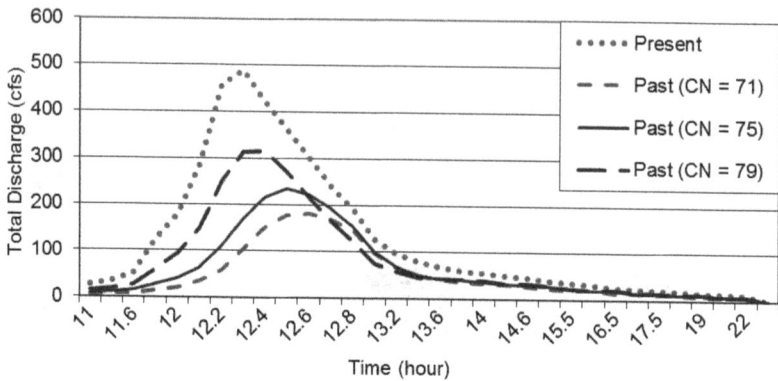

Figure 2.20:   Hydrograph for the past and present USM campus with 4 inches rainfall.

2.8.3   *Role of rainwater in eutrophication control*

Dredging TH to a depth of 4 m would remove most nutrients accumulated in the sediment and would significantly improve water quality instantly post dredging. The increased rainwater storage capacity would allow improved flushing to prevent buildup of nutrients in the sediment as a measure of eutrophication control. Further, increase in depth would provide a better buffer between the water column and the sediment, thereby helping to reduce eutrophication. Rainwater harvested on the campus can be used to provide the water needed for flushing TH at the rate of 0.01 per day or equivalent to a turn over time of 100 days. Post dredging, this flushing rate would maintain TH at an oligotrophic clear-water state with a low level of algae.

2.9   Eutrophication Control for TH

Lake eutrophication can be rehabilitated by several methods such as (a) application of macrophytes to remove nutrients and to suppress algae growth, (b) dredging of lake sediment to remove nutrients and pollutants, (c) flushing of lake water to remove nutrients and algae, and (d) reduction of external phosphorus loading. For USM, reduction of external nutrient loading is virtually impossible.

2.9.1 *Water hyacinth*

Water hyacinth has been used in eutrophication and wastewater treatment due to its high capacity for nitrogen and phosphorus absorption, and its ability to grow in heavily polluted water. However, water hyacinth is a highly invasive aquatic weed with a rapid growth rate and high adaptability to extreme conditions. It can survive in both tropical and temperate climatic conditions and is able to store nutrients for later stages of its life cycle. Because it is highly invasive, extreme care must be taken when water hyacinth is introduced in any water body. To understand the harm that water hyacinth can inflict to a lake, a set of three ordinary differential equations is used to model the dynamics between phosphorus P, algae A and water hyacinth W, the details of which can be found in Tay *et al.* (2020). Here, Eqs. (2.10)–(2.12) describe the growth of water hyacinth (WH) with harvesting h. Table 2.11 provides the units and parameter values for each of the symbols used in Eqs. (2.10)–(2.12). The growth of WH is phosphorus limiting, with a carrying capacity denoted by K kg/m^2. It is assumed that TH receives high phosphorus input from both external sources and from sediments to maintain a constant chl a level of 300 μg/L in the absence of WH. Water hyacinth is introduced into the highly eutrophic TH at time t = 0 day to control algal growth and to reduce chl a concentration. Figure 2.21 shows the algal concentration in TH after the introduction of WH. Two scenarios are simulated: (i) with no harvesting of WH and (ii) with constant harvesting of WH. WH suppresses the growth of algae by competing with algae for phosphorus and space. WH continues to growth until it reaches the carrying capacity of 70 kg/m^2, beyond which WH stops growing. The algae concentration A starts to decrease after the introduction of WH and continues to decrease to 250 μg/L of chl a, when WH reaches the carrying capacity of 70 kg/m^2. Then, the algae density begins to increase because the WH stops growing and stops taking up phosphorus. However, if WH is harvested constantly, then algal concentration will be suppressed to 120 μg/L of chl a. This optimal WH harvesting rate implies that WH must be removed from TH (with surface area of 6070 m^2) at the rate of 0.5 × 70 kg/m^2 × 6070 m^2 × 0.0506 d^{-1} = 10750 kg/day. This daily harvesting rate would require 6.5 workers per day, assuming a worker can remove 200 kg/hour for 8 hours

Table 2.11:   Definition and value of the parameters in Eqs. (2.10)–(2.12).

Symbol	Definition	Unit	Value	Source	Range
A	Algal concentration	μg/L	–	–	–
P	Phosphorus concentration	μg/L	–	–	–
C	DO concentration	μg/L	–	–	–
L	BOD concentration	μg/L	–	–	–
W	Water hyacinth wet weight per area	kg/m^2	–	–	–
b	Algal growth rate	d^{-1}	0.7	Curve fitting	0–1
h_a	Half saturation constant	μg/L	10	Jones (2018)	0–10
p_a	Phosphorus content percentage	–	1	Jones (2018)	0–1
h_1	Flushing rate	d^{-1}	0	–	0–0.8
g	Zooplankton grazing rate	d^{-1}	0.03	Genkai-Kato and Carpenter (2005)	0–0.91
s_a	Algal mortality rate	d^{-1}	0.085	Genkai-Kato and Carpenter (2005)	0–0.9
l_p	External P loading rate	μg/L/d	0.3	Curve fitting	–
r	Internal P sediment recycling rate	μg/L/d	0.3	Curve fitting	0–14
n	Half saturation value of recycling function	μg/L	2.4	Carpenter (2005)	0–2.4
q	Parameter for steepness of sigmoid function near n	–	20	Carpenter et al. (1999)	0–20
e	Phosphorus excretion associated with grazing	μg/μg	0.65	Genkai-Kato and Carpenter (2005)	0.4–0.8
k_a	Reaeration rate	d^{-1}	1.6	Curve fitting	0–7
C_s	Saturated DO concentration corresponding to each temperature	μg/L	7500	Rounds et al. (2013)	–
k_1	Deoxygenation rate	d^{-1}	0.29	Curve fitting	0.005–0.5

Table 2.11: (*Continued*)

Symbol	Definition	Unit	Value	Source	Range
p_{max}	Maximum oxygen production rate by photosynthesis at saturating lighting condition	d^{-1}	12.3	Computed	–
l_{BOD}	BOD loading	$\mu g/L/d$	2300	Curve fitting	–
g_w	Maximum water hyacinth growth rate	d^{-1}	0.11	Wilson *et al.* (2005)	0–0.11
h_w	Half saturation constant for phosphorus uptake by water hyacinth	$\mu g/L$	10	Eid and Shaltout (2017)	2–100
p_w	Phosphorus content per unit of water hyacinth	$\mu g \cdot m^2 /Lkg$	0.53	Su *et al.* (2018)	0–1
K	Maximum carrying capacity of water hyacinth	kg/m^2	70	Wilson *et al.* (2005)	50–76
h	Harvesting rate of water hyacinth	d^{-1}	0.0506	Computed	0–1

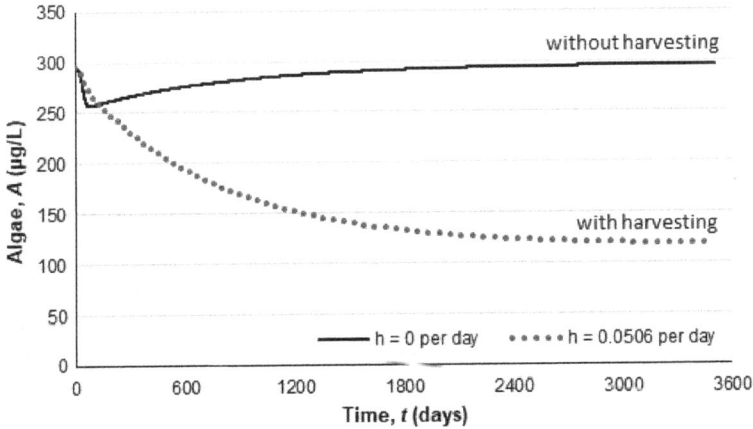

Figure 2.21: Simulated algal level, A ($\mu g/L$ chl a), with and without water hyacinth harvesting.

per day (Su *et al.*, 2018). This would impose a high financial cost, besides the problem of disposal of harvested WH. Hence, eutrophication control by WH is not feasible for TH. Other alternatives such as sediment dredging are explored. Dredging has two important contributions in controlling eutrophication. First, sediment dredging removes a significant quantum of nutrients from TH in an instant. Second, dredging TH to a depth of four meters would increase lake volume and would permit a higher rate of continuing flushing to remove more nutrients and algae from TH. Further, water harvested from RWH can be stored in TH to maintain a consistent rate of flushing.

$$\frac{dA}{dt} = bA\frac{P - p_a A}{h_a + P - p_a A} - (h_1 + g + s_a)A \tag{2.10}$$

$$\frac{dP}{dt} = l_p + r\frac{P^q}{P^q + n^q} + egA - bA\frac{P - p_a A}{h_a + P - p_a A}$$

$$- h_1 P - g_w p_w W\left(1 - \frac{W}{K}\right)\left(\frac{P}{P + h_w}\right) \tag{2.11}$$

$$\frac{dW}{dt} = g_w W\left(1 - \frac{W}{K}\right)\left(\frac{P}{P + h_w}\right) - hW \tag{2.12}$$

2.9.2   *Sediment dredging*

To control eutrophication in TH, three treatment methods have been examined: (a) mechanical aeration of TH, (b) EM solution and mudball treatment and (c) application of water hyacinth. All three treatment techniques have been proven to be not feasible or effective in controlling eutrophication in TH. Mechanical aeration does not remove nutrients, the main contributors to eutrophication. It merely increases DO concentrations, which are already high in TH during the daytime. Similarly, EM and mudball treatment does not remove nutrients from TH, the root cause of eutrophication, and hence is not effective in controlling eutrophication in TH. Further, EM and mudball treatments of up to 1.0 g L^{-1} had been reported to have increased chl *a* concentration from $\approx$120 to 325–435 μg L^{-1}, within 4 weeks of treatment. In another experiment, chl *a* concentration in EM and mudball treatments (52 μg L^{-1}) was significantly higher compared to that in controls (20 μg L^{-1}). Further, a high amount of clay introduced by mudballs would induce subsequent high turbidity in the water, further deteriorating water quality. In short, EM and mudball treatment is not effective in preventing cyanobacterial proliferation or in terminating algal blooms because they neither permanently bind to nor remove phosphorus from eutrophic systems. In summary, EM and mudballs have no inhibiting effect on cyanobacteria, and they could even be an extra source of nutrients and turbidity (Lurling *et al.*, 2010). Finally, highly invasive water hyacinth would not be effective in controlling eutrophication in TH. It would quickly and permanently chock up TH. Hence, a viable option to the rehabilitation of TH would involve sediment dredging to remove nutrients and other pollutants accumulated in the bottom sediment. To examine the impact of sediments on eutrophication, numerical simulation studies were performed to assess the contributions of sediment nutrients to eutrophication in TH (Teh *et al.*, 2009). The total source of P and N in the ratio of 1 part of P to 8 parts of N drives algal growth in TH. For a simple analysis, this nutrient source derived from the sediment layers is adjusted until the chl *a* concentration reaches 300 μg/l (Figure 2.22).

2.9.2.1 *Simulation results*

Figure 2.22 shows the simulated chlorophyll *a* and DO concentrations at the top 30-cm surface of TH before sediment removal. The simulation begins with a low level of chl *a* and with DO of 5 mg/L. Strong photosynthesis allows a steady state to be achieved after a month of simulation. After about a month of simulation time, chlorophyll *a* achieves steady state slightly below 300 μg/L. On the contrary, DO levels fluctuate between 4.5 and 11 mg/L, clearly indicating a highly eutrophic state. Simulation is then performed with the bottom sediment dredged to remove all nutrients, but subject to nutrient input into TH from the surrounding catchment, including sullage water from USM and sewage from outside. It is estimated that this nutrient input from the campus and the surrounding settlements contributes about 10% of the original nutrient input from the accumulated sediments. The simulated chl *a* levels after dredging are now reduced from 300 μg/L to about 40 μg/L, while DO level fluctuates around 6 mg/L (Figure 2.23), indicating a eutrophic condition. Over a long duration,

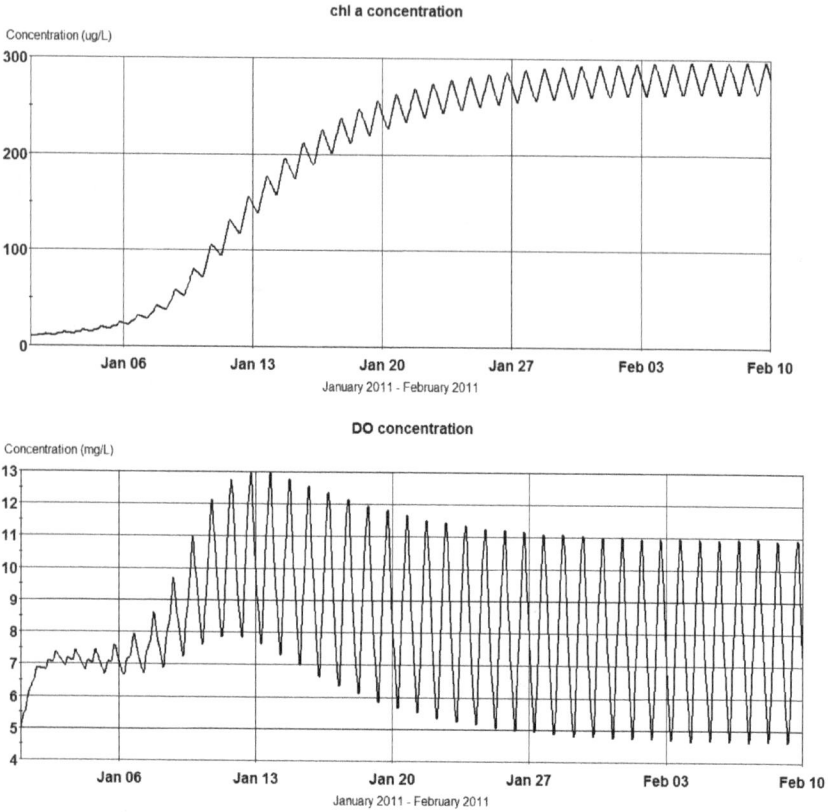

Figure 2.22:   Simulated chl *a* (top) and DO (bottom) in Tasik Harapan before sediment removal.

nutrients will build up in the sediment layer, thereby repeating the eutrophication cycle. Hence, flushing of water in TH is needed to constantly remove nutrients from TH. The source of this water may come from rainwater harvesting (RWH) (Furumai, 2008; Song *et al.*, 2009; Jones and Hunt, 2010). To create adequate storage volume in TH to retain additional water collected from RWH to support flushing, TH needs to be dredge to a depth of 4 m. This additional depth and volume will also reduce P and algal concentration further due to the dilution effect. The overall impact is a rehabilitation of TH to an oligotrophic state with chl *a* level below 10 μg/l and DO around 7 mg/L.

chl a concentration

Concentration (ug/L)

January 2011 - February 2011

DO concentration

Concentration (mg/L)

January 2011 - February 2011

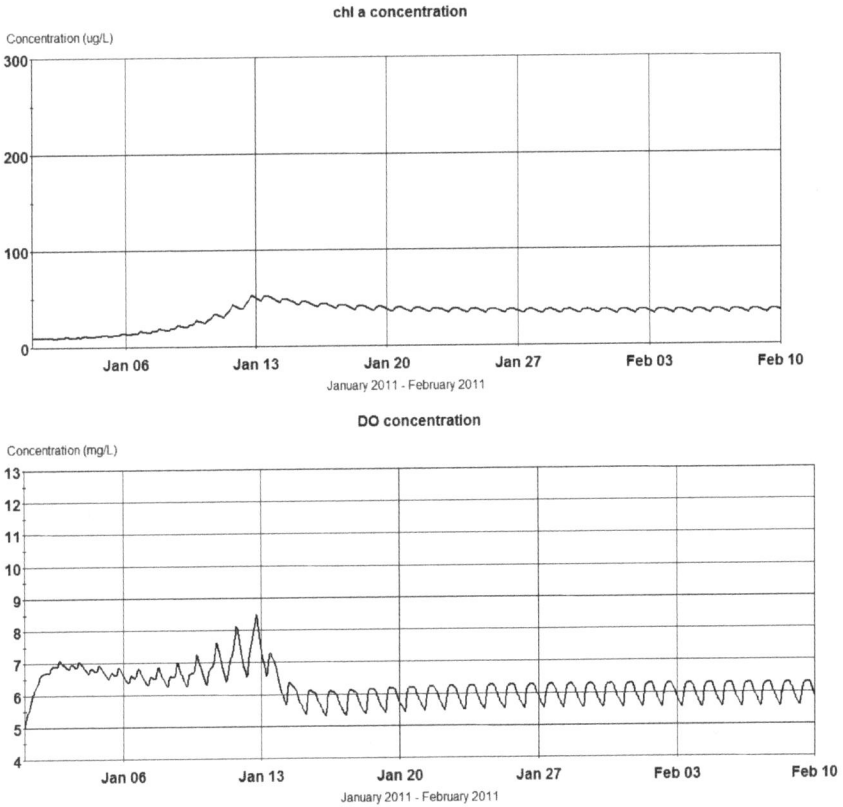

Figure 2.23: Simulated chl *a* (top) and DO (bottom) in Tasik Harapan after sediment removal.

2.9.2.2 *Dredging techniques*

There are two major techniques for sediment removal from freshwater lakes (Cooke *et al.*, 2005). The first is to drain the lake completely followed by excavation. The obvious limitation of this technique is that it is time consuming as water must be drained, and the basin must be allowed to dry sufficiently before excavation equipment can operate. The second and most common sediment removal technique is to dredge the sediments directly, which may cause temporary sediment resuspension and turbidity. Basically, a cutter head with steel blades dislodges the sediments, and a centrifugal pump sucks up the muck. This sediment and water mix (called a slurry) is piped to a holding basin on land where the water is drained off and the sediments are left to dry. For hydraulic dredging, a disposal basin must be designed and constructed that is large enough to not only hold all the dredged sediments but also allow the pumped sediment and water slurry enough time to settle out the solids and return relatively clear water to the lake. Further, lake sediments tend to increase in volume temporarily during and immediately after hydraulic dredging. Therefore, the disposal basin would have to be sized to hold 1.2 times of the lake sediment volume.

The sediments removed from the lake should be reused commercially to promote sustainability. Possible commercial usages of the sediments are agricultural soils, concrete production, brick fabrication, ceramic coatings, and coastal defense and rehabilitation (Bernardo *et al.*, 2000; Molino, 2002; Carone *et al.*, 2006). Landfills, croplands and pasturelands are commonly utilized for land application of trucked sediments. The geology of the disposal site must be considered. Some sites may be too porous and could allow nearby lakes, streams or groundwater tables to be impacted if the land is not lined. Costs increase when the sediments must be trucked far away, or the disposal site must be significantly modified to contain the sediments. If the lake sediments themselves are found to have elevated concentrations of contaminants (heavy metals, pesticides), special handling and disposal will be required. Nutrient-enriched sediments are not considered contaminated and therefore do not require special handling. Dredging costs can vary greatly, depending on several factors.

2.9.3   *Water flushing*

It may be concluded that sediment removal by dredging TH to a depth of 4 m is a viable means to control eutrophication in TH. Accumulated nutrients stored in the sediments would be removed instantly to prevent nutrient input into the water column from the sediment. However, nutrients coming from the campus and the surrounding settlements would contribute a high nutrient load to TH because of its shallow depth of 1.0 to 1.5 m. Dredging to a depth of 4 m will provide additional volume and dilution to reduce the impact of eutrophication. To prevent further accumulation of nutrients in the bottom sediment, it is essential to constantly flush out nutrients and

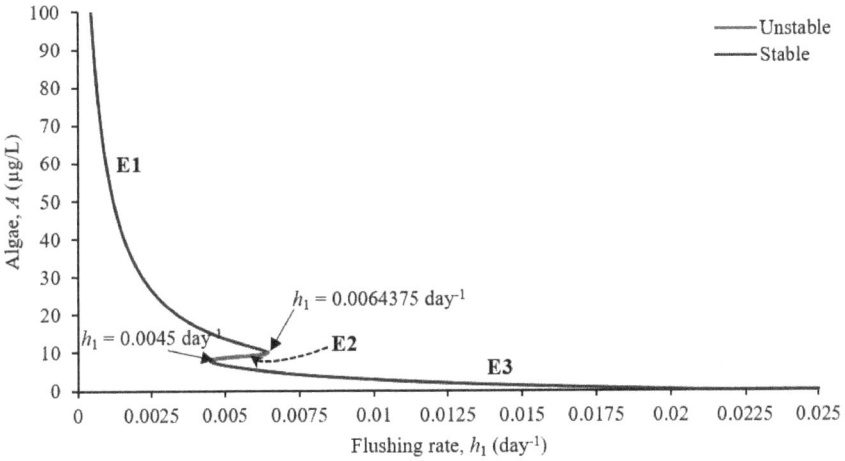

Figure 2.24: Bifurcation diagram of algae, A (μg/L), against flushing rate, h_1 (day^{-1}).

algae from TH. The water needed for this flushing can be obtained by harvesting and storing rainwater in TH with an increased depth. Water stored in TH can be utilized for landscape irrigation as the water quality is sufficiently good. This will reduce water and fertilizer consumption on campus, contributing to SDG 6. To be effective and efficient, an optimal flushing rate should be estimated as follows.

2.9.3.1 *Optimal flushing rate*

Flushing is one of the viable lake restoration methods for TH. In order to determine the optimal flushing rate required to restore TH to an oligotrophic state ($A < 2.6$ μg/L), a bifurcation analysis is performed. The bifurcation diagram linking algae A with flushing rate h1 is displayed in Figure 2.24. As expected, the algal concentration A decreases with increasing flushing rate, as flushing removes algae and P from TH. As may be seen from Figure 2.24, three equilibrium states E1, E2 and E3 (two stable and one unstable) exist when 0.0045 day^{-1} < h1 < 0.0064375 day^{-1}. In this range of h1, A will be attracted to either the stable steady-state E1 ($A > 10$ μg/L, eutrophic) or to the stable steady-state E3 ($A < 10$ μg/L, oligotrophic). If A is above E2 and below E1 (with 0.0045 day^{-1} < h_1 < 0.0064375 day^{-1}), it approaches the stable steady-state E1. An algal concentration A lower than E2 (with 0.0045 day^{-1} < h_1 <

0.0064375 day^{-1}) will decrease and approach the stable steady-state E3. The algal concentration will never approach the unstable steady-state E2. From Figure 2.24, a flushing rate of $h_1 > 0.0064375$ day^{-1} would restore TH to a mesotrophic state $(A < 9.735 \ \mu g/L)$. The oligotrophic state of TH can be achieved conservatively when $h_1 > 0.009$ day^{-1} $(A < 2.6 \ \mu g/L)$. During the flushing process, lake water laden with algae and phosphorus is flushed out of the lake. Based on the bifurcation diagram in Figure 2.24, a flushing rate of h_1 equal to or more than 0.01 day^{-1} would restore TH to an oligotrophic state $(A < 2.6 \ \mu g/L)$. A flushing rate of 0.01 per day would require 1% of the lake water to be replaced per day, equivalent to a residence time of 100 days. This suggests that the entire volume of TH needs to be replaced every 100 days to keep TH at an oligotrophic state. For this purpose, an adequate supply of water is needed to sustain the application of flushing. This can be achieved by rainwater harvesting on campus.

2.10   Conclusions

The following conclusions may be made. First, TH is highly eutrophic with wild fluctuation of DO over the diurnal cycle, reaching a high DO level of some 18 mg/L in the late afternoon. This highly eutrophic condition is detrimental to the provision of ecosystem services. A mechanical aerator is not an effective means to mitigate eutrophication in TH as it does not remove the sources of nutrients. Adding more DO by means of aerators into TH is meaningless in this situation since TH has excess dissolved oxygen due to high photosynthesis. Second, the application of a mudball and EM solution did not appear to reduce the degree of eutrophication in TH. The addition of mudballs may even increase turbidity and add additional nutrients that further complicate the eutrophication process. Third, the application of highly invasive water hyacinth is not a viable option for controlling eutrophication in TH, as its high growth requires constant removal, a costly process. Finally, removal of sediment from the lake bottom by dredging to a depth of 4 m appears to be a viable option. To remain in an oligotrophic state with a low level of algae, the water in TH must be flushed out at the rate of 0.01 per day. This is equivalent to replacing the lake water every 100 days. This sediment removal method is sustainable in the long run if in-campus rainwater harvesting can be utilized to provide the water needed for flushing to remove nutrients.

References

APHA (1985). Standard Methods for the Examination of Water and Wastewater, 16th edition, American Public Health Association, Washington, D.C.

Asha, S.C., Soo, L.C. and Chan, N.W. (2010). Tasik Harapan lake clean-up initiative. Abstract in the Proceeding of the 1st National Seminar on Environmental Humanities, 16–17 December 2010, Penang. Chan Ngai Weng (Ed.), p. 55.

Bennett, J.P. and Rathbun, R.E. (1972). Reaeration in Open-Channel Flow, US Geological Survey Professional Paper 737.

Benson, B.B. and Krause, D. Jr. (1984). The concentration and Isotopic Fractionation of Oxygen Dissolved in Fresh Water and Seawater in Equilibrium with the Atmosphere: I. Oxygen. *Limnology and Oceanography*, 29(3), 620–632.

Bernardo, G., Biscione, A., Marroccoli, M. and Molino, B. (2000). Reservoir rehabilitation by the sediment evacuation pipeline system and sediment utilization as raw material for cement industry. In: Proceedings of the International Conference New Trends in Water and Environmental Engineering for Safety and Life: Ecocompatible Solution for Aquatic Environments. Capri, Italy, 3–7 July, Balkema Publisher, The Netherlands.

Carone, M.T., Greco, M. and Molino, B. (2006). A sediment-filter ecosystem for reservoir rehabilitation. *Ecological Engineering*, 26, 182–189.

Carpenter, S.R. (2005). Eutrophication of aquatic ecosystems: bistability and soil phosphorus. *Proceedings of the National Academy of Sciences of the USA*, 102, 10002–10005.

Carpenter, S.R., Ludwig, D. and Brock, W.A. (1999). Management of eutrophication for lakes subject to potentially irreversible change. *Ecological Applications*, 9, 751–771.

Chapra, S.C. (1997). *Surface Water Quality Modeling*. McGraw-Hill, New York.

Chapra, S.C. and Di Toro, D.M. (1993). The delta method for estimating community production, respiration and reaeration in streams. *Journal of Environmental Engineering*, 117(5): 640–655.

Chapra, S.C. and Tarapchak, S.J. (1976). A Chlorophyll a model and its relationship to phosphorus loading plots for lakes. *Water Resources Research*, 12(6), 1260–1264.

Charumas, M. (2004). Strategies for sustainable management of fishery resources in the Pasad Jolasid Reservoir, Thailand through physio-chemical assessment. *Proceedings of the Second International Symposium on Southeast Asian Water Environment*, December 1–3, pp. 325–332, Ed. Shinichiro Ohgaki, Kensuke Fukushi, Hiroyuki Katayama and Satoshi Takizawa. University of Tokyo, AIT-Vietnam.

Chen, X., Ding, X. and Chen, S. (2004). The study of chlorophyll detection in coastal waters based on environmental variables. GIS and Remote Sensing in Hydrology, Water Resources and Environment (*Proceeding of ICGRHWE held at the Three Gorges Dam, China*, September 2003). IAHS Publication 289, pp. 316–321, Ed. Yangbo Chen, Kaoru Takara, Ian Cluckie and F. Hilaire De Smedt.

Cooke, G.D., Welch, E.B., Peterson, S.A. and Nichols, S.A. (2005). Restoration and management of lakes and reservoirs. Boca Raton, CRC Press, p. 616.

Eid, E.M. and Shaltout, K.H. (2017). Growth dynamics of water hyacinth (Eichhornia crassipes): A modeling approach. *Rendiconti Fisiche Accademia Lincei*, 28, 169–181.

Furumai, H. (2008). Rainwater and reclaimed wastewater for sustainable urban water use. *Physics and Chemistry of the Earth*, 33(5), 340–346.

Genkai-Kato, M. and Carpenter, S.R. (2005). Eutrophication due to phosphorus recycling in relation to lake morphometry, temperature, and macrophytes. *Ecology*, 86, 210–219.

Hashim, Z.H., Md. Shah, A.S., Mansor, M., Wan Omar, W.M., Md. Sah, S.A., Chong, A.S.C. and Kumar, K.S. (2006). Ecological assessment of Tasik Harapan, USM Main Campus. In: Healthy Campus Series 13, *Monitoring and Modelling of a University in a Garden*, (Eds.) Ahmad Izani Md. Ismail, Koh Hock Lye and Mohamed Izham Mohamed Ibrahim, Penerbit USM, pp. 48–70. ISBN: 983-3391-59-1.

Jones, M. (2018). Using a coupled bio-economic model to find the optimal phosphorus load in Lake Tainter WI. *Honors Research Projects*, 632.

Jones, M.P. and Hunt, W.F. (2010). Performance of rainwater harvesting systems in the southeastern United States. *Resources, Conservation and Recycling*, 54(10), 623–629.

Lurling, M., Tolman, Y. and van Oosterhout, F. (2010). Cyanobacteria blooms cannot be controlled by Effective Microorganisms (EM®) from mud- or Bokashi-balls. *Hydrobiologia*, 646, 133–143.

Mansor, M., Md Sah, S.A., Chong, S.C., Kumar, K.S., Wan Omar, W.M. and Md. Shah, A.S.R. (2004). Project Report: Assessment of Bioremediation Treatment of Harapan Lake Via Monitoring of Selected Physicochemical and Biological Parameters. Universiti Sains Malaysia.

Mara, D. (1998). *Sewage Treatment in Hot Climates*, 2nd Edition. John Wiley and Sons, Ltd.

Molino, B. (2002). Silting up of artificial reservoirs: Defence strategies—sediment management. In: *International Workshop Ecological, Sociological and Economic Implications of Sediment Management in Reservoirs*, Prignano, Cilento (Italy), 8–10 April.

Nielsen, E. and Pillay, C. (2006). Biological recovery of lakes employing the 'Bio Reco' effective aerator and 'Terra Biosa' beneficial microorganisms. In: *Healthy Campus Series 13, Monitoring and Modelling of a University in a Garden*, (Eds.) Ahmad Izani Md. Ismail, Koh Hock Lye and Mohamed Izham Mohamed Ibrahim, Penerbit USM, pp. 102–106.

Park, S.S. and Lee, Y.S. (2002). A water quality modeling study of the Nakdong River, Korea. *Ecological Modelling*, 152, 65–75.

Rounds, S.A., Wilde, F.D. and Ritz, G.F. (2013). Dissolved oxygen (ver. 3.0): U.S. Geological Survey Techniques of Water Resources Investigations, ch. A6.

Salas, H.J. and Thomann, R.V. (1978). A steady-state phytoplankton model of Chesapeake Bay. *Journal Water Pollution Council. Journal (Water Pollution Control Federation)*, 50(12), 2752–2770.

SCS (1975). Urban Hydrology for Small Watersheds, Tech. Release No. 20, Soil Conservation Service, US Department of Agriculture, Washington, D.C.

Song, J., Han, M., Kim, T. and Song, J. (2009). Rainwater harvesting as a sustainable water supply option in Banda Aceh. *Desalination*, 248, 233–240.

Su, W., Sun, Q., Xia, M., Wen, Z. and Yao, Z. (2018). The resource utilization of water hyacinth (Eichhornia crassipes [Mart.] Solms) and its challenges. *Resources*, 7, 46.

Tay, C.J., Teh, S.Y. and Koh, H.L. (2020). Eutrophication Bifurcation analysis for Tasik Harapan Restoration. *International Journal of Environmental Science and Development*, 11(8), 407–413. doi: 10.18178/ijesd.2020.11.8.1282.

Teh, S.Y., Koh, H.L., Chan, N.W. and Izani, A.M.I. (2006). Modeling Tasik Harapan as a Flood Retention Pond. *Proceedings of the Regional Conference on Ecological and Environmental Modeling* (ECOMOD 2004), 5–7 September 2004, Penang. In: Healthy Campus Series 13, Monitoring and Modelling of a University in a Garden, Ahmad Izani Md. Ismail, Koh Hock Lye and Mohamed Izham Mohamed Ibrahim (Eds.), Penerbit USM, pp. 84–101.

Teh, S.Y., Koh, H.L., Izani, A.M.I. and Mansor, M. (2008). Determining photosynthesis rate constants in Lake Harapan Penang. *Proceedings* of the First International Conference on Biomedical Engineering and Informatics (BMEI), 1, 27–30. Sanya, Hainan, China, Yonghong Peng and Yufeng Zhang (Eds.), IEEE Computer Society, USA, IEEE Computer Society, USA, pp. 585–590. doi: 10.1109/BMEI.2008.217.

Teh, S.Y., Koh, H.L., Zakaria, N.A. and Lee, L.M. (2009). Rehabilitation of Tasik Harapan: Issues and simulations. *Proceedings of the 5th Asian Mathematical Conference (AMC)*, 22–26. PWTC, Kuala Lumpur, Malaysia, pp. 435–441.

Thomann, R.V. and Mueller, J.A. (1987). Principles of surface water quality modeling and control. Harper and Row, Publishers, Inc., New York.

USDA NRCS (2009). Small Watershed Hydrology: WinTR-55 User Guide. Conservation Engineering Division, U.S. Department of Agriculture, Natural Resources Conservation Service.

USEPA (1985). Rates, Constants and Kinetics Formulations in Surface Water Quality Modeling (Second Edition). United States Environmental Protection Agency, Athens, Georgia 30613. EPA/600/3-85/040.

Wang, H., Hondzo, M., Xu, C., Poole, V. and Spacie, A. (2003). Dissolved oxygen dynamics of streams draining an urbanized and an agricultural catchment. *Ecological Modeling*, 160, 145–161.

Wilson, J.R., Holst, N. and Rees, M. (2005). Determinants and patterns of population growth in water hyacinth. *Aquatic Botany*, 81, 51–67.

Yacobi, Y.Z. and Schlichter, M. (2004). GIS application for mapping of phytoplankton using multi-channel fluorescence probe derived information. GIS and Remote Sensing in Hydrology, Water Resources and Environment (*Proceeding of ICGRHWE held at the Three Gorges Dam, China*, September 2003). IAHS Publication 289, pp. 301–307, Yangbo Chen, Kaoru Takara, Ian Cluckie and F. Hilaire De Smedt (Eds.).

Chapter 3

Regime Shift in Lake Eutrophication

3.1 Introduction

Eutrophication is a common feature in shallow tropical lakes surrounded by large human settlements. Eutrophic lakes are characterized by turbid water with an abundance of floating algae, induced by excessive nutrients such as phosphorus (P) and nitrogen (N) that enter the lakes from the surrounding settlements. On the contrary, oligotrophic lakes are characterized by clear water, with no or very little algae due to low nutrients. Natural oligotrophic lakes provide numerous essential services to ecosystems, wildlife and humans. However, many urban lakes are eutrophic due to pollutants and nutrient loading from domestic, industrial and agricultural activities in the surrounding neighborhood. These environmental stresses will undoubtedly impair the ecosystem services such as potable water supply, fishery and ecotourism provided by natural oligotrophic lakes. A lake may be classified into four categories of eutrophication as follows: (1) oligotrophic (Good), (2) mesotrophic (Fair), (3) eutrophic (Poor) and (4) hypertrophic (Very Poor) (Caspers, 1984). In a natural environment without human disturbances, a lake begins in a good and clear-water state known as oligotrophic with no visible algae growth and with low levels of Chl-a concentration of below 2 μg/L. However, nutrients produced from human settlements gradually turn the lake into a mesotrophic state, which is considered as fair, where algae may be mildly visible, with Chl-a levels ranging between 2 and 8 μg/L. Over time, nutrients continue to accumulate in the water column and in the sediments, eventually causing the lake to become turbid and eutrophic due to algae blooms, with Chl-*a* levels exceeding 9 μg/L. Further input of more nutrients will push the lake to a hypertrophic state, characterized by highly greenish-brownish colored water, with very high Chl-*a* levels exceeding tens or even hundreds of μg/L. However, we will consider only two states, namely, oligotrophic and eutrophic states, as the other two states may be considered as part of the continuum. The eutrophication of lakes is largely a consequence of human-induced nutrient enrichment and is hence frequently referred to as cultural eutrophication. To distinguish it from other more severe forms of anthropogenic pollution, eutrophication is legally defined as "the enrichment of water by nutrients causing an accelerated growth of algae and other forms of plant life to produce an undesirable disturbance to the aquatic environment". The natural progression of eutrophication symptoms from a clear-water (oligotrophic) state generally begins with an initial surge in phytoplankton biomass and micro-algal blooms. These turbid-water (eutrophic) state is then followed by more severe impacts, such as loss of submerged macrophytes, depleted oxygen levels, proliferations of harmful and toxic algal blooms, fish kills and the formation of dead zones (Conley *et al.*, 2009; Ferreira *et al.*, 2011). The transition from the clear-water oligotrophic state to the turbid eutrophic state is known as regime shift.

3.2　Literature Review on Regime Shift

A brief literature review on regime shift in ecosystems would provide useful scientific insights for this chapter. The stability of an ecosystem is subject to external perturbations and internal feedback, the study of which is a major topic in many fields in contemporary science (Müller *et al.*, 2009), particularly for climate change studies. Increasing concern has been directed to the scenario known as catastrophic regime shift, where a relatively small change in the environmental conditions can lead to a sudden quantum jump from one state to another (Scheffer *et al.*, 2012). Catastrophic regime shifts are known to pose a serious threat to ecosystems, the control of which requires a reliable set of early warning indicators or thresholds (Weissmann and Shnerb, 2016). Global climate change has the potential of inducing catastrophic regime shifts because of the complex nonlinear interactions across space and time involved in climate phenomena. Predicting the tipping point or threshold when a state begins to lose its stability and shifts to an alternative and undesirable state remains a daunting challenge. The mass conversion of heathland to grassland in north-western Europe during the past decades is a typical example of how crossing a tipping point may threaten the biodiversity and ecosystem services delivered by biota (van Voorn *et al.*, 2016). Many large shallow lakes in China, such as the Dianchi, Chaohu and Taihu, have already shifted from a clearwater oligotrophic state to highly turbid eutrophic conditions (Cao *et al.*, 2016) that have vastly impaired the valuable ecosystem services provided. These lakes are plagued by persistent turbid water and annual toxic algae blooms, caused by the excessive inputs of P and N from fertilizers over decades. Severe eutrophication in Lake Dianchi (Cao *et al.*, 2016) in Kunming Yunnan has impaired critical ecosystem services such as fresh water supply for 7 million people and ecotourism (Huang *et al.*, 2014). The ecosystem of Lake Chaohu began to show a regime shift from the clear-water oligotrophic state in the early 1970s to the turbid eutrophic state during the early 1980s. Since the 1990s, the lake has been confronted with persistent signs of further regime shift to severe eutrophication (Kong *et al.*, 2015). With a low trophic level of 2.92 out of a maximum of 6.00, the trophic status of Lake Taihu is considered undesirable, due to the serious and persistent eutrophication in the lake (Bai *et al.*, 2009; Wang *et al.*, 2011). At these three large eutrophic lakes, the Chinese government has invested billions trying to prevent the occurrence of toxic algae blooms and to sustain potable water supplies for the surrounding cities, totaling more than 20 million people. The following three subsections provide a brief overview of the important concept of regime shift and its associated thresholds, which are highly relevant to this chapter.

3.2.1 *Positive feedbacks and bi-stability post deforestation*

Availability of phosphorus (P) in soils is a major factor limiting vegetation growth for natural ecosystems and agricultural systems (Wardle *et al.*, 2004). Once it is lost from the soil, P is not replenished biologically (Vitousek *et al.*, 2010). Once removed from the earth, P becomes available for vegetation growth over very slow geologic time scales (Chadwick *et al.*, 1999). However, under conditions of limited P, soil microbes help in mitigating P losses and in enhancing P availability. Hence, the loss of microbial biomass due to deforestation increases P losses and decreases P availability. The P-vegetation-microbial ecosystem is driven by the interaction between vegetation growth, microbial abundance and P cycling. The dynamics of this P-vegetation-microbial ecosystem was incorporated into a simulation model to study vegetation regime shift. Simulation results suggest that the ecosystems most susceptible to regime shift are those that were previously deforested or those where the amount of P stored in the recalcitrant organic pool is low. For these ecosystems, a bi-stable equilibrium was observed after a significant loss of P exceeding a certain threshold. When deforestation occurred under a limited P availability threshold, both the vegetation and the microbial biomass were not able to recover, and the ecosystem shifted to the low-vegetation stable state for a long duration (Runyan and D'Odorico, 2013).

3.2.2   *Bi-stability of mangrove forests and freshwater plants*

Remote sensing imagery in southern Florida confirms the sharp boundaries, known as ecotone, between freshwater hardwood hammock communities and halophytic communities such as mangroves. Competition among plant species (Medina *et al.*, 2010; McKee, 2011) and self-reinforcing feedback between coastal vegetation and vadose zone salinity are involved in maintaining this ecotone (Teh *et al.*, 2008a). Along salinity gradients, this feedback creates a bi-stable equilibrium in which either the halophytic habitat or freshwater plant communities dominate as alternative stable states. The literature shows how transpiration in mangroves and freshwater plants responds differently to vadose zone salinity, thus altering the salinity through feedback. Simulation models demonstrate how self-reinforcing feedback, working together with a physical template including salinity gradients and sunlight, controls the bi-stability between halophytic and freshwater communities (Jiang *et al.*, 2015; Teh *et al.*, 2019). Regions of bi-stability along gradients of salinity have the potential for large-scale regime shifts following large pulse disturbances, such as hurricanes and tidal surges in Florida or tsunamis in other regions. The size of the region of bi-stability can be large for low-lying coastal habitats such as the Florida Everglades because of the high saline water table, which can extend several km inland due to salinity intrusion. This threat can be heightened potentially by climate change and sea level rise.

3.2.3 *Bi-stability in reef oyster subject to sediment accumulation*

Efforts over the past two decades to restore native oyster populations in the Chesapeake Bay have been extensive but largely ineffective (Ruesink *et al.*, 2005). However, recent restoration efforts and field experiments to revitalize reef oyster populations subject to sedimentation indicate that elevated reef height beyond certain thresholds can offset heavy sedimentation, promote oyster survival, encourage growth and improve disease resistance. This suggested the existence of alternative stable states for the ecosystem consisting of live oysters, reef height and sediment volume (Jordan-Cooley *et al.*, 2011). Bifurcation analysis performed on a system of three ordinary differential equations indicates that the initial reef heights dominate the outcomes of the equilibrium. This bifurcation analysis provided a theoretical framework for investigating a strategy for restoration of degraded reef oyster populations. Alternative stable states can be triggered by environmental disturbances in various ecosystems, such as soil salinization (Lal, 2015), kelp transition (Filbee-Dexter and Scheibling, 2014), coral reef degradation (Schmitt *et al.*, 2019) and lake eutrophication (Genkai-Kato and Carpenter, 2005). For lake regime shifts, a system consisting of two ordinary differential equations (ODEs) representing P and Chl-*a* (Genkai-Kato and Carpenter, 2005) is adopted in this chapter for analyzing regime shift thresholds in eutrophic lakes. Critical parameters determining the status of a lake regime shift are P loading into the lake (μg/L/day) and P flushing rate (per day) or its inverse, the P retention time (day).

3.3 Lake Regime Shift

Ecosystems, particularly aquatic ecosystems, are constantly subject to regime shifts among multiple locally stable and unstable states (Holling, 1973). Such regime shifts are fascinating ecological phenomena, involving multiple causes and many variables that change at different spatial–temporal scales, potentially altering the direction of feedbacks (Scheffer *et al.*, 2001; Scheffer and Carpenter, 2003). Lakes may undergo a regime shift between two alternative steady states, the oligotrophic and the eutrophic states. This regime shift is caused by external P input and internal interaction between water and sediment P. Mean depth, sediment P storage and temperature strongly influence the susceptibility of lakes to regime shifts, while water surface area plays a relatively insignificant role (Genkai-Kato and Carpenter, 2005). Many other factors, including lake morphometry, sediment quality and thickness, and dominance of macrophytes, also play a role, making the regime shift quite variable and difficult to predict before its occurrence. Regime shift from a desirable oligotrophic clear-water state to an undesirable eutrophic turbid condition will have adverse ecosystem impacts and may reduce important ecosystem services. Restoration back to the oligotrophic state after the shifts to eutrophic states is costly or may be impossible in certain situations. Hence, evaluation of the potential occurrence of regime shifts and the severity of the impacts is needed for good lake management. Abundant macrophytes may prevent P recycling from sediments and reduce the susceptibility of shallow lakes to regime shift. However, some lakes, such as the Sunway Lagoon, may be too deep relative to the surface area (depth of 8 m and surface area of 5 acres) to be protected by macrophytes in their limited littoral zones. Global warming may enhance eutrophication, and render lake restoration programs costly and less successful due to increased internal P recycling from the sediment at higher temperatures. Shallow lakes offer little buffer between the sediments and the water column. Hence, shallow lakes such as Lakes Dianchi, Chaohu and Taihu are most susceptible to regime shifts. Large lakes such as Dianchi, Chaohu and Taihu are difficult to restore back to the oligotrophic state once they become eutrophic. Frequent episodes of toxic Microcystis blooms occurred in these three large shallow lakes. At these three lakes, the China government has invested, and will continue to invest, billions to prevent the occurrence of these toxic algae blooms and to sustain reliable potable water supplies for the surrounding cities with a population of more than 20 million people. Reestablishing aquatic macrophytes in lakeside littoral zones and introducing artificially cultured filter-feeding carp (bighead and silver carp) to graze on phytoplankton successfully decreased the nutrient concentration to some extent (Lu *et al.*, 2012).

Recent research has revealed the important role played by lake bottom sediments in the internal nutrient dynamics that trigger and maintain regime shifts in lake eutrophication (McDonald *et al.*, 2010). The inorganic phosphorus (IP) at the lake bottom is bound to iron in the sediments under oxygenated conditions, but is released back to the water column when the sediment surface is deoxygenated. This is a main driver in regime shifts in shallow lakes, regardless of their surface areas. One of the main causes of oxygen depletion near the sediment layer is the decomposition of sinking phytoplankton, the density of which is related to lake's trophic status. High phytoplankton concentrations induced by eutrophication lead to more sedimentation that promotes bacterial respiration and anoxia. Anoxia, in turn, leads to additional P release from the sediments, triggering and sustaining a vicious cycle of positive feedback mechanism that promotes regime shift to eutrophication. An example is Harapan Lake, located in the USM campus at Penang, Malaysia, with a shallow depth of 1.5 m and a surface area of 1.5 acres. Persistent and frequent hypertrophic conditions occurred annually, with Chl-*a* concentration reaching 330-μg Chl $\cdot$ L-1 occasionally (Teh *et al.*, 2008b). The shallow depth enhances the feedback between the water column and the sediment P, promoting persistent eutrophication in the lake. The interplay of multiple abiotic, physical, chemical and biotic mechanisms in regime shifts observed in lakes may also arise in regime shifts of other types of ecosystems (Genkai-Kato and Carpenter, 2005), such as mangroves and grasslands, although the underlying causes are quite different. Insights gained from this research may have the potential of being applied to other ecosystems. We begin with a brief review of lake eutrophication and regime shifts between oligotrophic and eutrophic states in four lakes in China, namely, Lake Dianchi, Lake Chaohu, West Lake and a pair of reservoirs in Xiamen. This review will provide valuable insights on lake eutrophication remediation and management.

3.3.1 *Lake Dianchi*

Lake Dianchi is a plateau lake located in Kunming City, the capital of Yunnan province. It has a surface area of 330 km^2, with a mean depth of 4.4 m, and a watershed area of 2920 km^2. Over the decades since the 1970s, it has been receiving 90% of wastewater generated in Kunming City, with a population of seven million. The shallow depth coupled with a high population renders it prone to eutrophication. Indeed, Lake Dianchi is currently contaminated by severe eutrophication, with cyanobacterial blooms occurring annually over the whole lake. The lake water quality is rated as Grade V, the worst grade in China (Cao *et al.*, 2016). The recurrent algae blooms impair critical ecosystem services such as fresh water supply, fishery and ecotourism (Huang *et al.*, 2014). Most biomasses and trophic flows are primarily concentrated at the lower three trophic levels (TLs) out of a total of six TLs. The primary producer level (TL1) consists mainly of detritus and phytoplankton; while TL2 consists mainly of zooplankton and zoobenthos, and filter-feeding fish (silver carp and bighead carp) and herbivorous fish. Positive feedback within the lower two TLs locks the nutrients to the plankton communities and promotes the inflation of planktonic biomass. About 78% of the trophic flows from TL1 to TL2 originate from the detritus. High proportions of underutilized zooplankton biomass returned to the detritus, reflective of the low transfer efficiencies of 3% from TL2 to TL3. Located at the level of TL3, shrimp occupies about 67% of the total fishery, with the remaining consisting of zooplanktivory fish. The biomass of shrimp at TL3 in Lake Dianchi is large (29.85 t$\cdot$km^{-2}), indicating the importance of lake ecosystem services such as fishery and ecotourism, despite the eutrophic condition. However, the future scenarios for Lake Dianchi are not bright, unless the eutrophication can be effectively controlled. The mean trophic level of fish catch is estimated to be 3.06, which is reflective of the status for TL3. This TL value of 3.06 is slightly higher than the TL value of 2.92 for Lake Taihu (Li *et al.*, 2009a,b), and close to the TL value of 3.07 for Northern-Central Adriatic Sea (Coll *et al.*, 2007). Hence, Lake Dianchi, similar to Lake Chaohu and Lake Taihu, is considered a bottom-up control ecosystem (Shan *et al.*, 2014). The increasing trend of eutrophication in Lake Dianchi is consistent with the increasing trend of industrialization and urbanization in the surrounding city of Kunming. During the period from 1980 to 1985, the Chlorophyll-*a* (Chl-*a*) concentration was about 10-μg Chl$\cdot$L^{-1} (mesotrophic). By 1995, Chl-*a* gradually increased to 25-μg Chl$\cdot$L^{-1} (eutrophic). Within the next 15 years from 1996 to 2000, the levels increased exponentially to 160-μg

Chl$\cdot$L^{-1} (hypertrophic). From 2009 to 2010, the Chl-a levels maintained at 128.21 ± 35.32-μg Chl$\cdot$L^{-1} (hypertrophic). Total phosphorus (TP) varied between 0.2 and 0.4 mg $\cdot$L^{-1}, total nitrogen (TN) varied between 2.0 and 3.5 mg $\cdot$ L^{-1}, while Secchi depth varied between 13 and 36 cm (Shan *et al.*, 2014), indicating a status of hypertrophic condition.

3.3.2 *Lake Chaohu*

Lake Chaohu is the fifth largest lake in China, with a surface area of about 780 km^2, and mean depth of 3.06 m. Lake Chaohu provides important ecological services, including potable water supply, for the 7.6 million people living in the basin. The ecosystem of Lake Chaohu shifted from the clear-water to turbid-water state during the late 1970s. The lake began to face serious eutrophication since the 1980s. TP loading amounted to 1050 t $\cdot$ y^{-1} in the late 1980s (Tu *et al.*, 1990) and increased to 1550 t $\cdot$ y^{-1} during 2002 to 2010. About 85% of the TN and 77% of the TP loading originate from non-point sources. These loadings are higher than expected, thus suggesting the unspecified nutrient flux from the soil in the basin. Analysis by the DyN model indicates that the TP loading threshold that would shift the clear water to the turbid-water state is 631.8 ± 290.16 t $\cdot$ y^{-1}, which is equivalent to 0.73 μg$\cdot$L$^{-1}\cdot$d^{-1} of TP (Kong *et al.*, 2015). This external TP loading threshold will be used as a benchmark in simulations for Sunway Lagoon. Further, eutrophication of Lake Chaohu is more likely to be reversible (74.12%) than hysteretic (25.53%), based upon DyN simulation and analysis. To shift from the current turbid eutrophic state back to the clear-water oligotrophic state, the current TP loading must be reduced by two-third (Kong *et al.*, 2015). However, in real practice, the reduction of non-point sources is very difficult and costly. Hence, additional methods beyond nutrient input reduction, such as water level regulation, should be considered for lake restoration in Lake Chaohu.

3.3.3 *West Lake Hangzhou*

Located in the west of Hangzhou City, Zhejiang province, China, West Lake has an area of 6.5 km^2 and a mean depth of 2.27 m, giving a water volume of 1.49×10^7 m^3. Outbreak of algae blooms and ecosystem degradation caused by eutrophication have occurred frequently in West Lake since the 1950s. Abundant external nutrients, originating from industrial and domestic wastewater, agricultural activities in the basin coupled with nutrients from precipitation (Wang *et al.*, 2011; Zhang and Huang, 2011), are the important contributors to algal blooms (Najar and Khan, 2012). Submerged macrophyte restoration was successfully adopted to mitigate eutrophication and improve the aquatic ecosystem in West Lake (Zhang and Huang, 2011). After macrophyte restoration, macrophyte biomass and coverage were at a very high level, while nutrient levels and phytoplankton concentration significantly decreased. The decrease in nutrients is mainly attributed to the direct absorption by, and indirect release inhibition from, restored macrophytes. The reduction in phytoplankton might be the result of the growth inhibition due to decreased nutrients available for phytoplankton growth. Zooplankton, especially large-sized species, significantly increases after restoration, helping to graze down phytoplankton populations. The effective protection of zooplankton against fish predation provided by restored macrophytes might be the main reason for zooplankton increase. For the shallow West Lake, macrophyte restoration is an effective method to mitigate eutrophication and improve water quality, deserving the attention of lake management (Zhang and Huang, 2011). However, it remains an open question if macrophyte restoration is effective for Sunway Lagoon, as the lake is deep at 8 m relative to its small surface area of 5 acres, providing limited littoral zones for macrophyte restoration.

3.3.4 *Shidou and Bantou reservoirs*

The Shidou and Bantou Reservoirs are located in the headwater of Zhu River near Xiamen City, Fujian province, southeast China. A six-year study was conducted to examine the impacts of multiple disturbance events on the phytoplankton and cyanobacteria dynamics in these two subtropical reservoirs (Yang *et al.*, 2017). These disturbances consist of human resettlement, temperature change, rainfall and water-level fluctuations. The trophic states of these two reservoirs have changed from oligotrophic to eutrophic levels over time, providing an example of regime shift (Lv *et al.*, 2014;

Yang *et al.*, 2016). The reservoirs provide critical ecological services including water supply to one million people in Xiamen City. Similar to Sunway Lagoon, these reservoirs have almost no macrophyte cover, having depths between 8 and 15 m. Research showed that combined multiple environmental disturbances triggered two abrupt regime shifts between a cyanobacteria-dominated state and a non-cyanobacteria-dominated state. In late 2010, the combined effect of human resettlement (emigration) and natural disturbances (cooling, more rainfall, increase water level) led to a 60–90% decrease in cyanobacteria biomass. This was accompanied by the disappearance of cyanobacterial blooms, in tandem with an abrupt and persistent shift in the phytoplankton community. After summer 2014, however, combined weather and hydrological disturbances (warming, less rainfall, lower water level) occurred, leading to an abrupt and significant increase in cyanobacteria biomass, associated with a return to cyanobacteria dominance. These changes in phytoplankton community were strongly related to the nutrient concentrations and water-level fluctuations, as well as water temperature and rainfall. Sunway Lagoon might also be exposed to multiple disturbances, with consequences that are difficult to predict in advance. Extreme weather events and large human disturbances are anticipated to become more frequent and more severe during the twenty-first century under GCC scenarios. Hence, prudent sustainable lake management will require consideration of the background limnological conditions and the frequency and severity of disturbance events. In particular, sediment phosphorus plays an important role in shallow lakes because of its potential impacts on lake biodiversity and ecosystem functioning and services.

3.3.5 *Role of sediment phosphorus*

Research shows a persistent and positive correlation between the development of cyanobacterial blooms and the increase of soluble reactive phosphorus (SRP) in the lake water of Dianchi (Cao *et al.*, 2016). Cyanobacterial blooms are a driving force for phosphorus (P) mobility and exchange among sediments, water and cyanobacteria. The P in the sediments represents a significant supply for the growth of cyanobacteria. Although much effort has been made to reduce the external loading of P in lakes, P released from the sediments may act in such an intense and persistent manner that it prevents the improvement of water quality (Sondergaard *et al.*, 2003). Algal blooms have positive feedback on the release of P from sediment. Research in Lake Donghu shows that dissolved P released from sediment back to lake water could be enhanced by algal blooms Xie (2003). Generally, P is believed to be a limiting factor for cyanobacterial growth. The Chl-*a* and SRP in the overlying water are significantly related. Inorganic P (IP) is the main P fraction, with IP mainly consisting of NaOH-P in the heavily eutrophic sediment. IP and NaOH-P are relatively easier to migrate out of sediment back to the water. In another independent study, most of the P in the sediment consisted of IP (Bai *et al.*, 2009). The death and decaying cyanobacteria would promote the release of P from sediments back into the water body, and the released P would in turn help new cyanobacteria growth. The proliferation of cyanobacteria drives sediment P release, sustaining a vicious cycle. The development of cyanobacterial blooms caused an increase in the pH from 8 to 12. It has been demonstrated that an increase in pH may increase P release from sediments. The brief exposition on eutrophic conditions in several large and shallow lakes in China provides the background knowledge for the investigation of eutrophication in Sunway Lagoon.

3.4   Regime Shift in Sunway Lagoon

A small stagnant lake known as Sunway Lagoon (SL) located in the state of Selangor, Malaysia, has shown clear signs of regime shift from oligotrophy to eutrophication, due to high inputs of P from the surrounding township over the past decades. When the loadings of P into a lake reach a critical threshold (0.7 P μg/L/d), the lake shifts from an oligotrophic state to a eutrophic state. The regime shift from an oligotrophic to a eutrophic regime can be rapid once the concentration of P in a lake crosses a critical threshold. The tight coupling between sediment P and water column P renders the dynamics of eutrophication in shallow lakes highly nonlinear. Shallow lakes offer little buffer between the sediment and the water column. High nonlinearity offers ample opportunity for a sudden and abrupt regime shift that requires costly remediation action if the shift is not detected early (Cao *et al.*, 2016). It is therefore crucial to develop methodology, based upon monitoring and modeling, to enable early detection of such an abrupt shift Singh *et al.* (2015). Further, the model enables the development of effective mitigation measures for preventing further a shift to hyper eutrophication, a more severe form of eutrophication. This section aims to develop models suitable for rehabilitation of SL back to the desired clear-water oligotrophic condition. There are several approaches for developing lake regime shift models, some for tropical lakes, others for temperate lakes. It is preferable to have a model with model parameters (a) that are stable for a wide variety of limnological conditions and (b) that are readily estimated from available data (Genkai-Kato and Carpenter, 2005). Many factors can lead to a great variability among lakes in determining the thresholds for eutrophication and regime shifts. Among these factors, mean lake depth and temperature are the two most important parameters in determining the direction and strength of regime shift arising from eutrophication. Other factors such as the effects of macrophytes are closely associated with mean depth and are stronger in shallower lakes due to availability of adequate littoral plants in shallow lakes. However, there is little consensus within the literature concerning the precise mechanism on how macrophytes affect lake nutrient recycling between water column and sediments. Nevertheless, macrophyte cover is expected to decrease the recycling of P by reducing sediment resuspension. By providing refuge for zooplankton against fish predation, macrophytes can also help control algal blooms by enhancing zooplankton grazing of phytoplankton (Scheffer, 1998).

3.4.1 *Regime shift model*

A two-component model is developed to link algae concentration A to P concentration in SL and to investigate the conditions that govern regime shifts between oligotrophic and eutrophic states. As shown in Eqs. (3.1) and (3.2), the model consists of two dynamic variables A and P, where A is the concentration of algae, measured in Chl-*a* (μg Chl $\cdot$ L^{-1}), and phosphorus P is measured in μg P $\cdot$ L^{-1} (Genkai-Kato and Carpenter, 2005). The definitions and units of the model parameters are listed in Table 3.1. As may be observed in Eq. (3.1), Chl-*a* concentration A increases at a rate bPA due to algae

Table 3.1: Definition, units and value of variables and parameters for SL.

Symbol	Definition	Unit	Value
b	Algal growth rate per unit	L/μg P/d	0.900
g	Zooplankton grazing rate	d^{-1}	0.030
s	Algal sinking rate	m/d	0.085
h	Flushing rate	d^{-1}	(a) 0.01, (b) 0.005, (c) 0.0
e	P excretion rate associated with grazing	μg P/μg chl	0.650
l	External P loading rate	μg P/L/d	0.01–0.13
r	Internal P sediment recycling rate	μg P/L/d	0.1 Computed
z_e	Thickness of epilimnion	m	7.000

growth and decreases at a rate $- (g + s/z_e + h)A$, due to algae grazing g by zooplankton coupled with algae sinking at a rate s/z_e and flushing rate h. As observed in Eq. (3.2), P concentration increases by external loading l, internal sediment recycling r and excretion from zooplankton egA. On the contrary, P concentration decreases at the rate of — bAP due to algae uptake and — hP due to flushing.

$$\frac{dA}{dt} = bPA - \left(g + \frac{s}{z_e} + h\right) A \tag{3.1}$$

$$\frac{dP}{dt} = l + r + egA - bAP - hP \tag{3.2}$$

3.4.2   *Sunway Lagoon water quality*

SL receives water inflow containing nutrients via surface runoff and groundwater. It has no natural outflow, resulting in a persistent accumulation of P and N in the sediments. Currently, after each heavy rain, excess water in the lake is allowed to overflow a weir into the adjoining retention pond, to avoid flooding the vicinity. Some nutrients and pollutants are removed with the surface water being drained, but the actual quantum of nutrients and pollutants being drained from SL is unknown. As a control measure, we propose to regularly pump off water at a deeper water level of 7.0 m to remove as much nutrients as possible in order to reduce nutrient levels in the lake.

The rate of removal of water (flushing rate) and of key elements (P, N, BOD and Chl-*a*) will be recorded for regime shift simulation and analysis via Eqs. (3.1) and (3.2).

Several water and sediment quality surveys were conducted between 2010 and 2017. Monthly water quality sampling was conducted at three depths (0.5, 3.5 and 7.5 m below water surface) for six months from December 2016 to May 2017. Water quality parameters sampled included temperature, pH, BOD, DO, ammoniacal nitrogen, nitrate nitrogen, phosphorus, Chlorophyll-*a*, total coliform and *Escherichia coli*. Water quality and sediment quality surveys were conducted on 23 January 2017. Water quality samples were collected at three depths of 0.5 m (surface), 3.5 m (mid-depth) and 7.5 m (bottom) below water surface. A comprehensive sampling survey was conducted on 24 to 25 June 2010 to test water and sediment quality in SL. The water samples were collected at 10 spatially separated sampling locations at three water depths: surface (0.5 m), mid-depth (3.5 m) and lake bottom (7.5 m). Sediments were collected at 10 separate locations. Overall, the water quality was classified as Malaysia Standard Class IIB. The surface water dissolved oxygen (DO) levels were generally below 5 mg/L, indicating a mildly DO-stressed environment detrimental to aquatic lives including fish. The results of these water and sediment quality surveys are presented in the following three subsections.

3.4.2.1 *Monthly water quality 2016–2017*

Monthly water quality sampling was conducted at three depths (0.5, 3.5 and 7.5 m below water surface) for six months, from December 2016 to May 2017. Water quality parameters sampled included temperature, pH, BOD, DO, ammoniacal nitrogen, nitrate nitrogen, phosphorus, Chlorophyll-*a*, total coliform and *Escherichia coli*. As shown in Figure 3.1, high levels of BOD, NH_3N and Chl-*a* were recorded during the six months of sampling, indicating a eutrophic lake condition. More data are available in (Koh, 2017). A low-impact and sustainable solution to control SL eutrophication is proposed in this section. In general, a clear-water oligotrophic lake should have Chl-*a* level below 2 μg/L, while a meso-eutrophic lake should have Chl-*a* between 2 and 9 μg/L. Levels of Chl-*a* exceeding 9 μg/L are considered eutrophic and undesirable.

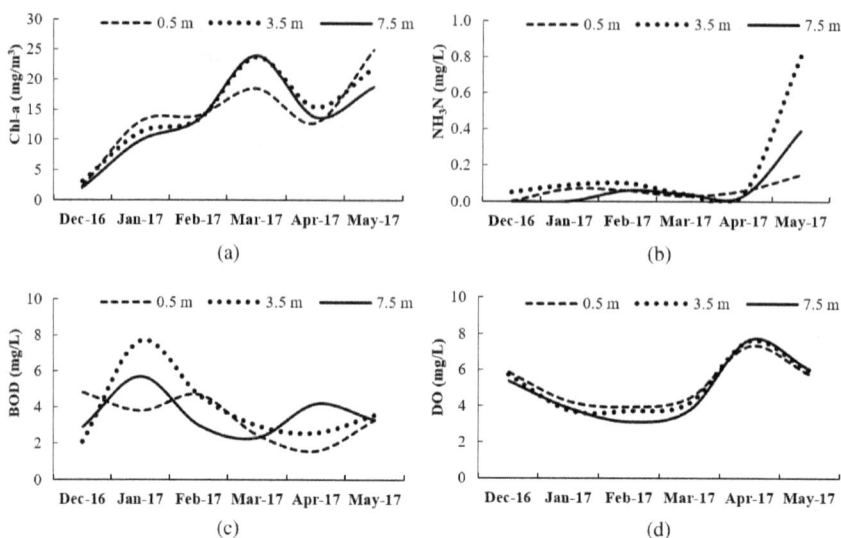

Figure 3.1:   (a) Chl-*a*, (b) NH$_3$N, (c) BOD and (d) DO sampled at three depths (0.5, 3.5 and 7.5 m below water surface) from December 2016 to May 2017.

3.4.2.2   *Water and sediment quality 2017*

Water quality and sediment quality surveys were conducted on 23 January 2017. The results indicated that the overall water quality had deteriorated from Class IIB to Class III and the lake had shifted from mesotrophic to eutrophic. Further, the N concentration in the sediments measured at 739 mg/kg, based upon an average of three sampling locations. The mean value of sediment N increased by 600%, compared to the mean value in 2010 based upon 10 sampling locations (between 68 and 246 mg/kg). Survey results for biochemical oxygen demand (BOD$_5$), DO and Chl-*a* at these three depths indicated high and undesirable levels, signifying eutrophic conditions. The sediment quality measured on 23 January 2017 is summarized in Table 3.2, indicating poor water quality. Therefore, there is an urgent need to improve water quality in SL to prevent it from shifting to the hyper eutrophic condition.

3.4.2.3   *Water and sediment quality in 2010*

A comprehensive sampling survey was conducted on 24 to 25 June 2010 to test water and sediment quality in SL. Overall, DO varied between 1.26 mg/L and 4.74 mg/L from bottom to surface,

Table 3.2:　Sediment quality in SL measured on 23 January 2017.

Component	Average value	Level
TP (mg/kg)	61.0	High
TN (mg/kg)	739	High
TOC (% w/w)	0.5	High

Table 3.3:　Summary of concentrations of five key elements in the sediments of SL sampled on 25 to 26 June 2010.

Component	Range of value	Level
TP (mg/kg)	22 to 81	High
TN (mg/kg)	68 to 246	High
TOC (% w/w)	0.4 to 1.5	High
Iron (mg/kg)	5225 to 13190	High
Aluminum (mg/kg)	8921 to 83890	High

indicating a DO-stressed condition detrimental to fish. Temperature was between $28°C$ and $31°C$, and pH was neutral at 7. Concentration of Chlorophyll-a (Chl-a) varied between 5 and 15 $\mu g/L$, suggesting a condition in between mesotrophic and eutrophic. With total kjeldahl nitrogen (TKN) concentration varying between 0.2 and 0.9 mg/L and total phosphorus (TP) between 0.1 and 0.4 mg/l, the water is considered as nutrient rich, capable of supporting vibrant Chl-a growth to a eutrophic level. The sediments were tested for 17 heavy metals at 10 locations. The heavy metal concentrations were generally low, except for iron, aluminum and manganese. However, the concentrations for P and N are high, reflecting a eutrophic lake condition capable of supporting vibrant algae growth. Table 3.3 is a summary of concentrations of five key elements in the sediments sampled on 25 to 26 June 2010, indicating an undesirable water quality condition in the sediments. Hence, regular removal of nutrient-rich lake water near the bottom sediment layer is proposed as an effective solution to control eutrophication in SL.

3.4.3　*Flushing rate F*

As discussed earlier, SL lake water should be flushed out regularly in order to prevent further eutrophication. Nutrient-rich water located near the sediment layers at a depth of 7.0 m should be removed daily to minimize the accumulation of nutrients in the sediments. We will demonstrate by regime shift model simulation that daily removal of nutrient-rich water from SL will help to control eutrophication. The water removed from SL can be used for landscape irrigation to water plants in the surrounding Sunway City, thereby saving costs of water and fertilizers consumed. To facilitate this regime shift analysis, we introduce the concept of retention time R (days) or its inverse, i.e., the concept of flushing rate F (day^{-1}). Thus, we define $F = 1/R$. Consider a lake of volume $V = 10^6$ m^3, with a pollutant T of concentration C_T kg/m^3. Assume that the water is flushed out at the rate of $W = (10^3$ m^3/day). Then, the hydraulic retention time R is given by $R = V/W = 10^6$ m$^3/(10^3$ m^3/day) $= 10^3$ days. The total mass M of pollutant T in the lake is $M = 10^6$ m$^3 \times C_T$ kg/m$^3 = 10^6$ C_T kg. Suppose now that the concentration of T near the bottom sediment is 10 times higher or 10 C_T kg/m^3. Further, suppose that we flush out the lake water near the bottom sediment (with concentration 10 C_T kg/m^3) at the same rate of $W = (10^3$ m^3/day). The mass of T flushed out per day is 10 C_T kg/m$^3 \times 10^3$ m^3/day $= 10^4$ C_T kg/day. The retention time R_T of pollutant T is $R_T = 10^6$ C_T kg/10^4 C_T kg/day $= 10^2$ day, while the hydraulic retention time R is still 10^3 days. Therefore, the hydraulic flushing rate F is $1/10^3$ day $= 0.001$ per day $= 0.001$ d^{-1}. On the contrary, the pollutant T flushing rate F_T is $1/10^2$ day $= 0.01$ per day $= 0.01$d^{-1}. This means that $F_T = 10$ F or the pollutant T flushing rate F_T is 10 times higher than the hydraulic flushing rate F. It is therefore beneficial to flush out water near the sediment to remove as much nutrients as possible with the same hydraulic flushing rate F.

3.4.4   *Simulation results*

To control SL eutrophication, a simple, sustainable and cost-effective method is proposed. Lake water at a depth of 7.0 m will be removed at daily frequency to flush out P-rich lake water (Koh, 2017). The system of two ODEs representing the evolution of Chl-a (3.1) and P (3.2) in SL is given in Eqs. (3.1) and (3.2), with the definitions, units and values of process parameters shown in Table 3.1. Further details are available elsewhere (Koh *et al.*, 2018a). The P flushing rate denoted by F_T (day^{-1}) is represented by h in Eq. (3.2). Three flushing rates,

(a)

(b)

(c)

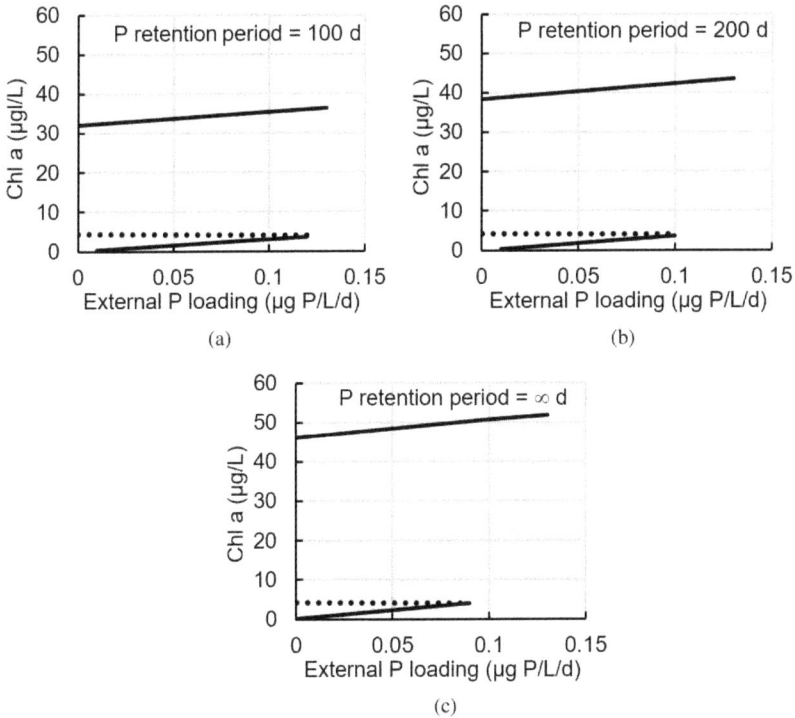

Figure 3.2: Simulated results for three scenarios of retention period: (a) 100 days $(h = 0.01 \text{ d}^{-1})$, (b) 200 days $(h = 0.005 \text{ d}^{-1})$ and (c) No flow $(h = 0.0 \text{ d}^{-1})$.

in decreasing order, are simulated: (a) $h = 0.01 \text{ day}^{-1} = 1/(100 \text{ days}$ retention time), (b) $h = 0.005 \text{ day}^{-1} = 1/(200 \text{ days retention time})$ and (c) $h = 0.0 \text{ day}^{-1} = 1/(\infty \text{ days retention time})$. The corresponding simulation results are demonstrated in Figure 3.2.

At the P flushing rate of 0.01 day^{-1}, corresponding to a retention time of 100 days, the regime shift tipping point is achieved at external P loading of $1.2 \text{ }\mu\text{g/L/d}$, and the Chl-$a$ concentration peaks at $35 \text{ }\mu\text{g/L}$, as may be seen from Figure 3.2(a). When the P flushing rate is reduced to 0.005 day^{-1}, corresponding to an increased retention time of 200 days, the regime shift tipping point is reduced to external P loading of $1.0 \text{ }\mu\text{g/L/d}$, and the Chl-$a$ concentration peaks at $40 \text{ }\mu\text{g/L}$ as illustrated in Figure 3.2(b). When the P flushing rate is further reduced to 0.0 day^{-1}, corresponding to an infinitely long retention time, the regime shift tipping point is further reduced to the external P loading rate of 0.8 μg/L/d, and the Chl-*a* concentration peaks at 50 μg/L, as may be seen from Figure 3.2(c). Hence, lowering flushing rates will result in lowering the regime tipping point and in increasing the Chl-*a* concentration. Given that the lake is shallow, and the temperature is high all year round, SL is vulnerable to eutrophication (Carpenter *et al.*, 1999). Hence, in addition to increasing P flushing, other measures should be considered, including reduction of fish density. For this purpose, a series of lab experiments was conducted to measure fish-specific respiration rates under lab conditions as discussed in the following subsection.

3.4.5   *Fish respiration rate*

A laboratory experiment in the staging facility at Sunway University was conducted to study the effect of fish respiration on dissolved oxygen (DO) level in water. For this purpose, 20 tilapias with different weights, ranging between 70 g and 335 g each, were selected. To prevent competition, one tilapia is placed into each tank for each experiment. Figure 3.3 shows the fish tanks used in this experiment. The water in each tank is aerated overnight to saturation. The dimension of the tanks, water depths and water volumes are presented in Table 3.4. At the start of the experiment on the following morning, the initial readings of DO, temperature and water level were recorded for each tank. Immediately, the aerator is turned off for the remaining part of the daily experiment. Subsequently, DO level in the water was measured at regular intervals of one hour until the end of the day. The experiment was continued until all 68 sets of

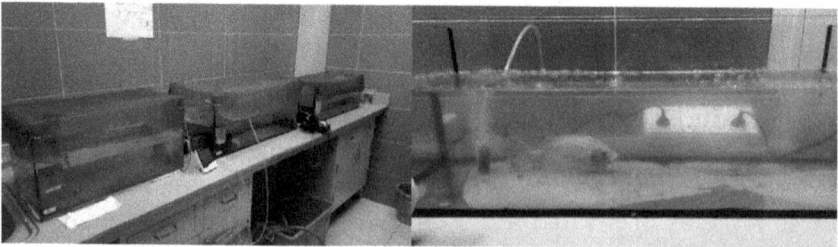

Figure 3.3:   Fish tanks used in the fish respiration experiment.

Table 3.4: Dimension, fish mass and specific respiration rates of the five tanks in the experiment.

Tank	Length (m)	Width (m)	Water depth (m)	Water volume (L)	Fish mass (g)	Specific respiration rate (g O_2/kg fish/day)
1	0.780	0.340	0.177	46.94	85	3.5
2	0.590	0.295	0.156	27.15	95	2.1
3	0.800	0.340	0.185	50.32	155	3.0
4	0.750	0.370	0.226	62.72	200	3.7
5	0.750	0.370	0.269	74.65	335	2.4

daily data comprising various combinations of tank dimensions and fish weights were obtained. Figure 3.4 demonstrates the decline in DO due to fish respiration in all five fish tanks. E2algae simulations indicated that the specific respiration rate varied between 2.1 and 3.7 g O_2/kg fish/day for sedentary fish at 23°C that are inactive most of the time. This range of specific respiration rates is less than the average value of about 5.0 g O_2/kg fish/day reported in the literature. For the regime shift model used for SL, the specific respiration rate of 5.0 g O_2/kg fish/day is used. In SL where fish compete for resources in an environment of higher temperature of around 29°C, the specific respiration rate should be higher at the rate 5 g O_2/kg fish/day.

Numerical simulations via E2algae were performed for a fish pond with a depth of 0.6 m, volume of 15.0 m^3 and atmospheric reaeration rate of 2.0 d^{-1}. The fish density was 0.8 kg/m^3, the specific respiration rate for fish in the pond was 5 g O_2/kg fish/day, while the fish feeding rate was kept at 2% per day wet weight. The simulation results indicated that fish will be killed because of low DO, without an aerator (Koh *et al.*, 2018b). Indeed, fish kills were observed frequently in the early part of the morning, presumably due to low DO. It is therefore important to reduce fish density in SL to 0.1 kg/m^3 or less to ensure adequate DO. Adequate DO near the sediment reduces P release from the sediment to the water column and hence mitigates eutrophication. With a depth of 8 m, the SL reaeration rate will be less than 2.0 d^{-1}, limiting oxygen replenishment from the atmosphere.

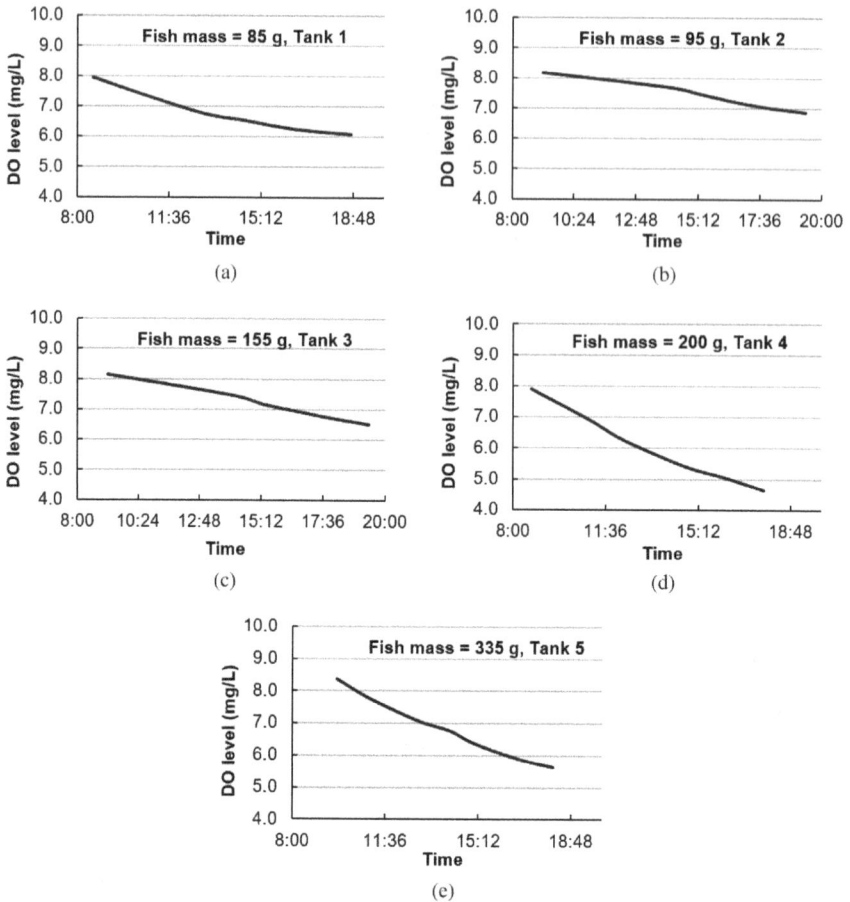

Figure 3.4:   Measured DO levels in the five experimental tanks with various fish masses.

3.4.6   *Discussion*

3.4.6.1   *Eutrophication control in sunway lagoon*

Lake eutrophication remains a daunting environmental challenge that is difficult to resolve, particularly for shallow tropical lakes such as Sunway Lagoon and Tasik Harapan, as well as more than 60% of the lakes and reservoirs in Malaysia. Although the causes of lake eutrophication regime shift have been broadly understood, the theoretical foundation is still under development. The main goal of this chapter is to attempt to clarify the underlying mechanism driving regime shift in lakes, with the hope to provide some insights for the management of lake eutrophication. With high annual temperature and abundant sunlight in the tropics, eutrophication is accelerated by large and persistent influx of nutrients such as P generated by anthropogenic activities. A high concentration of these nutrients in the lake stimulates algae to grow wildly, leading to depletion of DO in the lake water. The accumulation of P in the sediment and the release of P from the sediment back into the water column are the dominant processes driving the regime shift. The persistence of eutrophication is governed by important ecological thresholds and the existence of multiple equilibrium states or regimes within a lake. The knowledge on regime shift thresholds associated with eutrophication is a prerequisite in lake restoration efforts to improve the lake ecosystem. The analysis performed in this chapter would provide the fundamental knowledge and insights on the regime shift and its thresholds for SL, and offers sustainable and low-energy solutions for controlling eutrophication. It may be concluded that a sustainable and cost-effective rehabilitation program for SL is frequent daily removal of nutrient-rich lake water from a depth of 7.0 m. This daily flushing of P-rich water will delay the tipping point of the regime shift from 0.8 μg/L/d to 1.2 μg/L/d and significantly reduce P and Chl-*a* levels. Reduction of fish density to 0.1 kg/m^3 or less may also be beneficial in delaying the regime shift, in improving overall water quality and in increasing DO levels. Increased DO may inhibit release of P from sediment to water column, thereby helping to inhibit eutrophication. The methodology developed in this paper may be applicable to other tropical shallow lakes for the analysis and control of eutrophication, such as Tasik Harapan.

3.4.6.2   *Effects of temperature and depth*

Eutrophication in a tropical and shallow lake such as Tasik Harapan behaves in a fundamentally different manner compared to that of a temperate and deep lake such as Lake Fuxian in Kunming, China. Understanding the fundamental underlying mechanism that drives regime shift in both types of lakes would provide valuable insights for the management of lake water quality. For this purpose, Tay *et al.* (2021) have developed a theoretical framework on lake regime shift for Tasik Harapan and Lake Fuxian as representatives of both types of lakes. In the following two subsections, we will attempt to give a brief literature overview on their study. The aim of this overview is to provide guidance for eutrophication control and to elicit further research on regime shifts to aid eutrophication control.

3.5   Regime Shift in Tasik Harapan

Located in Universiti Sains Malaysia (USM), Penang, Malaysia, Tasik Harapan (TH) is a shallow tropical lake (Figure 3.5) with a surface area of about 6070 m^2 (1.5 acres), a volume of 10,000 m^3 and a mean depth of 1.0 to 1.5 m (Teh *et al.*, 2008b). Constructed for flood mitigation in USM in 1990, the lake has been undergoing a regime shift and has turned eutrophic due to high nutrients driving excessive algal growth. A lake is classified as eutrophic if the algal concentration is greater than 10 μg/L chl *a* (Chapra, 1997). With a reported high algal concentration occasionally exceeding 300 μg/L

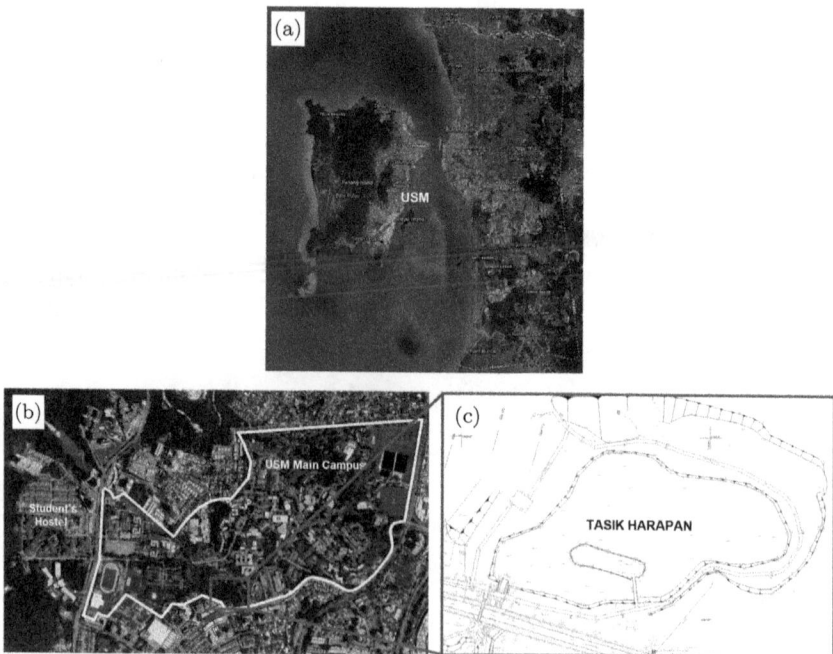

Figure 3.5:   Location of Tasik Harapan in USM, Penang, Malaysia.

chl *a*, TH is highly eutrophic (Teh *et al.*, 2008b). Mathematical models were developed by Tay *et al.* (2021) to provide a framework for understanding regime shift, and the associated bifurcation analysis. This section provides a brief summary of their findings on the regime shift in TH.

3.5.1   *Mathematical model for Tasik Harapan*

The mathematical model in Eqs. (3.3) and (3.4) was formulated to analyze the regime shift in the shallow tropical TH. Table 3.5 shows the definition and unit of the parameters for this algae-phosphorus model.

$$\frac{dA}{dt} = bA\frac{P - p_a A}{h_a + P - p_a A} - (h_1 + g + s_v)A \tag{3.3}$$

$$\frac{dP}{dt} = l_p + r\frac{P^q}{P^q + n^q} + egA - bA\frac{P - p_a A}{h_a + P - p_a A} - h_1 P \tag{3.4}$$

3.5.2   *Bifurcation analysis for Tasik Harapan*

Using the parameter values listed in Table 3.5 and the algae-phosphorus model in Eqs. (3.3) and (3.4), the regime shift threshold for lake eutrophication in TH is determined through bifurcation analysis. This bifurcation analysis aims (i) to provide information on the threshold value at which the lake became eutrophic, (ii) to identify the lake bifurcation behavior and (iii) to determine the effective restoration measures.

3.5.2.1   *Impact of external phosphorus loading l_p*

Bifurcation analysis of algae concentration A verses external phosphorus loading rate l_p is performed by means of XPPAUT. The analysis indicated that TH exhibits bifurcation behavior considered to be irreversible, because the lake remains eutrophic ($A = 131$ μg/L) even though the external loading l_p is reduced to zero (no P inflow). Therefore, other restoration methods such as dredging, flushing or bio-manipulation are needed in conjunction with drastic reduction in external loading in order to control eutrophication. Bio-manipulation

Table 3.5:   Definition and unit of the parameters in algae–phosphorus model for TH.

Variable/ Parameter	Definition	Unit	Value	Source	Range
A	Algal concentration	μg/L	—	—	—
P	Phosphorus concentration	μg/L	—	—	—
t	Time	d^{-1}	—	—	—
b	Algal growth rate	d^{-1}	0.7	Tay *et al.* (2020)	0–1
h_a	Half saturation constant	μg/L	10	Jones (2018)	0–10
p_a	Phosphorus content percentage	—	1	Jones (2018)	0–1
h_1	Flushing rate	d^{-1}	0	—	0–0.8
g	Zooplankton grazing rate	d^{-1}	0.03	Genkai-Kato and Carpenter (2005)	0–0.91
s_v	Algal mortality rate	d^{-1}	0.085	Genkai-Kato and Carpenter (2005)	0–0.9
l_p	External phosphorus loading rate	μg/L/d	0.3	Tay *et al.* (2020)	—
r	Phosphorus recycling rate	μg/L/d	0.3	Tay *et al.* (2020)	0–14
n	Half saturation value of recycling function	μg/L	10	Carpenter (2005)	0–10
q	Parameter for steepness of sigmoid function near n	–	20	Carpenter *et al.* (1999)	0–20
e	Phosphorus excretion associated with grazing	μg/μg	0.65	Genkai-Kato and Carpenter (2005)	0.4–0.8

refers to the reduction in algal concentration through an increase in zooplankton grazing (Jeppesen *et al.*, 2012).

The bifurcation diagram of algal concentration $A(\mu g/L)$ verses external phosphorus loading rate l_p indicated a low value of bifurcation tipping point threshold at $l_p = 0.01595 \ \mu g/L/d$. This low l_p threshold meant that TH is prone to eutrophication, as this low l_p threshold can easily be crossed. Three equilibrium states S1, S2 and S3 were identified, consisting of two stable (S1, S3) and one unstable (S2) equilibria. The lower stable state S3 is oligotrophic, while the higher stable state S1 is eutrophic. As l_p gradually increases from 0.0 $\mu g/L/d$, the algae concentration $A(\mu g/L)$ increases linearly along the lower stable oligotrophic state S3 until $A = 7.353 \ \mu g/L$ (mesotrophic state) is reached at the critical threshold point $l_p = 0.01595 \ \mu g/L/d$. Beyond this critical threshold point, a further increase in l_p would cause the equilibrium to "quantum jump" to another higher stable eutrophic state (S1) and remain there in a eutrophic state.

Since TH shows irreversible behavior, lake eutrophication cannot be reversed by merely reducing l_p. This implies that other lake restoration methods such as flushing and sediment dredging, coupled with drastic reduction in l_p, are required to significantly improve the eutrophic state of TH. The main driver of eutrophication in TH is strong internal P recycling driven by the strong exchange of P between sediments and the water column. The extreme shallowness of TH vastly promotes this sediment–water exchange, and hence promotes eutrophication. Dredging the sediment is needed (a) to remove accumulated sediments and P, (b) to increase the depth of TH for better dilution and (c) to increase the rate of flushing. Dredging will increase the threshold of eutrophication l_p, and hence will delay the onset of eutrophication. By decreasing the coupling between sediment and water column, dredging will decrease P concentration and decelerate algae growth. Dredging the nutrient-rich lake bottom sediment in TH to a depth of 4 m was therefore proposed in Chapter 2 as an effective measure to control eutrophication in TH.

3.5.2.2 *Impact of flushing rate* h_1

For the purpose of bifurcation analysis on the impact of flushing rate h_1, we assume that TH has a mean depth of 1 m, before dredging.

In order to determine the flushing rate required to restore TH to the oligotrophic state ($A < 2.6$ μg/L), a bifurcation diagram of algal concentration A against flushing rate h_1 was developed. As expected, the algal concentration decreases when the flushing rate increases. Three equilibrium states were identified consisting of two stable (S4, S6) and one unstable (S5) equilibria. The higher stable state *S4* with $A > 10$ μg/L is eutrophic, while the lower stable state *S6* with $A < 10$ μg/L is oligotrophic. A flushing rate of $h_1 > 0.02575$ d^{-1} would restore TH to the mesotrophic state ($A < 9.735$ μg/L), while the oligotrophic state ($A < 2.6$ μg/L) of TH can be achieved when $h_1 > 0.042$ d^{-1}. It should be noted that $h_1 = 0.02575$ d^{-1} and 0.042 d^{-1} means that the lake water in TH should be completely replaced every 40 days and 24 days, respectively. This high frequency of water replacement may not be feasible. Hence, it was proposed in Chapter 2 that TH be dredged to a deeper depth of 4 m to provide more water for flushing. As noted in Chapter 2, with a dredged depth of 4 m, eutrophication of TH can be controlled by flushing at the rate of 0.01 per day or equivalent to replacing the lake water every 100 days. This is possible by rainwater harvesting on campus.

3.6   Regime Shift in Lake Fuxian, China

As noted earlier, the depth and temperature of a lake play important roles in eutrophication. Both Sunway Lagoon and Tasik Harapan are shallow tropical lakes, and hence are prone to eutrophication. We now present an overview of a case study on a deep temperate lake to demonstrate the important role of depth and temperature in eutrophication (Tay *et al.*, 2021). As compared to SL and TH, Lake Fuxian has a large volume (189×10^8 m^3), a deep water column (mean depth 89 m) and a temperate climate. Hence, this overview will show that Lake Fuxian is more resilient to eutrophication (Zhang, 2015) than TH.

Located at Yunnan, China, Lake Fuxian (Figure 3.6) has an area of 211 km^2, a mean depth of 89 m, a maximum depth of 155 m and a water storage of 189×10^8 m^3 (Zhou *et al.*, 2018). As a large, oligotrophic lake and the second deepest lake in China, Lake Fuxian contributes 9.16% to freshwater storage in China (Wu *et al.*, 2017). It is 31.5 km long and 6.7 km wide on average. More than 20 rivers flow

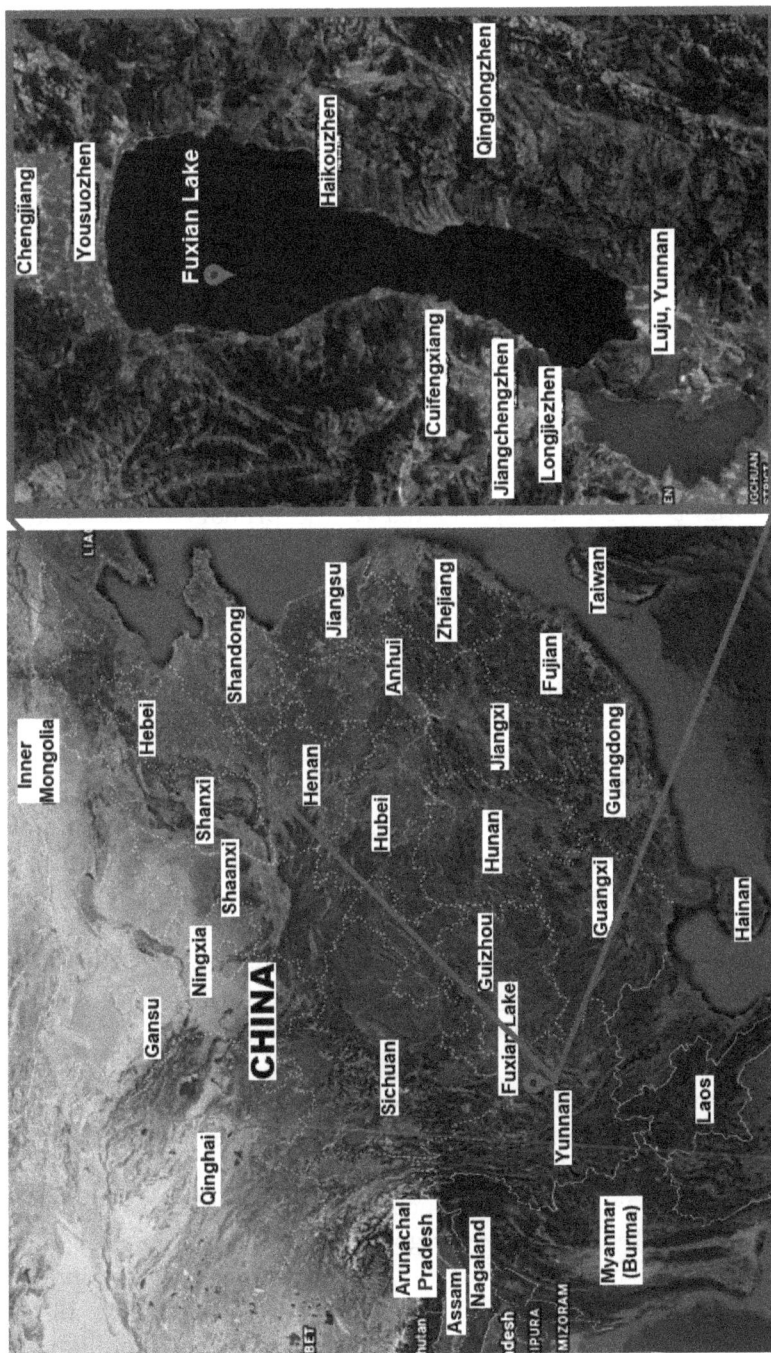

Figure 3.6: Location of Lake Fuxian in Yunnan, China.

into the lake, with the Gehe Watergate as the main inflow, while the Haikou River is the only outflow (Li *et al.*, 2011). Limited outflow has the propensity to increase the water retention time of Lake Fuxian to 167 years. The lake serves as a major drinking water source and is one of the most important freshwater resources in China. However, it has been reported that the lake's water quality has been deteriorating since the 1980s because of pollutants being discharged into the lake (Li *et al.*, 2011). The pollutants are derived from human activities such as agricultural and industry development, urbanization and tourism. A mathematical model was therefore developed to analyze the evolution of eutrophication in the deep temperate Lake Fuxian.

3.6.1   *Mathematical model for Lake Fuxian*

Phosphorus is commonly considered as the primary driver of lake eutrophication. Numerous studies have indicated the existence of a strong relationship between algal concentration and nutrient loading Li *et al.* (2014). Algal concentration is a common indicator of eutrophication, while phosphorus is the primary driver of eutrophication. Hence, algae and phosphorus are the two most important components of a eutrophication model as formulated in Eqs. (3.5) to (3.7) as proposed for the temperate Lake Fuxian (Genkai-Kato and Carpenter, 2005; Jones, 2018). The sources of phosphorus input into Lake Fuxian include external phosphorus loading (l_p), excretion associated with zooplankton grazing (egA) and recycling from sediment (r_1). The loss of phosphorus from the lake is due to flushing ($h_1 P$) and uptake by algae for growth ($bA(P - p_a A)/(h_a + P - p_a A)$). Phosphorus is a nutrient-limiting factor for algal growth and is represented by the term $(P - p_a A)/(h_a + P - p_a A)$. The loss of algae is due to flushing ($h_1 A$), grazing by zooplankton (gA) and mortality (sA/z_e).

In this model, the phosphorus recycling term (r_1 in Eq. (3.7)), suitable for temperate lake, is formulated based upon the phosphorus release rate from the sediment into the water column and the duration of anoxia during summer stratification (Genkai-Kato and Carpenter, 2005). The term r_1 is governed by the oxygen depletion rate, which is a function of algal concentration (A), hypolimnion thickness (z_h) and hypolimnion temperature (T) (Charlton, 1980). It is assumed that the phosphorus released from the hypolimnion into the epilimnion occurs at the maximal rate. This maximum rate is calculated as the product of 14 mg m^{-2} d^{-1} multiplied by the proportion of anoxic days in the hypolimnion during the stratified season, which is reported to be 150 days long (Genkai-Kato and Carpenter, 2005). Table 3.6 displays the definition and unit of the parameters for this algae-phosphorus model for the temperate Lake Fuxian.

$$\frac{dA}{dt} = bA\frac{P - p_a A}{h_a + P - p_a A} - \left(h_1 + g + \frac{s}{z_e}\right)A \tag{3.5}$$

$$\frac{dP}{dt} = l_p + r_1 + egA - bA\frac{P - p_a A}{h_a + P - p_a A} - h_1 P \tag{3.6}$$

where $r_1 = \frac{R}{z_e}$,

$$R = \frac{14}{150}\left[150 - \frac{DO \times (50 + z_h)}{3.8 \times \frac{1.15A^{1.33}}{9 + 1.15A^{1.33}} \times 2^{\frac{T-4}{10}} + 0.12}\right] \tag{3.7}$$

3.6.2 *Regime shift analysis for Lake Fuxian*

The analysis of the regime shift in Lake Fuxian is performed in three parts: (a) curve fitting to identify key parameter values relevant to Lake Fuxian eutrophication status, (b) bifurcation analysis to determine the types of lake responses for Lake Fuxian and (c) determination of external loading threshold that can trigger regime shift from the oligotrophic to eutrophic state in Lake Fuxian.

3.6.2.1 *Curve fitting for Lake Fuxian*

First, curve fitting was performed to identify the best set of parameter values that best reflect the eutrophication dynamic in Lake Fuxian. External phosphorus loading rate l_p is the most important parameter that governs the dynamics of lake eutrophication (Li *et al.*, 2014). The parameter value for external phosphorus loading rate l_p is estimated from curve fitting to reflect the recorded algal concentration in Lake Fuxian. Some of the other parameter values, such as zooplankton grazing rate and algal sinking rate, can be obtained from the literature review. The algal concentration data for the period 1990 to 2014 for Lake Fuxian are obtained from Zhou *et al.* (2018). The

Table 3.6:   Definitions and units of the parameters in algae–phosphorus model for the temperate Lake Fuxian.

Variable/ Parameter	Definition	Unit	Value	Source	Range
A	Algal concentration	μg/L	—	—	—
P	Phosphorus concentration	μg/L	—	—	—
t	Time	d^{-1}	—	—	—
b	Algal growth rate	d^{-1}	0.7	Genkai-Kato and Carpenter (2005)	0–1
h_a	Half saturation constant	μg/L	10	Jones (2018)	0–10
p_a	Phosphorus content percentage	—	1	Jones (2018)	0–1
h_1	Flushing rate	d^{-1}	1.6×10^{-5}	Li *et al.* (2011)	0–0.8
g	Zooplankton grazing rate	d^{-1}	0.03	Genkai-Kato and Carpenter (2005)	0–0.91
s	Algal sinking rate	m d^{-1}	0.085	Genkai-Kato and Carpenter (2005)	0–3
l_p	External phosphorus loading rate	μg/L/d	Different scenarios	—	—
DO	Saturated oxygen concentration corresponding to temperature	mg/L	10.3	Rounds *et al.* (2013)	—
z_h	Mean of hypolimnion thickness	m	88.55	Wang *et al.* (2017)	—
z_e	Epilimnion thickness	m	27.5	Wang *et al.* (2017)	—
T	Temperature at hypolimnion	°C	14.21	Wang *et al.* (2017)	—
e	Phosphorus excretion associated with grazing	μg/μg	0.65	Genkai-Kato and Carpenter (2005)	0.4–0.8

Table 3.7: Summary of fitted external phosphorus loading, $l_p(\mu g/L/d)$ into Lake Fuxian for year 1990. to 2014.

Year	Algal concentration, A (μg/L)	Fitted external phosphorus loading, l_p (μg/L/d)
1990–1999	0.5776 (average value)	0.0038
2000	0.8597	0.0140
2001–2014	1.6894–2.3636	l_p increases constantly by 0.00016 μg/L/d per year

algal concentration A in Lake Fuxian increased significantly in year 2001 from an average concentration of 0.5776 μg/L (1990–2000) to 2.1256 μg/L (2001). This sudden increase in algal concentration was attributed to increased sewage discharge into the lake as a result of the development of agriculture, industry and urbanization (Li *et al.*, 2011). Overall, the algal concentration data indicate that Lake Fuxian is presently classified as oligotrophic (Zhou *et al.*, 2018). However, it has been observed that Lake Fuxian is at the risk of becoming increasing polluted, as may be observed from the deteriorating water quality over the course of the past 25 years (Chen *et al.*, 2019).

For the curve-fitting simulation, the algae-phosphorus model in Eqs. (3.5) to (3.7) is used, with the definition and value of the parameters shown in Table 3.6. The best fit to the average values of algal concentration (0.5776 μg/L) for the years 1990 to 1999 is obtained with the external phosphorus loading rate of $l_p = 0.0038$ μg/L/d. In year 2000, the sudden increase in A is simulated by a sharp increase in the loading rate to $l_p = 0.0140$ μg/L/d. Subsequently, for the year 2001 and beyond, the increasing trend of algal concentration is simulated by annually increasing the external phosphorus loading rate linearly by 0.00016 μg/L/d per year from the initial loading of 0.0140 μg/L/d in 2001. Table 3.7 shows the summary of fitted l_p from year 1990 to 2014 for Lake Fuxian.

3.6.2.2 *Bifurcation analysis for Lake Fuxian*

Second, bifurcation analysis was then performed by using the model XPPAUT to determine the types of lake responses for Lake Fuxian, based upon the estimated l_p. Developed by Bard Ermentrout

Ermentrout (2002), XPPAUT is an open-source numerical tool for simulating, animating and analyzing dynamical systems. Since phosphorus is the main driver of eutrophication, l_p is selected as the bifurcation parameter in the analysis. No multiple equilibria are obtained and A increases linearly as l_p increases without any quantum jump. This is indicative of the reversible behavior of Lake Fuxian where eutrophication can be reversed by controlling the l_p alone.

3.6.2.3 *Phosphorus threshold for Lake Fuxian*

Third, the external phosphorus eutrophication loading threshold is determined by analyzing the effects of the external phosphorus loading rate (l_p) on lake water quality. Information on eutrophication threshold is vital in early warning analysis with the goal of preventing a regime shift in Lake Fuxian. Here, the external phosphorus loading threshold is defined as the external phosphorus loading above which eutrophication occurs ($A > 10$ μg/L). Two case studies of the numerical simulations are presented.

3.6.2.4 *Case study 1*

For case study 1, it was assumed that the current loading scenario for years 2001 to 2014 persisted beyond 2014. Hence, the external phosphorus loading rate continues to increase linearly by a moderate rate of 0.00016 μg/L/d per year beyond 2014. Simulation results suggested that Lake Fuxian would only become eutrophic by year 2380, if the current trend of a modest loading scenario persists. Since the lake has a large volume (189×10^8 m^3), a deep water column (mean depth 89 m) and a long retention time (167 years), Lake Fuxian is resilient to eutrophication and may change slowly in response to modest anthropogenic interference (Zhang, 2015).

3.6.2.5 *Case study 2*

For case study 2, the purpose was to determine the external phosphorus loading threshold that can trigger eutrophication in Lake Fuxian. Simulation results indicated that a sharp increase in l_p values to $l_p = 0.0765$ μg/L/d (which is five times the rate in year 2014), effective in year 2015 onward, could trigger a regime shift in Lake Fuxian in just three years. The value of A would increase to $A = 10$ μg/L in three years after l_p is increased to 0.0765 μg/L/d. This implies that $l_p = 0.0765$ μg/L/d may be identified as the external phosphorus loading threshold that can trigger eutrophication in Lake Fuxian. Any value of l_p which is equal to or greater than 0.0765 μg/L/d could result in eutrophication of Lake Fuxian. Hence, this threshold value of l_p serves as a guide for the relevant authorities in controlling or limiting the input of nutrients into the lake.

3.7 Comparative Analysis

Table 3.8 displays the comparison of the types of lake responses, regime shift thresholds and restoration methods for Lake Fuxian and Tasik Harapan. Lake Fuxian is identified as a reversible lake with a higher regime shift threshold of $l_p = 0.0765$ μg/L/d. On the contrary, Tasik Harapan is an irreversible lake with a lower value of regime shift threshold of $l_p = 0.01595$ μg/L/d. A lower value of regime shift threshold implies that eutrophication would occur $(A > 10$ μg/L) earlier as the lower threshold value of l_p can be crossed earlier. For a reversible lake, eutrophication can be reversed by controlling l_p alone. However, for the highly eutrophic and irreversible TH, model analysis suggests that a combination of reduction in l_p and other restoration methods such as flushing and sediment dredging is needed to restore the highly eutrophic and irreversible TH. Here, the differences in regime shift patterns for Lake Fuxian and TH are attributable to two critical factors, i.e., temperature and depth of lake as highlighted in Genkai-Kato and Carpenter (2005).

Table 3.8: Comparison of type of lake responses, regime shift thresholds and restoration methods between Lake Fuxian and Tasik Harapan.

Lakes	Lake Fuxian (temperate lake)	Tasik Harapan (tropical lake)
Type of lake response	Reversible	Irreversible
Regime shift thresholds	$l_p = 0.0765$ μg/L/d	$l_p = 0.01595$ μg/L/d
Restoration method(s)	Lake eutrophication can be reversed by controlling l_p alone	A combination of reduction in l_p with other restoration methods such as dredging and flushing

3.7.1   *Tasik Harapan*

Tasik Harapan is prone to eutrophication because of its *tropical* higher temperature and *shallow* depth. Located near the equator, Malaysia has a tropical climate which is hot and humid throughout the year (Tang, 2019). A higher temperature would promote eutrophication and render the restoration more difficult (Genkai-Kato and Carpenter, 2005). High solar irradiance and a high temperature of 26–28°C in Malaysia promote algal growth, which is optimal in the temperature range 20–30°C (Singh and Singh, 2015). A higher lake temperature, as in the case of TH, increases the bacterial activity, stimulates the mineralization of organic matter, diminishes oxygen solubility and reduces dissolved oxygen (DO) levels. As a result of reduced DO, more phosphorus is released from the sediment into the water column which is readily absorbed for algal growth. A shallow depth also facilitates the exchange of P between the sediments and the water column, thereby promoting eutrophication. Further, wind-induced resuspension of phosphorus from the sediment is more likely to occur in the shallow TH (Hickey and Gibbs, 2009). The phosphorus so released from the sediment is readily available to the photic zone because of its shallow depth. Hence, the removal of the lake bottom sediment via dredging is deemed the most effective restoration method, especially for shallow lakes such as TH (Carpenter, 2005; Hickey and Gibbs, 2009).

3.7.2   *Lake Fuxian*

Lake Fuxian is more resilient to eutrophication because of its *temperate* climate and *deep* depth. The epilimnion and hypolimnion temperatures of Lake Fuxian during summer are in the range of 22–25°C and 13–15°C, respectively (Wang *et al.*, 2017). This lower temperature inhibits algal growth. Further, Lake Fuxian is a deep lake with a mean depth of 89 m and a maximum depth of 155 m. This large depth buffers the upper water column from the bottom sediment layer, and protects the upper water column from excessive eutrophication. The phosphorus released from the sediment may not readily reach the upper epilimnion due to the thick middle hypolimnion that dilutes the recycled phosphorus, thereby inhibiting eutrophication in deep lakes (Genkai-Kato and Carpenter, 2005; Welch and Cooke, 2005).

3.8 Conclusions

The basic mathematical model consisting of algae and phosphorus for examining regime shifts in lakes is formulated for the temperate Lake Fuxian and the tropical TH, with distinctly different phosphorus recycling terms. For the temperate Lake Fuxian, the phosphorus recycling term used is the product of the maximal phosphorus release rate and proportion of anoxic days. In the study of regime shift in lakes, most of the mathematical models were developed for temperate lakes. In order to overcome this limitation and to examine regime shifts in tropical lakes, a sigmoid function of the phosphorus concentration is formulated to represent the phosphorus recycling in the tropical TH. This choice is justified based on three reasons: (i) the sigmoid function could represent the probability of anoxia in lakes, (ii) the alternative states of lakes can be represented by using the sigmoid function and (iii) the major recycling mechanism in shallow lakes is well represented by phosphorus resuspension that is proportional to the amount of phosphorus available in water. The following conclusions regarding the eutrophication status of Lake Fuxian and Tasik Harapan may be derived.

3.8.1 *Lake Fuxian is reversible*

Model simulations coupled with bifurcation analysis indicate that the current state of Lake Fuxian is reversible, with a higher regime shift threshold of phosphorus loading $l_p = 0.0765$ $\mu g/L/d$. Any value of l_p greater than this value could lead to lake eutrophication. Two distinct scenarios with distinct phosphorus loading rates l_p are used to estimate the time when Lake Fuxian could become eutrophic. First, a modest phosphorus loading rate l_p is used, i.e., the l_p is increased gradually at the rate of 0.00016 $\mu g/L/d$ per year. With this slow rate of l_p increase, Lake Fuxian is expected to slowly become eutrophic by the year 2380. Second, a higher phosphorus loading rate of $l_p = 0.0765$ $\mu g/L/d$ is applied. In this case, Lake Fuxian quickly becomes eutrophic in just three years. Hence, management of Lake Fuxian should ensure that this loading threshold is not exceeded in the future to prevent eutrophication.

3.8.2 *TH is irreversible*

Tasik Harapan may be characterized as an irreversible lake. It has a lower threshold value of regime shift at $l_p = 0.01595$ μg/L/d. This means that TH will easily reach a eutrophic state as the low regime shift threshold can be easily crossed. Further, once it becomes eutrophic, reversal of status is difficult as the lake is irreversible in characteristic. The main factors contributing to high eutrophication in TH are its shallow depth, its year-round high temperature and the accumulated P in the sediments.

This sharp contrast in regime shift behavior between Lake Fuxian and Tasik Harapan is attributed to the large contrast in the temperature and depth of the two lakes. Lake Fuxian is located in an area with temperate climate and has deep depth, while TH is located in an area with tropical climate and has shallow depth. Higher temperature favors algal growth. Moreover, the phosphorus released from the sediment will be better mixed in shallow waters than in deep waters, granting algae an easy access to sediment phosphorus in shallow lakes such as TH. Lake Fuxian, having a higher regime shift threshold of $l_p = 0.0765$ μg/L/d, is more resilient to eutrophication than TH, which has a lower regime shift threshold of $l_p = 0.01595$ μg/L/d. Model analysis suggests that for TH, reducing the external phosphorus loading to zero may not shift the eutrophic lake to the oligotrophic state. In its current eutrophic condition, a large reduction in the phosphorus input through dredging and flushing is needed to restore the irreversible state of TH.

References

Bai, X.L., Ding, S.M., Fan, C.X., Liu, T., Shi, D. and Zhang, L. (2009). Organic phosphorus species in surface sediments of a large, shallow, eutrophic lake, Lake Taihu, China. *Environmental Pollution*, 157, 2507–2513.

Cao, X., Wang, Y., He, J., Luo, X. and Zhang, Z. (2016). Phosphorus mobility among sediments, water and cyanobacteria enhanced by cyanobacteria blooms in eutrophic Lake Dianchi. *Environmental Pollution*, 219, 580–587.

Carpenter, S.R. (2005). Eutrophication of aquatic ecosystems: Bistability and soil phosphorus. *Proceedings of the National Academy of Sciences USA*, 102, 10002–10005.

Carpenter, S.R., Ludwig, D. and Brock, W.A. (1999). Management of eutrophication for lakes subject to potentially irreversible change. *Ecological Application*, 9(3), 751–771.

Caspers, H. (1984). OECD: Eutrophication of waters. Monitoring, assessment and control. 154 pp. Paris: Organisation for Economic Co-Operation and Development 1982. *Internationale Revue der gesamten Hydrobiologie und Hydrographie*, 69, 200–200 (1984).

Chadwick, O.A., Derry, L.A., Vitousek, P.M., Heubert, B.J. and Hedin, L.O. (1999). Changing sources of nutrients during four million years of ecosystem development. *Nature*, 397, 491–496.

Chapra, S.C. (1997). *Surface Water-quality Modeling*. McGraw-Hill, New York.

Charlton, M.N. (1980). Hypolimnion oxygen consumption in lakes: Discussion of productivity and morphometry effects. *Canadian Journal of Fisheries and Aquatic Sciences*, 37, 1531–1539.

Chen, J., Lyu, Y., Zhao, Z., Liu, H., Zhao, H. and Li, Z. (2019). Using the multidimensional synthesis methods with non-parameter test, multiple time scales analysis to assess water quality trend and its characteristics over the past 25 years in the Fuxian Lake, China. *Science of the Total Environment*, 655, 242–254.

Coll, M., Santojanni, A., Palomera, I., Tudela, S. and Arneri, E. (2007). An ecological model of the Northern and Central Adriatic Sea: Analysis of ecosystem structure and fishing impacts. *Journal of Marine Systems*, 67, 119–154.

Conley, D.J., Paerl, H.W., Howarth, R.W., Boesch, D.F., Seitzinger, S.P., Havens, K.E., Lancelot, C. and Likens, G.E. (2009). Controlling eutrophication: Nitrogen and phosphorus. *Ecology*, 323, 1014–1015.

Ermentrout, B. (2002). Simulating, analyzing, and animating dynamical systems: A guide to XPPAUT for researchers and students. SIAM, Philadelphia.

Ferreira, J.G., Andersen, J.H., Borja, A., Bricker, S.B., Camp, J., Cardoso da Silvaf, M., Garcés, E., Heiskanen, A.-S., Humborg, C., Ignatiades, L., Lancelot, C., Menesguen, A., Tett, P., Hoepffner, N. and Claussen, U. (2011). Overview of eutrophication indicators to assess environmental status within the European Marine Strategy Framework Directive. *Estuaries, Coastal and Shelf Science*, 93, 117–131.

Filbee-Dexter, K. and Scheibling, R.E. (2014). Sea urchin barrens as alternative stable states of collapsed kelp ecosystems. *Marine Ecology Progress Series*, 495, 1–25.

Genkai-Kato, M. and Carpenter, S.R. (2005). Eutrophication due to phosphorus recycling in relation to lake morphometry, temperature, and macrophytes. *Ecology*, 86(1), 210–219.

Hickey, C.W. and Gibbs, M.M. (2009). Lake sediment phosphorus release management — decision support and risk assessment framework. *New Zealand Journal of Marine and Freshwater*, 43(3), 819–856.

Holling, C.S. (1973). Resilience and stability of ecological systems. *Annual Review of Ecology and Systematics*, 4, 1–23.

Huang, C., Wang, X., Yang, H., Li, Y., Wang, Y., Chen, X. and Xu, L. (2014). Satellite data 9–15 regarding the eutrophication response to human activities in the plateau lake Dianchi in China from 1974 to 2009. *Science of the Total Environment*, 485–486, 1–11.

Jeppesen, E., Søndergaard, M., Lauridsen, T.L., Davidson, T.A., Liu, Z., Mazzeo, N., Trochine, C., Özkan, K., Jensen, H.S., Trolle, D., Starling, F., Lazzaro, X., Johansson, L.S., Bjerring, R., Liboriussen, L., Larsen, S.E., Landkildehus, F., Egemose, S. and Meerhoff, M. (2012). Biomanipulation as a restoration tool to combat eutrophication: Recent advances and future challenges. *Advances in Ecological Research*, 47, 411–488.

Jiang, J., Fuller, D.O., Teh, S.Y., Zhai, L., Koh, H.L., DeAngelis, D.L. and Sternberg, L.D.S.L (2015). Bistability of mangrove forests and competition with freshwater plants. *Agricultural and Forest Meteorology*, 213, 283–290.

Jones, M. (2018). Using a coupled bio-economic model to find the optimal phosphorus load in Lake Tainter WI. *Honors Research Projects*, 632.

Jordan-Cooley, W.C., Lipcius, R.N., Shaw, L.B., Shen, J. and Shi, J. (2011). Bistability in a differential equation model of oyster reef height and sediment accumulation. *Journal of Theoretical Biology*, 289, 1–11.

Koh, H.L. (2017). Sunway Lagoon Rehabilitation Study: Data, Simulation and Synthesis (Selangor: Sunway Lagoon Theme Park).

Koh, H.L., Tan, W.K. and Teh, S.Y. (2018a). Regime shift analysis and numerical simulation for effective ecosystem management. *International Journal of Environmental Science and Development*, 9, 192–199.

Koh, H.L., Teh, S.Y., Lee E, Tan, W.K., Sagathevan, K.A. and Low, A.A. (2018b). Derivation of optimal fish stocking density via simulation of water quality model E2Algae. *AIP Conference Proceedings*, 1974, 020042.

Kong, X., Dong, L., He, W., Wang, Q., Mooij, W.M. and Xu, F. (2015). Estimation of the long-term nutrient budget and thresholds of regime shift for a large shallow lake in China. *Ecological Indicators*, 52, 231–244.

Lal, R. (2015). Restoring soil quality to mitigate soil degradation. *Sustain*, 7, 5875–5895.

Li, M., Xie, G.Q., Dai, C.R., Yu, L.X., Li, F.R. and Yang, S.P. (2009a). A study of the relationship between the water body chlorophyll a and water quality factors of the offcoast of Dianchi Lake. *Yunnan Geographic Environment Research*, 21, 102–106.

Li, Y.K., Chen, Y., Song, B., Olson, D., Yu, N. and Chen, L.Q. (2009b). Ecosystem structure and functioning of Lake Taihu (China) and the impacts of fishing. *Fisheries Research*, 95, 309–324.

Li, Y., Gong, Z., Xia, W. and Shen, J. (2011). Effects of eutrophication and fish yield on the diatom community in Lake Fuxian, a deep oligotrophic lake in southwest China. *Diatom Research*, 26(1), 51–56.

Li, Y.P., Tang, C.Y., Yu, Z.B. and Acharya, K. (2014). Correlations between algae and water quality: Factors driving eutrophication in Lake Taihu, China. *International Journal of Environmental Science and Technology*, 11, 169–182.

Lu, J., Wang, H.B., Pan, M., Xia, J., Xing, W. and Liu, G.H. (2012). Using sediment seed banks and historical vegetation change data to develop restoration criteria for a eutrophic lake in China. *Ecological Engineering*, 39, 95–103.

Lv, H., Yang, J., Liu, L., Yu, X., Yu, Z. and Chiang, P. (2014). Temperature and nutrients are significant drivers of seasonal shift in phytoplankton community from a drinking water reservoir, subtropical China. *Environmental Science and Pollution Research*, 21, 5917–5928.

McDonald, C.P., Urban, N.R. and Casey, C.M. (2010). Modeling historical trends in Lake Superior total nitrogen concentrations. *Journal of Great Lakes Research*, 36, 715–721.

McKee, K.L. (2011). Biophysical controls on accretion and elevation change in Caribbean mangrove ecosystems. *Estuarine, Coastal and Shelf Science*, 91, 475–483.

Medina, E., Cuevas, E. and Lugo, A.E. (2010). Nutrient relations of dwarf Rhizophora mangle mangroves on peat in eastern Puerto Rico. *Plant Ecology*, 207, 13–24.

Müller, F., Burkhard, B. and Kroll, F. (2009). Resilience, integrity and ecosystem dynamics: Bridging ecosystem theory and management ed J. C. Otto and R. Dikau (Berlin: Springer) pp. 221–242.

Najar, I.A. and Khan, A.B. (2012). Assessment of water quality and identification of pollution sources of three lakes in Kashmir, India, using multivariate analysis. *Environmental Earth Sciences*, 66, 2367–2378.

Rounds, S.A., Wilde, F.D. and Ritz, G.F. (2013). Dissolved oxygen (ver. 3.0): U.S. Geological Survey techniques of water-resources investigations, book 9, chap. A6, sec. 6.2.

Ruesink, J., Lenihan, H., Trimble, A., Heiman, K., Micheli, F., Byers, J. and Kay, M. (2005). Introduction of non-native oysters: Ecosystem effects and restoration implications. *Annual Review of Ecology, Evolution Systematics*, 36, 643–689.

Runyan, C.W. and D'Odorico, P. (2013). Positive feedbacks and bistability associated with phosphorus-vegetation-microbial interactions. *Advances in Water Resources*, 52, 151–164.

Scheffer, M. (1998). *Ecology of Shallow Lakes*. Chapman and Hall, New York.

Scheffer M. and Carpenter, S.R. (2003). Catastrophic regime shifts in ecosystems: Linking theory to observation. *Trends in Ecology and Evolution*, 18, 648–656.

Scheffer, M., Carpenter, S., Foley, J.A., Folke, C. and Walker, B. (2001). Catastrophic shifts in ecosystems. *Nature*, 413, 591–596.

Scheffer, M., Carpenter, S.R., Lenton, T.M., Bascompte, J., Brock, W., Dakos, V., Van De Koppel, J., Van De Leemput, L.A., Levin, S.A., Nes, E.H.V. *et al.* (2012). Anticipating critical transition. *Science*, 338, 344–348.

Schmitt, R.J., Holbrook, S.J., Davis, S.L., Brooks, A.J., and Adam, T.C. (2019). Experimental support for alternative attractors on coral reefs. *Proceedings of the National Academy of Sciences USA*, 116(10), 4372–4781.

Shan, K., Li, L., Wang, X., Wu, Y., Hu, L., Yu, G. and Song, L. (2014). Modelling ecosystem structure and trophic interactions in a typical cyanobacterial bloom-dominated shallow Lake Dianchi, China. *Ecological Modelling*, 291, 82–95.

Singh, R., Reed, P.M. and Keller, K. (2015). Many-objective robust decision making for managing an ecosystem with a deeply uncertain threshold response. *Ecology and Society*, 20, 12.

Singh, S.P. and Singh, P. (2015). Effect of temperature and light on the growth of algae species: A review. *Renewable and Sustainable Energy Reviews*, 50, 431–444.

Sondergaard, M., Jensen, J.P. and Jeppesen, E. (2003). Role of sediment and internal loading of phosphorus in shallow lakes. *Hydrobiologia*, 506, 135–145.

Tang, K.H.D. (2019). Climate change in Malaysia: Trends, contributors, impacts, mitigation and adaptations. *Science of the Total Environment*, 650, 1858–1871.

Tay, C.J., Teh, S.Y. and Koh, H.L. (2020). Eutrophication bifurcation analysis for Tasik Harapan restoration. *International Journal of Environmental Science and Development*, 11(8), 407–413.

Tay, C.J., Teh, S.Y., Koh, H.L., Mohd Hafiz, M. and Zhang, Z. (2021). Managing regime shift in lake systems by modelling and simulation. In: *Modelling, Simulation and Applications of Complex Systems, Proceedings in Mathematics & Statistics*, Springer, in press.

Teh, S.Y, DeAngelis, D.L., Sternberg, L.D.L., Miralles-Wilhelm, F.R., Smith, T.J. and Koh, H.L. (2008a). A simulation model for projecting changes in salinity concentrations and species dominance in the coastal margin habitats of the Everglades. *Ecological Modelling*, 213, 245–256.

Teh, S.Y., Koh, H.L., Ismail, A.I.M. and Mansor, M. (2008b). Determining photosynthesis rate constants in Lake Harapan Penang. *International Conference on Biomedical*, 1, 585–590.

Teh, S.Y., Koh, H.L., DeAngelis, D.L., Voss, C.I. and Sternberg, L. (2019). Modeling $\delta^{18}O$ as an early indicator of regime shift arising from salinity stress in coastal vegetation. *Hydrogeology Journal*, 27(4), 1257–1276.

Tu, Q.Y., Gu, D.X., Yi, C.Q., Xu, Z. R. and Han, G.Z. (1990). *The Researches on the Lake Chaohu Eutrophication*. Publisher of University of Science and Technology of China, Hefei, China.

van Voorn, G.A.K., Kooi, B.W. and Bregt, A.K. (2016). Over-shading is critical for inducing a regime shift from heathland to grassland under nitrogen enrichment. *Ecological Complexity*, 27, 74–83.

Vitousek, P.M., Porder, S., Holton, B.Z. and Chadwick, O.A. (2010). Terrestrial phosphorous limitation: mechanisms, implications and nitrogen-phosphorous interactions. *Ecological Applied*, 20, 5–15.

Wang, F.E., Tian, P., Yu, J., Lao, G.M. and Shi, T.C. (2011). Variations in pollutant fluxes of rivers surrounding Taihu Lake in Zhejiang Province in 2008. *Physics and Chemistry of the Earth, Parts A/B/C* 36, 366–371.

Wang, L.J., Yu, H., Niu, Y., Niu, Y., Zhang, Y.L., Liu, Q. and Ji, Z.Y. (2017). Distribution characteristics of water temperature and water quality of Fuxian Lake during thermal stratification period in summer. *Huang Jing Ke Xue*, 38(4), 1384–1392.

Wardle, D.A., Walker, L.R. and Bardgett, R.D. (2004). Ecosystem properties and forest decline in contrasting long-term chronosequences. *Science*, 305, 509–513.

Weissmann, H. and Shnerb, N.M. (2016). Predicting catastrophic shifts. *Journal of Theoretical Biology*, 397, 128–134.

Welch, E.B. and Cooke, G.D. (2005). Internal phosphorus loading in shallow lakes: Importance and control. *Lake Reservoir Management*, 21(2), 209–217.

Wu, X., Deng, K., Ge, P., Wang, Y. and Xue, Y. (2017). Investigation on the ecological environment and resource protection management system of Fuxian lake. *Advanced Engineering Research*, 141, 290–295.

Xie, L.Q., Xie, P. and Tang, H.J. (2003). Enhancement of dissolved phosphorus release from sediment to lake water by Microcystis blooms — an enclosure experiment in a hyper-eutrophic, subtropical China lake. *Environmental Pollution*, 122, 391—399.

Yang, J., Lv, H., Yang, J., Liu, L., Yu, X. and Chen, H. (2016). Decline in water level boosts cyanobacteria dominance in subtropical reservoirs. *Science of the Total Environment*, 557–558, 445–452.

Yang, J.R., Lv, H., Isabwe, A., Liu, L., Yu, X., Chen, H. and Yang, J. (2017). Disturbance-induced phytoplankton regime shifts and recovery of cyanobacteria dominance in two subtropical reservoirs. *Water Research*, 120, 52–63.

Zhang, H. and Huang, G.H. (2011). Assessment of non-point source pollution using as spatial multicriteria analysis approach. *Ecological Modelling*, 222, 313–321.

Zhang, Y., Su, Y., Liu, Z., Chen, X., Yu, J., Di, X. and Jin, M. (2015). Sediment lipid biomarkers record increased eutrophication in Lake Fuxian (China) during the past 150 years. *Journal of Great Lakes Research*, 41, 30–40.

Zhou, Q., Zhang, Y., Li, K., Huang, L., Yang, F., Zhou, Y. and Chang, J. (2018). Seasonal and spatial distributions of euphotic zone and long-term variations in water transparency in a clear oligotrophic Lake Fuxian, China. *Journal of Environmental Science*, 72, 185–197.

Chapter 4

Role of Coastal Aquifers in Water and Food Security

4.1 Introduction

Rapid population growth coupled with fast economic development has imposed increasing and unsustainable demands on precious freshwater resources, both surface and subsurface groundwater. According to the United Nations World Water Development Report 2018, an estimated 3.6 billion people, nearly half the global population, currently live in areas that are potentially water scarce for at least one month per year (UN-Water, 2018). Worldwide, about 1.5 to 2 billion people rely on fresh groundwater stored in the aquifers as a main freshwater source for drinking, domestic and agricultural usage. However, declining groundwater quantity and quality that have been reported in many parts of the world have posed a severe threat to the resilience of fresh groundwater supply. This threat to fresh groundwater supply is further compounded by global climate change (GCC) and sea level rise (SLR). Temperature increase of between 1.4 and 5.8% and SLR of between 1.8 and 5.9 mm/year have been predicted by the GCC model. The adverse impacts of GCC and SLR may lead to salinization of coastal groundwater, resulting in substantial reduction in freshwater resources and drop in agricultural yields by the end of the 21st century. For impoverished coastal communities that rely on coastal aquifers for their freshwater supply, the consequence is water and food insecurity.

4.2   Salinization of Coastal Groundwater

SLR will increase surface seawater inundation because of increased seawater overtopping on coastal land. SLR will also increase subsurface saltwater intrusion because of an increase in underground seawater seepage through porous media. Increase in seawater inundation and saltwater intrusion due to the impact of SLR will reduce the thickness and volume of the aquifer lenses that sit on top of saline groundwater, thereby reducing freshwater stored in these coastal aquifers. The combined impact of surface seawater inundation and subsurface saltwater intrusion will increase the salinity in the freshwater stored in the coastal aquifers. The salinization of coastal fresh groundwater will render groundwater unsuitable for human consumption and agricultural use. The degradation of groundwater quality and reduction in quantity will have a profound impact on water and food security for coastal communities. This impact is particularly severe for those living in low-lying atoll islands throughout the Pacific and Indian Oceans, who depend critically on the fresh groundwater stored in the freshwater aquifers known as freshwater lenses (Anderson, 2002). To overcome this threat, adaptations such as alternative water resources (AWRs) including rainwater harvesting (RWH), water reuse, water conservation and desalination have received increasing acceptance as supplements to these scarce traditional water supplies to ensure water security (SDG6).

Recognizing the importance of sustainable water security to humanity, the United Nation endorses the Sustainable Development Goal 6 (SDG6) of "Ensuring availability and sustainable management of clean water and sanitation for all". Local SLR and GCC could further exacerbate pressure on already-stressed freshwater resources due to the reduction in fresh groundwater induced by SLR. The eradication of hunger (SDG2) is a major goal in the United Nations Millennium Development Goals adopted in 2000 and in the Sustainable Development Goals adopted in 2016. Hunger is tightly coupled to water and food security. Addressing the predicted water and food insecurity in many parts of the world is a daunting challenge in this century. Groundwater plays an important role in the challenge to achieve water and food security. Overextraction, seawater inundation and saltwater intrusion into groundwater aquifers will increase salinity in the soil and in the aquifers, posing serious threats to freshwater availability in coastal areas. Hence, understanding the complex and dynamic coupling between surface water and subsurface water is crucial in achieving sustainable freshwater resource utilization in coastal zones. Along the coastal plains worldwide, freshwater scarcity, soil degradation and loss of cropland will have a profound impact on water and food security, threatening SDG6 and SDG2. In wetlands worldwide, such as the Florida Everglades, increased seawater inundation and elevated saltwater intrusion into groundwater aquifers can have a pronounced impact on increasing soil salinity and on reducing available freshwater and therefore on reducing crop yields, directly threatening both water and food security. Similar remarks are pertinent to coastal communities living in the Pantai Aceh in Penang Island of Malaysia, for whom water security is a grave uncertainty. The low-elevation wetlands are subject to salinization because they are exposed to SLR, to increased seawater inundation and to elevated saltwater intrusion.

4.2.1 *Roles of coastal wetlands*

Wetlands are complex hydrogeological environments in which surface water and subsurface water interact strongly with the biotic and abiotic environment. Coastal vegetation, topography, climate change and tidal influence can interact to produce dynamic and complex responses to change soil salinity and to cause coastal vegetation regime shifts. Increasing salinity and the resulting vegetation regime shifts can post dire consequences to the coastal ecosystems and human settlements in and around coastal wetlands. Because wetlands serve as important habitats to many important species of plants and animals, profound changes in wetland structures due to GCC and SLR can induce undesirable consequences for the entire ecosystems. SLR has caused wetlands to show signs of adverse impact such as declining freshwater supply and degrading freshwater vegetation. The threat of SLR on fresh groundwater resources and coastal vegetation dynamics has been widely studied for wetlands at low-lying coastal plains in South Florida (Willard and Bernhardt, 2011), in North and South Carolina (Noe *et al.*, 2013), as well as in Louisiana (Teal *et al.*, 2012). Insights and projections of gradual (due to SLR) or abrupt (due to large storm surges) increase in groundwater salinity on vegetation are important in conserving and managing the coastal ecosystem. Two simulation models known as MANHAM and MANTRA (Teh *et al.*, 2013, 2015) have been developed by the authors and others to study the complex dynamics involving coastal vegetation competition and groundwater hydrology. In this chapter, we will use these two simulation models and SUTRA developed by the USGS to investigate the dynamic interaction between vegetation growth, groundwater salinity, freshwater precipitation and saltwater intrusion from tidal forcing. These model simulations will investigate the impacts of SLR on coastal vegetation succession and to highlight potential threats to typical coastal marshland ecosystem and human settlements in the USA and Malaysia. The potential threats include increasing soil salinity, undesirable vegetation regime shifts and loss of crop land because of precipitation changes, increased salinity intrusion and heightened seawater intrusion due to SLR induced by GCC.

4.2.2   *Impacts of SLR and GCC*

Global climate change has the potential of causing local sea level rise and local precipitation changes in vast extents of coastal areas. Local SLR and precipitation changes will have a significant impact on increasing soil salinization and decreasing agriculture production, immediately threatening water and food security. The adverse impact is particularly severe for low-lying coastal areas in South Florida, South and Southeast Asia, as well as in regions in the Sub-Saharan Africa that are currently water stressed. Reduced precipitation and SLR will decrease the availability of fresh surface water and subsurface water needed to sustain agriculture, human settlements and wildlife. In wetlands, interaction between coastal vegetation, groundwater salinity, SLR and increased tidal salinity intrusion may induce reinforcing positive feedback loops to further limit the freshwater supply by reducing the thickness and volume of the freshwater stored in aquifers or freshwater lens beneath the ground. Freshwater lens refers to the region, shaped like a lens, beneath the ground where freshwater is stored as an aquifer. Reduction in freshwater lens thickness and volume will reduce the quantity of fresh groundwater stored that is critical to agriculture, industrial and domestic consumption in many regions in the coastal plains. We will use three vegetation-groundwater simulation models SUTRA, MANHAM and MANTRA to demonstrate this dynamic and complex interaction between coastal vegetation and groundwater hydrology in response to SLR and precipitation changes in a typical coastal marshland in the USA and Malaysia to infer insights on the wider impacts of GCC and SLR at the local level. We then discuss implications on and adaptation to GCC impacts on water, agriculture and food security in several regions, including Malaysia, South Asia, the lower Mekong Delta and China to highlight the potential of catastrophic events of water crisis, droughts, famine and social-economic crisis.

4.3   Coastal Fresh Groundwater

Vast volumes of freshwater are available under the ground for domestic, industrial and agricultural purposes if the water is not depleted or contaminated by pollutants or salt. Fresh groundwater is therefore a critical component of the water cycle in all parts of the world. For example, many streams and lakes are fed primarily or in large part by groundwater, indicating the supreme importance of groundwater as an essential water resource. Groundwater supplies 39% of non-saline water used in the United States (Table 4.1). In the state of Florida, 62% of the entire water usage is supplied from fresh groundwater, contributing 12 million m^3 per day of fresh groundwater. Hence, the importance of preserving and enhancing the quantity and quality of groundwater resources under the scenarios of GCC and SLR is beyond doubt. A brief overview of groundwater hydrology and solute transport is essential for understanding the physics, biology and chemistry involved in solute transport simulation models. Models such as SUTRA, MANHAM and MANTRA are used in this chapter

Table 4.1:   Groundwater (GW) use in some states of the USA (Todd and Mays, 2005).

State	GW use intensity (m^3/day/km^2)	Total GW use ($10^6 m^3$/day)	% of GW use to total water use
California	180	73	49
Hawaii	160	2.6	53
New Jersey	150	3	42
Florida	86	12	62
USA	—	310	39

to simulate the impact of GCC and the associated SLR on coastal water resources.

4.3.1 *Coastal aquifer property*

The main source of fresh groundwater is rainwater that infiltrates into the aquifers beneath the ground. An aquifer is a water-bearing structure consisting of rocks and soils that can release its stored water in sufficient quantity to make it economically feasible to develop for water supply. However, saline seawater from the ocean may intrude into the freshwater aquifers to render them unusable because of the high salinity in the aquifer water. The rate of salinity intrusion into freshwater aquifers will increase with SLR. Freshwater from precipitation or artificial recharge infiltrates through the permeable ground surface and percolates into the underlying strata below. The water infiltrates through an unsaturated zone on its route to a saturated zone below. The two unsaturated and saturated zones are separated by the water table, where the pressure is atmospheric. However, a capillary fringe may exist to permit water to rise above the water table. This capillary rise provides much of the soil moisture utilized by plants. The extent of this capillary rise depends on soil particle sizes, with finer soil particles producing higher capillary rise. In fine silty sands, the capillary rise can be as high as 50 cm, while the rise in larger-sized gravel is much lower, typically only 2 to 3 cm. The water in this capillary fringe zone fluctuates in quantity and quality, depending on uptake by vegetation and percolation from infiltrated water. The aquifer can be either confined or unconfined, depending on whether a water table or free water surface exists under atmospheric pressure. It is possible for a groundwater aquifer to become overlain by impermeable material and thus be under pressure, in which case it is known as a confined aquifer, where the pressure is above atmospheric. Wells drilled into such confined aquifers are called artisan wells, which may provide free flowing water as the pressure is above atmospheric. Three aquifer properties are of paramount importance to the understanding of groundwater movement and its associated solute transport: (a) storage coefficient S (dimensionless), (b) hydraulic conductivity

K (meter per day) and (c) hydraulic transmissivity T (meter square per day).

4.3.2 Definitions of aquifer property

The storage coefficient S is defined as the volume of water ($V \, \mathrm{m}^3$) that an aquifer releases or takes into storage per unit surface area ($A \, \mathrm{m}^2$) of aquifer for each unit change in head ($h \, \mathrm{m}$) normal to that surface. The storage coefficient S is therefore a dimensionless quantity indicating a volume of water released per volume of aquifer ($\mathrm{m}^3/\mathrm{m}^2.\mathrm{m}$). For a vertical column of unit area extending through a confined aquifer, the storage coefficient S is equal to the volume of water released from the aquifer when the piezometric surface declines a unit distance. Typical values of S fall in a range of small numbers $0.00005 \leq S \leq 0.005$, indicating that large pressure changes over extensive areas are required to produce substantial water yields. A rule of thumb relationship $S = 3 \times 10^{-6} \, \mathrm{m}^{-1} \times b$ is often used to estimate S, where b is the saturated aquifer thickness in meters. Storage coefficient can best be estimated from pumping tests of wells or from groundwater fluctuation in response to atmospheric or ocean tide variations. Hydraulic conductivity K is defined as the volume of water (m^3) transmitted (at the prevailing kinematic viscosity) per unit time (day) per unit area of cross section (m^2) perpendicular to the direction of flow per unit of hydraulic gradient (dh/dL, dimensionless, where $L \, \mathrm{m}$ is the length of flow). The units are $K =$ volume/time/area $=$ m per day or md^{-1} indicates that hydraulic conductivity has the units of velocity md^{-1}. Hydraulic conductivity K can vary significantly, depending on particle sizes (Table 4.2). Hydraulic transmissivity T is defined as the volume of water transmitted (at the prevailing kinematic viscosity) per unit time per unit width of aquifer per unit of hydraulic gradient. The units are $T =$ volume/time/width $= \mathrm{m}^2$ per day or $\mathrm{m}^2\mathrm{d}^{-1}$, indicating that hydraulic transmissivity T has the units of diffusion or dispersion $\mathrm{m}^2\mathrm{d}^{-1}$. Note that $T = Kb$, where b is the saturated thickness of the aquifer in m. We next present two simple analytical models to allow direct and quick interpretation of groundwater hydraulic responses subject to external tidal forcing and rainwater recharge.

Table 4.2:   Values of hydraulic conductivity and particle size (Todds and Mays, 2005).

Material	Hydraulic conductivity (m/day)	Particle size (mm)
Gravel, coarse	150	16.0–32.0
Gravel, medium	270	8.0–16.0
Gravel, fine	450	4.0–8.0
Sand, coarse	45	0.5–1.0
Sand, medium	12	0.25–0.5
Sand, fine	2.5	0.125–0.25
Silt	0.08	0.004–0.062
Clay	0.0002	<0.004

4.4   Groundwater Model SUTRA

The model SUTRA developed by the USGS is commonly used to simulate saturated–unsaturated density-dependent groundwater flow with solute or energy transport (Voss and Provost, 2003). Equations (4.1) and (4.2) show the coupled fluid and solute mass balance equations used in SUTRA. The definition and unit of the symbols used in Eqs. (4.1) and (4.2) are listed in Table 4.3.

$$
\left(S_w \rho S_{op} + \varepsilon \rho \frac{\partial S_w}{\partial p} \right) \frac{\partial p}{\partial t} + \varepsilon S_w \frac{\partial \rho}{\partial C} \frac{\partial C}{\partial t}
$$

$$
- \nabla \left[\left(\frac{k k_r \rho}{\mu} \right) \cdot (\nabla p - \rho g) \right] = Q_p \tag{4.1}
$$

$$
\frac{\partial (\varepsilon S_w \rho C)}{\partial t} + \nabla \cdot (\varepsilon S_w \rho v C) - \nabla \cdot [\varepsilon S_w \rho (D_m I + D) \cdot \nabla C] = Q_p C^* \tag{4.2}
$$

The standard Galerkin finite element method is used to discretize in space with quadrilateral elements, and a weighted difference method is used for temporal discretization. Construction and preparation of model input are facilitated by Argus Open Numerical Environments (Argus ONE) using the SUTRA Plug-In Extension (PIE). The simulation results are then visualized using Model Viewer, a USGS post-processing graphic utility (Hsieh and Winston, 2002).

Table 4.3:   Definition of symbols used in Eqs. (4.1) and (4.2).

Symbol	Unit	Definition
S_w	—	Water saturation
ρ	kg m^{-3}	Fluid density
S_{op}	m s^2 kg^{-1}	Specific pressure storativity
ε	—	Porosity
p	kg m^{-1} s^{-2}	Fluid pressure
t	s	Time
C	kg kg^{-1}	Mass-based solute concentration
k	m^2	Permeability tensor
k_r	—	Relative permeability
M	kg m^{-1} s^{-1}	Fluid viscosity
g	m s^{-2}	Gravitational acceleration
Q_p	kg m^{-3} s^{-1}	Fluid mass source
v	m s^{-1}	Fluid velocity
D_m	m^2 s^{-1}	Molecular diffusion
I	—	Identity tensor
D	m^2 s^{-1}	Dispersion tensor
C^*	kg kg^{-1}	Solute concentration of the fluid sources

4.4.1   *Analytical solution*

As part of model validation, we develop an analytical formula to quantify the thickness and volume of freshwater lens in coastal aquifers (Teh *et al.*, 2013). For simplicity of presentation, we consider a circular island of radius R(m) receiving a freshwater recharge from rainfall at a rate W(m s^{-1}). A freshwater aquifer or lens sits on top of the saline groundwater, as illustrated in Figure 4.1. Darcy's law is applied to groundwater discharge Q(m^3 s^{-1}) in the radial r(m) direction (Dupuit, 1863) to yield the following equation:

$$Q = -2\pi r K(z+h)\frac{dh}{dr} = \pi r^2 W \qquad (4.3)$$

In Eq. (4.3), K (ms^{-1}) is the hydraulic conductivity, h(m) is the water table elevation above sea level and z(m) is the depth of the freshwater–seawater interface below sea level. The relationship between h and z is expressed by the Ghyben–Herzberg Eq. (4.4), which assumes a sharp interface between fresh and saline

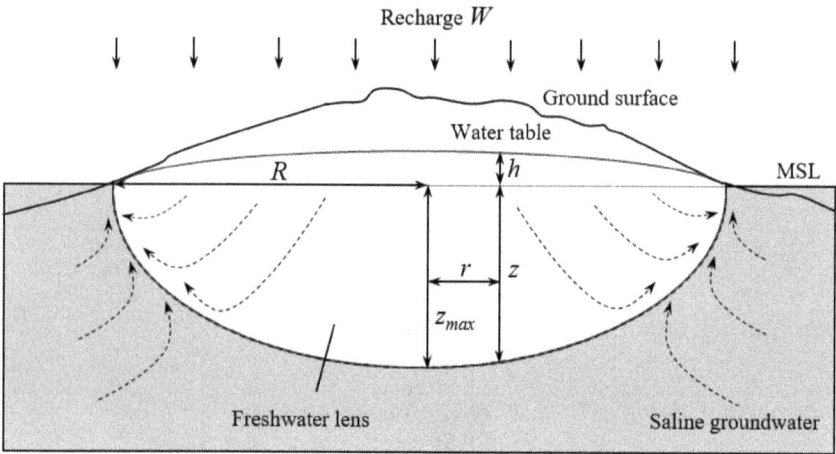

Figure 4.1:   Freshwater lens in a circular atoll island.

groundwater (Arnold, 1968):

$$z = \frac{\rho_f}{\rho_s - \rho_f} h = \frac{\rho_f}{\Delta\rho} h \tag{4.4}$$

The freshwater and seawater densities are denoted by ρ_f and ρ_s (kg m^{-3}), respectively. Combining Eqs. (4.3) and (4.4) gives Eq. (4.5):

$$-2\pi r K \left(z + \frac{\Delta\rho}{\rho_f} z \right) \frac{\Delta\rho}{\rho_f} \frac{dz}{dr} = \pi r^2 W \tag{4.5}$$

The rearrangement of Eq. (4.5) yields Eq. (4.6):

$$z\, dz = -\frac{Wr}{2K(1 + \Delta\rho/\rho_f)\Delta\rho/\rho_f} dr \tag{4.6}$$

The solution of Eq. (4.6) is given as Eq. (4.7):

$$z^2 = -\frac{Wr^2}{2K(1 + \Delta\rho/\rho_f)\Delta\rho/\rho_f} + C \tag{4.7}$$

The integration constant C can be determined using the boundary condition of $z = 0$ when $r = R$. With this, Eq. (4.7) can then be expressed as Eq. (4.8):

$$z = \sqrt{\frac{W(R^2 - r^2)}{2K(1 + \Delta\rho/\rho_f)\Delta\rho/\rho_f}} \tag{4.8}$$

Thus, the depth z to seawater at any location r is a function of the recharge rate W, island radius R and aquifer hydraulic conductivity K. The maximum thickness of freshwater lens z_{max} is obtained by substituting $r = 0$, at the center of the island. Then, the volume of extractable water in the freshwater lens V (m^3), which has the shape of a semi-ellipsoid, is given by Eq. (4.9) (Gualbert, 2001):

$$V = \frac{2}{3}\pi(1 + \Delta\rho/\rho_f)z_{max}R^2\varepsilon \tag{4.9}$$

4.4.2 Relative importance of W, R and K

The analytical solutions presented in Eqs. (4.8) and (4.9) indicate that the thickness z(m) and volume V(m^3) of the freshwater lens are governed by three properties: (1) recharge rate W, (2) island radius R and (3) aquifer hydraulic conductivity K. It is useful to quantify the relative importance of each of the three parameters W, R and K for governing aquifer lens thickness z. The sensitivity index of z relative to a parameter p is the ratio of the percentage change in z to the percentage change in the parameter p (Chitnis *et al.*, 2008) as defined in Eq. (4.10):

$$S_p^z = \frac{\partial z}{\partial p} \times \frac{p}{z} \tag{4.10}$$

Table 4.4 shows the sensitivity indices of z relative to each of the parameters W, R and K. The sensitivity indices show how significantly each of these parameters contributes to lens thickness z. A positive index means that an increase in that parameter value would result in an increase in the z value. Conversely, a negative index means that an increase in that parameter value would result in a decrease in the z value. The sensitivity analysis shows that z is most sensitive to the island radius R, with $S_R = +1.0$. This means that an increase (decrease) of 10% in R will increase (decrease) z by 10%. The sensitivity indices for the recharge rate W and aquifer hydraulic conductivity K are +0.5 and −0.5, respectively. Hence, an increase (decrease) of 10% in R will cause a decrease (increase) in z by 5%.

Table 4.4:   Sensitivity indices of freshwater lens thickness relative to the geo-hydrologic parameters.

Geo-hydrologic parameter	Sensitivity index $(-)$
Recharge rate (m s^{-1})	$+0.5$
Island radius (m)	$+1.0$
Hydraulic conductivity (m day^{-1})	-0.5

4.4.3   *Relationship between z and W, R and K*

It is useful to quantify the relationship between the freshwater lens thickness z and each of the parameters W, R and K, as shown in Figure 4.2. Figure 4.2(a) shows that lens thickness increases with increasing recharge rate W. Any decrease in groundwater recharge W may allow more intrusion of saline water into the aquifer, and cause a reduction or thinning of the freshwater lens thickness. As illustrated in Figure 4.2(b), the size of the island as measured by its radius R is linearly related to the lens thickness z. Small islands ($R < 400$ m) generally have thin lenses ($z < 5$ m), whereas larger islands ($R > 400$ m) generally have thicker lenses (5 m $\leq z \leq 25$ m). Small islands receive limited volumes of surface recharge W and consequently have small freshwater lenses. Further, small islands are also subject to larger saltwater intrusion that increases the soil salinity and that reduces the freshwater lens thickness. Figure 4.2(c) shows a logarithmic decrease in lens thickness with increasing aquifer hydraulic conductivity K. The higher the hydraulic conductivity K, the higher will be the rate of saline groundwater intrusion from the seawater. This higher salinity intrusion from the sea will cause in an increased inland migration of the seawater edge and hence an increase in ground salinity. Sensitivity analysis indicated that the freshwater lens thickness is most significantly affected by island radius R, followed by surface recharge rate W, and aquifer hydraulic conductivity K. Numerical simulations are carried out to quantify the impact of these three geo-hydrologic parameters on fresh groundwater sustainability. The thinning of freshwater lens is observed to be linearly correlated with decreasing recharge rate W and decreasing width of the island R. Further, the lens thickness z decreases logarithmically with increasing aquifer hydraulic conductivity K.

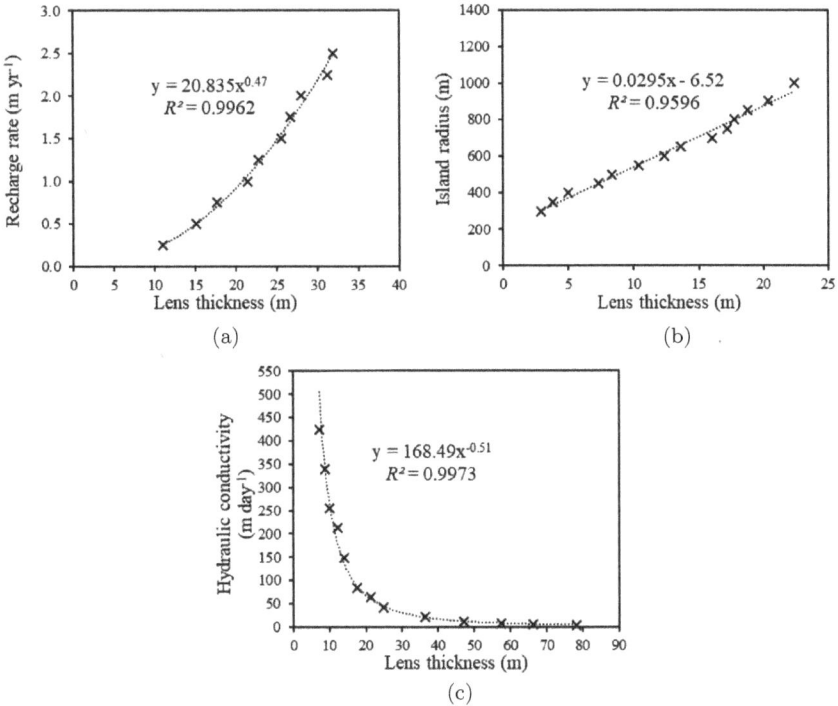

Figure 4.2: Sensitivity analysis results for each of the three parameters examined: (a) recharge rate W, (b) island radius R and (c) hydraulic conductivity K.

4.4.4 Comparison between analytical and numerical solutions

As part of the validation exercise, we compare the aquifer lens thickness obtained from numerical simulation results with that from the analytical solution. A three-dimensional groundwater flow and solute transport model SUTRA is used to simulate the dynamics of freshwater lens in response to SLR. Good agreement is found between the analytical and numerical solutions for lens thickness. For this purpose, the lens boundary is the interface between freshwater and seawater with 50% of seawater concentration (Lu *et al.*, 2016). Figure 4.3 shows that the lens thickness derived from analytical solutions is in a good agreement with that obtained from the numerical results. ($R^2 = 0.9957$). This indicates that the analytical solutions provide valid and comparable results even with the simplifications

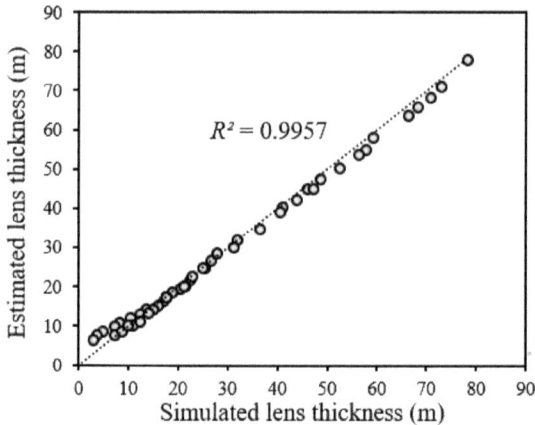

Figure 4.3:   Comparison between analytical and numerical solutions for freshwater lens thickness.

and assumptions imposed on the analytical mode. It is essential in sustainability study to have available a simple analytical model, solution and methodology that are readily understood and used by all.

4.5   SLR Reduces Fresh Groundwater Supply

The rate of global SLR is estimated to have accelerated to 3.2 ± 0.4 mm yr^{-1} during 1993–2003, from the average rate of 1.8 mm yr^{-1} during 1961–2003. The Intergovernmental Panel on Climate Change (IPCC) has predicted an SLR of 0.5 to 1.0 m by the end of the 21st century, under the plausible worst-case RCP 8.5 scenario, (IPCC, 2013). An SLR of such a magnitude could drastically reduce the size or radius R of atoll islands across the Pacific and Indian Oceans. This reduction in atoll island radius R will have a profound impact on increasing groundwater salinity. The result is a profound reduction in the freshwater aquifer lens thickness and volume V as shown in Eq. (4.9). Hence, in this subsection, we use SUTRA to examine the impact of SLR on the extent of salinization of coastal aquifers and on the corresponding reduction in freshwater resources.

4.5.1   *SUTRA simulation domain*

The computational setup for the three-dimensional simulation domain is illustrated in Figure 4.4 in 2D cross-sectional view and

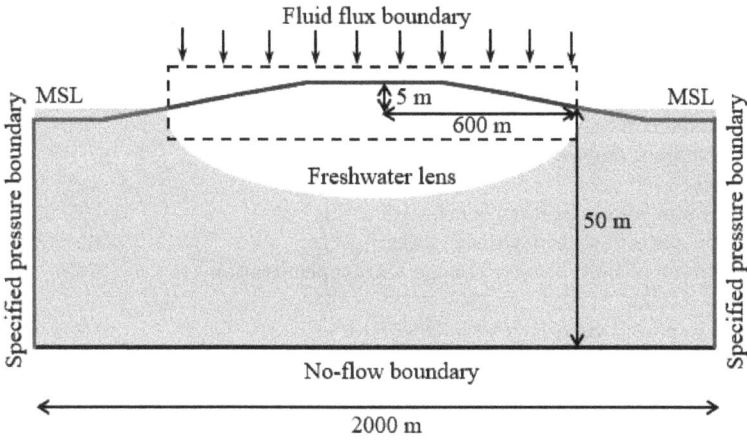

Figure 4.4:   Model domain and assigned boundary conditions.

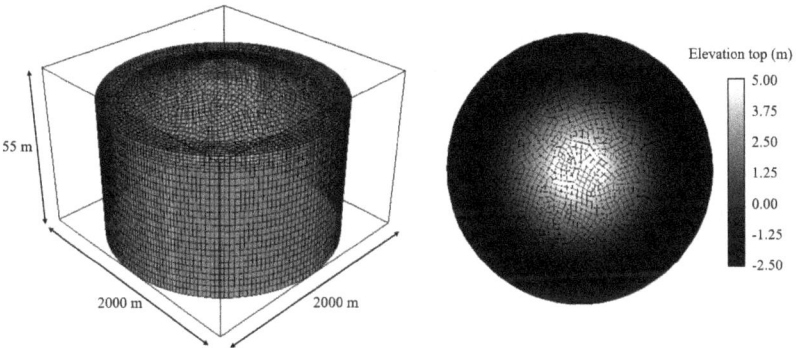

Figure 4.5:   Oblique (left) and top (right) view of the 3D finite element mesh for the island model.

in Figure 4.5 in 3D view. A hypothetical circular island of radius 600 m is used as the baseline model. The model domain is extended to a distance of 400 m offshore. The model domain is discretized into 65,364 nodes and 60,250 quadrilateral elements, with 25 vertical layers that extend from the land surface to a uniform depth of 50 m below mean sea level (MSL). The element grid size in the radial direction is 40 m, whereas the element grid size in the vertical direction varies from 1 to 2.5 m. The applied spatial and temporal discretization satisfies the Peclet (Pe) and Courant (Cr) number criterion for numerical stability of the SUTRA code. For the initial condition, the model domain is assumed to be saline throughout the entire domain.

Table 4.5:   Input parameters used for SUTRA model simulations.

Parameter	Value
Hydraulic conductivity (m day^{-1})	85
Longitudinal and transverse dispersivity (m)	10, 0.1
Porosity (−)	0.3
Freshwater and seawater concentration (kg kg^{-1})	0, 0.0357
Freshwater and seawater density (kg m^{-3})	1000, 1025
Coefficient of fluid density change with concentration (kg m^{-3}kg^{-1})	700
Fluid dynamic viscosity (kg m^{-1} s^{-1})	0.001
Molecular diffusion (m^2 s^{-1})	10^{-9}
Fluid and matrix compressibility (m s^2 kg^{-1})	4.47×10^{-10}, 10^{-8}
Land surface slope (m m^{-1})	0.0125

This choice of initial condition will have no impact on the simulation results if the simulation is run over a long period. For boundary conditions, specified pressure boundaries are assigned to all nodes below sea level as well as to the seaward vertical boundary. Inflowing saltwater at these seaward nodes has the concentration of seawater ($C = 0.0357$ kg kg^{-1}). An inflow of freshwater recharge through the atoll land surface via rainfall infiltration is set at the recharge rate W of 0.75 m yr^{-1}. This recharge rate W is applied evenly over the entire land surface, with inflowing fresh rainwater having zero salinity. This corresponds to an average annual rainfall of 1.875 m reduced by 60% runoff loss. A no-flow boundary is specified along the solid bottom of the domain. Table 4.5 summarizes the input parameters used for this set of SUTRA model simulations.

4.5.2   *SUTRA simulation results*

Three scenarios are considered consisting of (i) existing condition (pre-SLR), (ii) SLR of 0.5 m and (iii) SLR of 1.0 m. The simulations are run for 20 years to achieve steady-state condition with a time step of 0.2 year until the system reaches equilibrium. Twenty years of simulation time is adequate for the simulation to achieve equilibrium state as the island is small with a radius of only 600 m. For a large island or peninsula, a longer simulation time is required to achieve equilibrium state, often in the order of 100 years.

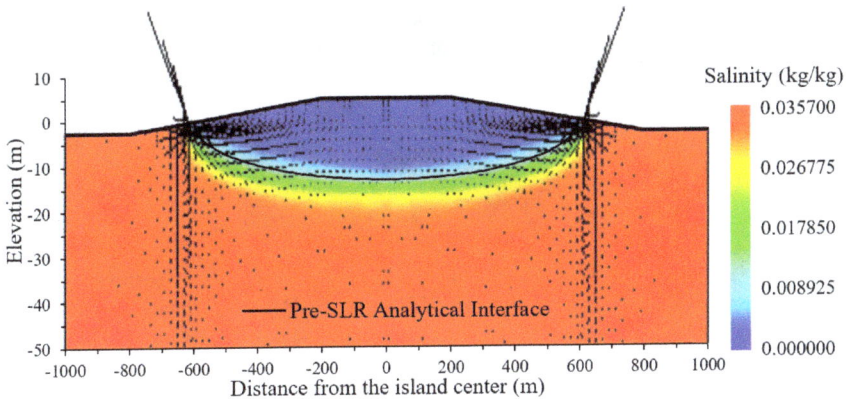

Figure 4.6:   Steady-state salinity and velocity distributions before SLR.

Figure 4.6 shows the steady-state salinity and velocity distributions under scenario (1) with 0.0 m SLR (existing condition), in which freshwater is indicated in blue and seawater in red. A mixing boundary zone is seen between the freshwater and seawater regions. The sharp interface between freshwater and seawater derived from the analytical solution (solid line) is compared with that predicted by the numerical model. The numerically simulated lens thickness is 13.3 m, which is in a reasonable agreement with the analytical result of 13.05 m (2% difference). The separation between freshwater outflow in the upper layer and seawater inflow in the lower layer is clearly visible. Changes in current direction in the mixing region along the sharp interface indicate that the water from both regions is drawn into the mixing region and transported toward the sea. This movement causes an increase in the shearing velocity at the interface.

Figure 4.7 shows the simulated freshwater lenses and island land boundary for the three scenarios of SLR of (a) 0.0 m, (b) 0.5 m and (c) 1.0 m. The radius of the island is reduced by SLR from 600 m at no SLR to 560 m at 0.5 m SLR and to 520 m at 1.0 m SLR. The freshwater lens thickness is reduced from 13.3 m with no SLR to 12.3 m at 0.5 m SLR and to 11.3 m at 1.0 m SLR. The loss in lens thickness is caused by increase in soil salinity due to seawater intrusion and saltwater inundation induced by SLR. The losses in simulated lens thickness and volume are given in Table 4.6. The maximum freshwater lens thickness is reduced by 8% and 15%, corresponding to SLR of 0.5 and 1.0 m, respectively. For SLR of 1.0 m,

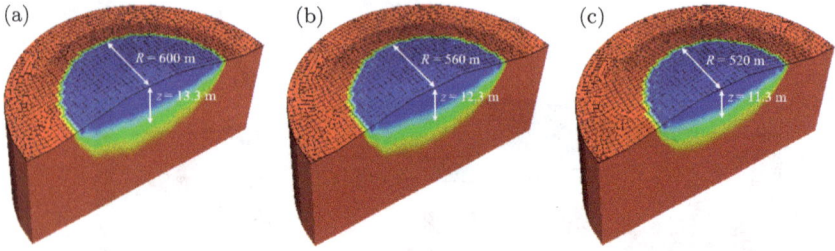

Figure 4.7: The change in the radius of the freshwater lens in response to sea level rise: (a) existing condition (pre-SLR), (b) SLR of 0.5 m and (c) SLR of 1.0 m.

Table 4.6: Groundwater availability and freshwater requirement for RWH for the two SLR scenarios.

SLR (m)	Lens thickness (m)	Lens volume (m^3)	Recharge rate required to maintain lens volume (m yr^{-1})	Freshwater requirement for RWH (%)
0.0	13.30	3,083,596.25	0.75	—
0.5	12.30	2,484,188,42	1.18	38
1.0	11.30	1,967,834.17	1.84	97

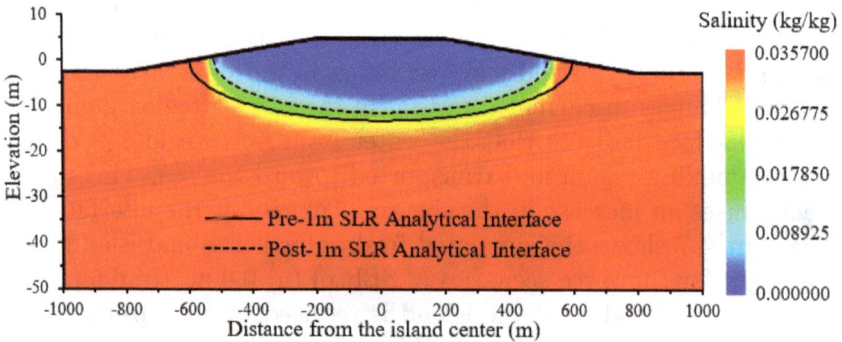

Figure 4.8: Steady-state salinity distribution in response to SLR of 1.0 m.

the interface between the freshwater and seawater moves landward by 80 m, causing increased salinization of the soil and aquifer. Simulation shows that SLR of 1.0 m can cause ~25% inundation of land area and ~36% loss in freshwater volume (see Figure 4.8), which will pose severe hardships to atoll islanders.

4.6   Propagation of Tidal Signal

We next derive a simple mathematical description of changes in groundwater head h in a confined aquifer in response to tidal forcing. For simplicity, consider the one-directional flow of groundwater in a confined aquifer as shown in Figure 4.9.

The differential equation governing the flow is given as Eq. (4.11), where h is the rise or fall of the piezometric surface with reference to the mean sea level, x is the distance inland from the outcrop, S is the storage coefficient of the aquifer, T is transmissivity and t is time. The amplitude of the tide is h_0 in meter (see Figure 4.9).

$$\frac{\partial^2 h}{\partial x^2} = \frac{S}{T}\frac{\partial h}{\partial t} \tag{4.11}$$

The solution of Eq. (4.11) subject to the boundary conditions (i) $h = h_0 \sin(\omega t)$ at $x = 0$ and (ii) $h = 0$ at $x = \infty$ is given by (4.12). The frequency ω is related to the tidal period t_0 by $\omega = 2\pi/t_0$.

$$h = h_0 e^{-x\sqrt{\pi S/t_0 T}} \sin\left(\frac{2\pi t}{t_0} - x\sqrt{\pi S/t_0 T}\right) \tag{4.12}$$

From Eq. (4.12), it can be seen that the amplitude h_x of groundwater fluctuations at a distance x from the shore equals $h_x = h_0 e^{-x\sqrt{\pi S/t_0 T}}$. The time lag t_L between the time series of ocean tide and the response of groundwater at the location x is given by the formula $t_L = x\sqrt{t_0 S/4\pi T}$. This wave that propagates inland has a travelling velocity of $v_w = x/t_L = \sqrt{4\pi T/t_0 S}$ and a wavelength

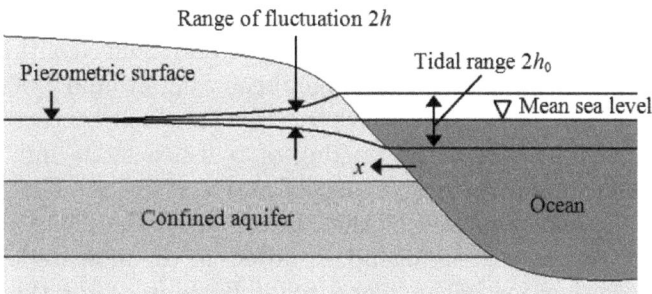

Figure 4.9: Groundwater level fluctuations in a confined aquifer produced by ocean tides.

of $L_w = v_w t_0 = \sqrt{4\pi t_0 T/S}$. The amplitude decreases by a factor $e^{-2\pi}$ for each wavelength. Water flows into the aquifer during half of each cycle and flows out during the other half with a volume given by $V = h_0\sqrt{2t_0 ST/\pi}$. The above analysis is also applicable to a good approximation to water table fluctuations of an unconfined aquifer if the range of fluctuations is small in comparison to the saturated thickness. The simple analytical formulation provides a basis for the estimation of aquifer parameters such as hydraulic transmissivity T. It also provide the insight that the tidal signal or influence increases with decreasing distance from the sea. Hence, for small islands such as atolls in the Indian Ocean or the Pacific, tidal influences on groundwater hydrology can be strong. Under GCC and local SLR, this strong tidal influence will decrease freshwater storage and supply in coastal aquifers in atoll islands, as shown in Eqs. (4.9) and (4.12). To adapt to this adverse impact, islanders might have to rely on harvesting rainwater stored in various types of water cisterns for their water needs.

4.6.1   *History of water cisterns*

The use of water cisterns to store water has a long history of more than 10,000 years. Robust and versatile, water cisterns have ranged in construction from small simple clay pots to large belowground and aboveground structures. From prehistoric times to the present, water cisterns have been used for the purpose of meeting water needs through seasonal variations, to store rainfall runoff water and aqueduct water originating in springs and streams. It has been traced back to the Neolithic Age, when waterproof lime plaster cisterns were built into the floors of houses in villages of the Levant, an area that occupied a large part of Southwest Asia and beyond. The Neolithic Age is the cultural period of the Stone Age, beginning around 8,000 BC in the Middle East and later elsewhere, characterized by the development of agriculture and the making of polished stone implements. The Romans made extensive use of water cisterns. The cultural habits and social customs of the ancient Romans, along with the needs of growing craftsmanship, resulted in an increased demand for water, either for bathing and toilet flushing or for the various workshops. These needs led to the construction of large water collection cisterns where previously the water demand was managed with small-scale rainwater collection and water reuse (Antoniou, 2010). Pompeii had an extensive water distribution system including both aqueduct water and well water. In addition, the roofs of houses collected rainwater that flowed through terracotta pipes down to cisterns where water was stored for domestic use. In Pompeii, the aqueduct and well water were contaminated by the volcano fallouts, requiring cisterns to be used for storing drinking water (Crouch, 1993). The three-aisled vaulted Roman cistern in the ancient town of Aptera, Chania, had a volume of about 3000 cubic meters. Large Roman cisterns were also built in Spain, Southern Italy, Crete, Asia Minor and North Africa, where the largest number of cisterns can be found. As pointed out by Wilson (2001), in Roman North Africa, vast cistern complexes were used in conjunction with aqueducts to meet community water demands. These community cisterns had capacities that were several thousand cubic meters, and were much larger than domestic cisterns. The large cisterns in North Africa were typically located where the aqueducts reached the edge of towns. Most cisterns of these periods were supplied by rainfall runoff, but there were many examples in which spring water was stored. The sustainability of water cisterns has been proven beyond doubt.

4.6.2 *Sustainability of water cisterns*

The increased number of people during the Roman era led to an increase in cistern sizes and to the innovative combinations of various cisterns used as integral parts of the impressive water conveyance and storage technology. The characteristics of durability, environmental adaptability and sustainability of cisterns built in the past are the key to their sustainable use in the future. Rapid technological progress in the twentieth and the twenty-first centuries has, however, created a misplaced disregard, indeed a total disrespect, for these proven past water technologies because these past water technologies were mistakenly considered to be inferior to the present water management systems. By the late fourth millennium BC, cisterns were essential elements of water management techniques used in dryland farming, such as in Jawa in northeastern Lebanon (Roberts, 1977). Carved out of solid rock, lined with large stones and sealed with clay to keep water from leaking out, these cisterns became the essential feature of a well-designed city. These past water technologies can provide valuable insights for improving current water engineering systems for long-term sustainability. These past dryland farming and cistern technologies may be adapted for use in Pantai Acheh and elsewhere in Malaysia. These alternative or complementary water resources, such as rainwater harvesting, are critical to local sustainability where water is a scarcity or where soil salinization impairs crop cultivation under the impact of GCC and SLR.

4.7   Rainwater Harvesting

Simulation models as given by Eq. (4.9) indicate that SLR could reduce fresh groundwater storage in coastal aquifers due to seawater inundation and saltwater intrusion into the aquifers. This will pose a significant risk to atoll islanders in achieving SDG6 that aims to ensure universal access to clean water and sanitation. RWH receives increasing attention as an alternative water supply for long-term water sustainability. The runoff water collected from roofs or ground surfaces can be stored in tanks for later domestic use or directed to recharge groundwater aquifers. In this section, we provide an estimation of RWH capacity or potential needed to compensate the loss of fresh groundwater stored in aquifers due to SLR.

4.7.1   *Water harvesting potential*

The water harvesting potential V (m^3 yr^{-1}) can be calculated via Eq. (4.13):

$$V = C \times P \times A \times E \qquad (4.13)$$

Here, C (dimensionless) is the runoff coefficient, P (m yr^{-1}) is the annual rainfall, A(m^2) is the catchment area and E (dimensionless) is the collection efficiency after accounting for losses due to evaporation and catchment retention (Alegre and Tynan, 2014). Based on the simulation results presented in Section 4, for an SLR scenario of 0.5 m, only 87% of the pre-SLR land area is available to receive surface recharge from rainfall. To keep the lens volume at a pre-SLR level, a recharge rate of 1.18 m yr^{-1} (Table 4.6) is needed to maintain the lens volume. With the existing recharge rate of 0.75 m yr^{-1} from precipitation, an additional recharge rate of $1.18 - 0.75 = 0.43$ m yr^{-1}

is required to recharge the groundwater aquifer. Assuming that 60% of the total rainfall is lost by runoff (Rachwal and Holt, 2008) and collection efficiency is 100%, the amount of rainwater that can be effectively harvested can be calculated as follows:

Island catchment area $= \pi \times 560^2 \, \text{m}^2 = 985203 \, \text{m}^2$

Total annual rainfall $= 1.875 \, \text{m} \, \text{yr}^{-1}$

Runoff loss $= 0.6 \times 1.875 \, \text{m} \, \text{yr}^{-1} \times 985203 \, \text{m}^2 \times 1$

$\quad = 1108353 \, \text{m}^3 \text{yr}^{-1}$

Additional recharge requirement $= 0.43 \, \text{m} \, \text{yr}^{-1} \times 985203 \, \text{m}^2$

$\quad = 423637 \, \text{m}^3 \text{yr}^{-1}$

Percentage of required RWH $= 423637/1108353 \times 100\% \approx 38\%$

Hence, for this hypothetical island, 38% of rainwater runoff should be collected to compensate the loss of fresh groundwater due to seawater inundation and saltwater intrusion under an SLR scenario of 0.5 m. Based on an average annual rainfall of 1.875 m, 423,637 m³ of water can potentially be collected from a total catchment area of 423,637/1.875 = 225,940 m². This is equivalent to 1506 roof tops with an average roof surface area of 150 m². For each house, a roof catchment area of 150 m² that receives annual rainfall of 1.875 m will be able to capture 1.875 × 150 = 281.25 m³ of rainwater per year. The roof-collected rainwater is then diverted into storage tanks for later use.

4.7.2 *Storage tank size*

It is generally estimated that the storage tank size should be around 5% of the annual rainwater supply (Sarada, 2015). The required storage tank size can be calculated as follows:

Total annual rainfall $= 1.875 \, \text{m} \, \text{yr}^{-1}$

Roof catchment area $= 150 \, \text{m}^2$

Storage requirement $= 1.875 \, \text{m} \, \text{yr}^{-1} \times 150 \, \text{m}^2 \times 0.05 = 14.06 \, \text{m}^3$

Height of tank $= 2 \, \text{m}$

Radius of tank $= \sqrt{14.06 \, \text{m}^3/(\pi \times 2 \, \text{m})} = 1.5 \, \text{m}$

Size of tank $= \pi \times 1.5 \, \text{m} \times 1.5 \, \text{m} \times 2 \, \text{m} = 14.14 \, \text{m}^3$

Therefore, a cylindrical tank with radius 1.5 m and height 2 m is needed for the required water storage capacity of 14.06 m^3. For an SLR scenario of 1 m, a higher additional recharge rate of 1.84 − 0.75 = 1.09 m yr^{-1} is required to recharge the groundwater aquifer. As a result, the island is greatly in need of an effective RWH system in order to collect 97% of rainwater runoff, which would require a cylindrical storage tank with radius 1.9 m and height 3 m.

4.7.3   *Summary on RWH*

An increase in sea level could reduce the island's width or radius R, and thereby could reduce the available volume of extractable freshwater V stored within the freshwater aquifer lens. Our analysis suggests that for an SLR of 0.5 m, 38% of rainwater needs to be collected and directed to recharge the groundwater aquifer in order to maintain the freshwater volume stored in the groundwater aquifer at a pre-SLR level. This level of rainwater harvesting can be achieved by having each of 1506 houses on the island fitted with a 150 m^2 roof top collection area. The collected rainwater from each household can be stored in a cylindrical tank of radius 1.5 m and height 2 m. For an SLR of 1 m, it is estimated that about 97% of rainwater should be harvested to compensate the loss of fresh groundwater. The scale of the rainwater harvesting facility for this scenario of an SLR of 1.0 m would be more than double those for the scenario of 0.5-m SLR. This analysis suggests that food security in small islands or in low-elevation coasts such as Pantai Acheh is uncertain under the impact of GCC and SLR.

4.8   Food Security SDG2

Along low-elevation coastal and estuarine regions, the impact of GCC and SLR can cause sharp increases in soil salinity and profound decreases in fresh groundwater supply due to seawater inundation due to seawater overtopping on coastal land and saltwater intrusion via groundwater transport. Water security and food security in these regions are highly vulnerable to continuously changing climatic patterns under GCC and SLR. Crop may respond to increase in temperatures in several ways. For rice, the yield may be reduced by up to 10% for every 1°C increase in nighttime temperature. Being sensitive to high temperatures beyond their normal range, most vegetables may not be viable for commercial production (Lal, 2013). A combination of compelling factors has heightened the global, regional and national sensitivity to water and food security. These factors include poor soils, declining freshwater resources, soil salinization and land degradation. Other factors are population growth, inadequate agricultural infrastructures, plant disease and unfavorable climate. The world demand for cereals is predicted to increase from 1.84 billion tons in 1997 to 2.50 billion tons in 2020. To meet this cereal demand, sustainable eco-hydrology principles and best practices are crucial in achieving climate-resilient agriculture intensification. It is crucial to produce more food from the same area of land and from the same amount of water used, while reducing the environmental impact and negative externalities. An important criterion in sustainable agriculture is water conservation and prevention of soil loss.

Soil losses from agricultural land are currently occurring at an alarming rate of over 6 ton ha^{-1} yr^{-1} globally. This is 15 times more than the average soil loss rate of 0.4 ton ha^{-1} yr^{-1} over the geological history of the earth. This global environmental carnage caused by agriculture intensification and poor soil management needs to be addressed urgently. In the long run, this loss in cropland will not be adequately compensated by incremental improvements in food production alone. Water conservation and soil loss management are important in promoting resilient and sustainable agriculture. Simple onsite and in-field crop residual retention can increase crop production, minimize soil loss and reduce water consumption. For example, corn grain yield can be increased to 10.5 ton/ha with 100% residue retention compared to 6.3 ton/ha without crop residue retention, resulting in a significant 67% increase (Lal, 2013). However, climate model HadCM3 developed by the IPCC indicated that regional imbalance in cereal production is likely to persist and widen through time. Major food insecurity will occur in low-latitude regions and in arid and semi-arid grasslands.

4.8.1   *Climate model projection*

Arid and semi-arid regions cover 37% of the earth's terrestrial area. Sustainable management of these grasslands is important in improving water and food security in the regions. The shortfalls in meeting the food security in the regions may be met by abundant and sustainable productions of the four major cereals elsewhere. The four major crops for human consumptions are wheat, rice, maize and soybean, which together account for 85% of the world cereal exports. Abundant and sustained yields from these four cereals will contribute significantly toward balancing regional differences in crop productions and relieving the shortages in the arid and semi-arid grasslands. It is critical to ensure abundant and sustainable production of these four cereals subject to various scenarios of GCC. To examine this adequacy, a global climate model known as HadCM3 has been developed by the IPCC Special Report on Emissions Scenarios (SRES) (Nakićenović *et al.*, 2000). Global production in the future appears secure and stable with sustainable practices in agriculture. However, regional imbalances in cereal production are likely to persist and widen through time, with major vulnerabilities occurring in low-latitude regions and in arid and semi-arid grasslands (Parry *et al.*, 2004). Elevated CO_2 by itself may promote plant growth and increase yield of most agricultural plants, leading to an average increase of 4% to 5% in cereal yield for every 100 ppm CO_2 increase. However, the realization of these modeled beneficial effects of CO_2 in the field remains uncertain and complex due to the unknown crop interactions with temperature, nutrients, water, salinity and other stresses (Derner *et al.*, 2003). Generally, the SRES scenarios reported would result in crop yields decreasing in developing countries and yields increasing in developed countries. Yield decreases are especially significant in Africa and parts of Asia with expected losses of up to 30%. In these locations, the beneficial direct effects of CO_2 increase are not sufficient to compensate for the adverse effects of soil degradation, cropland loss, temperature rise and precipitation changes on crop yields. Cereal yields dramatically decrease in developing countries as a result of regional decreases in precipitation and large temperature increases. Soil degradation and loss of cropland will have a profound impact on the food security in Africa and many parts of Asia and possibly in China in the long run.

Simulation results strongly suggest that the present-day production capacity under the current management regimes will not be able to sustain the long-term needs of a sustained growing population. Increasing urban affluence is likely to create increasing demands for a richer diet that can only be met by increasing demands on resources beyond the capability of the planet to sustain. Behavioral, institutional and technological innovations are crucially needed to enhance agricultural infrastructures, to improve land and water management, and to reduce consumption and minimize wastes during production and supply chain processes. Strengthening the integrated utilization of agricultural resources based on local eco-hydrological conditions is particularly important. In subregions facing water and food insecurity, early warning systems are needed to alleviate the impact of major catastrophic events on local water and food adequacy and supplies (Ye and Ranst, 2009). An understanding of the impact of salinity on crop yield is essential in the alleviation of food crisis.

4.8.2 *Salinity impact on crop yield*

A crop yield model has been developed to predict crop yields subject to the interactions and feedback mechanisms in the plant–water–nitrogen–salinity regime (Pang and Letey, 1998). Model results clearly indicate that salinity can increase nitrogen leaching from soil and can decrease corn yields. For this study, available nitrogen is fixed at 200 kg/ha, while salinity levels are kept at three levels (EC = 0.2, 2 and 10 dS/m), where EC denotes electrical conductivity. It investigates the impact of available water (cm) on corn relative yield (%) and on nitrate leaching. For each of the three salinity levels, the corn relative yield gradually increases with increasing water availability, following a sigmoid S curve, up to a maximum, which is achieved at 50–60 cm of available water. The relative yield subsequently declines with further increase in water availability exceeding 60 cm, implicating nitrate leaching with excess available water. The maximum relative yield decreases with increasing salinity levels (70% at 0.2, 65% at 2 and 50% at 10 dS/m). For each of the three salinity levels, nitrate leaching does not occur below 80 cm of available water, at which point available water balances the plant uptake capacity. Beyond 80 cm of available water, nitrate leaching then sharply increases with increasing available water, following a sigmoid S curve, eventually reaching a maximum when available water exceeds 100 cm. The maximum level of nitrate leaching is 18 kg/ha at EC = 0.2 dS/m, increases to 20 kg/ha at EC = 2 dS/m and peaks at 30 kg/ha at EC = 10 dS/m. It is clear from this study that salinity stress increases nitrogen leaching and decreases corn yields (Wang and Baerenklau, 2014), implicating the dire consequences of SLR. This finding is highly relevant to food production as over 800 million hectares of land worldwide is affected by salinity (Alvarez *et al.*, 2015) via a so-called "primary salinization", where salts are transported by capillary flow of brackish water from groundwater into the root zone. The underlying mechanisms of salinity stress on plants are well explained by Pang and Letey (1998) as follows: "Salinity leads to reduced plant growth, which leads to reduced evapotranspiration, which leads to more leaching, which leads to removal of N from the root zone. Reduced N leads to reduced plant growth, which leads to less evapotranspiration, which leads to more leaching, which leads to even less N in the root zone" (p. 1426). Based upon the results of crop yield models, it may be surmised that the current system of rice production in Malaysia in unsustainable. On the contrary, the average paddy yield in another experimental area, consisting of 40 farms, is well above the national average. The government should therefore devise policies supportive of sustainable paddy cultivation in Malaysia.

4.8.3   *Non-sustainable rice production in Malaysia*

At 3817 kg/ha, the national average productivity level for rice in Malaysia is low, as compared to the world average of 4527 kg/ha. In 2013, the total paddy production in Malaysia was 2.63 million tons cultivated from 688,207 ha of farmland. However, the potential is vast for improving rice productivity in Malaysia. In a recent study, Muazu *et al.* (2015) report that the average paddy yield in their study area, consisting of 40 farms, is 7625 kg/ha, which is well above the national average. This yield is close to the average rice yield in America (7616 kg/ha), although it is less than the rice productivity levels in Australia (9896 kg/ha) and in China (8098 kg/ha). The rice productivity level in the study area is, incidentally, higher than that in India (3800 kg/ha), in Bangladesh (4870 kg/ha) and in Japan

(4650 kg/ha). The vast discrepancy in regional rice yields carries the implication that rice productivity worldwide has the potential of vast improvement if the appropriate cultural, institutional and technological innovations are implemented in the regions with low rice productivity. Farmers in the study area reap merely 7.76 times the energy they invested, with a minute monetary benefit–cost ratio of 1.37. This low level of rice productivity deserves close attention. Further, 84% of the total energy input used by the farmers for the entire cultivation period is derived from fossil-based non-renewable resources. This high level of fossil non-renewable fuel input and low benefit–cost ratio is detrimental to the sustainability of paddy cultivation in Malaysia in the long run. The government should devise policies supportive of sustainable paddy cultivation by channeling incentives to farmers who adopt best sustainable cultivation practices that promote environment health and, more importantly, that conserve water to ensure water security as in SDG6.

4.9 Water Security SDG6

Water and land are the main limiting factors for crop cultivation and food production. Producing more crop per drop of water used and more crop per unit of land cultivated is therefore critical in meeting the ever-increasing global demands for food. The internationally accepted definition of water scarcity is set at 1000 m^3 per capita. Any nation or region that has water availability below 1000 m^3 per capita is deemed to confront water scarcity. China's endowment of land and water resources is notably below the world average on a per capita basis. China's annual water supply is equivalent to 1856 m^3 per capita, merely a notch above the water scarcity level, or merely 25% of the world's average of 7425 m^3 per capita. Hence, for China, sustainable management of water distribution at various temporal and physical scales must be maintained, in conjunction with equitable and sustainable cropland allocations and other natural resources' utilization to ensure water and food security. Proper utilization and conservation of precious fresh surface water, groundwater and other natural resources must be practiced at the scales of the farm, district and region. This requires a thorough understanding of the spatial-temporal linkage between food security, water security and land use.

Integrated strategies and effective management options are required to address the complexity in water distributions, land management and food production of agricultural systems. A recent model study in northwest China indicates that a 15% water saving could be achieved by improved regulation of canal water conveyance and distribution, coupled with improved irrigation scheduling and proper land leveling (Todorovic *et al.*, 2015). Further improvements would contribute significantly toward addressing China's water and food security. An estimated 81% of water resources of China are found in the south, while most of China's arable crop land (64%) is located in the arid and semi-arid north. Further, the average supply of fresh groundwater in China is four times greater in the south than in the north. This north–south disparity in water resources is addressed by the large South-North Water Transfer project that could deliver 50 km^3 per year of water from the Yangtze River Basin in the south to the North China Plain, benefiting 300 to 325 million people. Further, saltwater intrusion and water pollution in heavily polluted cities such as Chongqing have led to a reduction of GDP in local areas by about 1.2%, with damages in agriculture production constituting the largest share (56%) of the costs (Khan *et al.*, 2009). Hence, environmental management is another important dimension in securing water and food security in China, as in many other countries located in arid and semi-arid areas facing salinity intrusion into groundwater.

4.9.1 *Saline groundwater*

Unprecedented escalation in water demand from domestic, irrigation and industrial sectors is depleting freshwater resources at such a fast rate that more nations will face a severe and prolong water crisis. It is important to search for additional sustainable freshwater resources from non-conventional sources such as wetlands and marshes. It is even more important to conserve water usage to a minimum. Water scarcity is a particularly serious problem for agricultural production in arid and semi-arid areas. A majority of the arid and semi-arid countries are not able to fulfill the required demand for water and food under the scenarios of climate change. Shallow groundwater is potentially a valuable source of additional water supply to partially meet crop water requirements in these arid and semi-arid regions. Saline groundwater is often found at a shallow depth in irrigated areas of arid and semi-arid regions and is associated with problems of soil salinization and land degradation. This shallow fresh groundwater, via capillary rise, is a valuable resource that can be utilized directly. In this way, it is possible to meet part of the crop water requirement, even where the groundwater is mildly saline, thus reducing the demand for irrigation water and alleviating the burden of disposal of saline drainage effluent (Gowing *et al.*, 2009). However, this is accomplished at the expense of reduced yields. In Iran, where groundwater contributes more than 50% of the total water requirement, groundwater with salinity levels of 2.9 dS/m or higher results in pronounced yield losses (Talebnejad and Sepaskhah, 2015). With limited major sources of water, Arabic countries depend on natural precipitation, water conservation and groundwater to fulfill their water needs. Excessive reliance on groundwater has resulted in continuous unsustainable drawdown of aquifers (Misra, 2014). Artificial recharge of the dried aquifer systems with partially treated water may be one solution. The unsaturated lithology (vadose zone) acts as a natural filter that can remove essentially all pollutants before the water reaches the freshwater lens if the vadose zone has the desirable quality and property. Rain-fed agriculture systems can be upgraded through enhanced rainwater harvesting and augmented by artificial recharge systems. Being one of the world's driest continents, Africa continues to face a severe water crisis. Climate change will directly affect African countries, in particular, the Sub-Saharan region, with declining crop yield and escalating water demand. To meet the water demand from arid and semi-arid regions would require substantial and sustained investments in technology and innovations. It is essential to prevent salinization of water as saline water is detrimental to human health, and is an impediment to good health and well-being (SDG3).

4.9.2　Health impact of saline water

For safe drinking water, the USEPA stipulates that the maximum allowable salinity concentration is limited to 250 mg/L. SLR-induced salinity intrusion into coastal rivers and aquifers may result in increased salinity in drinking water far exceeding this safety limit. For example, the IPCC study indicates that marine and coastal ecosystems in South and Southeast Asia (SSA) will be affected by

SLR, imparting severe impact on water security in those developing countries located along deltaic regions. The impacts are particularly severe for the regions which are subjected to frequent large storm surges from the sea that can inundate up to 100 km inland.

While the global SLR rate is estimated to be in the range of 1.8 to 5.9 mm/year (IPCC), for Bangladesh, the rate is higher (4.6–7.8 mm/year). SLR-induced salinity intrusion in Bangladesh will pose immense pressure on existing water resources. Salinity intrusion into surface and subsurface water is determined by a combination of factors, including rainfall, river flow, tidal elevations and groundwater extraction, as well as the influence of SLR and other climatic variables. A combination of (1) reduced rainfall and river flows, (2) increased tidal elevations and (3) elevated groundwater extraction will contribute to vastly increased salinity intrusion along coastal regions, as indicated by several model simulation studies. In addition, large-scale shrimp farms have contributed significantly to increased groundwater salinity, soil degradation, and a lower yield and lower acreage of rice in Bangladesh. Of critical concern is the salinity during the dry seasons. To provide perspective, the average level of river salinity in the Passur River in Khulna (a coastal area) is estimated at 8210 mg/L in the dry season, resulting in an average human salt intake of up to 16 g/day from river water alone, which is 10 times higher than the recommended daily salt intake (Vineis *et al.*, 2011). Similarly, in the critically dry 1991 water year, electrical conductivity at the Sacramento and San Joaquin rivers averaged 589 μS/cm, corresponding to 920 mg/L, posing dire medical consequences with regard to blood pressure. A study in Massachusetts, USA, reported that systolic and diastolic blood pressures of high-school pupils from towns with a high sodium content (272 mg/L) in public drinking water were significantly higher than those in matched cohorts in the lower-sodium towns (20 mg/L) by 3 to 5 mmHg after controlling dietary salt intake (Calabrese and Tuthill, 1981). The problem of saline intrusion due to climate change and SLR can potentially affect 11 Asian mega-deltas, and other large deltas or estuaries elsewhere, such as the Nile and the Mississippi. This salinity intrusion may in turn have adverse impact on sustainable agriculture in the long run.

4.9.3 *Sustainable agro-aquaculture*

Producing one-third of the world's food from 240 million ha, irrigated agriculture is the largest user of the world's freshwater (Dugan *et al.*, 2006). To improve food security, efforts should be devoted to enhancing water productivity in irrigated agriculture, by improving crop yield per drop of water used and by the harvest of fish, both cultured and catch, from irrigation systems. Out of the 240 million ha irrigated agriculture, an estimated 600,000 km of large channels and 2.4 million km of small channels are potentially available for fish culture. Fish culture in flooded rice fields can increase water productivity substantially. Culture fish production can achieve 600 kg/ha/year in shallow-flooded areas and up to 1,500 kg/ha/year in deep-flooded areas, without reduction in rice yield and wild fish catch. Culture-based fisheries in small reservoirs can further augment fish yield in a substantial way as they do in China (743 kg/ha/year), in Sri Lanka (300 kg/ha/year) and in Cuba (125 kg/ha/year). In Asia, potential yields from small reservoir fisheries are estimated to be 500 to 2,000 kg/ha/year. In these enhanced agro-aquaculture ecosystems, the opportunity exists to build large fenced-in areas, up to several hectares, by creating enclosed water bodies and stocking them with fish. Substantial additional benefits can be obtained by combining intensive aquaculture with irrigated crop production in this way. At national, community and family levels, these systems of agro-aquaculture are critically important in sustaining food security by improving efficiency of water utilization in food production. This improvement in water productivity by crop–fish agro-aquaculture mediated via cultural, technological and institutional innovations will benefit large numbers of low-income families in Africa, Asia and Latin America, where freshwater fisheries are a crucially important resource for poor rural families (Dugan *et al.*, 2006). Rich in protein and minerals, these fish are high-value food, which can be and are harvested using a range of simple, low-cost technologies.

Rivers and their associated floodplains are particularly important in sustaining these low-cost fish harvests. For example, in Cambodia, fish harvested primarily from the Mekong river system constitutes 65%–75% of total protein in the diet. The total direct use value of the fishery resources of the Lower Mekong Basin has been estimated as USD 1478 million (Sverdrup-Jensen, 1999). Studies in Vietnam and Bangladesh show that socially and financially viable approaches for integrating fish into rice culture systems are possible, with potential extension to the Sub-Saharan Africa (Delgado *et al.*, 2003). Agro-aquaculture is widely regarded as playing a crucial role in meeting regional food requirements over the coming decades. Unfortunately, a growing number of rivers run dry along parts of their course for part of the year, including the Colorado (USA) and the Huang Ho (China). Restoring a vibrant hydrology of these river systems is critical in sustaining food and fish production. Also critical to a sustained fishery in rivers and flood plains are water level and river flow, the timing, duration and regulation of the floods, characteristics of the flooded zones, fish migration routes and dry season refuges. Sufficient investment in research and development will go a long way toward achieving food security in these local regions. Salinization in the lower reaches of river systems as well as reduced flows at the mouth of many big rivers, as in the Lower Mekong, is of grave concern. In addition to the adverse effects on agriculture, this salinization drives a change in the natural vegetation structure, leading to reduced biodiversity, impacting subsequently the livelihood of local populations. Higher salinity in estuaries encourages the invasion of many marine predators whose presence would decimate the safe nursery for fishery. Hence, restoring a healthy and vibrant hydrology of the riverine wetland ecosystems is critical in sustaining water and food security. This is the major goal of the Comprehensive Everglades Restoration Plan (CERP), authorized by the US Congress in 2000 for implementation over some 30 years with a budget of USD 10 billion. This CERP inspires a case study to improve water and food security for the state of Penang in Malaysia under the impact of GCC and SLR (Kh'ng *et al.*, 2021), by means of the simulation model MANTRA.

4.10   Case Study in Penang

A coupled hydrology–salinity–vegetation model MANTRA has been developed for simulating the impact of SLR on the sustainability of groundwater and on the mangrove succession and zonation at Pantai Acheh, Penang Island. The simulated soil salinity agrees reasonably well with the measured salinity in the mangrove forest of

Pantai Acheh. MANTRA also successfully reproduces the observed vegetation zonation pattern in Pantai Acheh. Simulation results indicate that the response of coastal wetlands to SLR is sensitive to the topography of the study area and to the availability of areas suitable for vegetation migration landward. The simulation analysis reveals that fresh groundwater is non-viable in Penang Island (Kh'ng *et al.*, 2021), which is confirmed by JMG KPP through personal communication. It is estimated that a 1 m rise in sea level could lead to the loss of 22% of the mangrove forest within the study domain in Pantai Acheh. The landward migration of mangroves is restricted by upland topographic profile and human developments.

SLR and the associated increase in surface seawater inundation would reduce freshwater infiltration to many existing seaward mangrove areas in P. Acheh, Penang, leading to mangrove degradation and wetland habitat loss. Mangrove ecosystems are able to adapt to changes in sea level by expanding landward into areas of higher elevation, or even by growing upward in place (McLeod and Salm, 2006). However, the landward migration of mangroves in P. Acheh has been restricted by high upland topographic profile and by extensive coastal developments around the area, resulting in mangrove loss at lower intertidal elevations. The upland freshwater vegetation remains relatively intact and yet, their biomasses are slightly reduced with increasing soil salinity due to SLR-induced saltwater intrusion. For an SLR of 1.0 m, the mangrove area within the study domain in Pantai Acheh will be reduced to 2.08 km^2, representing a mangrove loss of 22% (Kh'ng *et al.*, 2021). This soil salinization and wetland habitat loss will have significant implications for the sustainability of local water security (SDG6) and food security (SDG2). It is therefore critical to seek alternatives that are viable in the long run for enhancing SDG6 and SDG2 at the local level. Coastal zone communities in developing nations such as Malaysia and other nations around Southeast Asia are vulnerable to the impact of GCC and SLR. Poor socio-economic conditions will constrain mitigation and adaptation options and measures. Large and dense populations living near low-lying coastal zones would amplify any adverse impacts to disproportional levels. Yet, awareness among multi-stakeholders on the grave GCC and SLR challenges confronting the region is still weak. Collaborative partnerships are rare among the government, business, civil society and local communities. This weakness leads to a failure in incorporating climate change mitigation and adaptation measures (SDG13) into national and regional coastal zone development policy and processes. In view of this daunting challenge in integrated coastal zone management strategy at the national and regional levels, an adaptation strategy for developing alternative water resources tailored to local conditions might be more feasible. For example, rainwater can be harvested and stored in water cisterns to enhance water security at the local level to mitigate the vulnerability of Penang to water insecurity (SDG6).

4.10.1   *Vulnerability of Penang to SDG6*

Penang has the lowest domestic water tariff in Malaysia. Consequently, Penang has recorded the highest per capita domestic water consumption in Malaysia at 278 L/c/d in 2018. This consumption is 70% over the 165 L recommended by the World Health Organization (WHO). Currently, Penang's water supply is heavily dependent (80%) on its only major raw-water source from Sg. Muda in Kedah with its quality and quantity constantly threatened by logging in Ulu Muda (Chan, 2012). Based upon the simulation analysis, there is no viable groundwater in Penang Island as a backup resource, diminishing the sustainability of SDG6 (Kh'ng *et al.*, 2021). Any occurrence of persistent and severe droughts driven by climate change would further increase the risk of water insecurity in Penang, with dire consequences. The following subsections provide some findings and recommendations on alternative water resources (AWR) available for ensuring water security and sustainability of SDG6 in Penang. Further, elevated soil salinity due to SLR will reduce crop yields, undermining SDG2 for Penang. AWR must be developed to shield Penang from impairment to SDG6 and SDG2. A major threat to water security in Penang is the potential contamination of water source by logging in the Ulu Muda water catchment in Kedah.

4.10.2   *Threats from logging in Ulu Muda*

Penang receives 80% of its water from the Ulu Muda reservoirs. Any disruptions in water supply from the Ulu Muda catchment would spell disaster for water provision in Penang. The water quality in the Ulu Muda reservoirs is constantly threatened by heavy runoffs after rain, which carry heavy loads of suspended solids into the reservoirs. Hence, high suspended soil concentrations in the Ulu Muda reservoirs had caused frequent water disruption to Penang. Rainfall intensity exceeding 100 mm in 24 h occurs frequently in the Ulu Muda catchment. This high rainfall volume and high intensity in the tropical climate of Malaysia can render soil easily eroded during a heavy rainfall. Stream suspended solid concentrations immediately after a high-intensity storm often exceed 5000 mg/L. Felling, removing and transporting trees can further disturb the forest floor and expose the soil to more erosion. Logging activities in the Ulu Muda forest covering an area of 122,798 ha in Kedah are accompanied by high intensity of roads and skid trails that expose the forest floors to bare soil. The creation of these access roads and skid trails exposes bare soils to erosion during rainfall, resulting in a sharp increase in suspended sediment (SS) concentration in the receiving water bodies (Baharuddin *et al.*, 1995). Logging activities have the tendency to create conditions that may contribute to increased bank failures, and may vastly increase soil loss and sediment yield by one order of magnitude or more. This exposed soil drastically changes the fundamental characteristics of the forest floor. Crawler and tractors are used to haul the felled trees via extensive skid trails to collection points. From here, additional access roads are constructed to transport the collected logs for further processing outside the forest. These extensive series of disturbances to the forest floor in Ulu Muda will cause severe soil erosion during a rain, and will grossly contaminate the water source of Penang. Therefore, logging in Ulu Muda must be banned.

Logging Ban in Ulu Muda

The cumulative impacts of these series of disturbances to the forest floor cause a very high degree of soil loss from the disturbed surfaces. In the Batangsi River catchment, logging activities using skid trails have been observed to have generated high sediment yield of 28.3 t/ha/year (Lai, 1993). Similarly, logging in the Sg Wang experimental watersheds was noted to have resulted in a sharp increase in SS. In Plot W3 in the study area where conventional logging was employed, peak concentration of SS increased dramatically from 1980 mg/L in 1997 before logging to 92,514 mg/L in 1999 soon after logging began, representing an increase of 47 times. In a nearby Plot W2, where reduced impact logging was practiced, the peak concentration was 38,904 mg/L in 1999 soon after logging began. Furthermore, SS levels were noted to exceed 50,000 mg/L in other experimental plots that employed conventional logging methods. A soil loss rate of 50 t/ha/year was estimated for Plot W3, which is 40 times higher than the levels before logging (Lai, 2002). These high levels of soil loss persisted until the end of the fifth year after logging. The absence of any canopy or ground cover coupled with a long slope length along the logging roads leads to sharp increases in sediment yield. For a terrain with a high relative soil erodibility index of 3.34, consisting of silty clay soil, with a 150 m-long uniform slope of 20% and canopy and ground cover of 85%, simulation results indicated that the soil loss amounted to 24 t/ha for a single event runoff volume of 100 mm (Lee *et al.*, 2004). This is equivalent to a very high soil loss rate in excess of 100 t/ha/year, as there are several storms of this intensity or greater in any one year. The mean sediment concentration in the runoff at the lowest segment of the 150 m slope is 24,270 mg/l. The total length of the logging roads in the forest reserve in Ulu Muda was estimated to be 406 km, with a width of 24 m. Simulations indicated that sediment yields from these permanently and totally exposed roads can be extremely high. In 2004, the Malaysian Federal Cabinet had issued a ban order on logging at the Ulu Muda forest, to preserve water quality there. It is critical that this logging ban in Ulu Muda be strictly enforced. Further, in view of the grave uncertainty regarding the reliable supply of clean raw water from Ulu Muda, it is equally critical for Penang to develop AWR such as RWH.

4.10.3   *Rainwater harvesting RWH*

RWH is a process of collecting, diverting and storing rainwater collected from roof catchment area for later domestic use or for recharging of groundwater aquifers. RWH has the potential to be implemented in Malaysia due to its high rainfall volume. The National Hydraulic Research Institute of Malaysia (NAHRIM) has successfully implemented several RWH projects in Shah Alam and Sandakan and has developed software named Tangki NAHRIM to estimate the size of rainwater tanks (PBAPP, 2019).

Table 4.7: Optimum rainwater storage tank size (m^2) for Penang.

Demand (L/d)	Roof catchment area (m^2)					
	50	100	200	300	400	500
100	0.5	0.5	0.5	0.5	0.5	0.5
200	1.8	1.0	0.8	0.7	0.7	0.7
400	–	3.6	2.0	1.6	1.6	1.5
600	–	12.5	4.0	3.1	2.7	2.5
800	–	–	7.0	4.8	4.0	3.8
1000	–	–	12.0	7.2	5.9	5.1

The effectiveness of RWH to provide water for each household is determined using the reliability ratio (E_T), which is defined in Eq. (4.14) as the rainwater yield (Y_t) over the total water demand (D_t):

$$E_T = \frac{\sum_{t=1}^{T} Y_t}{\sum_{t=1}^{T} D_t} \times 100\% \tag{4.14}$$

Modeling simulation is carried out with Tangki NAHRIM software to estimate the optimum rainwater storage tank size for Penang Island, as presented in Table 4.7. Tank size is considered sufficient and acceptable once the E_T value has reached 75%. For example, for a given four-person household with water demand of 1000 L/d (250 L/c/d) and roof area of 200 m^2, a storage tank size of 12 m^3 could reduce usage of treated water by about 75% per month, representing a significant economic saving and a meaningful improvement in achieving water security. Therefore, governmental agencies, research entities and policymakers should play an important role in promoting the installation of RWH systems as well as in providing financial and technical assistance for concerned authorities and entities. RWH is an essential element in the chain of water supply in Penang in view of the certainty regarding water insecurity in Penang.

4.10.4 *Water insecurity in Penang*

The 2009 "Masterplan Study for Potable Water Supply in Penang until year 2050" has proposed the implementation of the Sungai

Perak Raw Water Transfer Scheme (SPRWTS) to tap Sungai Perak as a second raw-water source for Penang (NAHRIM, 2014). Based on the Masterplan Study, the total potential yield of the SPRWTS for Penang is 1,000 million liters per day (MLD) by 2050, to be delivered in four phases of 250 MLD each from 2020. Besides ensuring water security for Penang and North Perak until year 2050, the SPRWTS will also mitigate the risks of Penang's overdependency on Ulu Muda as a primary raw-water resource. Conventional treatment of river water is more cost efficient than wastewater recycling and desalination. Penang's average domestic water tariff for the first 35,000 L per month is RM 0.32 per 1,000 L. In Singapore, a similar consumption volume would cost RM 7.44 (SGD 2.47) per 1,000 L. This cost includes water conservation tax and waterborne fee to account for additional costs incurred in desalination and water recycling technologies currently employed alongside conventional water treatment operations in Singapore. This simple comparison indicates that water tariff in Penang is vastly below the true cost of production, treatment and delivery, and does not reflect the grave uncertainty in delivering water with such a low tariff. The true cost of water tariff should reflect the actual cost that might be added in order to increase water security in Penang. For example, the true cost of alternative water sources should be factored into the water tariff. However, the weak and incompetent political leadership in Penang appears to be incapable of evaluating and implementing a water tariff scheme that would protect Penang from future water crisis.

4.10.5   *Groundwater abstraction*

Information on groundwater availability in Penang region is limited. Consistent with the information available from JMG KPP, simulation results indicated that fresh groundwater is practically non-existent in Penang Island (Tay *et al.*, 2021). However, several water resource studies have revealed the presence of potential groundwater aquifers in Seberang Prai, Penang. For example, Ladang Bertam in Kepala Batas has been utilizing groundwater to irrigate oil palm and rubber plantations during the dry season (Kamaludin, 1990). In addition, JMG had conducted exploratory drilling in Nibong Tebal during the 9th Malaysia Plan (2006–2010) period and found that the water quality is good except for the elevated iron content (Ranhill Consulting, 2011). Hence, further monitoring and modeling research should be conducted to evaluate the feasibility of groundwater extraction and artificial recharge by RWH, for example, in Seberang Prai and Nibong Tebal as potential sources of potable water to enhance water security in Penang.

4.11 Conclusions and Recommendations

Simulation models such as MANTRA are capable of providing insights on the impact of seawater inundation and saltwater intrusion due to sea level rise and storm surge-induced inundation on coastal vegetation subject to the influence of various stochastic and deterministic factors. This chapter outlined the risks of soil salinization, soil degradation and loss of arable land due to climate change and population growth. The implications on water and food security are then discussed. We conclude this chapter with the following remarks.

4.11.1 *More crop per drop*

We must produce more crop from less drops of water through sustainable agriculture in order to feed a growing population that is projected to reach 9 billion by 2050. Integrating hydrology and agro-ecosystem technology to optimize on the relationship between soil, water, nutrients and crops is important to ensure food security. Judicious choice of efficient management systems can minimize losses of water by reducing surface runoff and evaporation and by maximize storage of soil water in the root zone. In this way, we can significantly reduce water scarcity. High soil erosion exacerbates pollution and reduces soil organic carbon, leading to diminished yield. This must be controlled. Adaptation of agricultural, water and food systems to climate change necessitates appropriate economic and policy interventions at the national and international levels. Accelerated soil erosion, by water and wind, and soil salinization are the dominant processes that have reduced the global per capita arable land area from 0.42 ha in 1960 to 0.22 ha in 2004 (Lal, 2013). Restoration of degraded and decertified soils is therefore a high priority, particularly in SSA and densely populated Asia. An increasing preference for a meat-based diet by large populations of developed and emerging economies exerts a stronger demand on grain consumption.

Energy and water consumed and greenhouse gas emitted in animal-based food are 10–20 times more than that based on grains and plants. Adoption of plant foods will therefore go a long way toward water and food security. The yield gap between the global average yield and the maximum attainable yield must be reduced by use efficiency of resources. Ample opportunities exist to vastly improve productivity in mixed water–crop–livestock systems practiced worldwide by small-scale farmers who collectively produce half of the world's food. Producing more crop from less drops of water through sustainable agriculture is not just a promise. It is a mandatory necessity.

4.11.2   *Unsustainable water consumption in Penang*

Penang relies heavily on very limited and unsustainable water sources. Its current water management practices are not sustainable to meet growing demand and to overcome climate change impact. Hence, alternative water supply options such as rainwater harvesting, inter-state water transfer and groundwater exploitation deserve careful consideration. In combination with consumer behavioral education and research, these alternative water resources are critically needed to ensure a sustainable water future for Penang Island toward achieving SDG6 and SDG2. Key elements for achieving these two SDGs include (a) adapting integrated coastal zone management to counter GCC and SLR, (b) adapting integrated water management to mitigate increased risk of floods and droughts due to GCC and SLR, and (c) protecting mangrove forests and their ecosystem services from further degradation. Above all, the time is ripe for Penang to embark on a journey toward water and food security at the local level by taking a first step in encouraging and implementing RWH technology in achieving SDG6 and SDG2. In the following subsections, several recommendations based upon successes achieved in several countries regarding this journey would be presented. The goal is to inspire and aspire.

4.11.3   *Recommendations on RWH*

4.11.3.1   *Sustainable RWH technology*

Rainwater can be a good supplement for treated tap water for certain usages, such as garden irrigation, toilet flushing and laundry, as it has reasonably good water quality with zero hardness, no sodium and nearly neutral pH, provided proper harvesting and storage technology is applied. RWH can reduce local water demand. With the water saved, neighboring townships can mutually support each other during water shortage, thereby enhancing local water security. Yet, despite facing chronic water scarcity, many urban cities, such as Georgetown of Penang, treat rainwater as a risk rather than as a valuable resource. Skepticism regarding RWH technologies and sustainability persists, in both low- and high-precipitation areas. Questions still remain, such as how reliable this water source is, how large a storage is required and whether RWH is sustainable. Toilet flushing and laundry water demand of a single-family house can be met with a relatively small tank, even with moderate precipitation. Stored rooftop rainwater can meet more than 60% of the landscape irrigation demand in single- and multi-family buildings. The main drawback and deterrence, however, is the long payback period for most RWH systems currently in use (Domènech and Saurí, 2011).

4.11.3.2 *Policies, regulations and incentives*

Policies, regulations and incentives are good strategies to overcome this financial deterrence in order to advocate RWH systems in residential areas. It is worth noting that the rapid expansion of RWH systems in Australia had coincided with a period of growing interest in community-based, grassroots and self-help development (White *et al.*, 2006). Further, the development of alternative sources for water has benefited significantly from individual and community efforts (Turner *et al.*, 2007), beyond policy, regulations and incentives. Hence, regulations and incentives should be strengthened by social community education aimed to reinforce the perceived and proven benefits of RWH, and to inform the technicality and proper use of RWH systems. In addition, policies and regulations must be effective in overcoming the shortage of space and high land cost that may make the installation of large surface storage systems expensive and impractical (DNRMW, 2006).

4.11.3.3 *Global perspectives*

The adoption of RWH systems worldwide can be characterized as lethargic in most countries and encouraging in others. Nevertheless, success stories abound due to new regulations and incentives being developed worldwide to foster the use of rainwater. As an example, 48% of households in South Australia had installed rainwater tanks (Australian Bureau of Statistics, 2006). The Metropolitan Area of Barcelona had started to promote the use of rainwater through specific regulations and incentives, to conserve drinking water and to save costs. In Catalonia of Spain, several municipalities had approved water-saving regulations that mandate new buildings with a certain garden area to install RWH systems. Jordan and Sri Lanka had approved RWH policies at the national level (Goonetilleke *et al.*, 2005). At the local levels, many other regulations and incentives had been approved to support RWH. Similar regulations and incentives had also been approved in many states and cities of India. In Brazil, the government is supporting a program that aims to install one million water cisterns in the semi-arid areas of Brazil. In Australia, several initiatives at the national and regional level promote the use of alternative water sources, including RWH. The Australian Government offered rebates of up to $500 to all houses installing an RWH system (Australian Government, 2009). In South Australia, all new buildings have to install an RWH system or some other alternative sources of water supply to supplement piped water. In Germany, rainwater harvesters are exempted from paying stormwater taxes as they are considered to have contributed to stormwater attenuation systems (Hermann and Schmida, 2000). In Belgium, new buildings with a roof area greater than 100 m^2 are required to install RWH and stormwater attenuation systems. Persistently facing severe water shortages and excessive storm water inundations for decades, the Penang government can and must learn from the experiences and perspectives of these progressive urban cities on the wisdom of promoting and implementing RWH and stormwater attenuation systems. Penang must immediately begin to plan and implement rainwater harvesting and stormwater attenuation systems for a sustainable future dominated by uncertainty and threats posed by climate change and sea level rise (SDG13). In particular, the enactment of various state laws in Texas, USA, to promote RWH and alternative water resources is a role model for Penang to emulate.

4.11.3.4 *Texas as role model*

The harvesting of urban rainwater to supply non-potable water demands is emerging as a viable option, as a means to augment increasingly stressed urban water supply systems. There has been a fundamental shift in the approach to the provision of urban water services in many cities and countries around the world, mainly due to water stress. In America, rainwater harvesting is mandatory in new buildings of Tucson (Arizona), Santa Fe County (New Mexico) and several Caribbean islands. In particular, Texas enacted several state laws to support rainwater harvesting, such as the requirement of certain new state facilities to incorporate RWH systems in their design. The Texas Tax Code §11.32 allows the exemption of part or all of the assessed value of the property on which approved water conservation initiatives, such as RWH, are implemented. The Texas Tax Code §151.355 exempts rainwater harvesting equipment and supplies from state sales tax. The Texas Property Code §202.007 prevents a homeowner's association from prohibiting the use of RWH systems. The Texas Local Government Code §580.004 supports RWH at residential, commercial, industrial and educational facilities through incentives such as discounts for rain barrels or rebates for water storage facilities. The Texas Water Development Board (TWDB) provides useful information on all aspects of RWH, including the popular Texas Manual on RWH on designing residential and small-scale commercial systems. Through their website and through other outreach and education efforts, TWDB supports concerted community-based efforts to promote RWH, including providing financial support for rainwater harvesting research studies (TWDB, 2005). The Penang government should learn from the Texas experience.

References

Alegre, X.V. and Tynan, N. (2014). Preliminary project plan for a rainwater harvesting system for the complexe de L'Etoile du Sahel (Tunisia) (preliminary Report). Sousse: Municipality of Sousse.

Alvarez, M.P., Carol, E., Hernandez, M.A. and Bouza, P.J. (2015). Groundwater dynamic, temperature and salinity response to the tide in

Patagonian marshes: Observations on a coastal wetland in San Jose Gulf, Argentina. *Journal of South American Earth Sciences*, 62, 1–11.

Anderson, W.P.Jr. (2002). Aquifer salinization from storm overwash. *Journal of Coastal Research*, 18(3), 413–420.

Antoniou, G.P. (2010). Ancient Greek lavatories: Operation with reused water. In: Mays, L.M. (ed.) *Ancient Water Technologies*, Springer Science and Business Media, Dordrecht, pp. 67–86. doi: 10.1007/ 978-90-481-8632-7.4.

Arnold, V. (1968). A note on the Ghyben-Herzberg formula. *Hydrological Sciences Journal*, 13(4), 43–46.

Australian Bureau of Statistics (2006). Water account Australia 2004–2005. Australian Government.

Australian Government (2009). Water for the Future. National Rainwater and Greywater Initiative. http://www.environment.gov.au/water/publications/action/pubs/nrgi.pdf (accessed 27 November 2020).

Baharuddin, K., Mukhtaruddin, A.M. and Nik Muhamad, M. (1995). Surface runoff and soil loss from a skid trail and a logging road in a tropical forest. *Journal of Tropical Forest Science*, 7(4), 558–569.

Calabrese, E.J. and Tuthill, R.W. (1981). The influence of elevated levels of sodium in drinking water on elementary and high school students in Massachusetts. *Science of Total Environment*, 18, 17–33.

Chan, N.W. (2012). Managing urban rivers and water quality in Malaysia for sustainable water resources. *International Journal of Water Resources Development*, 28(2), 343–354. doi: 10.1080/07900627. 2012.668643.

Chitnis, N., Hyman, J.M. and Cushing, J.M. (2008). Determining important parameters in the spread of Malaria through the sensitivity analysis of a mathematical model. *Bulletin of Mathematical Biology*, 70(5), 1272–1296.

Crouch, D.P. (1993). Water management in ancient Greek cities. Oxford: Oxford University Press.

Delgado, C., Nikolas, W., Rosegrant, M.W., Meijer, S. and Ahmed, M. (2003). Fish to 2020: Supply and Demand in Changing Global Markets. International Food Policy Research Institute, Washington, DC and WorldFish Center, Penang, Malaysia, pp. 1–226.

Derner, J.D., Johnson, H.B., Kimball, B.A., Pinter Jr, P.J., Polley, H.W., Tischler, C.R., Bouttons, T.W., LaMorte, R.L., Wall, G.W., Adam, N.R., Leavitt, S.W., Ottman, M.J., Matthias, A.D. and Brooks, T.J. (2003). Above-and below-ground responses of C3-C4 species mixtures to elevated CO_2 and soil water availability. *Global Change Biology*, 9, 452–460.

DNRMW (2006). Department of Natural Resources Mines and Water. Water for southeast Queensland: A long-term solution. Department of Natural Resources, Mines and Water, Australian Government.

Domènech, L. and Saurí, D. (2011). A comparative appraisal of the use of rainwater harvesting in single and multi-family buildings of the Metropolitan Area of Barcelona (Spain): Social experience, drinking water savings and economic costs. *Journal of Cleaner Production*, 19(6–7), 598–608.

Dugan, P., Dey, M.M. and Sugunan, V.V. (2006). Fisheries and water Productivity in tropical river basins: Enhancing food security and livelihoods by managing water for fish. *Agricultural Water Management*, 80, 262–275.

Dupuit, J. (1863). Études Théoriques et Pratiques sur le Mouvement des Eaux dans les Canaux Découverts et à Travers les Terrains Perméables [Theoretical and Practical Studies of Water Movement in Open Channels and Permeable Rocks], Dunod, Paris.

Goonetilleke, A., Thomas, E., Ginn, S. and Gilbert, D. (2005). Understanding the role of land use in urban stormwater quality management. *Journal of Environmental Management*, 74, 31–42.

Gowing, J.W., Rose, D.A. and Ghamarnia, H. (2009). The effect of salinity on water productivity of wheat under deficit irrigation above shallow groundwater. *Agricultural Water Management*, 96, 517–524.

Gualbert, D.E. (2001). Density Dependent Groundwater Flow: Salt Water Intrusion, Interfaculty Centre of Hydrology Utrecht, Utrecht.

Hermann, T. and Schmida, U. (2000). Rainwater utilisation in Germany: Efficiency, dimensioning, hydraulic and environmental aspects. *Urban Water*, 1(4), 307–316.

Hsieh, A. and Winston, R.B. (2002). User's Guide to Model Viewer, a Program for Three-dimensional Visualization of Ground-water Model Results, Open File Report 02–106, U.S. Geological Survey.

IPCC (2013). Climate Change 2013: The Physical Science Basis, Working Group 1 Contribution to the Fifth Assessment Report of the International Panel On Climate Change, Cambridge University Press.

Kamaludin, H. (1990). A summary of the quaternary geology investigations in Seberang Prai, Pulau Pinang and Kuala Kurau. Geol Soc Malaysia, 26, 47–53. doi: 10.7186/bgsm26199005.

Khan, S., Hanjra, M.A. and Mu, J.X. (2009). Water management and crop production for food security in China: A review. *Agricultural Water Management*, 96, 349–360.

Kh'ng, X.Y., Teh, S.Y., Koh, H.L. and Shuib, S. (2021). Sea Level Rise Undermines SDG2 and SDG6 in Pantai Acheh, Penang, Malaysia. *Journal of Coastal Conservation*, 25, 9. doi: 10.1007/s11852-021-00797-5.

Lai, F.S. (1993). Sediment yield from logged steep upland catchments in Peninsular Malaysia. In: *Hydrology in Warm Humid Regions*, 219–230. IAHS Publ. 216. IAHS Press, Wallingford, UK.

Lai, F.S. (2002). Suspended sediment yield changes resulting from forest harvesting in the Sg Wang experimental watersheds, Kedah, Peninsular Malaysia. In: IHP-VI Technical Document in Hydrology no. 1, UNESC O Jakarta Office, 210–215.

Lal, R. (2013). Food security in a changing climate. *Ecohydrology and Hydrobiology*, 13(1), 8–21.

Lee, H.L., Koh, H.L. and Al'Rabia'ah, H.A. (2004). Predicting soil loss from logging in Malaysia. GIS and Remote' Sensing in Hydrology, Water Resources and Environment (Proceedings of ICGRHWE held at the Three Gorges Dam, China, September 2003). IAHS Publ. 289.

Lu, C., Xin, P., Kong, J., Li, L. and Luo, J. (2016). Analytical solutions of seawater intrusion in sloping confined and unconfined coastal aquifers. *Water Resources Research*, 52(9), 6989–7004.

McLeod, E. and Salm, R.V. (2006). Managing mangroves for resilience to climate change gland. International Union for Conservation of Nature and Natural Resources (IUCN), Switzerland.

Misra, A.K. (2014). Climate change and challenges of water and food security. *International Journal of Sustainable Built Environment*, 3, 153–165.

Muazu, A., Yahya, A., Ishak, W.I.W. and Khairunniza-Bejo, S. (2015). Energy audit for sustainable wetland paddy cultivation in Malaysia. *Energy*, 87, 182–191.

NAHRIM (2014). NAHRIM technical guide no. 2: The design guide for rainwater harvesting systems. *National Hydraulic Research Institute of Malaysia (NAHRIM)*, Kuala Lumpur.

Nakićenović, N., Alcamo, J., Davis, G. *et al.* (2000). *Emissions Scenarios: A Special Report of Working Group III of the Intergovernmental Panel on Climate Change*, Cambridge Univ. Press, New York.

Noe, G.B., Krauss, K.W., Lockaby, B.G., Conner, W.H. and Hupp, C.R. (2013). The effect of increasing salinity and forest mortality on soil nitrogen and phosphorus mineralization in tidal freshwater forested wetlands. *Biogeochemistry*, 114, 225–244.

Pang, X.P. and Letey, J. (1998). Development and evaluation of ENVIRO-GRO, an integrated water, salinity, and nitrogen model. *Soil Science Society of America Journal*, 62(5), 1418–1427.

Parry, M.L., Rosenzweig, C., Iglesias, A., Livermore, M. and Fischer, G. (2004). Effects of climate change on global food production under SRES emissions and socio-economic scenarios. *Global Environmental Change*, 14, 53–67.

PBAPP (2019). Sungai Perak: The best future raw water resource for Penang and north Perak. Perbadanan Berkalan Air Pulau Pinang Holdings Bhd (PBAPP), Penang.

Rachwal, A.J. and Holt, D. (2008). Urban rainwater harvesting and water reuse: Review of potential benefits and current uk practices (technical report FR/G0006). Marlow: Foundation for Water Research.

Ranhill Consulting (2011). Review of the national water resources study (2000–2050) and formulation of national water resources policy (Final Report Volume 9-Pulau Pinang). Department of Irrigation and Drainage (JPS), Kuala Lumpur.

Roberts, N. (1977). Water conservation in ancient Arabia. *Proceedings of the Seminar for Arabian Studies*, 7(1977), 134–146.

Sarada, M. (2015). Roof-top harvesting of rainwater: An answer to present day water crisis of India (with special reference to west Bengal). *Journal of Environmental Science, Toxicology and Food Technology*, 9(6), 36–42.

Sverdrup-Jensen, S. (1999). Policy issues deriving from the impact of fisheries on food security and the environment in developing countries. In: Ahmed, M., Delgado, C., Sverdrup-Jensen, S., Santos, R.A.V. (Eds.), Fisheries Policy Research in Developing Countries: Issues, Priorities and Needs, ICLARM, Manila, Philippines, pp. 73–91.

Talebnejad, R. and Sepaskhah, A.R. (2015). Effect of different saline groundwater depths and irrigation water salinities on yield and water use of quinoa in lysimeter. *Agricultural Water Management*, 148, 177–188.

Teal, J.M., Best, R., Caffrey, J., Hopkinson, C.S., McKee, K.L., Morris, J.T., Newman, S. and Orem, B. (2012). Mississippi River Freshwater Diversions in Southern Louisiana: Effects on Wetland Vegetation, Soils, and Elevation. Lewitus, A.J., Croom, M., Davison, T., Kidwell, D.M., Kleiss, B.A., Pahl, J.W., Swarzenski, C.M. (Eds.). Final Report to the State of Louisiana and the U.S. Army Corps of Engineers through the Louisiana Coastal Area Science & Technology Program; coordinated by the National Oceanic and Atmospheric Administration, p. 49.

Teh, S.Y., Koh, H.L., DeAngelis, D.L. and Turtora, M. (2013). Interaction between salinity intrusion and vegetation succession: A modeling approach. *Theoretical and Applied Mechanics Letters*, 3, 1–032001.

Teh, S.Y., Turtora, M., DeAngelis, D.L., Jiang, J., Pearlstine, L., Smith, T.J.III and Koh, H.L. (2015). Application of a coupled vegetation competition and groundwater simulation model to study effects of sea level rise and storm surges on coastal vegetation. *Journal of Marine Science and Engineering*, 3, 1149–1177.

Todd, D.K. and Mays, L.W. (2005). *Groundwater Hydrology*, Third Edition. John Wiley & Sons, Inc., New York, USA, p. 636.

Todorovic, M., Lamaddalena, N., Jovanovic, N. and Pereira, L.S. (2015). Agricultural water management: Priorities and challenges. *Agricultural Water Management*, 147, 1–3.

Turner, A., Hausler, G., Carrard, N., Kazaglis, A., White, S., Hughes, A. and Johnson, T. (2007). Review of water supply-demand options for south east Queensland. Sydney and Cardno, Brisbane, Australia: Institute for Sustainable Futures.

TWDB (2005). The Texas manual on rainwater harvesting, 3rd edn. Texas Water Development Board (TWDB), Austin, Texas, USA.

UN-Water (2018). Nature-based Solutions for Water, United Nations World Water Development Report, UNESCO.

Vineis, P., Chan, Q. and Khan, A. (2011). Climate change impacts on water salinity and health. *Journal of Epidemiology and Global Health*, 1, 5–10.

Voss, C.I. and Provost, A.M. (2003). A Model for Saturated-unsaturated, Variable-density Ground-water Flow with Solute or Energy Transport, Water-resources Investigation Report 02–4231, U.S. Geological Survey.

Wang, J. and Baerenklau, K.A. (2014). Crop response functions integrating water, nitrogen, and salinity. *Agricultural Water Management*, 139, 17–30.

White, S., Campbell, D., Giurco, D., Caroline, S., Kazaglis, A., Fane, S., Deen, A. and Martin, J. (2006). Review of the metropolitan water plan: Final report, Institute for Sustainable Futures, Sydney, ACIL Tasman and SMEC Australia.

Willard, D.A. and Bernhardt, C.E. (2011). Impacts of past climate and sea level change on Everglades wetlands: Placing a century of anthropogenic change into a late-Holocene context. *Climatic Change*, 107, 59–80.

Wilson, A. (2001). Urban water storage, distribution, and usage. In: Koloski-Ostrow, A.O., (ed). Water use and hydraulics in the Roman city. Dubuque: Kendall/Hunt Publishing Company. pp. 83–125.

Ye, L.M. and Ranst, E.V. (2009). Production scenarios and the effect of soil degradation on long-term food security in China. *Global Environmental Change*, 19, 464–481.

Chapter 5

Comprehensive Everglades Restoration Plan

5.1 Everglades: A Fragile Ecosystem

The Everglades is the most well-known wetlands in the United States
and the most distinct in the world. Located in south Florida, USA,
the Everglades stretches some 200 km from the southern edge of
Lake Okeechobee in the north to the Florida Bay in the south, drop-
ping in elevation by only 5.3 m over this 200 km stretch. Covering an area of 10,000 km^2, the Everglades, popularly known as the
River of Grass (Douglas, 1988), consists of many land types. They
include freshwater marshes, tropical hardwood hammocks, pinelands
and tree islands, besides cypress heads and domes, mangrove swamps
and coastal saline flats. Freshwater flows through the Everglades
have been sharply reduced because of hydrological reengineering and
severe human intervention by pioneer and subsequent settlements
in the past decades. Coupled with rapid urbanization of the area,
these developments have decimated the Everglades to patches of pool
waters and seasonally dry lands, incapable of supporting biodiversity
and vibrant wildlife. Further, many biological species in the Everglades are heavily contaminated by toxicants through the food consumed, although the contamination level in the water column may
not be that high. The bioaccumulation of toxicants, such as mercury
and PCBs, in the Everglades fish via the food web has an adverse
impact on the health of the entire ecosystem. Moreover, seasonally

varying hydrological regimes in the Everglades constrain the distribution of fish communities that provide foraging resources for wading birds. Subject to seasonal hydrological regimes, the constrained spatial-temporal distribution of fish community in the Everglades exerts its impact on wading bird ecology and on the overall biodiversity. Decimated habitats threatened by environmental toxicants and constrained by compromised hydrology pose a daunting challenge to the Everglades ecosystem. Because of its low elevation, the Everglades landscapes and ecosystem are vulnerable to sea level rise and its associated salinity intrusion, leading to increased soil salinity. Frequent and severe storm surges induced by hurricanes inundate the coastal areas in the Everglades and cause sharp salinity changes in the inundated soils for extended durations. Sharp salinity changes over long durations may have irreversible impacts on the sensitive plant ecosystems in the Everglades, such as the glycophytic (salt-intolerant) hardwood hammocks and halophytic (salt-tolerant) mangroves. The degradation of plant species will in turn adversely affect many other biological species, such as fish, wading birds and mammals. To protect the fragile ecosystems in the Everglades, the Everglades National Park was formed in 1947 and it has been a World Heritage Site since 1976. To provide a framework to restore, protect and preserve the water resources of the larger central and southern Florida, including the Everglades, the Comprehensive Everglades Restoration Plan (CERP) was enacted by the U.S. Congress in 2000. The CERP is anticipated to be implemented over a period of 30 plus years, with a budget of around USD 10 billion.

5.1.1   *The Everglades CERP*

The fragile Everglades in southern Florida is a unique network of subtropical coastal wetlands that are facing daunting challenges for survival. Past human developments have halved its original acreage. The U.S. federal government had a long history of active involvement in the Everglades development. Beginning in the 1940s, the U.S. Army Corps of Engineers (USACE) have been implementing flood control projects that drained water away from the Everglades. These flood control projects, coupled with agricultural and urban development, have contributed to the shrinking and degradation of the Everglades water resources and wetland ecosystem. CERP was therefore enacted

by the U.S. Congress in 2000 for the restoration of the Everglades
ecosystem. With an initial budget of \$8.2 billion, the federal govern-
ment is expected to pay half of that, with an array of state, tribal
and local agencies paying the other half (Stern *et al.*, 2010). More
recent estimates, however, indicated that the CERP plan would take
more than the anticipated 30 years to implement and would cost an
additional \$1.63 billion. CERP focuses on increasing storage of excess
water in the rainy season to provide more water during the dry season
for the ecosystem and for urban and agricultural users in the greater
Everglades region. Figure 5.1 shows the historical (left), current (middle) and restored (right) water flows through the Everglades. Histor-
ically, ample water would flow from Lake Okeechobee through the
Everglades before the water would drain into the Atlantic Ocean and
the Gulf of Mexico. These slow-moving waters nourished the lands
and supported a vibrant ecosystem. As indicated in Figure 5.1 (middle), much of the water is diverted to the eastern coastal cities. Very
little water gets into the Everglades, resulting in damaged hydrology
and degraded ecosystems. The CERP plan is to redirect the flows to
go through the Everglades to restore its hydrology and ecosystems.

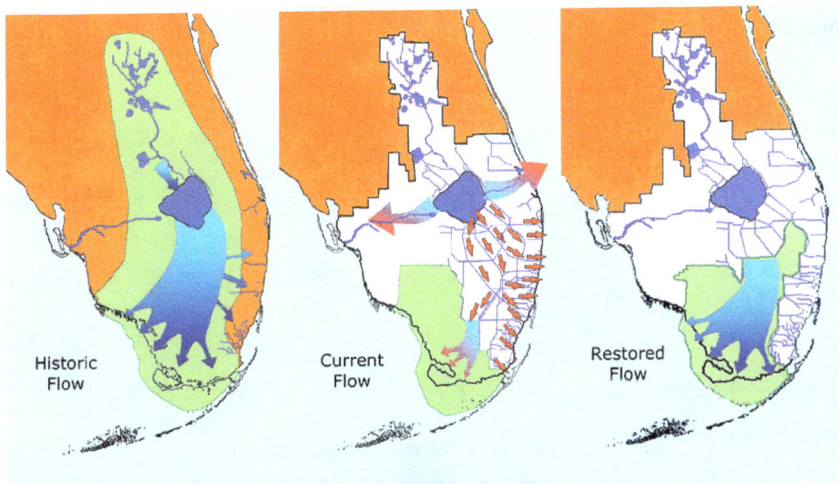

Figure 5.1: Historic, current and restored water flow of the Everglades
ecosystem.
Source: Institute of the Environment, University of Minnesota.

The CERP thus provides a framework to restore, protect and preserve the water resources of central and southern Florida, including the Everglades, covering 16 counties over a 47,000 km^2 area. The revitalized water resources will in turn restore and preserve the degraded ecosystems. The State of Florida (via the South Florida Water Management District) and the USACE are responsible for undertaking various projects under the CERP to help ensure the proper quantity, quality, timing and distribution of waters to the Everglades and all of south Florida.

The primary aspiration of CERP is to capture freshwater that now flows unused to the Atlantic Ocean and the Gulf of Mexico and redirect the water to areas that need the water most, including the Everglades. The restored water will benefit the ecosystems as well as cities and farmers by enhancing water supplies for the south Florida economy (Sklar *et al.*, 2005). The beneficial effects of increased freshwater flow resulting from the CERP, however, may be compromised in some places by increased saltwater intrusion and salinity overwash events associated with SLR. The economic benefits of this restoration effort appear to justify its expense. A 2012 economic study reported that for every penny spent on the CERP, the local economy will gain at least four cents. A series of biennial reports from the U.S. National Research Council have reviewed the progress of CERP. The fourth report in the series, released in 2012, found that little progress has been made in restoring the core of the remaining Everglades ecosystem. Instead, most project constructions so far have occurred along its periphery. The report noted that to reverse and restore ongoing ecosystem degradation, it is necessary to expedite restoration projects that target the central Everglades. It is essential to improve both the quality and quantity of the water that is critical to the benefit of the Greater Everglades Ecosystem Restoration (GEER).

5.1.2   *The Everglades GEER*

The CERP underscores the deep concern among government agencies, conservation groups, scientists and the public about the uncertain fate of the GEER. The greater Everglades was once a vast and free-flowing "river of grass" stretching from the Kissimmee Chain of Lakes near Orlando in the north through Lake Okeechobee in central Florida and finally into the southern Everglades (Figure 5.1, left). Early in the 20th century, the region's hydrology was reengineered by building a dike around Lake Okeechobee that continued to evolve into its current form in the 1960s. The dike prevented the lake's historical spillovers during the rainy seasons and protected downstream residents and farmlands from floods. Water from the entire northern watershed is pooled in Lake Okeechobee. Currently, very little water is permitted to travel its natural southward path to the sensitive and protected Everglades. Most of the water is drained into the eastern and western coasts of the Atlantic Ocean and the Gulf of Mexico (Figure 5.1, middle). As a result, areas like the Everglades National Park receive too little water, while Lake Okeechobee gets too much. The same lands and waterways in the Everglades are now drained, diked, segmented and controlled. The CERP aims to redirect the water to flow through the Everglades (Figure 5.1, right) to revitalize the ecosystems. The pooled water in Lake Okeechobee is nutrient rich, containing 600 parts per billion (600 ppb) of phosphorus. Most of the nutrient loads in the Everglades come from phosphorus that has accumulated in soils and in sediments from past human activities, called "legacy" phosphorus. Consequently, nutrients abound almost everywhere, driving regime shifts in plant communities, in aquatic life and in wildlife, potentially impairing ecosystem services. Excessive nutrients, particularly phosphorus, have fueled both the loss of native sawgrass and the spread of invasive cattail, driving associated regime shifts. The dense cattails often prevent movements of alligators, turtles, wading birds and other native wildlife. Small fish and aquatic macroinvertebrates, the primary foods of wading birds, are the immediate victims. And, populations of large fish that prey on small fish have been diminished, causing a decline in alligators and other large predators in the Everglades. The disturbed ecosystems may undergo further disturbances under the scenario of climate change and sea level rise. Hence, climate change adaptation (SDG 13) is a central part of the CERP.

5.1.2.1 *Climate change adaptation*

Excess water pooled in Lake Okeechobee is channeled to two estuaries (Figure 5.1, middle), on Florida's east and west coasts, causing coastal eutrophication there. To reduce the overwhelming water from entering Lake Okeechobee and to relieve hydraulic pressure building on its dikes, water to the tune of around 1.2 million acre-feet has to be retained in the upper watershed above Lake Okeechobee (Fisher, 2014). The influx of additional and adequate freshwater during the dry seasons, released from storage above Lake Okeechobee, will help prevent saltwater intrusion and mitigate the potential impact of sea level rise. This is arguably the best strategy for southern Florida to adapt to climate change and its associated sea level rise. If successful, these restoration efforts will protect subterranean aquifers from saltwater intrusion, will delay the impacts of sea level rise along the coast and will buy precious time for wildlife to adapt to the changing environment induced by GCC. Wading birds, for example, have declined 90 percent after drainage canals were constructed in the 1930s, because of degradation of aquatic habitats critically needed by the birds. Delivery of adequate clean water to the right places at the right times of the year is the prerequisite for the restoration of these degraded aquatic habitats. It is also a prerequisite for adaptation to GCC and SLR in the Everglades. This prerequisite is expected to restore the ecological vitality of the greater Everglades ecosystem, from periphyton to fish to wading birds. These restoration efforts will also help maintain the natural source of freshwater for drinking, farming and industry. More importantly, restored water flows will mitigate the impact of SLR and delay a vegetation regime shift from hammocks to mangroves in the Everglades.

5.1.3 *Vegetation regime shift in Everglades*

Over the last century, the sea level has risen approximately 15 to 20 cm worldwide. Along most of the US Atlantic and Gulf of Mexico coasts, including the Everglades, a sea level rise (SLR) at the rate of about 2 to 3 mm per year has been reported, and the total SLR is likely to reach 55 to 60 cm over the next century (Titus and Richman, 2000). It has been projected that the global temperature might increase by 1.4% to 5.8% between 1990 and 2100 due to greenhouse gas emissions. Rising global temperature might lead to more severe and more frequent hurricanes. This is because warmer temperatures heat up the ocean surface and provide additional energy for storms to intensify as they swirl through the ocean

(Webster *et al.*, 2005). More intense and more frequent hurricanes in combination with SLR will increase the salinity of the vadose zone. Increase in salinity in the vadose zone is particularly worrisome in the Everglades, as the area is flat and low-lying. Increase in salinities of the vadose zone beyond a certain threshold level might eradicate the salinity-intolerant hardwood hammocks at higher elevations and promote landward migration of mangroves, triggering a regime shift. Stratigraphic evidence indicates that mangroves have replaced freshwater marsh along coastal southern Florida throughout the Holocene (Gleason *et al.*, 1974). The regime shift of former pineland on Key Largo to the current mangroves had been attributed to SLR (Alexander and Crook, 1974). This invasion of mangroves to replace freshwater plants occurs in other parts of the world as well. For example, in the central part of the Bragança peninsula in Brazil, increasing salinity due to SLR had caused the invasion of the black mangroves (*Avicennia germinans*) into the higher elevation plain previously dominated by freshwater grasses (*Sporobulus virginicus*) and herbs (*Sesuvium portulacastrum*), as reported by Lara *et al.* (2002). Based upon data covering the study period from 1972 to 1992, they noted that the areas dominated by grasses and herbs had shrunk from 8.8 km^2 in 1972 to 5.6 km^2 in 1997. Rising sea level would also mean higher storm surges, even if the intensity and frequency of the storm surges do not change. A calculation made by Najjar *et al.* (2000) indicates that a 100 year flood will occur 3 to 4 times more frequently by the end of 21st century in the mid-Atlantic coastal region due to GCC. The mangrove and hammock forests in the flat and low-lying Everglades are particularly vulnerable to regime shifts caused by the projected increase in hurricanes and SLR induced by climate change. Using hurricane data from 1886 to 1990, Doyle *et al.* (2003) investigated the impact of hurricanes on mangrove communities in southern Florida by means of the model HURASIM-MANGRO. Their study concluded that hurricanes are responsible for the distribution of mangrove communities across southern Florida today. The vulnerability and susceptibility of forests to hurricanes have also been well documented in the field (Gilman *et al.*, 2006). If the storms intensify over the coming century, the structure and composition of mangroves may be altered, thereby shifting the ecotone boundaries between the two competing forest types, such as mangroves verses hammocks.

5.1.4   *Shifting ecotone boundary*

The ecotone boundaries between two competing forest types, such as mangroves verses hammocks, may shift because of SLR coupled with major storm surges and inundations. For example, Baldwin and Mendelssohn (1998) studied the effects of soil salinity and sea-water inundation coupled with clipping of aboveground vegetation on two adjoining plant communities, *Spartina patens* and *Sagittaria lancifolia*. The study concluded that the vegetation might shift to a salt-tolerant or a flood-tolerant species, depending on the level of flooding and salinity at the time of disturbance. Using artificial flooding, Person and Ruess (2003) studied the effects of tidal inundation on the elevation zonation of the three dominant plant communities in a subarctic saltmarsh. The study reported an 83% decline of woody vegetation of the "slough levee" community at the higher elevation (ca 25 cm above mean tidal range) because of the flooding. Many examples of vegetation regime shift and ecotone changes reported in the literature suggest that rapid shifts in vegetation between hardwood hammock and mangrove might be caused by large salinity disturbances such as hurricanes or persistent salinity shocks induced by SLR. The feedback dynamics between salinity increase and vegetation species competition among mangroves and hardwood hammocks is the central mechanism of vegetation regime shift (Scheffer *et al.*, 2001). Understanding how the coastal ecosystems respond to salinity perturbations is important for proper coastal management to minimize losses of coastal infrastructures, to minimize losses of valued ecosystems and to maximize available management options (Gilman *et al.*, 2006). For this purpose, simulation models have been developed to determine the magnitude of a perturbation that might precipitate and sustain such a regime shift from hammocks to mangroves.

5.2   Hammock–Mangrove Regime Shift

Many species of trees such as hardwood hammocks, mangroves, pine and cypress are indigenous to the Everglades. Some of these species have become fragile due to excessive interference by humans and frequent violent disruptions caused by Southern Florida weather systems (Baldwin *et al.*, 2001). In particular, the hardwood hammock and mangrove are affected by the frequent occurrence of hurricanes that alter the salinity structures of the Everglades. Sea level rise under the scenario of global climate change will intensify the impact of salinity on the trees. Hardwood hammock trees grow in areas where the ground level is high enough to avoid seasonal flooding following summer rains. Being salinity intolerant, hammocks could not possibly thrive in coastal areas with high salinity. Tropical hardwood hammocks are rich in species diversity. The dominant species are live oak, West Indian mahogany, Florida royal palm, gumbo limbo and strangler fig. These trees do not grow very tall in the Everglades due to frequent damage and interruption of growth inflicted by occasional extreme cold weather events and destructive hurricanes. Being salinity tolerant, mangroves cover the entire edge of the western and southern coast of the Everglades where salinity is high. This extensive ecosystem provides good protection for the shore land from rising sea water, storm surges and hurricanes. They provide an excellent habitat for many species of fish, birds and other invertebrates to thrive in. The three prominent species of mangroves in Everglades are the red, black and white mangroves. Red mangroves (*Rhizophora mangle*) usually occupy the areas closest to the sea. They are distinguished by their extensive system of prop roots that protrude downward from the tree trunk. Black mangroves (*Avicennia germinans*) are found slightly inland from the coast, while the white mangrove (*Laguncularia racemosa*) is found further inland and usually in drier areas than the other two mangrove species. The sharp differences in the tolerance of mangroves and hammocks to salinity determine the competitive edge of one species over the other in the Everglades. This provides a compelling motivation to investigate the role of salinity in driving regime shifts between mangroves and hammocks under the influence of GCC and SLR in the Everglades.

5.2.1 *Role of salinity in regime shift*

In the Greater Everglades region of southern Florida, regime shift has been observed in some areas, where one plant species has been colonized by the other plant species because of salinity perturbation that favors one species over the other. Mangroves and hardwood hammocks could occupy overlapping landscapes (Sklar and van der Valk, 2003). Typically, mangrove and hardwood hammock trees are not well intermixed in one common area. Rather, one species would occupy an area that is separated by a sharp boundary (ecotone) from an area occupied predominantly by the other species. The ecotone boundary between the two vegetation types is relatively stable, except being induced by drastic salinity changes. A perturbation in salinity may induce changes in the distribution of hardwood hammocks and mangroves because of the differential sensitivity to salinity, resulting in a shift from one vegetation type to the other. For example, salt deposited by hurricanes may increase salinity in the soil to a level that is lethal or sub-lethal to the glycophytic hardwood hammock but that is readily tolerated by the halophytic mangroves. This increase in salinity favors the mangrove at the expanse of hardwood hammocks, thus facilitating the mangroves to outcompete hammocks and eventually causing a regime shift by colonizing areas previously occupied by salinity-intolerant hammocks. Besides hurricanes, gradual SLR may induce sufficient salinity alteration over a prolonged period that can cause a regime shift in the Everglades.

5.2.2 *Role of SLR in regime shift*

In the past half a century, mangroves have migrated about 3.3 km inland in the southern Everglades, largely at the expense of freshwater marshes and swamp forests (Gaiser *et al.*, 2006). This mangrove invasion is caused by SLR and upstream hydrological manipulation that favor mangroves over the freshwater species. The possibility of regime shift caused by salinity alteration due to SLR stimulates research on identifying what specific physiological mechanisms and what environmental perturbations might trigger such regime shifts. For example, GCC and its associated SLR and changes in hurricane occurrence frequency and intensity are environmental perturbations that might induce the existing vegetation to shift from one type to the other. As noted earlier, rising global temperature might lead to more severe hurricanes as warmer temperatures ultimately heat up the ocean water, which in turn would supply more thermal energy for storms to intensify as the storms swirl through the ocean (Webster *et al.*, 2005). In combination with salinity gradient, coastal plant physiology plays an important role in sustaining a positive plant feedback mechanism to maintain a sharp ecotone boundary and to promote a regime shift.

5.2.3 *Role of plant positive feedback*

A combination of positive feedback and salinity gradients is primarily responsible for the sharp ecotone between mangrove and hammock vegetation types in coastal habitats such as the Everglades in south Florida. For example, model simulations performed by SEHM (Spatially Explicit Hammocks and Mangroves) showed that positive feedback between plants and soil salinity contributed to the sharp ecotone between hammocks and mangroves (Jiang *et al.*, 2012). Coastal vegetation in south Florida typically comprises salinity-tolerant (halophytic) mangroves bordering salinity-intolerant (glycophytic) hardwood hammocks (Sklar and van der Valk, 2003), in constantly evolving competition. These two competing species are normally not interspersed. Rather, sharp transition zones or ecotones typically separate the halophytic mangroves from the glycophytic hardwood hammock species. The strength of the positive feedback is mainly determined by the rate of changes in the salinity of soil pore water. Salinity of soil pore water in turn is mainly determined by the alternating physical processes of capillary rise of saline water from the saline water table versus infiltration of freshwater precipitation. A combination of high elevation and high salinity of the water table would create a strong positive feedback that would result in a sharp ecotone or a high aggregation index. In nature, high salinity in the water table is usually associated with low elevation at the intertidal zone. This would imply a strong positive feedback would prevail at the intertidal zone, where mangroves would dominate over hammocks. At high elevations associated with low water table salinity, weak positive feedback would prevail, allowing hammocks to dominate over mangroves. This is because hammock vegetation is competitively dominant over mangroves under low-salinity conditions, while mangrove vegetation is dominant under high salinity. Therefore, it would be expected that the ecotone boundaries of these two competing vegetation types would occur where the salinity level is equally favorable to both in their competition. At the equilibrium ecotone boundary, another plant physiological mechanism, i.e., relative plant water uptake rates, would determine the outcome of the competition.

5.2.4 *Role of plant water uptake*

At the equilibrium ecotone boundary, the relative rates of plant water uptake would determine the outcome of this competition between mangroves and hammocks. Where a high level of water uptake (e.g., >2.3 mm/day) prevails, mangroves would dominate the area, with higher salinity. Where a low level of water uptake (e.g., <1.0 mm/day) prevails, hammocks would dominate much of the area, with lower salinity. For example, hardwood hammocks, being restricted to areas with low salinity below 7 ppt (Sternberg and Swart, 1987), generally occupy areas of higher elevation, where salinity is low. Mangroves, being able to tolerate a wide range of salinities, ranging from low to high, may have an edge against hardwood hammocks along the salinity gradients of higher salinity. Hence, sharp ecotone boundaries typically separate uniform patches of the two vegetation types, each of the order of a hectare. Three primary ecological factors govern the maintenance of the sharp ecotones between mangroves and hammocks and oppose changes that might occur due to disturbances. These three factors are (1) salinity gradient, (2) positive feedback between plant response and soil salinity, and (3) relative water uptake rates that create an environment favorable to one species. Each vegetation community would influence its local water environment to favor itself, thereby reinforcing the boundary between communities and sharpening the ecotones. Salinity gradient caused by tidal flux is the key factor separating vegetation communities. But, salinity gradient alone would not give rise to sharp ccotones. Positive feedback involving competition between vegetation types in response to vadose zone salinity sharpens the ecotones and maintains the ecological resilience of mangrove–hammock ecotones to oppose small disturbances. Relative water update rates then further reinforce the strength of the ecotone stability. The effects of precipitation on positive feedback are the strongest during the dry season because its low precipitation creates sharp salinity gradient (Jiang *et al.*, 2012). Hardwood hammock species are competitively superior to mangroves in low-salinity areas, while mangrove species are competitively superior to hammocks in high-salinity areas. Hence, the ecotones between patches of the two competing species tend to occur at intermediate values of salinity along a gradient, with salinity ranging between 7 and 15 ppt. Influx of more freshwater into the Everglades under CERP would maintain low salinity and hence would favor hammocks over mangrove in the Everglades. Conversely, reduced freshwater flows combined with SLR in the Everglades could induce a salinity increase in the vadose zone that might trigger a vicious cycle of regime shifts.

5.3 Vicious Cycle in Regime Shift

It has been documented that mangroves have migrated 3.3 km inland in the southern Everglades during the past half-century, at the expense of freshwater marshes and swamp forests (Gaiser *et al.*, 2006). This inland invasive shift of mangroves is a consequence of a vicious cycle caused by three factors: (1) salinity increase induced by SLR and hydrological manipulation that reduced the amount of freshwater flowing through the Everglades, (2) positive feedback between plants and soil salinity, and (3) relative water update rate of competing plants governed by soil salinity. The combination of these three factors reinforces the shift from hardwood hammocks to mangroves in an environment of rising salinity. An environment of rising salinity can be created by a combination of disturbances such as SLR, storm surge inundations or reduced precipitations. If the disturbances are sufficiently large, the mangrove–hammock equilibrium may be pushed beyond a threshold or tipping point to trigger a regime shift. Strong feedback exists between salinity of the unsaturated (vadose) zone of the soil and wetland vegetation. The feedback dynamics between vegetation and salinity of the vadose zone of the soil determine the spatial pattern of hardwood hammocks and mangroves. The boundaries between the two competing vegetation types are relatively stable, under mild environmental disturbances. The intolerance of the hardwood hammocks to high salinity helps to maintain the resilience of the mangrove–hammock coexistence. However, a severe disturbance to the salinity in the vadose zone might cause a shift from one vegetation type to the other. For example, numerical simulation by the MANHAM model (Teh *et al.*, 2008) has indicated that a regime shift from hardwood hammock to mangrove can be induced by a heavy storm surge inundation that deposits large amounts of seawater into the wetlands. For instance, a sufficiently strong storm surge inundation that completely saturates the vadose zone with 30 ppt salinity for one day might lead to a regime shift from hammocks to mangroves and cause areas previously dominated by hardwood hammocks to be taken over by mangroves. Lighter storm surge inundations that merely saturate the vadose zone with less than 7 ppt salinity would not induce vegetation shifts because of the resilience of mangrove–hammock ecotones to oppose small disturbances (Teh *et al.*, 2008). The thickness of the vadose zone determines the residence time of high salinity in the vadose zone and the rate of mangrove domination. Therefore, vadose zone salinity plays an important role in initiating and maintaining regime shift.

5.3.1   *Role of vadose zone salinity*

Salinity in the vadose zone plays a significant role in determining the distribution of mangroves and hammocks in southern Florida. The vadose zone is the aerated zone of soil above a saline water table, overlying a brackish water lens. Salinity in the vadose zone is a function of precipitation (P), evaporation (E), transpiration (T) via plant water uptake and infiltration (I) from the saline water table below (Figure 5.2). The top two curves in Figure 5.2 show the transpiration rates of mangroves (left) and hammocks (right) as a decreasing function of salinity, indicating that both transpiration rates decrease with increasing salinity. However, the transpiration of mangroves decreases gradually with increasing salinity, while that of hammocks drops sharply with each increase in salinity. Equations (5.1) and (5.2) are the transpiration rates of hammocks $R_1(S_v)$ and mangroves R_2 (S_v), respectively, as a function of salinity S_v(ppt) in the vadose zone. At salinity of 0.0 ppt, R_1 and R_2 are 2.12 and 2.60, respectively. However, at salinity of 10 ppt, R_1 and R_2 are both reduced to 0.50 and 2.43, respectively. At salinity of 15 ppt, R_1 and R_2 are 0.325 and 2.34, respectively. It can be seen that hammocks virtually stop transpiring at high salinity above 10 ppt, while mangroves continue to transpire albeit at slightly reduced rates at higher salinity. Salinity is further increased by aboveground seawater overtopping (O) and belowground salinity diffusion (D) from the highly saline sea. GCC can have significant impact on salinity in the vadose zone through changes in P and E. SLR can increase O and D and hence increase I. Plant transpiration T will be affected by a combination of GCC

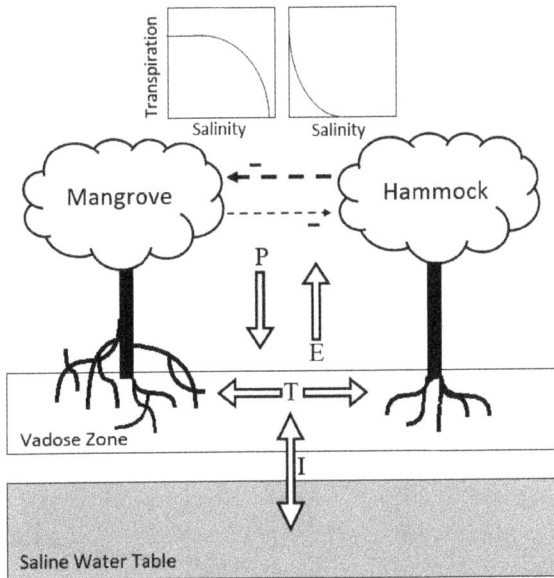

Figure 5.2: Hardwood hammock and mangrove transpiration and water uptake (T) as a function of salinity of the vadose zone (S_v).

and SLR.

$$R_1(S_v) = 2.6 - \frac{2.73 S_v}{3 + S_v} \tag{5.1}$$

$$R_2(S_v) = 3.9 \left(\frac{60 - S_v}{90 - S_v} \right) \tag{5.2}$$

Both mangrove and hammock species obtain their water from the vadose zone, the unsaturated zone between the soil surface and the top of the water table. In coastal areas, plant evapotranspiration can deplete water in the vadose zone during the dry season, which can lead to infiltration by more saline ground water, leading to elevated salinity in the vadose zone. When vadose zone salinities increase, hardwood hammock trees tend to decrease their evapotranspiration to limit the salinization of the vadose zone. But, mangroves can continue to transpire even at higher salinities, thereby increasing vadose zone salinity in the process, although transpiration may slow down at sufficiently high salinity (Passioura *et al.*, 1992). Therefore, each vegetation type promotes local salinity conditions in the vadose zone that favor itself in the competition. In the absence of large abiotic disturbances, this positive feedback between vegetation and soil salinity tends to maintain the stability of sharp ecotones or boundaries between the two types. However, in the presence of certain large abiotic disturbances, such as heavy storm surge inundations, or persistent SLR, the positive feedback could also lead to the possibility of a rapid large-scale vegetation shift. This shift can occur if the disturbance is large enough to push the vegetation–salinity feedback system over a threshold and initiate a vicious cycle to destabilize the resilience of the mangrove–hammock equilibrium. A sharp and persistent increase in salinity in an area initially dominated by hardwood hammocks will reduce hardwood tree growth and favor the invasion by mangroves. The invading mangroves in turn will further increase vadose zone salinity, which in turn promotes further invasion by mangroves. This positive feedback mechanism promotes the vicious cycle of increasing salinity and increasing mangrove invasion and could lead to a regime shift in vegetation over a large spatial scale. SLR, GCC and heavy storm surges are potential perturbations that could increase salinity beyond a threshold, leading to vegetation shift from hardwood hammocks to mangroves. Examining the mechanism, the potential outcomes and their implications for SDGs of this regime shift is the main goal in this chapter. The three reinforcing factors that may cause regime shift consist of (1) salinity gradient, (2) positive feedback between plant response and soil salinity, and (3) relative water update rates that create an environment favorable to one species. These three factors may render it possible to detect a potential regime shift early enough to facilitate adaptation measures. Of these three factors, the water update rate operating at the vadose zone appears to play the most important role in this early detection of regime shift.

5.3.2   *Early detection of regime shift*

Along vulnerable coastal regions, such as the Everglades, which are potentially subject to large-scale regime shift, coastal conservation efforts may benefit from early detection of potential regime shift well before the shift occurs. Landward invasion of salinity-tolerant (halophytic) mangroves at the expense of salinity-intolerant (glycophytic) freshwater vegetation has been the subject of fascinating literature on coastal ecosystems (e.g., Nicholls and Cazenave, 2010; Ross *et al.*, 1994, 2000, 2009; Saha *et al.*, 2011). Such landward invasion or regime shifts from freshwater vegetation to mangroves may be gradual because of gradual SLR (Doyle and Girod, 1997; Krauss *et al.*, 2011). On the contrary, large salinity pulses caused by large and intensive storm surge overwash covering extensive contiguous areas may initiate rapid regime shifts (Baldwin and Mendelssohn, 1998). Intense storm surge overwash covering extensive contiguous areas may lead to large salinity pulses in the vadose zones of freshwater vegetation habitats (Steyer *et al.*, 2010). Along low-lying southern Florida including the Everglades and other regions in south-eastern United States, a combination of SLR and frequent storm surges presents the most severe threats to coastal fresh groundwater and vegetation. GCC and SLR may induce a combination of elevated soil salinization, reduced precipitation and severe storm surges. Over time, this combination may lead to regime shift from salinity-intolerant vegetation to salinity-tolerant species, because of the intolerance of glycophytic vegetation to salinity. Coastal conservation efforts along these vulnerable coastal regions may benefit from early detection of potential regime shift due to salinity stress.

Good indicators for salinity stress are the ^{18}O and $\delta^{18}O$ values of water in the xylem of trees, which can be used as a surrogate for salinity in the rooting zone of plants. Measured and simulated $\delta^{18}O$ values in the tree xylem are useful in investigating the competitive responses of glycophytic verses halophytic trees. Coupled with simulated salinity in the vadose zone, these $\delta^{18}O$ values can be used to predict the outcomes of this vegetation competition. MANTRA-O18 simulations suggest that the impacts of long-term soil salinization on diminishing the resilience of salinity-intolerant trees can be detected up to 25 years before the glycophytic trees are threatened with regime shift to halophytic species. This early detection provides critical lead time and valuable information and insights useful for planning adaptation strategy to mitigate the adverse impacts of SLR and GCC. The fifth Intergovernmental Panel on Climate Change (IPCC) Assessment Report (AR5) has developed four global scenarios of GCC according to different representative concentration pathways (RCPs). The worst-case scenarios may have the potential impact of inducing vegetation regime shifts that could potentially lead to devastating consequences to those coastal communities that are ill equipped and short in vital resources that are essential for climate adaptation. The combination of SLR, reduced precipitation and socio-economic development can generate the greatest risks to low-lying coastal communities, with these risks increasing over time (Abadie *et al.*, 2017). New Orleans, Guangzhou and Miami are three coastal regions that are expected to be badly impacted by SLR that may cause population migrations, abandonment or social chaos. Early detection of potential regime shifts in these regions might provide useful insights to plan and implement climate change adaptation measures (SDG13). Two simulation models known as MANHAM-O18 and MANTRA-O18 have been developed for this purpose.

5.4   Model MANHAM-O18

The invasion of salinity-tolerant mangroves into freshwater marshes and into salinity-intolerant hammocks in southern Florida in the past decades (Ross *et al.*, 2000) is mainly caused by the intrusion of brackish seawater into previously freshwater areas (Saha *et al.*, 2011). This seawater intrusion is facilitated by several factors including (a) the flat landscape with low elevation and porous soil substrate (Hoffmeister, 1974), (b) sea level rise (Guha and Panday, 2012) and (c) hydrologic management along the coastal area that reduced freshwater in the area (Fourqurean and Robblee, 1999). For example, diversion of freshwater flows from the freshwater areas in the Everglades by canal systems (Sklar *et al.*, 2002) causes intrusion of saline ocean water into the Everglades. Further, seawater intrusion from the saline water table causes an increase in the salinity of the pore water in the vadose zone (Stephens, 1995), located between the soil surface above and the saline water table below. The salinity of the vadose zone has a great impact on plant community distributions, as plants vary in their salinity tolerance (Ross *et al.*, 1992). The intrusion of brackish seawater inland is accompanied by an inland spatial vegetation shift of the boundary or ecotone between salinity-tolerant halophytic vegetation and salinity-intolerant glycophytic vegetation (Ross *et al.*, 2000). To improve wetland management in the Everglades, it is beneficial to develop a reliable indicator and a robust model to provide early detection of vegetation shift. As noted earlier, a good indicator for plant salinity stress is the relative abundance

δ^{18}O of the stable isotope ^{18}O of water in the xylem of trees. This indicator δ^{18}O can be developed into a modeling tool for early detection of potential regime shift.

5.4.1 *Early detection of regime shift*

Vegetation shifts from salinity-intolerant vegetation to salinity-tolerant vegetation can be monitored by satellite imagery. However, it is hard to predict a particular area, or a particular tree, that is vulnerable to such a shift. The relative abundance δ^{18}O of the stable isotope ^{18}O is a good indicator of the potential vegetation shift for three main reasons (Zhai *et al.*, 2016). First, there is a curvilinear relationship between the δ^{18}O value and salinity in the vadose zone in mangrove-dominated areas. Second, hammock trees with higher probability of being replaced by mangroves had higher δ^{18}O values in their plant stem water. This difference could be detected two years before the trees reached a tipping point, beyond which future replacement became certain. Third, individual hammocks that were eventually replaced by mangroves with a 50% replacement probability had higher stem water δ^{18}O values three years before their replacement became certain, compared to those from the same population which were not replaced. It is promising to track the yearly δ^{18}O values of oxygen isotope ^{18}O in plant stem water in hammock forests to predict impending salinity stress leading to mortality of hammocks (Zhai *et al.*, 2016). It is worth noting that, besides oxygen isotope ^{18}O, various other isotopes have been extensively used in many hydrogeological and environmental studies.

5.4.2 *Isotopes in hydrogeological study*

Hydrogeological and environmental isotopes provide a useful tool to monitor and partition mixing processes in water supply networks and to provide substantial knowledge on urban groundwater systems (Grimmeisen *et al.*, 2017). These isotopes are frequently used to investigate the impact of climate variability (Howard, 2007; Lerner, 2002). The isotopic compositions are diagnostic tools that have been used for tracing specific natural and anthropogenic water sources, their mixtures and their transportation (Adar and Nativ, 2003; Lecuyer *et al.*, 2012). The δ^2H and δ^{18}O values of waters are extensively used to monitor and understand the hydrologic cycle in natural environments (Gat, 1996), to quantify evaporative losses (Li *et al.*, 2015) and to distinguish mixtures of source waters (Ehleringer *et al.*, 2016; Tipple *et al.*, 2017). In the headwaters of the Condamine River, a catchment in Southeast Queensland, Australia, the δ^2H and δ^{18}O values of waters have been utilized for three important purposes: (a) to determine groundwater–surface water interactions (Martinez *et al.*, 2015), (b) to identify a clear evaporation trend during baseflow conditions and (c) to confirm that creek flows are sustained by groundwater input. In Jordan, a multiple stable isotope approach has been used to investigate water network leakages, indicating excessive water loss from the networks and calling for improved water supply management (Grimmeisen *et al.*, 2017). The δ^{18}O and δ^2H ratios in the groundwater samples in Guarani Aquifer System (GAS) in São Paulo State, Brazil, have been used to identify a paleo-climatic signature emplaced under arid conditions (Elliot and Bonotto, 2017). Combined with carbon isotope systematics (δ^{13}C, ^{14}C), these signatures support the hypothesis that the groundwater represents fossil water. In the GAS, groundwater isotope and environmental tracer analysis provides the means to water resource managers to maintain the sustainability of abstraction (Chang *et al.*, 2013; Elliot *et al.*, 2014; Foster *et al.*, 2009). Literature review indicates that the oxygen isotope composition (δ^{18}O value) of plant stem water is a reliable indicator of salinity stress in plants. Hence, oxygen isotope composition δ^{18}O is used by Zhai *et al.* (2016) for monitoring and analyzing the impact of salinity on vegetation shift in southern Florida.

5.4.3   *Oxygen isotope $\delta^{18}O$*

Stable isotopes are isotopes that do not decay over time. They have been used extensively in many scientific investigations. Of the several isotopes normally present in natural compounds, the lighter ones are more abundant than the heavier and rarer ones. For instance, in water, the lighter ^{16}O represents the larger part (typically 99.757%) of the oxygen, whereas the heavier ^{18}O represents the smaller part (0.205%). Stable isotope data are generally expressed as "δ, delta-values", which represent the deviation of the rare to the abundant isotope ratio (X_{sample}) of a sample substance from that of commonly accepted standard reference materials ($X_{standard}$) as defined as Eq. (5.3):

$$\delta = \left(\frac{X_{sample}}{X_{standard}} - 1 \right) 1000 \text{ ‰} \qquad (5.3)$$

In the case of oxygen isotope, $X_{sample} = {}^{18}O/{}^{16}O$ is the ratio of the rare to common isotope of oxygen in the sample, while $X_{standard}$ is the standard reference value following the recognized "Vienna Standard Mean Ocean Water" (VSMOW), established by the International Atomic Energy Agency as $X_{standard} = 0.0020052$. The $\delta^{18}O$ value in plant stem water computed by Eq. (5.3) is a reliable indicator of salinity stress in plants.

5.4.4 $\delta^{18}O$ as indicator of salinity stress

The oxygen isotope composition ($\delta^{18}O$ value) of plant stem water is a reliable indicator of salinity stress in plants (Vendramini and Sternberg, 2007), and hence is a good indicator for potential regime shift of salinity-intolerant vegetation. The $\delta^{18}O$ value is computed using Eq. (5.3). There are two main reasons for the good correlation between plant salinity stress and the $\delta^{18}O$ value. First, in southern Florida and in other places where measurements are available (e.g., Waccamaw River, South Carolina; Zhai *et al.*, 2018), the $\delta^{18}O$ value of water is a good indicator of its salinity. Freshwater shows less ^{18}O enrichment ($\delta^{18}O \approx -3‰$) than saline seawater does ($\delta^{18}O \approx +4‰$) (Sternberg and Swart, 1987; Sternberg *et al.*, 1991). Second, $\delta^{18}O$ tends to increase in tree stem water when the soil becomes increasingly more saline, even though the salt ions are excluded from the tree roots of mangroves. Therefore, this makes $\delta^{18}O$ value of water in the xylem of trees an ideal indicator for salinity in the rooting vadose zone of plants, and hence a good indicator of potential vegetation shift from glycophytic to halophytic plants. Therefore, the $\delta^{18}O$ value is used to track salinity and vegetation shift in the simulation model known as MANHAM-O18.

5.4.5   *MANHAM-O18 simulation results*

Zhai *et al.* (2016) incorporated $\delta^{18}O$ into the previously published model of competition between hardwood hammocks and mangroves (Sternberg *et al.*, 2007) and developed the model referred hereafter as MANHAM-O18. Simulations by MANHAM-O18 showed that glycophytic hammock trees that had accumulated higher $\delta^{18}O$ values in the plant stem water had a higher probability of being replaced by halophytic mangroves. This higher $\delta^{18}O$ value could be detected in the glycophytic trees and in the MANHAM-O18 model simulations, three years before the glycophytic trees began to approach a tipping point toward the halophytic species, beyond which future replacement by halophytic trees became certain (Zhai *et al.*, 2016). These simulation results suggest that it is promising to use the $\delta^{18}O$ value of plant stem water in glycophytic hammock forests or the simulated $\delta^{18}O$ value to predict the probability and timing of hammock replacement by halophytic species. The timing of this replacement of hammocks by mangroves is a result of vegetation response to vadose zone salinity arising from a combination of SLR, storm surges and reduced precipitation. The inherent limitation in the model MANHAM-O18 can be improved to make it more robust and accurate.

5.4.6   *Limitation of MANHAM-O18*

However, the MANHAM-O18 model used by Zhai *et al.* (2016) has a limitation. It had incorporated some simplifying assumptions regarding the hydrology and solute transport dynamics in the vadose zone, rendering the simulation results less accurate. It is therefore desirable to improve on the accuracy of MANHAM-O18 by removing the simplifying assumptions regarding the hydrology and solute transport dynamics, in order to obtain better simulation results. A more realistic and accurate simulation on $\delta^{18}O$ as a monitoring agent for regime shift can be accomplished by the development of a new simulation model known as MANTRA-O18 to improve on MANHAM-O18. The original MANTRA (Teh *et al.*, 2015) is the result of dynamically coupling the spatially explicit model of two competing vegetation types (MANHAM) with a realistic hydrology and a solute dynamics model (SUTRA). Further details of the MANHAM model can be found in Teh *et al.* (2008). The new and enhanced model MANTRA-O18 is developed by the inclusion of oxygen isotope (^{18}O) transport in MANTRA. In brief, MANTRA-O18 simulates the groundwater flow and the transport of salt and oxygen isotope ^{18}O, including plant water uptake and salinity adsorption in the vadose zone. MANTRA-O18 will be employed here to study the effects of a combination of abiotic processes including SLR, reduced precipitation and storm surges on soil salinity, subject to plant δ^{18}O uptake from the vadose zone along a coastal salinity gradient. Indication of salinity stress and a potential regime shift from hammocks to mangroves can be detected by MANTRA-O18 some 25 years before the actual occurrence of the regime shift. This early warning provided by MANTRA-O18 presents precious time for planning and implementing climate change mitigation and adaptation measures (SDG13).

5.5   MANTRA-O18 Model

The original MANTRA (Teh *et al.*, 2015) is formed by combining the vegetation competition model MANHAM with the groundwater and solute transport model SUTRA. MANTRA provides an integrated model that simulates the possible effects of three abiotic processes: (a) SLR, (b) reduced precipitation and (c) storm surge overtopping on coastal groundwater and on vegetation competition between freshwater hammocks and saline-water mangroves. The feedback between the abiotic and biotic parameters is incorporated in MANTRA via the biotic plant water uptake process that is salinity dependent. GCC and SLR have the potential of altering these three abiotic process parameters and hence of changing the biotic response of plants via plant water update in the vadose zone. Facilitated by strong positive feedback between abiotic and biotic parameters, GCC and SLR can induce vegetation regime shift in the Everglades and southern Florida. The effect of SLR is gradual over decades. Reduced precipitation during the dry seasons can drastically increase vadose salinity to levels toxic to hammocks leading to death of hammocks and initiating regime shift of hammocks to mangroves. Storm surges can be a single large event or a series of smaller events. Saline water from the storm surges will infiltrate into the vadose zone to increase vadose salinity detrimental to hammocks. Further details of MANTRA are described in Appendix 2 of Teh *et al.* (2015). Finally, to provide an early indicator of vegetation shift, $\delta^{18}O$ is incorporated into MANTRA to become MANTRA-O18 by inclusion of a computational module for a mass balance of oxygen isotope composition signature $\delta^{18}O$.

5.5.1 *Oxygen isotope signature $\delta^{18}O$*

Consistent with the original MANTRA modeling framework, the enhanced MANTRA-O18 can simulate vegetation competition between glycophytic and halophytic vegetation along salinity gradients. It simulates the temporal changes in vegetation biomass subject to the groundwater salinity conditions. Therefore, MANTRA-O18 delivers the output of SUTRA, including fluid pressure and solute concentration. More importantly, it also produces as output the simulated oxygen isotope signature $\delta^{18}O$ to permit the analysis of an onset of vegetation shift. The transport of the isotopic fraction $^{18}O/^{16}O$ of the water is incorporated into MANTRA-O18 by defining an additional solute mass balance equation for ^{18}O. The simulated ^{18}O values are then expressed in delta notation $\delta^{18}O$. As in MANTRA, the fluid pressure varies spatially and temporally due to variations in salinity that change the fluid density. But, variations in the isotopic fraction, being very small, will not affect the fluid density and fluid pressure. Variations in fluid density and fluid pressure drive the flow of groundwater, which in turn drives the solute transport model. Currently, MANTRA-O18 uses spatial discretization, available in MANTRA, called "fishnet-type" meshes of regularly connected quadrilateral finite elements connected four to each internal node. Further details regarding MANTRA-O18 are available in Teh *et al.* (2019). The following subsections are devoted to a brief description of the biotic and abiotic processes central to the dynamics of MANTRA-O18. Competition between hammocks and mangroves, driven by salinity in the vadose zone, is the most important biotic process governing regime shift.

5.5.2 *Hammock–mangrove competition*

Salinity in the vadose zone plays an important role in driving the competition outcomes between hammocks and mangroves, governed by the relative sensitivity of hammocks and mangroves to salinity. In a saline groundwater environment, plant transpiration increases soil salinity by removing water from the vadose zone via plant water update. But, there is a fundamental difference between hammocks and mangroves in their water uptake in saline environment. As the salinity increases, glycophytes reduce their transpiration, while halophytes persist in their transpiration. In this manner, glycophytes attempt to resist increasing salinity by reducing transpiration in order to increase their survival in rising salinity. And, when the salinity reaches a critical level, glycophytes stop transpiring altogether in order to maintain the soil salinity to a level low enough for its survival. In short, glycophytes' growth is reduced in an environment of high salinity in order to reduce soil salinity and to increase their survival. This is a fine example of trade-off between growth and survival. On the contrary, halophytes continue to transpire even in an environment of high salinity, thereby continuing to increase the soil salinity to a higher level. In this manner, the salinity will continue to increase, ultimately to a level beyond the tolerance of glycophytes, allowing the halophytes to gain competition advantage over the glycophytes. In short, in a saline environment, halophytes have a competitive advantage over glycophytes. However, under low-salinity conditions, the halophytes are outcompeted by glycophytes, which are superior in their ability to grow by superior competition for light in low salinity as compared to halophytes. The growth of the vegetation biomass depends on the plant gross productivity, respiration and litterfall. The gross productivity is determined by the competition for light and the effect of salinity. The salinity effect on plants is expressed through their relative plant water uptake rate subject to salinity influence (Teh *et al.*, 2008).

5.5.3 *Plant water uptake*

Water uptake by plants for productivity drives the infiltration of saline water from the water table into the vadose zone and causes the increase in soil salinity. A field study by Ewe and Sternberg (2005) in southern Florida indicated that in mangrove-dominated forests, the salinity in the upper soil (vadose zone) is higher than that in the water table. This is because mangrove transpiration removes water from and excretes salt into the vadose zone, thereby further increasing vadose salinity. In MANTRA-O18, the abiotic groundwater flow model SUTRA interacts with the biotic vegetation dynamics model

MANHAM through plant water uptake. This SUTRA–MANHAM interaction drives the positive feedback between the abiotic and biotic processes. Plant water update rate, being salinity dependent, essentially reinforces the sharp ecotone boundary between hammocks and mangroves, as simulated in MANTRA and observed in southern Florida and elsewhere. This positive feedback via relative plant water uptake is responsible for vegetation regime shift from hammocks to mangroves, subject to salinity influence. The scenarios of GCC and SLR may strengthen salinity influence and may accelerate regime shift under certain conditions. But, what are the conditions that trigger and sustain regime shift? And, what is the indicator suitable for the early detection of the onset of regime shifts? The composition of the oxygen isotopes as discussed earlier allows the identification of early signals of regime shift. The transport of the oxygen isotopic fraction $^{18}O/^{16}O$ accompanying the plant water uptake is incorporated into MANTRA-O18 by defining an additional solute mass balance equation for oxygen isotopes as follows.

5.5.4   *Mass balance equation for $\delta^{18}O$*

The fluid mass source–sink term Q_p [M/s] in Eq. (5.2) accounts for addition or subtraction of fluid including pure water mass or the mass of any solute or isotopes dissolved in the fluid. Following the notation used in the documentation of SUTRA, [M] is fluid mass units, [M_s] is solute mass units, [L] is length units and [s] is seconds. This Q can be used to characterize the uptake of water by the plants ($Q = Q_p$). This Q_p term in the plant mass balance equation is used to represent the addition (source, $+$) or extraction (sink, $-$) from the mass balance system. As a function of salinity, the total water uptake $R = f(C)$ [L/s] by plants is determined by the salinity concentration $C[M_s/M_f]$ calculated by SUTRA. Here, L is the dimension of vertical distance or depth. The salinity, C, is derived from SUTRA solution of the solute mass balance equation, including fluid flow, diffusion and dispersion. We assume a closed canopy so that evaporation is negligible. The fluid mass per unit time (Q_p) required by the plants in a certain horizontal cell for transpiration can be estimated by

$$Q_p = R \cdot A_s \cdot \rho \text{ [M/s]}. \tag{5.4}$$

Here, A_s represents the horizontal surface along the depth dimension $[L^2]$ and ρ is the fluid density of freshwater $[M/L^3]$. For the two-dimensional cross-sectional model, the width of each cell is assumed to be 1.0 m. Thus, A_s depends on the length or horizontal element size. Further details of the changes made in SUTRA to link it with MANTRA (incorporating ^{18}O) are described in the electronic supplementary material (Teh *et al.*, 2018). The salinity C computed in SUTRA is a function of three factors: plant water uptake, saline tidal influx and freshwater precipitation percolating into the soil.

5.5.5 *Role of freshwater precipitation*

Freshwater precipitation is one of the main drivers in determining soil salinity and vegetation shift in southern Florida and elsewhere. For example, limited freshwater precipitation in the Moreton Bay of Australia, leading to high soil salinity, is identified as the driving factor promoting the landward mangrove expansion (Eslami-Andargoli *et al.*, 2009). Changes in regional rainfall may therefore have a great influence on mangrove and hammock distribution (Snedaker, 1995). By seasonally averaging the daily precipitation data generated based on monthly mean and standard deviation using a uniformly distributed random variable, the seasonally averaged precipitations in south Florida during the wet and dry seasons are 0.0038 m/day and 0.0017 m/day, respectively, as shown in Figure 5.3. During the wet season, the vadose zone is recharged by fresh rainwater and not by saline groundwater, resulting in low salinity as fresh rainwater can dilute soil salinity. Soil salinity during the wet season should therefore be less than that during the dry season. This observation is supported by the field study of Ewe and Sternberg (2005) in southern Florida that indicated that salinity at 50 cm belowground is higher during the dry season and lower during the end of the wet season. This observation is also consistent with the simulation results (Zhai *et al.*, 2016) indicating that the maximum salinity of the water table occurred in April, near the end of the dry season, and that the minimum salinity occurred in October, near the end of the wet season. The timings of the maximum salinity during the dry seasons and the minimum salinity during the wet seasons also matched salinity measurements of a coastal hardwood hammock water table in the Everglades (Saha *et al.*, 2014). The higher soil salinity during the dry season may be

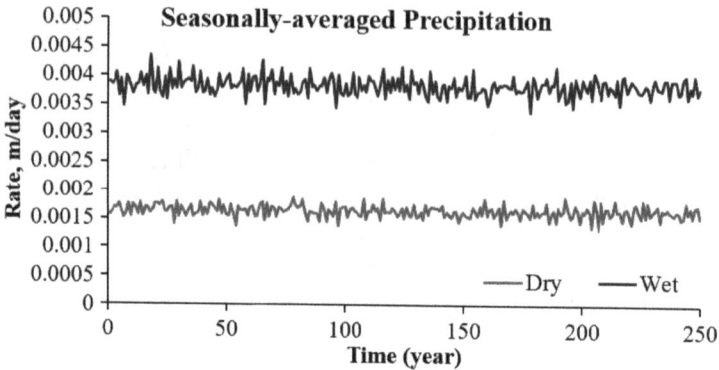

Figure 5.3:   Seasonally averaged precipitation during wet (top) and dry (bottom) seasons. Daily precipitation data were generated based on monthly mean and standard deviation using a uniformly distributed random variable.

sufficiently high in some areas to exclude hammocks because of hammocks' fatal sensitivity to high salinity. SLR, proximity to seawater and low precipitation (during the dry seasons), as observed in the Everglades and in southern Florida, may present conditions suitable to drive vadose salinity to high levels lethal to hammocks. Hence, in an area with such conditions, a moderately high initial mangrove concentration might be sufficient to trigger and sustain a regime shift. This may be the reason why a random mixture of hammocks and mangroves is rare in southern Florida, as this state is unstable under the influence of strong abiotic–biotic feedback, in an environment with low precipitation, low elevations and high tidal influx.

5.5.6   *Role of tidal influx*

Tidal influx and flooding, precipitation, evaporation and plant transpiration drive water movement in the vadose zone. Tidal flooding carries saline seawater with relatively high $\delta^{18}O$ values to the vadose zone, thereby increasing both salinity and $\delta^{18}O$ values. Both evaporation and plant transpiration can cause a water deficit in the vadose zone and drive infiltration of saline water upward into the vadose zone from the water table below. Tidal influx water with a high $\delta^{18}O$ value of $+4\%o$ (Sternberg and Swart, 1987) may move water with high $\delta^{18}O$ values and high salinity into the water table and then up into the vadose zone. At the start of simulation, the vadose zone pore water is given an initial $\delta^{18}O$ value of negative 3‰, but this value will change as a function of precipitation, infiltration, transpiration and tidal input. The positive linear relationship between salinity and $\delta^{18}O$ in the water table, without transpiration, is a consequence of mixing of freshwater having low $\delta^{18}O$ values with saline seawater having high $\delta^{18}O$ values. In a tidal creek system of Australia, Wei *et al.* (2012) observed a familiar trend, in which the water table had higher salinities and higher $\delta^{18}O$ values in the creek site, due to tidal influx. Further inland, both salinity and $\delta^{18}O$ values are lower due to reduced tidal flux. Increased tidal flooding has been noted to have played a strong role in inland migration of salt marsh at the Elkhorn Slough in California (Wasson *et al.*, 2013). The influence of tidal flooding may be enhanced in the future under GCC, as tidal flooding frequency and strength can be increased by SLR (Ezer and Atkinson, 2014). Hence, SLR might contribute sufficient tidal flooding to promote inland invasion of mangroves in southern Florida, including the Everglades. The CERP proposes to return freshwater flow to the Everglades to historic conditions (Perry, 2004). The implementation of this CERP plan will increase freshwater flows, particularly during the dry seasons, to mitigate the impact of tidal forcing and to reduce salinity of the coastal habitats in southern Florida (Herbert *et al.*, 2011). This anticipated pathway may reverse the current trend of soil salinization and inland invasion of mangroves into hammock habitats via integrated effects of strong feedback between abiotic and biotic process.

5.5.7 *Abiotic and biotic integration*

MANTRA will be used to simulate the possible combined effects of strong feedback between biotic and abiotic processes on salinity in coastal groundwater and vegetation competition outcomes between freshwater hammocks and saline-water mangroves. The three main abiotic processes driving soil salinity and hence vegetation shift are (1) precipitation, (2) tidal seawater intrusion and (3) storm surge overtopping. The main biotic feedback is the plant relative transpiration that increases salinity in the vadose zone. Hammocks have the tendency to slow or even stop transpiration at moderate salinity to enhance their survival at the expense of growth. On the contrary, mangroves have the unique adaptation to continue to transpire in saline environment and therefore to continue to increase vadose salinity. In this manner, mangroves would drive regime shift and invade hammock habitats by creating high salinity. The adaptation of mangroves to salinity such as salt excretion in mangrove roots that occurs during water uptake can result in salt accumulation in the vadose zone (Tomlinson, 1994). This in turn can lead to higher salinity in the vadose zone than that in the water table, consistent with observation noted by Ewe and Sternberg (2005) in southern Florida. This observation is also consistent with the simulation results of Zhai *et al.* (2016) that indicate that in mangrove-dominated areas, the salinity in the vadose zone is higher than that in the water table. The abiotic and biotic process integration that may lead to regime shift will be simulated by MANTRA-O18 to provide early detection of potential regime shift.

5.6   MANTRA-O18 Simulation Results

5.6.1   *Early detection of regime shift*

Regime shift from glycophytic hammocks to halophytic mangroves is induced by persistently increasing soil salinity due to higher tides, lower precipitation, more storm surges or SLR, a process that might take several decades. It would be beneficial to have an early indicator to signal an impeding regime shift before its actual occurrence. The tipping point of regime shift is indicated by two parameters: (1) the density of mangroves reaching 50% and (2) the increase in $\delta^{18}O$ value (henceforth annually averaged) over the previous year reaching 0.1‰. Both parameters are deemed as indicators of impeding regime shift, caused by persistent and prolonged salinity stress in the plants due to high salinity in the soil. The salinity, $\delta^{18}O$ values and the respective density of the hammocks and mangroves simulated by the model MANTRA-O18 are monitored in every spatial cell. Simulation results indicate that a strong linear relationship exists between the time to regime shift tipping point (from hammocks to mangroves) and the time when the annual $\delta^{18}O$ increase in the plant xylem water exceeds 0.1‰, with a high value of $R^2 = 0.976$. A $\delta^{18}O$ annual increase exceeding 0.1‰ can be observed on average 26 years before the onset of regime shifts, with a standard deviation of 5 years

(Teh *et al.*, 2019). Further research is warranted to verify if the $\delta^{18}O$ annual increase rate can serve as a reliable early indicator of coastal vegetation shift caused by persistent and prolonged salinity stress in the plants and in the soil. An irreversible regime shift may be identified by a combination of three parameters: (a) consistent yearly $\delta^{18}O$ increase exceeding 0.1‰, (b) consistent yearly decrease in the biomass of freshwater plant and (c) consistent increase in biomass of halophytic plants. In short, the increase in oxygen isotope signature $\delta^{18}O$ in the stem water of coastal vegetation and in the vadose zone may serve as an early indicator of regime shift from glycophytic to halophytic species. The intricate relationship between salinity and $\delta^{18}O$ in the vadose zone deserves further attention.

5.6.2 *Relationship between salinity and $\delta^{18}O$*

Plant transpiration withdraws saline water from the vadose zone, removing both salt and $\delta^{18}O$ from the soil. However, mangroves can exclude salt, thereby can cause the accumulation of salt and increase salinity in the vadose zone. To further analyze the effect of mangrove plant transpiration on the relationship between salinity and $\delta^{18}O$ in the unsaturated vadose zone, normalized values are employed. The $\delta^{18}O$ values are normalized by dividing the $\delta^{18}O$ values by the range of $\delta^{18}O$ values (i.e., the difference between the maximum and minimum $\delta^{18}O$ values $= 4-(-3) = 7$‰). Similarly, salinity is normalized by dividing salinity by the range of salinities (i.e., the difference between the maximum and minimum salinities $= 0.030$–0.000 kg/kg $= 0.030$ kg/kg). The relationship between normalized salinity and normalized $\delta^{18}O$ in the unsaturated vadose zone is shown in Figure 5.4 for two scenarios: (a) with salt accumulation effect due to mangrove transpiration and (b) without salt accumulation effect as in the case of hammock transpiration. In a simulation without plant transpiration effect, the relationship between the normalized salinity and normalized $\delta^{18}O$ is linear, as shown in the full line in Figure 5.4, as expected. However, when plant water uptake with salt accumulation is simulated, this relationship becomes curvilinear. As the mangrove plants withdraw water, the salinity increases as the salt accumulates in the unsaturated zone. However, because $\delta^{18}O$ is withdrawn by the plants along with the water, the $\delta^{18}O$ values will not accumulate in the unsaturated zone. Hence, in mangrove-dominated

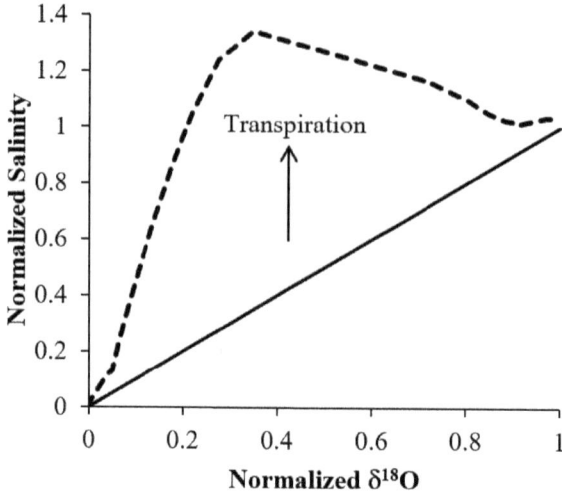

Figure 5.4:   Relationship between normalized salinity and normalized $\delta^{18}O$ with transpiration (dashed line) and without transpiration (solid line).

areas, normalized salinity increases in a curvilinear manner relative to normalized $\delta^{18}O$ increase, as shown in the dotted line in Figure 5.4. In a similar manner, the response of salinity after a storm surge in a mangrove-dominated cell differs significantly from that in a hammock-dominated cell because mangrove transpiration adds salt to the cell in the vadose zone, while hammocks do not. However, the response of $\delta^{18}O$ after a storm surge in a mangrove-dominated area is similar to that in a hammock-dominated area. MANTRA-O18 simulation results to be presented in the section that follows would help to cement this basic understanding.

5.6.3   *Salinity and $\delta^{18}O$ after a storm surge*

After a storm surge, the salinity in the vadose zone located in a mangrove-dominated area differs significantly from that in a hammock-dominated area. To assess the impact of storm surge on salinity and $\delta^{18}O$, we begin the storm surge overwash event with inundating all cells with seawater to a depth of 1 m, with salinity of 0.030 kg kg^{-1} and $\delta^{18}O$ value of $+4\%_0$ to mimic seawater. Simulation results for the next 200 days are plotted in Figure 5.5 for salinity (left) and $\delta^{18}O$ value (right). In Figure 5.5, at Day 0 following the

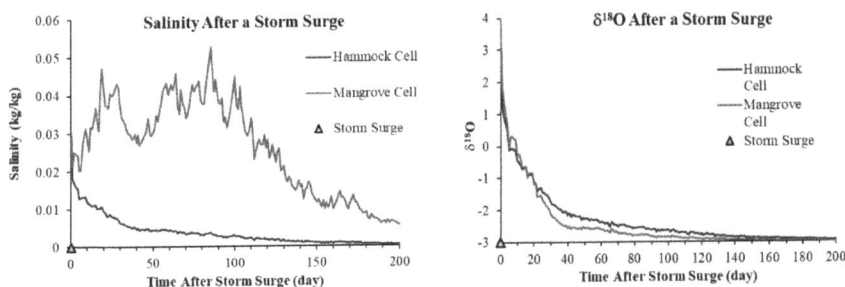

Figure 5.5: Simulated salinity (left) and $\delta^{18}O$ (right) in hammock and mangrove cells after a storm.

storm surge, all the mangrove and hammock cells are saturated with seawater at 0.030 kg kg^{-1} salinity and a $\delta^{18}O$ value of $+$ 4‰. The subsequent evolution of salinity in the mangrove cell is vastly different from that in the hammocks (Figure 5.5 left). In a mangrove cell, salinity continues to increase, peaking in value around Day 80. During mangrove transpiration, salt is excluded from the root zone into the vadose zone, thereby increasing the salinity there. Subsequently, the salinity begins to decline gradually because of increasing contribution from freshwater dilution and flushing from freshwater precipitation imposed after the storm surge. By Day 200, the salinity in the mangrove cell is reduced by precipitation to 0.008 kg kg^{-1}. On the contrary, in the hammock cell, salinity declines exponentially over time because of freshwater dilution and flushing from precipitation. By Day 200, salinity in the hammock cell has declined to the pre-storm level. This simulation serves to illustrate the feedback mechanism that helps to promote mangroves at the expense of hammocks in an environment of elevated salinity.

Figure 5.5 (right) shows the $\delta^{18}O$ values in a hammock cell and in a mangrove cell. At Day 0, both mangrove and hammock cells have the $\delta^{18}O$ value of seawater of $+$4‰. Subsequently, the $\delta^{18}O$ values in both the mangrove cell and in the hammocks cell decline exponentially, following almost a similar pattern, due to dilution and flushing by freshwater precipitation. Dilution and flushing by freshwater precipitation play a dominant role in causing the exponential decline in the $\delta^{18}O$ values in both types of mangrove and hammock cells. Transpiration appears to play a relatively insignificant role.

Eventually, the $\delta^{18}O$ value in both types of cells returns back to $-3‰$, the value for freshwater.

5.7   Impact of Temperature, Toxicity and Hydrology

The Everglades is the habitat for more than 11,000 species of plants, fish and marine mammals. A complex ecosystem with a wide variety of species of plants, fish, birds, pythons and alligators, the Everglades has undergone many changes in the past century. These changes and alterations at various scales have threatened its species abundance and biodiversity. Managing such a complex ecosystem with its biodiversity requires an integrated approach that involves many different trophic levels of organisms. Subjected to many environmental and ecological stresses over the last century, the Greater Everglades ecosystems have been the focus of major restoration projects (GEER) in the US. Fish communities play an important role in this complex ecosystem, functioning as an intermediary between the lower producers and the higher consumers. Hence, fish community ecology has been a core area of research in many monitoring and modeling projects in GEER, as a tool in ecological risk assessment and mitigation. Freshwater fish and invertebrates are important primary components in the Everglades ecosystem. They function at several trophic levels in the wetlands, from being primary consumers of plant materials and detritus, such as crayfish, to being among the top aquatic predators, such as the largemouth bass. Factors that influence fish and invertebrate numbers, biomass and composition will ultimately affect the entire ecosystem in the Everglades. Model simulations show that temperature-dependent starvation mortality exerts significant influences on fish population densities. Other important temperature-dependent functions include food resources and water-level fluctuations (Al-Rabai'ah *et al.*, 2002). Further, fish communities in the Everglades are subject to acute and chronic stresses from toxic chemicals such as mercury and PCB. The accumulation of mercury in fish is a problem in the Everglades (Atkeson and Axelrad, 2003). The Florida Department of Health has recommended limited consumption of several species of sport fish, in particular largemouth bass, in which mercury concentration had averaged 2.5 mg/kg, which exceeds all health-based standards (FDEP, 2003).

The long-term impact of these toxicants on the fish community and the entire ecosystem is a major concern in GEER. Hydrology is a major driving force in the Everglades ecosystems. Precipitation and evapotranspiration are the major factors affecting the timing and extent of water-level fluctuations (Duever *et al.*, 1994). The impact of temperature, toxicity and hydrology on fish population is a major concern in GEER and will be the focus of the next three subsections.

5.7.1 *Impact of temperature on fish community*

Temperature variation between 17°C and 32°C recorded in the Everglades has a significant impact on seasonal fish population dynamics and on related wetland ecosystems such as the wading birds that feed on fish. Hence, climate warming might have a significant impact on fish in the Everglades and on other ecosystems that are dependent on fish such as wading birds and alligators. Temperatures affect fish biological activities such as growth and size, which in turn influence size-dependent fecundity (Anastàcio *et al.*, 1999), mortality and survival. This temperature variation has more direct impact on small fish with short life cycles. To simulate the impact of temperature on fish populations in the Everglades, the small Everglade fish populations in the marsh areas are aggregated into two functional groups (FGs): (a) FG1, the killifish (Cyprinodontiods), and (b) FG2, the livebearer (Poecilids). Each FG is subdivided into 5 day age classes during their life cycles. To reflect the impact of temperature on fish and their food prey, temperature-dependent fish biological processes are formulated, including fish growth rate, fecundity, consumption rates and mortality as well as prey availability (Al-Rabai'ah *et al.*, 2002). Fish consumption rate is weight and temperature dependent. The food consumption rate function has the skew bell shape that increases slowly until it reaches a maximum at the optimum temperatures of consumption, then decreases rapidly as temperature increases above the optimum (McDermot and Rose, 2000). Fish mortality sub-modules are formulated in the model, including starvation mortality which is proportional to the ratio of prey needed to prey available. The model results show that temperature-dependent starvation mortality is an important factor that influences fish population densities. High fish population densities tend to occur around temperature ranges at which food consumption requirement

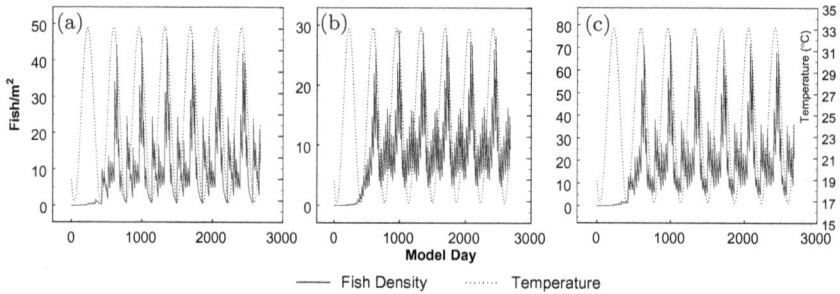

Figure 5.6: Fish density in marsh areas for (a) FG 1, (b) FG 2 and (c) Total small fish, with temperature effects.

is moderate (Al-Rabai'ah *et al.*, 2002). Sensitivity analysis involving food resources, water levels and variations in temperature-dependent items is conducted to determine their relative importance compared to temperature dependence. Sensitivity analysis of the model simulation shows that increased fish starvation mortality, induced by temperature via decreased food availability, is an important limiting factor for fish density. This result is consistent with similar research done on the Everglades (Loftus *et al.*, 1990).

Figures 5.6(a) and 5.6(b) show temperature-dependent densities of functional groups FG1 and FG2 of age larger than 30 days in the marsh areas of the Everglades. Very young fish of less than 30 days are excluded to avoid sharp and transient changes in density due to large numbers of births. These sharp increases in baby fish and eggs are moderated by predation over a period of 30 days. Figure 5.6(c) show the total number of fish in the marsh area, indicating a mean population density ranging from 17 to 20 fish m^{-2} (Al-Rabai'ah *et al.*, 2002). This overall density is consistent with the observed data on long-term dynamics of the Everglades small fish assemblage (Loftus and Eklund, 1994), which reported a mean value of 15.5 to 17.1 fish m^{-2}. To further evaluate the effect of temperature on fish populations in the model, all temperature terms were removed from the model, keeping other parameters with the same values. The results of this scenario analysis as plotted in Figure 5.7 reaffirm the important role of temperature. Significant difference exists between the two outputs of Figure 5.6 (with temperature) and Figure 5.7 (without temperature), especially between the peak density values. The implications of

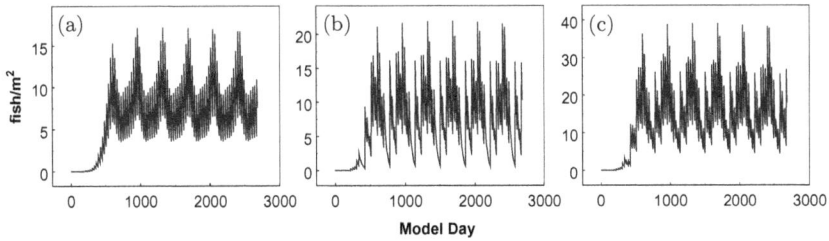

Figure 5.7: Fish density in marsh areas for (a) FG 1, (b) FG 2 and (c) Total small fish, after eliminating all temperature parameters from the model.

climate change on fish populations in the Everglades deserve further research.

5.7.2 *Impact of PCB on fish community*

Fish community in the Everglades is regularly exposed to a number of toxic chemicals, such as mercury and PCBs, whose harmful effects, both acute and chronic, are of keen interest to GEER researchers, including the authors. PCBs have been widely distributed in the south Florida Everglades since the early 70s, probably as a result of volatilization and transport by aerosols and fallout with dust or rain (Pfeuffer, 1991). PCBs may increase fish mortality (Boese *et al.*, 1995), and disrupt sexual development in fish females (Matta *et al.*, 1998), leading to negative effect on fish reproduction (Orn *et al.*, 1998). The fish survival fraction or probability $S(t, c)$ at exposure time t and exposure concentration c is given in Eq. (5.5). The slope parameters μ and β in Eq. (5.5) were calibrated so that toxicity values at different exposure durations agree with PCB toxicity data available in the literature (Johnson and Finley, 1980; Mayer and Ellersieck, 1986; AQUIRE database of EPA). The LC_{50} and LT_{50} are, respectively, the median lethal concentration and lethal exposure time needed to cause 50% mortality of the homogeneous fish population. Fish in the Everglades is aggregated into five functional groups (FGs). The survival probability for each FG depends, as indicated in Eq. (5.5), on LT_{50} and LC_{50} of the fish FG. For example, at age 200 days and exposure concentration c = 10.0 μg/L, the survival probability is given by S (200, 10) = 0.1 for FG1. Figure 5.8 (left) shows fish survival fraction of FG1 as functions of exposure duration t for eight PCB concentrations of 20, 10, 5, 3, 2, 1, 0.5 and

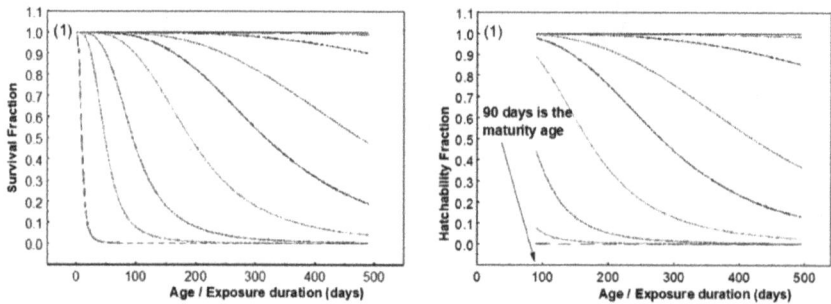

Figure 5.8:   Fish Survival Fraction and Hatchability Fraction of FG1 as functions of exposure duration at 8 PCB concentrations of 20, 10, 5, 3, 2, 1, 0.5 and 0.1 μg/L (from bottom left to top right).

0.1 μg/L (from bottom left to top right). Figure 5.8 (right) shows fish hatchability of FG1 as functions of exposure duration t for eight PCB concentrations of 20, 10, 5, 3, 2, 1, 0.5 and 0.1 μg/L (from bottom left to top right). Note that the hatchability of eggs or larvae subject to toxicity is also given by the same type of effect function. At high values of c, say $c = 20$ μg/L, both survival fraction and hatchability fraction decay sharply to 0.0 over time (both bottom curves in Figure 5.8).

$$S(t, c) = \frac{\left(\dfrac{LC_{50}}{c}\right)^{1/\alpha} \cdot \left(\dfrac{LT_{50}}{t}\right)^{1/\beta}}{1 + \left(\dfrac{LC_{50}}{c}\right)^{1/\alpha} \cdot \left(\dfrac{LT_{50}}{t}\right)^{1/\beta}} \tag{5.5}$$

A time–concentration–effect model has been developed to assess the long-term effect of PCBs on multiple generations of fish community in the Florida Everglades, subject to combined effects of mortality and hatchability impairment (Al-Rabai'ah *et al.*, 2005). Fish is aggregated into five functional groups (FGs), and partitioned into small and large fish classes. The model predicts the population density time series for each of the five FGs, based upon two factors, namely, exposure concentration c and exposure duration t. Simulation is performed for fish communities in the Everglades freshwater marsh areas, subject to exposure duration and exposure concentration, according to Eq. (5.5). Each successive increase in the toxicant levels would cause a successive decrease in fish population

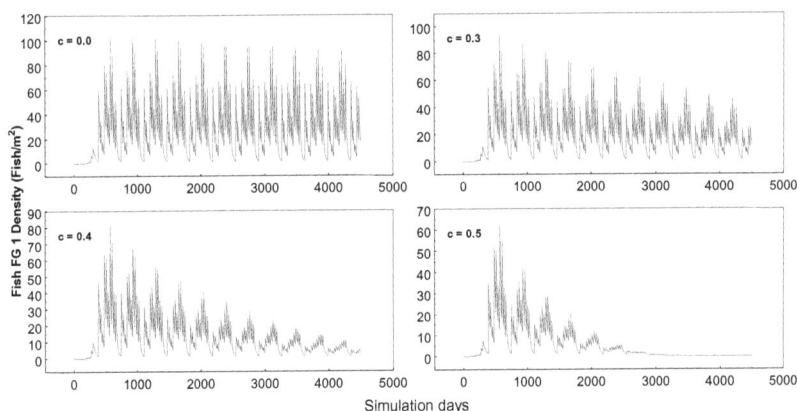

Figure 5.9: Fish FG1 density time series exposed to PCB concentrations of 0.0, 0.3, 0.4 and 0.5 μg/L.

density (Figure 5.9). For FG1, fish population density declines 50% to 50 fish/m^2, within 115 months at c = 0.3 μg/L (top right curve of Figure 5.9). Fish functional group FG1 populations would ultimately crash at toxicant concentrations of 0.4 and 0.5 μg/L (Figure 5.9, bottom curves). The results further show that a PCB concentration as low as 0.1 μg/L appears to have an adverse impact on fish population. The U.S. EPA has recommended the water quality criteria for PCB to not exceed a concentration of 0.014 μg/L in order to protect freshwater organisms from chronic effects and the standard of 1.0 μg/L from acute effects (USEPA, 1999). Figure 5.9 shows fish FG1 density time series exposed to four PCB concentrations of 0.0, 0.3, 0.4 and 0.5 μg/L. The model results indicate that a concentration of 0.3 μg/L could have severe long-term effects on FG1 and indeed for all five FGs.

5.7.3 *Impact of water levels on fish community*

Precipitations in southern Florida exhibit a bimodal pattern of wet summer and dry winter seasons, with little change in the total annual rainfall over the last 100 years. Precipitations are correlated with thunderstorms and tropical cyclones, with over 60% of them occurring in the summer months between June and September. This has resulted in late summer high water depths, which fall gradually in the winter, with a rapid decline in spring due to high evapotranspiration

(Duever *et al.*, 1994). These seasonal fluctuations in water depths have a profound impact on fish populations and hence on the ecosystems that feed on fish. The Everglades ecosystem is a complex subtropical marshland, a habitat harboring several endangered species and large rookeries of wading birds. The underlying food web consisting of small fish and invertebrates is the foundation of this biodiversity. The ecosystem shows strong hydrological seasonality with distinct dry and wet seasons that dramatically alter the distribution of flooded and dry landscapes. To persist in such a seasonally fluctuating ecosystem, biota must adapt to these annual cyclical conditions. A substantial decline of the traditional bird communities in the Everglades has occurred over the past several decades (Ogden, 1994). To comprehend the decline in the wading bird populations, it is essential to understand the dynamics of the fish and invertebrate populations subject to piscivorous predation (Figure 5.10) in response to seasonal changes in hydrology and water depth. The fish and invertebrate community in the Florida Everglades marshes is subject to varying seasonal fluctuations in water levels in the marsh areas.

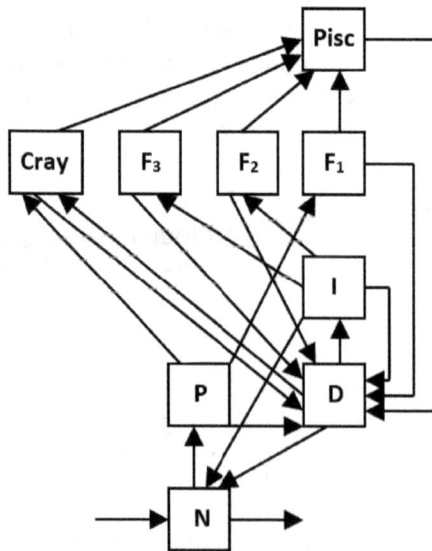

Figure 5.10: Organization of trophic levels of the basic food web structure of the Southern Florida fish community model.
Notes: Pisc = piscivorous fish, Cray = crayfish, F1 = intermediate fish species, F2 = good disperser fish species, F3 = good resource exploiter fish species, I = invertebrates, P = periphyton, D = detritus and N = nutrients.

We investigated the effect of seasonal fluctuations in water levels on the fish communities in the marsh areas. For this purpose, we developed a spatially explicit landscape simulation model of the lower trophic level food web in the Everglades marshes that includes the small fish and invertebrate community (Jopp *et al.*, 2009a). We focused on an analysis of a baseline scenario with annual water level amplitude of 0.6 m. In an accompanying paper, we extended the analysis to cover several representing scenarios in which the water-level fluctuation amplitudes vary between 0.2 m and 0.8 m (Jopp *et al.*, 2009b). During the wet season, following the flooding water, the fish community spreads across the flooding landscape and grows in population size. During the drying season, the retreating fish are crowded into ever smaller areas of permanent water. In the shallow and crowded areas, they are hunted down by higher trophic levels, such as wading birds, piscivorous fish, reptiles and mammals. With the annual reflooding of extensive wetland areas, the diminished fish species disperse and grow by exploiting the different habitat types depending on their individual traits. Will they grow adequately during the wet seasons to compensate for the high loss during the dry seasons? What are the combined impacts of hydrological fluctuations and toxic chemicals such as mercury and PCB on the viability of fish community and the wading birds feeding in the marsh areas? To what extent should water levels be restored under CERP to facilitate fish recovery to adequate levels? What would be the impact of climate change on the structures of the community? Opportunistic fish species will be the first to disperse into and exploit reflooded areas. Gleaner species, which are better at exploiting resources at low trophic levels, are more successful in dominating permanently flooded wetland areas. This combination of elevation heterogeneity and water-level fluctuations enabled a robust community of many fish and invertebrate species to coexist while feeding on the same resource. Any dramatic disruption in this delicate balance would spell disaster to the entire ecosystem, particularly the high-level wading birds that depend crucially on the marsh fish. This is a major concern in GEER.

5.7.3.1 *Top-down cascading effects*

Some top-down cascading effects are observed in the simulations (Jopp *et al.*, 2009a,b), where high density of piscivores is associated with reduced fish and crayfish density, and with the associated high density of invertebrates. The invertebrate's peak density observed in the high elevation and shallower waters at the beginning of the wet reason is the result of being free from fish predation. There is a decrease in invertebrate population in the lower elevation deeper water following the start of the dry seasons. This is because of an increase in fish predation that follows the retreating water into the deeper water. The fish themselves are also freed from piscivore top-down predation. The fish and crayfish show distinctive pulses at the edge of the drying fronts as water levels decline. These pulses are the primary food sources for the wading birds and higher trophic levels that feed on the trapped fish. This sharp gradient in species composition during the dry season eventually will smooth out during the wet season with rising water levels inducing the movement of all living species up the elevation gradient. The elevation heterogeneity is one such process that leads to a patchy distribution of permanent and temporary retention of water that provides refuge for fish. These refuge regions are important for population recovery and biomass production in newly flooded regions (DeAngelis *et al.*, 2005). This fish model should incorporate a sub-model for mercury methylation in areas that undergo periodic flooding and drying as well as including a mercury bioaccumulation sub-model for the food web.

References

Abadie, L.M., Galarraga, I. and Sainz de Murieta, E. (2017). Understanding risks in the light of uncertainty: Low-probability, high-impact coastal events in cities. *Environmental Research Letters*, 12, 014017. doi: 10.1088/1748e9326/aa5254.

Adar, E. and Nativ, R. (2003). Isotopes as tracers in a contaminated fractured chalk aquitard. *Journal of Contaminant Hydrology*, 65, 19–39. doi: 10.1016/S0022-1694(98)00275-3.

Alexander, T.J. and Crook, A.G. (1974). Recent vegetational changes in southern Florida. In: P.J. Gleason (ed.) Environments of Southern Florida: Present and Past. Memoir 2: Miami Geological Society, Miami, Florida, pp. 61–72.

Al-Rabai'ah, H.A., Koh, H.L., DeAngelis, D. and Lee, H.L. (2002). Modeling fish community dynamics in the Florida everglades: Role of temperature variation. *Water Science and Technology*, 46(9), 71–78.

Al-Rabai'ah, H., Koh, H.L., DeAngelis, D. and Lee, H.L. (2005). Modeling the long-term effect of PCBs on Everglades fish communities. *Wetlands Ecology and Management*, 13, 73–81.

Anastàcio, P.M., Nielsen, S.N. and Marques, J.C. (1999). CRISP (crayfish and rice integrated system of production): 2. Modelling crayfish (Procambarus clarkii) population dynamics. *Ecological Modelling*, 123, 5–16.

Atkeson, T. and Axelrad, D. (2003). Mercury monitoring, research and environmental assessment 2003. Everglades consolidated report, South Florida Water Management District and Florida Department of Environmental Protection, January 2003.

Baldwin, A.H. and Mendelssohn, I.A. (1998). Effects of salinity and water level on coastal marshes; an experimental test of disturbance as a catalyst for vegetation change. *Aquatic Botany*, 61, 255–268.

Baldwin, A., Egnotovich, M., Ford, M. and Platt, W. (2001). Regeneration in fringe mangrove forests damaged by Hurricane Andrew. *Plant Ecology*, 157(2), 149–162.

Boese, B.L., Winsor, M., Lee II, H., Echols, S., Pelletier, J. and Randall, R. (1995). PCB congeners and hexachlororbenzene biota-sediment accumulation factors for Macoma nasuta exposed to sediments with different total organic carbon contents. *Environmental Toxicology and Chemistry*, 14, 303–310.

Chang, H.K., Aravena, R., Gastmans, D., Hirata, R., Manzano, M., Vives, L., Rodrigues, L., Aggarwal, P.K. and Araguás, L. (2013). Role of isotopes in the development of a general hydrogeological conceptual model of the Guarani aquifer system (GAS). In: 27 March–1 April 2011, IAEA Vienna, STI/PUB/1580 Isotopes in Hydrology. *Marine Ecosystems and Climate Change Studies: Proceedings of the International Symposium held in Monaco*, 2, 281–290.

DeAngelis, D.L., Trexler, J.C. and Loftus, W.F. (2005). Life history trade-offs and community dynamics of small fishes in a seasonally pulsed wetland. *Canadian Journal of Fisheries and Aquatic Sciences*, 62, 781–790.

Douglas, M.S. (1988). *The Everglades: River of Grass* (revised edition). Pineapple Press, Sarasota, Florida, USA.

Doyle, T.W. and Girod, G.F. (1997). The frequency and intensity of Atlantic hurricanes and their influence on the structure of South Florida Mangrove communities. In: Diaz, H.F. and Pulwarty, R.S. (eds.) *Hurricane, Climate and Socioeconomic Impact*. Springer, Berlin, Heidelberg, pp. 109–120.

Doyle, T.W., Girod, G.F. and Books, M.A. (2003). Chapter 12: Modeling mangrove forest mitigation along the southwest coast of Florida under climate change. In: Integrated Assessment of the Climate Change Impacts on the Gulf Coast. Ning, Z.H., Turner, R.E., Doyle, T. and Abdollahi, K.K. (lead authors). Region. Gulf Coast Climate Change Assessment Council (GCRCC) and Louisiana State University (LSU) Graphic Services, pp. 211–221.

Duever, M.J., Meeder, J.F., Meeder, L.C. and McCollom, J.M. (1994). The climate of South Florida and its role in shaping the Everglades ecosystem. In: Everglades: the System and its Restoration (ed. by S.M. Davis and J.C. Ogden). St. Lucie Press, Delray Beach, Florida, USA.

Ehleringer, J.R., Barnette, J.E., Jameel, Y., Tipple, B.J. and Bowen, G.J. (2016). Urban water—a new frontier in isotope hydrology. *Isotopes in Environmental and Health Studie*, 52, 477–486. doi: 10.1080/10256016.2016.1171217.

Elliot, T., Bonotto, D.M. and Andrews, J.N. (2014). Dissolved uranium, radium and radon evolution in the continental intercalaire aquifer, Algeria and Tunisia. *Journal of Environmental Radioactivity*, 137, 150–162. doi: 10.1016/j.jenvrad.2014.07.003.

Elliot, T. and Bonotto, D.M. (2017). Hydrogeochemical and Isotopic Indicators of Vulnerability and Sustainability in the GAS Aquifer São Paulo State, Brazil. *Journal of Hydrology: Regional Studies*, 14, 130–149. doi: 10.1016/j.ejrh.2017.10.006.

Eslami-Andargoli, L., Dale, P., Sipe, N. and Chaseling, J. (2009). Mangrove expansion and rainfall patterns in Moreton Bay, Southeast Queensland, Australia. *Estuarine, Coastal and Shelf Science*, 85, 292–298.

Ewe, S.M.L. and Sternberg, L. (2005). Water uptake patterns of an invasive exotic plant in coastal saline habitats. *Journal of Coastal Research*, 23, 255–264.

Ezer, T. and Atkinson, L.P. (2014). Accelerated flooding along the U.S. East Coast: On the impact of sea-level rise, tides, storms, the Gulf Stream, and the North Atlantic Oscillations. *Earth's Future*, 2, 362–382.

FDEP (2003). Integrating Atmospheric Mercury Deposition with Aquatic Cycling in South Florida. Revised edition, November 2003. Florida Department of Environmental Protection, Florida, USA.

Fisher, M. (2014). Restoring the Greater Everglades Ecosystem. CSA Science News February 2014. doi: 10.2134/csa2014-59-2-1.

Foster, A., Hirata, R., Vidal, A., Schmidt, G. and Garduño, H. (2009). The Guarani Aquifer Initiative–towards realistic groundwater management in a transboundary context. Sustainable Groundwater Management Series, GW-MATE/World Bank, Washington.

Fourqurean, J.W. and Robblee, M.B. (1999). Florida Bay: A history of recent ecological changes. *Estuaries*, 22, 345–357.

Gaiser, E.E., Zafiris, A., Ruiz, P.L., Tobias, F.A.C. and Ross, M.S. (2006). Tracking rates of ecotone migration due to salt-water encroachment using fossil mollusks in coastal South Florida. *Hydrobiologia*, 569, 237–257.

Gat, J.R. (1996). Oxygen and hydrogen isotopes in the hydrologic cycle. *Annual Review of Earth and Planetary Sciences*, 24(1), 225–262. doi: 10.1146/annurev.earth.24.1.225.

Gilman, E.L., Ellison, J. and Coleman, R. (2006). Assessment of mangrove response to projected relative sea-level rise and recent historical reconstruction of shoreline position. *Environmental Monitoring and Assessment*, 124, 105–130.

Gleason, P.J., Cohen, A.D., Brooks, H.K., Stone, P., Smith, W.G. and Spackman, Jr., W. (1974). The environmental significance of Holocene sediments from the Everglades and saline tidal plain. In: P.J. Gleason (ed.) Environments of Southern Florida: Present and Past. Memoir 2: Miami Geological Society, Miami, Florida, pp. 61–72.

Grimmeisen, F., Lehmann, M.F., Liesch, T., Goeppert, N., Klinger, J., Zopfi, J. and Goldscheider, N. (2017). Isotopic constraints on water source mixing, network leakage and contamination in an urban groundwater system. *Science of the Total Environment*, 583, 202–213. doi: 10.1016/j.scitotenv.2017.01.054.

Guha, H. and Panday, S. (2012). Impact of sea level rise on groundwater salinity in a coastal community of south Florida. *Journal of the American Water Resources Association*, 48, 510–529.

Herbert, D.A., Perry, W.B., Cosby, B.J. and Fourqurean, J.W. (2011). Projected reorganization of Florida Bay seagrass communities in response to the increased freshwater inflow of Everglades restoration. *Estuaries and Coasts*, 34, 973–992.

Hoffmeister, J.E. (1974). Land from the sea: The geologic story of South Florida. Coral Gables: University of Miami Press.

Howard, K.W. (2007). Urban Groundwater, Meeting the Challenge. IAH Selected Papers on Hydrogeology 8. CRC Press.

Jiang, J., DeAngelis, D.L., Smith, T.J., Teh, S.Y. and Koh, H.L. (2012). Spatial pattern formation of coastal vegetation in response to external gradients and positive feedbacks affecting soil porewater salinity: A model study. *Landscape Ecology*, 27(1), 109–119. doi: 10.1007/s10980-011-9689-9.

Johnson, W.W. and Finley, M.T. (1980). Handbook of Acute Toxicity of Chemicals to Fish and Aquatic Invertebrates. Resource Publication 137. U.S. Department of the Interior, Fish and Wildlife Service, Washington, DC, 4–17.

Jopp, F., DeAngelis, D.L., Koh, H.L., Teh, S.Y., Trexler, J.C. and Jiang, J. (2009a). Modeling Food Web in Florida Everglades Marshland I. *Proceedings of the 5th International Conference on Asian and Pacific Coasts (APAC 2009)*, 13–16 October 2009, Nanyang Technological University (NTU), Singapore, Volume 1, Soon Keat Tan and Zhenhua Huang (eds.), World Scientific Publishing, Singapore, pp. 258–264. doi: 10.1142/9789814287951_031.

Jopp, F., DeAngelis, D.L., Koh, H.L., Teh, S.Y., Trexler, J.C. and Jiang, J. (2009b). Modeling Food Web in Florida Everglades Marshland II. *Proceedings of the 5th International Conference on Asian and Pacific Coasts (APAC 2009)*, 13–16 October 2009, Nanyang Technological University (NTU), Singapore, Volume 1, S.K. Tan and Z. Huang (Eds.), World Scientific Publishing, Singapore, pp. 265–271. doi: 10.1142/9789814287951_0032.

Krauss, K.W., From, A.S., Doyle, T.W., Doyle, T.J. and Barry, M.J. (2011). Sea-level rise and landscape change influence mangrove encroachment onto marsh in the Ten Thousand Islands region of Florida, USA. *Journal of Coastal Conservation*, 1–10.

Lara, R., Szlafsztein, C., Cohen, M., Berger, U. and Glaser, M. (2002). Implications of mangrove dynamics for private land use in Bragança, North Brazil: A case study. *Journal of Coastal Conservation*, 8, 97–102.

Lecuyer, C., Bodergat, A.M., Martineau, F., Fourel, F., Gurbuz, K. and Nazik, A. (2012). Water sources, mixing and evaporation in the Akyatan Lagoon, Turkey. *Estuarine, Coastal and Shelf Science*, 115, 200–209. doi: 10.1016/j.ecss.2012.09.002.

Lerner, D.N. (2002). Identifying and quantifying urban recharge: A review. *Hydrogeology Journal*, 10, 143–152. doi: 10.1007/s10040-001-0177-1.

Li, S., Levin, N.E. and Chesson, L.A. (2015). Continental scale variation in ^{17}O-excess of meteoric waters in the United States. *Geochimica et Cosmochimica Acta* doi.org/10.1016/j.gca.2015.04.047.

Loftus, W.F. and Eklund, A.M. (1994). Long-term dynamics of an Everglades small fish assemblage. Pages 461–484 in Everglades: The ecosystem and its restoration (Davis, S.M. and Ogden, J.C. Eds). Delrey Beach, Fl: St. Lucie Press, p. 826.

Loftus, W.F., Chapman, J.D. and Conrow, R. (1990). Hydroperiod effects on Everglades marsh food webs, with relation to marsh restoration efforts, pp. 1–22 IN G. Larson and M. Soukup (Editors). Fisheries and Coastal Wetlands Research. *Proceedings of the 1986 Conference on Science in National Parks*, Vol. 6.

Martinez, J.L., Raiber, M. and Cox, M.E. (2015). Assessment of groundwater — Surface water interaction using long-term hydrochemical data and isotope hydrology: Headwaters of the Condamine River, Southeast Queensland, Australia. *Science of the Total Environment*, 536, 499–516. doi: 10.1016/j.scitotenv.2015.07.031.

Matta, M., Cairncross, C. and Kocan, R. (1998). Possible effects of polychlorinated biphenyls on sex determination in rainbow trout. *Environmental Toxicology and Chemistry*, 17(1), 26–29.

Mayer, F.L. and Ellersieck, M.R. (1986). Manual of acute toxicity: Interpretation and database for 410 chemicals and 66 species of freshwater animals. US Department of the Interior, Fish and Wildlife Service, 160, 506–553.

McDermot, D. and Rose, K.A. (2000). An individual-based model of lake fish communities: application to piscivore stocking in Lake Mendota. *Ecological Modelling*, 125, 67–102.

Najjar, R.G., Walker, H.A., Anderson, P.J., Barron, E.J., Bord, R.J., Gibson, J.R., Kennedy, V.S., Knight, C.G., Megonigal, J.P., O'Connor, R.E., Polsky, C.D., Psuty, N.P., Richards, B.A., Sorenson, L.G., Steele, E.M. and Swanson, R.S. (2000). The potential impacts of climate change on the mid-Atlantic coastal region. *Climate Research*, 14, 219–233.

Nicholls, R.J. and Cazenave, A. (2010). Sea-level rise and its impact on coastal zones. *Science*, 328, 1517–1520. doi: 10.1126/science.1185782.

Ogden, J.C. (1994). A comparison of wading bird nesting colony dynamics (1931–1946 and 1974–1989) as an indication of ecosystem conditions in the southern Everglades, pp. 533–570, in: S.M. Davis and J.C. Ogden (eds.), Everglades: The system and its restoration, St. Lucie Press, Delray Beach, Florida.

Orn, S., Andersson, P.L., Forlin, L., Tysklind, M. and Norrgren, L. (1998). The impact on reproduction of an orally administered mixture of selected PCBs in zebrafish (Danio rerio). *Achieve Environmental Contamination and Toxicology*, 35(1), 52–57.

Passioura, J.B., Ball, M.C. and Knight, J.H. (1992). Mangrove may salinize the soil and in so doing limit their transpiration rate. *Functional Biology*, 6, 476–481.

Perry, W. (2004). Elements of South Florida's comprehensive everglades restoration plan. *Ecotoxicology*, 13, 185–193.

Person, B.T. and Ruess, R.W. (2003). Stability of a subartic saltmarsh: Plant community resistance to tidal inundation. *Ecoscience*, 10(3), 351–360.

Pfeuffer, R.J. (1991). Pesticide residue monitoring in sediment and surface water within the South Florida Management District, Volume 2. Technical Publication 91-01, South Florida Water Management District, West Palm Beach, Florida.

Ross, M.S., O'Brien, J.J. and Flynn, L.J. (1992). Ecological site classification of Florida Keys terrestrial habitats. *Biotropica*, 24, 488–502.

Ross, M.S., O'Brien, J.J. and Sternberg, L. (1994). Sea-level rise and the reduction in pine forests in the Florida Keys. *Ecological Applications*, 4, 144–156. doi: 10.2307/1942124.

Ross, M.S., Meeder, J.F., Sah, J.P., Ruiz, P.L. and Telesnicki, G.J. (2000). The southeast saline Everglades revisited: 50 years of coastal vegetation change. *Journal of Vegetation Science*, 11, 101–112. doi: 10.2307/3236781

Ross, M.S., O'Brien, J.J., Ford, R.G., Zhang, K. and Morkill, A. (2009). Disturbance and the rising tide: the challenge of biodiversity management for low island ecosystems. *Frontiers in Ecology and the Environment*. doi: 10.1890/070221.

Saha, S., Bradley, K., Ross, M.S., Hughes, P., Wilmers, T., Ruiz, P.L. and Bergh, C. (2011). Hurricane effects on subtropical pine rocklands of the Florida Keys. *Climate Change*, 107, 169–184. doi: 10.1007/s10584-011-0081-1.

Saha, S., Sadle, J., Heiden, C. and Sternberg, L. (2014). Salinity, groundwater, and water uptake depth of plants in coastal uplands of Everglades National Park (Florida, USA). *Ecohydrology*, 8, 128–136.

Scheffer, M., Carpenter, S., Foley, J. A., Folke, C. and Walker, B. (2001). Catastrophic shifts in ecosystems. *Nature*, 413, 591–596.

Sklar, F.H. and van der Valk, A. (Eds.) (2003). Tree Islands of the Everglades. Kluwer Academic Publishers, Boston.

Sklar, F., McVoy, C., VanZee, R., Gawlik, D.E., Tarboton, K., Rudnick, D. and Miao, S. (2002). The effects of altered hydrology on the ecology of the Everglades, pp. 39–82. In: The Everglades, Florida Bay, and Coral Reefs of the Florida Keys: An Ecosystem Sourcebook. J.W. Porter and K.G. Porter (eds.). CRC Press, Boca Raton, Florida.

Sklar, F.H., Chimney, M.J., Newman, S., McCormick, P., Gawlik, D., Miao, S., McVoy, C., Said, W., Newman, J. and Coronado, C. (2005). The ecological-societal underpinnings of Everglades restoration. *Frontiers in Ecology and the Environment*, 3, 161–169.

Snedaker, S.C. (1995). Mangroves and climate change in the Florida and Caribbean region: Scenarios and hypotheses. In: Wong Y-S, Tam NY Eds. *Asia-Pacific Symposium on Mangrove Ecosystems*. Dordrecht: Springer, pp. 43–49.

Stephens, D.B. (1995). Vadose Zone Hydrology. Boca Raton: CRC Press, pp. 1–2.

Stern, C.V., Sheikh, P.A. and Carter, N.T. (2010). Everglades restoration: The federal role in funding. *Congressional Research Service*, 7–5700 www.crs.gov RS22048. CRS Report for Congress.

Sternberg, L. and Swart, P.K. (1987). Utilization of freshwater and ocean water by coastal plants of southern Florida. *Ecology*, 68, 1898–1905. doi: 10.2307/1939881.

Sternberg, L., Ish-Shalom-Gordon, N., Ross, M. and O'Brien, J. (1991). Water relations of coastal plant communities near the ocean/freshwater boundary. *Oecologia*, 88, 305–310. doi: 10.1007/BF00317571.

Sternberg, L., Teh, S.Y., Ewe, S., Miralles-Wilhelm, F. and DeAngelis, D. (2007). Competition between Hardwood Hammocks and Mangroves. *Ecosystems*, 10(4), 648–660. doi: 10.1007/s10021-007-9050-y.

Steyer, G.D., Cretini, K.F., Piazza, S., Sharp, L.A., Snedden, G.A. and Sapkota, S. (2010). Hurricane influences on vegetation community change in coastal Louisiana, US Geological Survey Open-File Report 2010-1105, US Geological Survey, Reston, VA, p. 21.

Teh, S.Y., DeAngelis, D., Sternberg, L., Miralles-Wilhelm, F.R., Smith, T.J. and Koh, H.L. (2008). A simulation model for projecting changes in salinity concentrations and species dominance in the coastal margin habitats of the everglades. *Ecological Modelling*, 213(2), 245–256. doi: 10.1016/j.ecolmodel.2007.12.007.

Teh, S.Y., Turtora, M., DeAngelis, D.L., Jiang, J., Pearlstine, L., Smith, T.J. and Koh, H.L. (2015). Application of a coupled vegetation competition and groundwater simulation model to study effects of sea level rise and storm surges on coastal vegetation. *Journal of Marine Science and Engineering*, 3, 1149–1177. doi: 10.3390/jmse3041149.

Teh, S.Y., DeAngelis, D.L., Voss, C.I., Sternberg, L. and Koh, H.L. (2018). MANTRA-O18: An Extended Version of SUTRA Modified to Simulate Salt and $\delta18O$ Transport amid Water Uptake by Plants. *E3S Web of Conferences*, 54, 00039. doi: 10.1051/e3sconf/20185400039.

Teh, S.Y., Koh, H.L., DeAngelis, D.L., Voss, C.I. and Sternberg, L. (2019). Modeling $\delta^{18}O$ as an early indicator of regime shift arising from salinity stress in coastal vegetation. *Hydrogeology Journal*, 27(4), 1257–1276. doi: 10.1007/s10040-019-01930-3.

Tipple, B.J., Jameel, Y., Chau, T.H., Mancuso, C.J., Bowen, G.J., Dufour, A., Chesson, L.A. and Ehleringer, J.R. (2017). Stable hydrogen and oxygen isotopes of tap water reveal structure of the San Francisco Bay area's water system and adjustments during a major drought. *Water Research*, 119, 212–224. doi: 10.1016/j.watres.2017.04.022.

Titus, J. and Richman, C. (2000). Maps of lands vulnerable to sea level rise: Modeled elevations along the U.S. Atlantic and Gulf Coasts. *Climate Research*, 18, 205–228.

Tomlinson, P.B. (1994). The botany of mangroves. Cambridge: Cambridge University Press.

USEPA (1999). National Recommended Water Quality Criteria – Correction, US Environmental Protection Agency, Office of Water, EPA 822-Z-99-001, Washington DC.

Vendramini, P.F. and Sternberg, L. (2007). A faster plant stem-water extraction method. *Rapid Communications in Mass Spectrometry*, 21, 164–168. doi: 1002/rcm.2826.

Wasson, K., Woolfolk, A. and Fresquez, C. (2013). Ecotones as indicators of changing environmental conditions: Rapid migration of salt marsh-upland boundaries. *Estuaries and Coasts*, 36, 654–664.

Webster, P.J., Holland, G.J., Curry, J.A. and Chang, H.R. (2005). Changes in tropical cyclone number, duration, and intensity in a warming environment. *Science*, 309, 1844–1846.

Wei, L., Lockington, D.A., Poh, S.-C., Gasparon, M. and Lovelock, C.E. (2012). Water use patterns of estuarine in a tidal creek system. *Oecologia*, 172, 485–494.

Zhai, L., Jiang, J., DeAngelis, D.L. and Sternberg, L. (2016). Prediction of Plant Vulnerability to Salinity Increase in a Coastal Ecosystem by Stable Isotope Composition (d18O) of Plant Stem Water: A Model Study. *Ecosystems*, 19, 32–49.

Zhai, L., Krauss, K.W., Liu, X., Duberstein, J.A., Conner, W.H., DeAngelis, D.L. and Sternberg, L. (2018). Growth stress response to sea level rise in species with contrasting functional traits: A case study in tidal freshwater forested wetlands. *Environmental and Experimental Botany*, 155, 378–386. doi: 10.1016/j.envexpbot.2018.07.023.

Chapter 6

Mangrove Current Status and Future Trends

6.1 Introduction

Mangroves are some of the most carbon-rich ecosystems in tropical and subtropical regions with a mean carbon density of 1,023 Mg carbon/ha (Donato *et al.*, 2011). Mangroves account for 14% of carbon sequestration in the world's oceans, although they cover merely 0.5% of the total coastal ocean area. Covering less than 1% of the total tropical forest area, mangroves account for about 3% of carbon sequestered by the world's tropical forests (Alongi, 2002). Environmental factors, including temperature, salinity and rainfall, are key determinants that have a strong influence over the growth, survival and distribution of mangroves. Mangroves are found in the intertidal zones of tropical and subtropical regions of the world. Mangroves exist on every continent except Antarctica. Consisting of halophytic trees and shrub species, mangroves are intertidal wetland forests that live in the tropical and subtropical regions between latitude 30% north (e.g., Florida) and 30% south (e.g., southern Australia). In the literature, mangrove also refers to the tidal forest that includes trees, shrubs, palms, epiphytes and ferns (Tomlinson, 1986). The distinctive community of plants and animals associated with mangroves is sometimes referred to as the "mangal". A plethora of coastal and terrestrial fauna, including fish, crustaceans, snakes and mammals, share the wetland habitats with some 70 vegetation species of mangroves. Hence, mangroves provide invaluable ecosystem services to humans and wildlife. These forests grow around the mouths of rivers, in tidal swamps and along coastlines. Mangroves are regularly inundated by saline or brackish water and are subjected to constant salinity stresses due to vast variations in salinity over the diurnal and seasonal cycles. Mangroves must adapt to constant salinity stresses, in addition to high temperature and oxygen deprivation in a water-logged environment, at the expense of growth and development. This has resulted in the very low species diversity of the mangrove vege-tation today, compared to the high biodiversity found in, e.g., coral reefs and tropical rainforests (Ricklefs and Latham, 1993). For exam-ple, 223 tree species per hectare have been recorded in the lowland tropical rainforest in Sarawak (Proctor *et al.*, 1983). On the con-trary, there may be only two or three mangrove species per hectare. The distribution of mangroves, covering 118 countries, is described in detail by Tomlinson (1986). Physiologically adapted to the tropics and subtropics and requiring minimum temperatures, mangroves are not expected to expand poleward, being limited by frost frequency and severity (Twilley, 1998; Saintilan *et al.*, 2014). They can tolerate salinity regimes ranging from freshwater terrestrial habitats to hyper-saline coasts with salinity exceeding 100 parts per thousand. Their abundance among terrestrial plant communities is subject to com-petition with other plant species that are better adapted to the less saline terrestrial environment. Around the globe, mangrove forests have been acknowledged to provide numerous benefits and ecosys-tem services crucial to human populations, wildlife and the habitats they occupy. Human disturbances and natural causes have long been recognized as threats to mangrove persistence. For the protection of mangroves, actions have been taken by national governments in collaboration with international organizations through mutual agree-ments. However, losses of mangrove habitats continue, after having lost more than 50% worldwide. Around 35% of the mangrove for-est coverage area was lost worldwide during the 1980s and 1990s alone (Valiela, 2001). Overall, the global decline of mangroves has slowed; however, consistent and coordinated actions are required to ensure their long-term survival (Alongi, 2002). The global loss of mangroves can be attributed largely to human population growth and land conversion to agriculture such as rice farming, aquaculture and overexploitation of mangrove timber. Occupying some 18 million ha worldwide as reported by Spalding *et al.* (1997), mangrove global coverage was revised downward to 14 million ha by Giri *et al.* (2011), and further reduced to 8 million ha by Hamilton and Casey (2016).

In this chapter, we review the status of mangrove forests and the conservation efforts taken to protect them. Threats to the persistence of mangroves and potential solutions for their effective conservation will be critically examined. Case studies from various regions of the world will be evaluated. The integration of human livelihood needs into mangrove conservation goals in a key to long-term sustainability of mangrove forests throughout the world. We provide brief overviews of the following: (1) Current Status of Mangroves, (2) Failures of Current Conservation Approaches, (3) Potential Threats for Mangroves in the Future and (4) Innovation and Success in Mangrove Conservation.

6.2   Mangrove Conservation and Restoration

Mangrove losses over the past decades have created a major concern, as their loss may reduce the numerous critical services and benefits provided to nature and to people. Mangroves play an important role in buffering coastlines against storm surges and tsunamis through wave reduction induced by the presence of mangrove forests with extensive prop root systems (Kathiresan and Rajendran, 2005; Wolanski, 2007; Barbier *et al.*, 2008; Teh *et al.*, 2009). Loss of mangroves will lead to less protection from coastal flooding, storm surges and high winds such as hurricanes. The protection provided by the 6 km to 30 km zones of mangroves had been reported to have reduced by 70% the flooded area in southwestern Florida caused by Hurricane Wilma in 2005 (Zhang *et al.*, 2012a; Liu *et al.*, 2013). More importantly, mangroves reduced the loss of human life from the 1999 cyclone that struck Orissa, India (Barbier, 2016). To help protect mangrove forests, some countries such as Guyana have engaged in educational outreach to alert the public about the potentially catastrophic consequences to mangrove deforestation. Mangroves provide nursery habitats for juvenile coral reef fishes of many species (Nagelkerken *et al.*, 2000). Mangroves, especially the Rhizophora with extensive prop roots, provide structural heterogeneity that is favorable both to prey attempting to avoid predators and to predatory fish searching for invertebrate prey hiding within the root structure (Laegdsgaard and Johnson, 2001). Hence, mangrove forests are efficient aquaculture farms in nature. Mangroves also serve as sinks for carbon, through the accumulation of living biomass, through the deposition of litter and dead wood and through the trapping of sediments delivered from the uplands. Carbon in mangrove sediments build up vertically in response to sea level rise (SLR), thereby helping to prevent soil erosion (McLeod *et al.*, 2011) and serving as a carbon and nutrient sink. Mangroves can improve coastal water quality contaminated by wastewater inputs from uplands via denitrification in the anaerobic environment (Ewel *et al.*, 1998) and through nitrogen fixation by certain bacteria and cyanobacteria associated with mangrove mud and the prop root systems (Kimball and Teas, 1975; Pelegri and Twilley, 1998). To reverse the ongoing losses of mangrove forests, an integrated approach to mangrove conservation and management is essential. The development of mangrove and wetland resources falls under a complex interaction between various levels of governing institutions, management, private ownership and community engagement (Berkes, 2004). Actions are taken at the national, regional and international levels for the conservation of mangrove and for the sustainable use of wetland resources, coordinated via interlocking international conservation authorities. Of these, the most well known is the Convention on Wetlands of International Importance especially as Waterfowl Habitat, known popularly as the Ramsar Convention. Working closely with Ramsar is the Convention on the Conservation of Migratory Species of Wild Animals and the Convention on International Trade in Endangered Species of Wild Fauna and Flora (CITES). Organizations working under the umbrella of the United Nations include the United Nations Forum on Forests (UNFF) and the United Nations Framework Convention on Climate Change (UNFCCC). Others include the Convention on Biological Diversity (CBD) and the Convention for the Protection of World Cultural and Natural Heritage. Working in concert, these agreements and conventions have provided mechanisms for the protection of large areas of mangrove forests globally. In addition to conservation agreements at the international level, individual nations have contributed to efforts for the protection, conservation and restoration of mangrove forests, via mandatory protection by governments and through locally initiated voluntary measures and actions. Coordination between governments at various levels and local community utilization of these resources are critical to ensuring long-term success of these conservation actions.

Occupying 42% of world mangrove areas, Asia has the largest land coverage of mangroves. The U.S. has been the focus of many decades of published research on mangrove forests. While the status of mangrove forests varies by country and by region, many mangrove forests experience similar patterns of disruption and threats to their persistence from encroachment of urban development, from rampant conversion of mangrove land to agriculture and aquaculture, and from unsustainable mangrove timber harvest. The following four sections will present (a) the current status of mangroves, (b) their degradation and (c) their restoration in four countries: Malaysia, Vietnam, China and USA. This literature overview may reveal gaps in knowledge, divergence in approaches and social, cultural imperatives supporting or opposing mangrove conservation efforts.

6.3   Mangroves in Malaysia

6.3.1   *Current status of mangroves in Malaysia*

There are currently six Ramsar sites in Malaysia: three in Johor, two in Sabah and one in Sarawak. Malaysia has a total mangrove coverage of 575,000 ha, down from 695,000 ha in the 1970s. The main causes of this 17% mangrove loss include urbanization and the associated infrastructure development, land conversion and coastal reclamation for agriculture and aquaculture, and natural disturbances due to coastal land erosion. Designated as a Permanent Forest Reserve since 1904, the Matang Mangrove Forest (MMF) is the largest mangrove forest in Malaysia, covering an area of about 40,000 ha in the state of Perak. A sustainable and well-managed forest system, the MMF is the oldest mangrove reserve in Malaysia. The MMF is tasked to produce a sustainable and consistent yield of renewable forest resources while maintaining ecosystem biodiversity and richness. The MMF is not designated as a Ramsar site. Rich in mangroves and intertidal mudflats, the three Ramsar sites in Johor are in Pulau Kukup, Sungai Pulai and Tanjung Piai. Located about 1 km offshore from the southwestern region of Johor, Pulau Kukup is an uninhabited and pristine mangrove island of approximately 650 ha surrounded by extensive

800 ha of intertidal mudflats. The island is important for flood control and protection from storm surges and coastal erosion. Covering an area of 9,000 ha, Sungai Pulai is Peninsular Malaysia's largest estuarine mangrove system, consisting of seagrass beds, intertidal mudflats and an inland freshwater riverine forest. It provides shoreline stabilization from coastal erosion and flood prevention for adjacent villages. With an area covering 500 ha, Tanjung Piai is an 8 km strip of coastal mangroves and intertidal mudflats at the southernmost tip of continental Asia. The mangroves along Tanjung Piai prevent excessive sediment input from entering the waterways and help stabilize the shoreline against high-energy wave action. The Sarawak Mangrove Forest Reserve of the Kuching Wetlands National Park is a mangrove forest covering approximately 6,600 ha, having been reduced from its former area of 17,000 ha. The Lower Kinabatangan-Segama Wetlands of Sabah is Malaysia's largest Ramsar site, with an area of 78,000 ha consisting of mangrove forest and peat swamp. It is formed by three protected forest reserves: Trusan Kinabatangan Forest Reserve, Kuala Maruap–Kuala Segama Forest Reserve and Kulamba Wildlife Reserve. The 24 ha Kota Kinabalu (KK) Wetland of Sabah is the most recent site being designated as a Ramsar site in March 2017. Located just two kilometers away from the KK city, the patch of mangrove forest is what remains of the original forest along the entire coastline of the KK city before urbanization, making it a site of cultural heritage significance.

6.3.2   *Mangrove degradation in Malaysia*

Loss and degradation of mangrove forests in Malaysia is alarming. About 1,200 ha or about 1% of mangrove coverage in Peninsular Malaysia was lost every year between 1990 and 2010 (Hamdan *et al.*, 2010, 2012). The main reasons for such losses were identified as land conversion for agriculture and aquaculture, as well as coastal erosion. Richards and Friess (2016) reported that more than 15,800 ha (or 2.83%) of mangrove area was lost between 2000 and 2012. Over that period, deforestation of mangrove forests for aquaculture, rice and oil palm plantations was identified as the major cause of the loss. They attributed 38% loss of this mangrove in Malaysia to conversion of mangrove forests to oil palm plantations. Malaysia's Matang Mangrove Forest, well known for its sustainable forest management, is susceptible to pollution from upstream industrial areas. For the Marudu Bay Mangrove in Sabah, overexploitation of scarce marine resources and mangrove trees, in combination with pollution from nearby oil palm plantations, is a major concern that threatens the mangrove. A combination of threats could potentially undermine the integrity of the Lower Kinabatangan-Segama Wetlands, one of the Ramsar sites in Malaysia. Logging and the associated loss of wildlife and habitat loss pose a major hazard. Soil erosion and agricultural waste from nearby oil palm plantations and mills cause degradation in water quality. These are the problems identified by the Sabah Biodiversity Centre (2011) as problems that threaten the long-term sustainability of the mangrove ecosystem. The Kuching Wetlands National Park, located downstream of a township with high population density and large development area, is at risk for environmental degradation. The main sources of this environmental concern are wastes from land-clearing activities, residuals from untreated solid and liquid wastes, as well as discharge from a nearby stone quarrying operation. Non-native plant invasions displacing native species are likely to have a significant ecological impact. For example, non-native plants such as *Acrostichum aureum* and *Acanthus* species often take over deforested mangrove areas, and may restrict the regrowth of mangroves. Land clearing activities in urban development and overexploitation of forests exacerbate soil erosion. Soil erosion is a common threat faced by wetlands in Johor such as those of Pulau Kukup, Sungai Pulai and Tanjung Piai, which are mainly scattered along the coastal areas and rivers.

6.3.3 *Mangrove restoration in Malaysia*

Restoration and replanting of degraded mangroves have achieved little success. Restoration and replanting programs were undertaken by the Malaysian Government under the Ninth Malaysian Plan (2006–2010) with a budget of RM40 million. Unfortunately, this mangrove replanting in Malaysia achieved limited progress due to three main reasons: (a) inappropriate choice of mangrove species, (b) poor technical execution and (c) unsuitable replanting locations. First, the appropriate mangrove species must be chosen to provide good coastline protection, to yield good economic returns as firewood and be suitable to the local tidal conditions. Mangrove species differ in their ability to reduce wave impacts because of their drag force and hydraulic resistance. For tsunami wave reduction, the palm *Pandanus odoratissimus*, and *Rhizophora apiculata*, is more effective than other common vegetation, such as the mangrove *Avicennia alba*. *Rhizophora apiculata* is a favorite choice because of its commercial value as poles and logs for charcoal. On the contrary, *Avicennia sp.* has limited commercial value as firewood, but is suitable as protective green shelterbelts and coastal bio-shields against wave surges and tsunamis. Mangroves also differ in their ability to recolonize new or degraded habitats. This observation implies the importance of selecting appropriate species (a) that can act as good wave barriers to offer sufficient shoreline protection and (b) that are viable and robust to recolonize the zones intended for them. These two criteria may not be mutually compatible at some locations. Second, innovative and suitable mangrove replanting efforts require technical inputs from several specialists from various fields, such as engineers, hydrologists, ecologists and botanists, to plan bio-technical options for initial ground stabilization and subsequent mangrove replanting works. This usually would require the construction of a hard and a flexible breakwater that (i) can facilitate seawater flushing of the restoration area during high tide and (ii) can promote sediment accretions during low tide (Tamin et al., 2011). These two requirements often contradict each other, rendering the technical work challenging. At about USD 142,000 per ha, the cost of constructing hard breakwater is substantial and therefore requires significant justifications. Innovative pre-planting trials, species selection and enrichment planting are crucial for the success of the restoration efforts. The conflicting performance of each species does not help in the selection process. Third, the choice of a suitable replanting location is crucial in order to promote growth of young propagules. Within the restoration area in Sg Haji Dorani Selangor, Malaysia, sedimentation remains an ongoing problem that threatens the survival of young propagules. *Avicennia marina* appears to show greater tolerance toward root burial than *Rhizophora apiculata*. Nevertheless, within a year after the construction of the hard breakwater, attempts to establish *Avicennia marina* have failed due to active sediment accretion that resulted in root burial and ultimately death of the planted seedlings (Tamin et al., 2011). After 90 days post planting, the seedling mortality reached

75%, being covered by 9.3 cm of sediment burial at the Sg Haji Dorani site in Selangor Malaysia (Affandi *et al.*, 2010). After 120 days post planting, the mortality was even higher at 93%, being covered by 11.5 cm of sediment burial. Much needs to be done before sustained rehabilitation becomes the norm rather than the exception. In the Philippines, *Rhizophora sp.* is also the preferred choice for mangrove restoration because of their good commercial return from the harvest of firewood (Primavera and Esteban, 2008). Currently, *Avicennia sp.* is widely planted in degraded mangrove sites throughout the Philippines, however.

6.4 Mangroves in Vietnam

6.4.1 *Current status of mangroves in Vietnam*

An excellent and extensive review of mangrove status, deforestation and restoration in Vietnam is provided in Hai *et al.* (2020). This subsection attempts to briefly summarize the superb review by Hai *et al.* (2020) to provide valuable education and insights on mangrove restoration and reforestation worldwide. Out of 408,500 ha of mangroves existing in Vietnam in 1943, 124,000 ha was destroyed during the Second Indochina War from 1965 to 1970 (Westing, 1983). Further, mangrove deforestation continued at the rate of about 0.25% loss per year during the period from 2000 to 2012 (Richards and Friess, 2016). Currently, natural mangrove forests occupy only 21% of the existing mangrove forests in Vietnam. The remainder are replanted (McNally *et al.*, 2011). Mangroves in Vietnam are distributed into four zones, from Zone I in the north to Zone IV in the south. Located in the southern and warmer coastal region of Vietnam, Zone IV has nearly 80% of the total mangrove coverage in Vietnam (MARD, 2014). This Zone IV receives nourishing alluvium and fresh water from the Cuu Long and Dong Nai river systems that sustain the mangroves here. The favorable environmental, climatic, oceanographic and topographic characteristics of Zone IV positively influence its robust mangrove distribution. It is less impacted by storms than the other three zones. With many sunny days and high radiation, Zone IV supports the highest rates of growth of mangrove species.

6.4.2 *Mangrove degradation in Vietnam*

Mimicking the situation in other countries in Southeast Asia, Vietnam's mangroves have experienced rapid conversion to agriculture and aquaculture (Richards and Friess, 2016). The Vietnam Doi Moi Renovation actively promoted large-scale conversion of mangrove forests to shrimp farms for export and local consumption. Fine weather in Zone IV located in warmer southern Vietnam in the Mekong Delta favors rapid growth of aquaculture. The area of shrimp farming in Zone IV increased from 3,000 ha in 1980 to 40,000 ha in 1987 and reached 60,000 ha by 1992 (De Graaf and Xuan, 1998). The total area of shrimp aquaculture in Vietnam by 1994 reached an astonishing 200,000 ha, second only to Indonesia (Tobey *et al.*, 1998). Vietnam's Ministry of Agricultural and Rural Development has a plan to further increase the shrimp farming area from the current 650,000 ha to 670,000 ha by 2030 (MARD, 2015). Aquaculture is not the only pathway for mangrove destruction and loss in Vietnam. Large areas of mangroves were converted for sea dike protection and for the provision of ecosystem services for local communities (Lebel *et al.*, 2002). The mangroves in Vietnam were further destroyed by land reclamation to cater to agricultural expansion, infrastructure development, urbanization and associated industry, and tourism development (Hawkins *et al.*, 2010). More recently, further mangrove loss was driven by artisanal endeavors such as clam farming, timber and fuelwood harvesting, fishing and shellfish collection (McNally *et al.*, 2011). Mangrove degradation and loss were further aggravated by urban waste discharge in coastal areas (14 million tons annually), by oil spill accidents (Nguyen, 2012) and by coastal erosion and destructive impacts from waves and storms (Cat *et al.*, 2006). Depending on the coastal topography, sea water can invade up to 3 km inland, with depths of 4 to 5 m. This salinity intrusion vastly alters the local salinity regimes, thereby decreasing the resilience of mangroves. The combined and cumulative impacts of these detrimental disturbances have significantly contributed to a loss of nearly 60% of the total mangrove area since 1943. Over the 30 years from 1983 to 2013, another 84,000 ha of mangroves was lost, much of which occurred in Zone IV. However, the loss of mangrove coverage has been slowed by restoration programs since 1975 undertaken by the

State and non-government organization sponsorships. The current mangrove degradation and deforestation in Vietnam will likely be exacerbated by future impacts of climate change (Ward *et al.*, 2016). Because of the low tidal range and low sediment output, local sea level rise may pose the greatest threat to mangroves in Vietnam, except for the Mekong and Red River Deltas (Alongi, 2008).

6.4.3 *Mangrove restoration in Vietnam*

The three main objectives and motivations in the development of mangrove restoration programs in Vietnam are as follows: (1) the legal requirement to implement mangrove restoration for protection of inland properties and economic production activities, (2) the clear ecosystem and economic benefits provided from mangrove restoration activities and (3) the aspiration to utilize mangroves to mitigate the significant loss and damage due to coastal disasters and conflicts (Buckingham and Hanson, 2015). To mitigate against tropical typhoons and adapt to climate change, mangrove restoration projects had been concentrated in northern Zone II and central Zone III of Vietnam (Thu and Populus, 2007). Recently, mangrove restoration in Zone IV has received more impetus due to concern about the impacts of climate change and SLR (Thuc *et al.*, 2016). The concern is particularly discernible along the Mekong River Delta, which is exposed to the adverse impact of SLR. Mangrove restoration objectives are divided into five categories: (1) forest products, (2) fisheries, (3) coastal protection and stabilization, (4) climate mitigation and adaptation, and (5) ecosystem preservation and ecosystem service provision (Ellison, 2000). Around half of all mangrove restoration projects in Vietnam are focused on coastal protection and stabilization, especially in Zones II and III, which are more vulnerable to natural disasters (Buckingham and Hanson, 2015). By contrast, mangrove restoration efforts in Zone IV in southern Vietnam were aimed at poverty alleviation and livelihood diversification (Buckingham and Hanson, 2015) in a region which is less vulnerable to typhoons and storms (Takagi *et al.*, 2014). From the mid-2000s, all mangrove restoration projects have set their main or supplemental objectives on climate change mitigation and adaptation.

Mangrove restoration efforts in Vietnam have a long history and can be divided into three periods: (1) the first period, from 1975 to 1980, (2) the second period, from 1981 to 1990, and (3) the third period from the 1990s to the present. Approximately 52,000 ha of mangrove was replanted during each period, the success or otherwise of which was not recorded. In addition to state-funded projects, mangroves in Vietnam have been restored with support from international NGOs beginning from 1990, where projects ranged in scope from small to large. With the total area of mangrove restoration in Vietnam reaching 197,000 ha, restorations have played a significant role in maintaining mangrove forest cover in Vietnam. As there are no assessments of survival rates after restoration projects are finished, it is not clear how the restoration has contributed to the current total mangrove cover in Vietnam. In Zone IV located in southern Vietnam, mangrove restoration activities have addressed poverty alleviation and livelihood diversification for food and income generation (Buckingham and Hanson, 2015). These projects had brought opportunities for nearly 8,000 households as they receive land leases and benefits from the development of infrastructure, such as schools, roads and health care (Powell *et al.*, 2011b). Economic values generated from 150 ha of restored mangroves were estimated to be worth USD 1.1 million (priced in 2010), which is higher than investment in agriculture with returns of only USD 0.55 million (priced in 2010) over a 22 year period at a discount rate of 10% (Tuan and Tinh, 2013).

The success of restoration can be assessed by four main indicators: (1) survival rate of planted trees measured within months of planting to the third year, (2) forest growth comprising tree growth, stand density and stem form of timber trees and (3) environmental success, including vegetation structure, ecosystem function and species biodiversity; and (4) socioeconomic success as measured by local income or local employment opportunities (Lebel *et al.*, 2002). However, the success of these restoration projects in Vietnam cannot be properly evaluated because of inadequate data. One of the main reasons leading to the failure of mangrove restoration programs is the lack of incentives for long-term management. Despite been governed by formal contracts with local communities to plant and protect mangrove plantations, the Red Cross restoration program was not able to achieve a high level of success. This is because the contracts did not provide long-term incentives for local people to protect mangroves from human impacts, such as illegal cutting or grazing when the program ended in 2006 (Powell *et al.*, 2011a). Co-management is an emerging approach in which the government engages local communities in long-term mangrove restoration projects. In this approach, government agencies share decision-making, responsibility and accountability with local communities whose livelihoods depend on ecosystem services provided by mangroves (Schmitt and Duke, 2014). Through participatory processes, resource users and resource governance authorities negotiate a formal agreement on their respective management roles, responsibilities and rights to establish a pluralistic governance body (Borrini-Feyerabend *et al.*, 2012).

6.5 Mangroves in China

6.5.1 *Current status of mangroves in China*

China has one-third of the world's mangrove species (Wang, 2007), although the mangrove forests in mainland China make up only 0.14% of the world's mangrove area. Although China may have had as much as 250,000 ha in the past, the total area of mangrove forests currently is 25,000 ha, having lost much due to deforestation. Mangroves experienced a 50% loss in China, from 40,000 ha in 1957 down to 18,800 ha in the mid-1980s (He and Zhang, 2001). Therefore, mangrove conservation in China plays a vital role in biodiversity conservation of the world's mangrove forests. The distribution of mangroves extends from Yulin Port in Hainan (18° 9′ N) to Fuding of Fujian (27° 20′ N), and covers the Hong Kong, Guangdong and Guangxi provinces. Mangroves are categorized as true mangrove and semi-mangrove species based upon their living environment (Wang *et al.*, 2011). True mangroves are species that are strictly distributed in the intertidal zone, while semi-mangroves are species that could grow in the intertidal zone and on land. China has 37 species of mangroves, representing 20 families and 25 genera, including 26 true mangrove species and 11 semi-mangrove species (Li and Lee, 1997; Wang, 2007).

6.5.2 *Mangrove degradation in China*

China has experienced three stages of mangrove forest loss and degradation due to land conversion for the purposes of (1) transformation to agricultural areas in the 1960s and 1970s, (2) development of aquaculture in mangrove forests during the 1980s and (3) recent urbanization via development of ports, docks and business districts (Wang, 2007). Significant loss and degradation of mangrove areas in China have been recorded for the three decades since 1950. Upstream and higher intertidal habitats are more vulnerable than downstream and lower intertidal habitats, with alteration in habitats to cater to human settlements being the main threats. Alteration and modification of habitats and its coastal hydrology may induce a shift of vegetation structures, undermine ecosystem sustainability and ultimately cause the loss and degradation of mangroves. In southern China, mangrove cover has been reduced from 40,000 ha in 1950 to 15,000 ha in 1990. Sea walls and dikes stretching over 1576 km in Guangdong built to protect coastal infrastructures adversely impacted over 80% of mangroves. With the longest coastlines and largest mangrove covers in China, Guangdong has suffered a loss of 81% of its mangrove forests from 1956 to 1990s. However, mangrove areas in Guangdong have been increasing over the past two decades, due to conservation efforts, incentivized by the knowledge that mangroves are important ecosystems that provide numerous benefits to humans and their habitats. Three to five exotic species have been used in many afforestation projects since 1985. Many replanting efforts are undertaken with these fast-growing species that can adapt well to the habitats and that can provide protection against storm surges. However, this is achieved at the expense of species diversity and richness that may undermine the long-term sustainability of the conservation and restoration efforts if this trend persists. Another case of mangrove loss occurred in the Tieshan port of Beihai in Guangxi province, where urban development has resulted in mangrove forest loss of 370 ha. The mangrove loss from this urban development project alone may exceed the total area of mangrove trees replanted in all of China in one year. Further, invasive species also threaten the persistence of natural indigenous mangrove species. For example, the species *Spartina alterniflora* is one of the most successful invasive species in mangrove wetlands and has successfully dispersed to virtually all mangrove sites in China. Another species *Sonneratia apetala*, which is used in reforestation efforts, has aroused big concern because of its potential to invade native mangrove forests (Ren *et al.*, 2009). In addition to habitat loss, human urban activities have also caused serious coastal pollution, increased disease spread and caused pest damage in mangrove wetland ecosystems. Animal waste and residuals from the application of pesticides and fertilizers are major concerns to mangrove health. For example, pollution from domestic duck farming had resulted in the *Sphaeroma* blooms that occurred in 2010 in Hainan and Guangxi, leading to the demise of a mangrove forest nearby (Fan *et al.*, 2014). Severe pest damage to the *Aegiceras corniculatum* mangrove forest in the Quanzhou Bay of Fujian was likely caused by pollution from aquaculture development (Zuang *et al.*, 2011; Li, 2012a,b).

6.5.3 *Mangrove restoration in China*

The area with available land suitable for mangrove restoration is often constrained by land tenure conflicts in the coastal zone in many countries such as China and the Philippines. Hence, restoration of abandoned fishponds for mangrove replanting is difficult in Guangdong as in most of southern China. The main reason is unresolved land tenure issues due to the ownership of massive ponds being held by numerous private stakeholders (Peng *et al.*, 2016). Nevertheless, restoring mangroves and semi-mangroves in abandoned ponds is feasible in most nature reserves in Guangdong and in southern China (Peng *et al.*, 2013). To restore such abandoned fishponds successfully would require the restoration of the altered hydrological conditions back to the normal condition with the suitable salinity and sedimentation regimes needed by mangroves. The mangrove plantations with introduced species currently account for approximately 16% of the total mangrove area in China. The fast growth and high adaptability of introduced species ensure their continuing expansion in the future. But, this "artificial" growth has aroused much criticism over the long-term ecological impacts on native mangroves. Further, such mangrove "plantations" tend to be dominated by low diversity, often with only one or two dominant species. From an ecological perspective, mixed-cultured mangroves with higher diversity are preferred as they can deliver higher ecosystem services and higher carbon sequestration potential (Chen *et al.*, 2012) in addition to improved nursery functions. Further, mixed mangrove cultures are also more resilient to human and natural perturbations due to their intrinsic "portfolio effects" of diversification. Currently, there are 35 mangrove conservation areas in China, which are managed by the central or local governments (Chen *et al.*, 2009). Among the 49 Ramsar Convention sites in China, some focus specifically on protecting mangroves, such as the Dongzhaigang Mangrove Nature Reserve (5,400 ha), Mai Po Marshes and Inner Deep Bay (1,540 ha), Shankou Mangrove Nature Reserve (4,000 ha), Fujian Zhangjiangkou National Mangrove Nature Reserve (2,360 ha) and Guangxi Beilun Estuary National Nature Reserve (3,000 ha).

Mangrove conservation initiated by private citizens is often more innovative. For example, in Guangxi Province of China, a mangrove area was developed into an eco-farming aquaculture system that benefits both mangroves and the community. This innovative mangrove-aquaculture system does not require the cutting of mangrove trees or the conversion of mangroves. It has been perfected to a point that no industrial feed input is required to feed the aquaculture, following field trials over a five-year period between 2007 and 2012 conducted in onsite mangrove areas. This unique innovation succeeded in addressing the conflict between mangrove conservation and the economic returns of aquaculture (Fan *et al.*, 2013). A network of underground tubes and pipes is buried in between mangrove prop roots to augment benthic habitats for fish. This mangrove-aquaculture ecosystem can bring in a mean return of between USD 27,000, and 45,000 per hectare per annum. The environmentally friendly eco-farming site is accessible by boardwalks that facilitate ecotourism and public education, thereby increasing the income of the farmers and promoting UNSDG. The system has been shown to be wave resistant, easy to operate and has low management cost, while the products are of high quality and high value.

6.6   Mangroves in the USA

6.6.1   *Current status of Mangroves in the USA*

Florida contains the largest area of mangrove swamp in the US (Odum *et al.*, 1982), of which the Everglades is the most popular one.

Covering an area of 10,000 square kilometers, the Greater Everglades of Florida is one of the most well-known wetlands in the United States and the most distinct in the world. Several species of trees, such as hardwood hammocks, mangroves, pine and cypress, are indigenous to the Everglades' ecosystems. They have become extremely fragile due to past excessive interference by humans and constant disruptions induced by Southern Florida weather systems, which can be rather violent (Swiadek, 1997; Baldwin *et al.*, 2001). The hardwood hammock and mangrove ecosystems may be affected by the frequent occurrence of hurricanes that alter the salinity structure of the Everglades. Hardwood hammock trees grow in areas where the ground level is sufficiently high to avoid seasonal flooding following summer rains. Among the most appealing and fascinating features in the Everglades, the tropical hardwood hammocks are rich in species diversity, with the dominant species being Live Oak, West Indian mahogany, Florida royal palm, Gumbo Limbo and Strangler Figure. These trees do not grow very tall in the Everglades due to frequent damage and interruption of growth inflicted by occasional extreme cold weather events and destructive hurricanes. Mangroves cover the entire edge of the western and southern coast of the Everglades. This extensive ecosystem provides good protection for the shore land from rising sea water, storm surges and hurricanes. Further, it is an excellent habitat for many species of fish, birds and other invertebrates to thrive in. The three prominent species of mangroves in Everglades are the red, black and white mangroves. Red Mangroves (*Rhizophora mangle*), which usually occupy the areas closest to the sea, are distinguished by their extensive system of prop roots that protrude downward from the tree trunk. Black Mangroves (*Avicennia germinans*) are found inland from the coast, while the White Mangrove (*Laguncularia racemosa*) is found further inland and usually in drier areas than the other two mangrove species.

The distribution of mangroves in the US stretches north to Cedar Key on the Gulf Coast and Ponce de Leon Inlet on the Atlantic Coast, both at about 29° 10′ N latitude (Odum *et al.*, 1982). Northward expansion is occurring because of decreasing frequency of freezes due to global warming (Doughty *et al.*, 2016). The area of mangroves for Texas on the northern Gulf Coast was 2,181 ha in 1990; but, it has increased to an estimated 3,790 ha recently (Armitage *et al.*, 2015). In Louisiana, chill-tolerant black mangrove (*Avicennia germinans*) shrubs have interspersed and have rapidly recolonized and expanded into saltmarshes at the northern extreme of their current range in the last three decades (Michot *et al.*, 2010; Osland *et al.*, 2013, 2015). Mangroves are not native to Hawaii, but have been introduced, with both positive and negative impacts (Allen, 1998). Because Florida contains the greatest spatial extent of mangroves and greatest research emphasis in the U.S., it is the focus here. The mangrove community consists of mainly three species, namely, black mangrove, red mangrove (*Rhizophora mangle*) and white mangrove (*Laguncularia racemose*). A fourth species, buttonwood (*Conocarpus erecta*), is not a true mangrove because it displays neither the viviparous reproduction strategies nor the specialized root structures of any true mangrove species (Nelson, 1994). Sea level rise has enabled many mangrove species to expand their habitats from mangrove swamps into upland forests (Tomlinson, 1986). Most of the area occupied by mangroves lies within the Everglades National Park, which borders on Florida Bay and the Gulf of Mexico. Much of the remaining mangrove area lies within land owned by the Federal, state and county governments, or by non-profit organizations such as the National Audubon Society.

6.6.2   *Mangrove degradation in the USA*

Many mangroves in the Everglades have been severely degraded by excessive human exploitation and rapid urbanization over the past century. The Everglades mangrove ecosystem is vulnerable to changes in sea level and salinity because of its low-lying landscape. Frequent and severe storm surges induced by hurricanes inundate coastal areas and cause sharp salinity changes in the inundated soils. These salinity pulses may have irreversible impacts on the sensitive plant ecosystems in the Everglades, including the freshwater glycophytic hardwood hammocks and saltwater-tolerant halophytic mangroves. The degradation of plant species may in turn adversely affect many other biological species, such as fish and birds. Mangroves are now protected by law in some parts of the United States. However, several threats remain, including a legacy of clear-cutting to make way for human development, as has occurred in Florida's southerly island chain and in the Upper Florida Keys. Mangrove habitats were reduced and fragmented through dredging and filling for residential canals (Kruczynski and McManus, 2002), which increases the vulnerability to further destruction or invasion by exotic plants (Strong and Bancroft, 1994). In Florida's vast Everglades wetland, a primary threat has been the altered freshwater flows into the mangrove estuaries. The reduced freshwater flow has increased salinity in some locations. Although mangroves can tolerate high salinity, it is known to have a negative impact on seedlings (Koch and Snedaker, 1997). In some instances, mangrove die-off in Florida Bay may be due to salinity outside the range of tolerance (McIvor *et al.*, 1994). Black mangroves are vulnerable to overflooding or hypersaline conditions due to changes in tidal fluxes, caused by the type of marsh impoundment used for mosquito control (Rey *et al.*, 1990). As hurricanes are a natural environmental disturbance in southern Florida, mangroves have adapted to their periodic occurrence. However, an increase in frequency and intensity of hurricanes due to global warming poses a grave threat to mangroves in south Florida. Thousands of hectares of mangrove forest were reported to have suffered catastrophic disturbances from Hurricane Andrew in 1992 (Doyle *et al.*, 1995) and from Hurricane Wilma in 2005 (Smith *et al.*, 2009). Repeated and frequent intensive disturbances have turned some former mangroves into mudflats, which show no sign of returning to mangroves (Smith *et al.*, 2009). Another constant threat is the possibility of a large oil spill in the Gulf of Mexico, which is a site of a major petroleum industry. For example, an oil spill from a land tank in Bahia Las Minas of Panama killed 69 ha of mangroves, or 6% of the mangroves in the bay, and inflicted damage to a further 34% (Duke *et al.*, 1997). Other concern is that invasive species, particularly Brazilian pepper (*Schinus terebinthifolius*) and colubrine (*Colubrina asiatica*), have invaded some mangrove areas (Davis *et al.*, 2005). Pollutants, herbicides and other runoff constituents from agricultural and urban areas also pose risks. But, sea level rise may be a greater future threat.

6.6.3 *Mangrove restoration in the USA*

Severe human intervention and rapid development around the Everglades area through pioneer and subsequent settlements have reduced the Florida Everglades to patches of pool waters and seasonally dry lands, incapable of supporting vibrant wildlife. To stop the deteriorating situation and to reverse this undesirable state, conservation efforts have been put in place. These efforts may take many years to eradicate some of the damage done in the past. To protect the fragile ecosystems in the Everglades, the Everglades National Park was formed in 1947 and has been a World Heritage Site since 1976. In the year 2000, the Comprehensive Everglades Restoration Plan (CERP) was approved by the US Congress to provide a framework to restore, protect and preserve the water resources of central and southern Florida, including the Everglades. With an initial budget of $7.8 billion that covers a period of 30 years, CERP is the most expensive and comprehensive environmental restoration attempt in history. Other smaller conservation programs have been initiated by local communities.

For example, an experimental conservation program has been initiated in south Florida for early detection of mangrove forest degradation, involving the rehabilitation and monitoring of 220 ha of dead and stressed mangrove forest. This mangrove forest has succumbed to slow degradation over the last three decades. The clear associations between the observed mangrove die-offs and modifications to local hydrology have been scientifically established. This observation generated great interest among residents and provided the incentive and methodology for the protection of mangroves. The local coastal management plan is developed to stop further mangrove losses through preemptive action. This prototype mangrove "heart attack" prevention model is needed globally to rehabilitate vast areas of dead or stressed mangroves over the coming decades. Prevention involves three elements: (1) monitoring for early detection of small degrees of degradation, (2) identification of thresholds that may trigger acute losses and (3) a sustained interest in ameliorating stresses in advance of acute losses (Lewis *et al.*, 2016).

6.7   Integration of Community and Science

Current management and policy for mangrove conservation worldwide have not been fully successful in ensuring the conservation and sustainable use of these resources. There are a myriad of reasons for this failure, including the failure to integrate the differing spatial and temporal scales at which the mangrove ecosystem services are provided. The management, policy and regulatory institutions operate at the larger scale of ecological services which are typically long term covering large spatial extents. On the contrary, local community resource users operate at the smaller demand scale, which is typically short term covering a limited spatial extent at the community level. The local communities lack the broader incentives and institutional imperatives to integrate their social demands into the ecological scales of mangrove ecosystem provisions. This lack of coordination and integration across spatial-temporal scales has resulted in a profound mismatch between the social benefits demanded at the local scale by the local communities and the ecological costs imposed at the regional scale. Research conducted in Vietnam, Indonesia, Brazil, Ghana and the Mekong Delta reported that social opposition has occurred frequently for two reasons: (1) the failure in managing the conflict between mangrove resource provision and resource utilization and (2) the failure in integrating science with local community knowledge (Máñez *et al.*, 2014). These two failures are in fact closely related. Success of scientific knowledge transfer from the scientific sectors to the local communities in conservation projects depends strongly on how successful the science integrates with the local community knowledge base. The following five subsections will present the status of integrating science with local community knowledge in five countries: Vietnam, Indonesia, Brazil, Ghana and the Mekong Delta.

6.7.1 *Vietnam*

In Vietnam, the lack of integration and coordination between local communities and the scientific sectors had led to limited success of mangrove conservation. The science-based approaches and the local knowledge-based approaches have been separately designed to manage mangrove-dominated muddy coasts (MDMCs), according to their respective perspectives. This segregated methodology achieved limited success for both parties, due to inadequate integration of goals and misplaced suspicion between the two distinctly different stakeholders with distinctly disparaged approaches. New mechanisms are therefore needed to promote a high level of integration and coordination between local knowledge and scientific methods to effectively manage MDMCs in a sustainable way (Nguyen *et al.*, 2017a). The science-based approaches require deep specialist knowledge, which are laboratory based and inventory driven, typical of science. This scientific approach is beyond the capacity of local communities to comprehend and appreciate, leaving little opportunity for their active participation and meaningful consultation during design, planning and implementing processes (Nguyen and Parnell (2017b)). Local knowledge-based approaches have faced two key challenges, namely, (1) limited local knowledge and non-committal participation, and (2) doubtful long-term sustainability due to financial constraints faced by the local communities. In short, active involvement and meaningful consultation between local communities and the science sectors is urgently needed for effective implementation of integrated science–community-based approaches (Stojanovic *et al.*, 2004).

6.7.2   *Indonesia*

In Indonesia, similar problems are encountered in many projects such as the project known as the Segara Anakan Conservation and Development Project (SACDP), funded by the Asia Development Bank (ADB), in cooperation with the government of Indonesia. Grassroots communities pose a grave challenge to the project as the communities perceived that the scientific sectors did not incorporate problem-solving options that can benefit the local communities. The programs were perceived as oriented toward the wishes and benefits of the established elite communities. This perception led to strong social opposition before the implementation of the mangrove conservation project. As a result, all programs were compromised, with science contributing little to success (Dharmawan *et al.*, 2016). Hence, integrating community knowledge and aspirations with good science and relevant research into mangrove conservation efforts is critical for its success (Dharmawan *et al.*, 2016).

6.7.3   *Brazil*

In Brazil, mangrove conservation is hampered by frequent opposition from the indigenous communities, is impaired by the overarching economic interests that are detrimental to mangrove conservation and by the lack of operative conservation policies and best practices. The future of mangrove conservation in Brazil depends on the integration of indigenous communities, who depend on mangrove goods and services, with science and other society sectors through integrated science–community-based co-management. Government agencies, academic institutions, NGOs and funding agencies can play a vital role as agents of action and awareness generation (Datta *et al.*, 2012). For example, local communities should be educated on how and why abandoned shrimp ponds and degraded mangrove forests can cause enormous damage to artisanal fishermen and regional fisheries. This damage to local fisheries can be further compounded by impairment to the natural functioning of local hydrological and soil integrity. Brazilian shrimp farming could benefit from adaption to sustainable utilization of bountiful natural services provided by mangrove forests and the adjacent estuaries to produce shrimp in a sustainable way (Ferreira and Lacerda, 2016). Mangroves are constantly under grave threat by conversion of mangrove habitats to land for aquaculture, agriculture, industry and urban development. This land use conversion has destroyed more than 50,000 ha, or about 4%, of the total mangrove cover over the past three decades. Restoration efforts have somewhat minimized these losses, but only 5% of the total degraded area has been restored.

6.7.4 *Ghana*

In Ghana, a study in the Volta estuary embraced the concept of community–science-based co-management of mangrove restoration and management operated at the local scales. Resource users are the key stakeholders who play significant roles and take full responsibility in maintaining mangrove conservation within their jurisdiction (Aheto *et al.*, 2016). Resource users' livelihoods, food security, economic benefits and wellbeing are the primary motivation to sustainably exploit, restore and manage the resources (Wagner, 2001). Financial motivations are the basis for sustainable mangrove reforestation and prevention of unsustainable exploitation of mangrove stands in Volta. Success in community–science-based mangrove management has also been achieved in several countries around the world, especially in South and Southeast Asia, South America, and Eastern and Southern Africa (Datta *et al.*, 2012). This finding is consistent with the observation of Melana *et al.* (2000) that "people first and sustainable mangrove forest management will follow". In this context, replanting and restoration communities are not primarily motivated by ecological interests alone but rather in conjunction with compensatory financial rewards. Achieving clarity on ownership and the rights of access to coastal resources could lead to more effective mangrove resource management through local institutional arrangements. Local communities alone cannot typically manage coastal ecosystems for reasons such as limited research capacity, restricted area of jurisdiction, budget constraints and local politics (Sorensen and McCreary, 1990). With objectives clearly established at the outset, co-management is an approach in which governments can enforce certain authority and share responsibilities with local communities in coastal resource management with resource users, academic institutions and non-governmental organizations as partners.

6.7.5   *Mekong Delta*

In the Mekong Delta, a study in the Ca Mau province supports the hypothesis that sufficient economic incentives, coupled with adequate legal rights protection and taking full responsibility over their allocated mangrove forests, is the key to success in mangrove conservation. Shrimp farmers may be able to plant, protect and sustainably manage mangroves if they are given these assurances and protections. Mangrove conservation and shrimp farming can coexist in harmony in an environment of economic incentives assisted by good science and supported by appropriate legislation. This exertion calls for a reevaluation of the wrong perception that shrimp farming invariably leads to deforestation and degradation of mangrove habitat (Ha *et al.*, 2012). As income from mangrove timber harvests is very low compared to shrimp harvests, an equitable profit-sharing scheme has been devised to ensure that maximizing aquaculture production will not compromise the mangroves' functional integrity. While complying with the regulations stipulated by the state government regarding the forest-to-pond ratio, shrimp farmers must also devote attention to maintaining the quality and integrity of mangroves under their care. This seemingly conflicting requirement poses a constant challenge to the sustainability of this arrangement (Ha *et al.*, 2012; Bush *et al.*, 2010).

6.8   Ecosystem Service Economic Valuation

Concern over the loss of mangrove ecosystems has gradually shifted to focusing on the disruption or loss in the provision of ecosystems services by mangrove forests. These ecosystem services and benefits include protection of coastal infrastructures and human settlements from storms and surges that damage property and that cause death and injury. Other benefits such as buffering climate change impacts induced by sea level rise, and reducing saltwater intrusion and coastal erosion are equally important, although appearing rather distant. However, up until recently, no economic valuation in dollar terms had been performed to examine the economic contribution of these ecosystem services. In the last decade, economic valuation of ecosystems services has increasingly been developed and utilized in mangrove conservation management and policy deliberation and decisions. The United Nations Sustainable Development Goals agenda (UN, 2015) has listed mangrove conservation and human well-being as two primary goals. These two apparently conflicting aspirations can be met by innovations such as ecologically sustainable aquaculture that relies on eco-farming within mangrove forests. This eco-farming can achieve harmony between human wealth and mangrove health, as mangroves coexist in harmony with aquaculture. This harmony is possible because mangroves and aquaculture are given equitable economic valuations, known as ecosystem service economic valuation (ESEV).

6.8.1 *Objectives of ESEV*

Ecosystem services may be defined broadly as the well-being provided to humans by natural ecosystems. The Millennium Ecosystem Assessment (MA, 2003, 2005) differentiates between four major service areas provided by ecosystems, namely, (a) supporting, (b) provisioning, (c) regulating and (d) cultural services. This division of ecosystem services was designed to promote conflict resolution regarding the provision of ecosystem services among the diverse stakeholders. For mangroves, this conflict resolution is difficult to achieve because of several reasons: (1) large and diverse distributions of stakeholders, (2) less defined systems and functions, (3) overlapping marine and terrestrial areas assigned to mangroves and (4) different regulations and jurisdictions. Ecosystem services must therefore be linked with human well-being in terms of social-economic values, a process known as ESEV. The development of ESEV approaches has been fueled by the growing need to deal with ecosystem management globally, by giving an economic valuation in Dollar terms to ecosystems. Economic Valuation studies have been advocated to support decision-making and management of ecosystems. ESEV provides useful information about the social economic benefits and costs associated with alternative policies and facilitates the assessment of trade-offs and synergies inherent in ecosystem-based management. To inform decision-making, ESEV can be used to connect science and community into effective and equitable policy. Translating good science, effective governance and strong community engagement into effective policy can be difficult, however. This difficulty arises because scientists, communities and decision-makers often operate with different time frames, in diverse areas of expertise and knowledge, and with divergent objectives and incentives (Waite *et al.*, 2015). The science–community–policy gap can be narrowed by improving communication among scientists, decision-makers and other stakeholders through capacity building, outreach and engagement. This communication can be improved if the diverse stakeholders have a common economic platform to exchange and trade their views and a common valuation system to evaluate alternatives and trade-offs. This common platform is an ESEV. Scientific research would more likely be used in management decisions if it can explicitly respond to the needs of decision-makers, and if it can collect economic valuation (EV) data that are targeted at specific policy issues to address specific ecosystem services (ESs). The ESEV is the EV of the ESs provided by mangrove forests. Usefulness and issues pertinent to ESEV in three geographic areas, India, Southeast Asia and the Caribbean, will be discussed in the following three subsections.

6.8.2   *Mangrove ESEV in India*

An example of the usefulness of ESEV is the EV of the ESs provided by mangrove forests in India. A significant rise in mangrove cover in the state of Gujarat post 1993 has been attributed to mangrove plantation or regeneration activities in the state (Sahu *et al.*, 2015; FSI, 2011). These mangrove conservation efforts have been motivated by the knowledge that mangrove-related fish species contribute 10 to 32% to total marine fisheries catch in economic returns. The change in fish production due to changes in mangrove cover is examined by comparing fish production in Gujarat to that in other coastal regions of India, under the initiative known as The Economics of Ecosystems and Biodiversity (TEEB) India. Under this TEEB India, the contribution of mangroves to the total marine fish output in India is estimated at 1.86 tons per hectare per year. In economic terms, the EV of mangrove contributes INR 68 billion in the state of Gujarat alone (Anneboina *et al.*, 2017; values are standardized to 2012–2013 prices). This ESEV of mangroves in India can provide a means to compare and trade off alternative policy options in a rational manner if this ESEV is extensive enough to cover a broad range of policy trade-offs.

However, ESEV is confronted with daunting challenges in developing countries because of lack of data, lack of funding and lack of institutional trust (Torres and Hanley, 2017). ESEV can be used to identify who gets the benefits, and who faces the costs, particularly across the social, economic divide. Estimation of non-use values should be transparent and be performed with collaboration among social, natural and political scientists to build trust and to reduce conflicts of interest (Torres and Hanley, 2017). The non-use values include (1) local users' complex perceptions of landscape, (2) their holistic sense of well-being and (3) their context-specific socio-cultural valuation of mangrove ESEV. These non-uses that transcend the monetary value are indispensable criteria to be incorporated into conservation policies (Queiroz *et al.*, 2017). It is crucial to have a good understanding of the complex interrelationships between social and natural systems. It is equally important to understand the multiple dimensions and different time scales of ecosystem services. Such an approach is consistent with the United Nations Sustainable Development Goals of improving human well-being while promoting the conservation of marine ecosystems (UN, 2015).

6.8.3 *Mangrove ESEV in Southeast Asia*

In lower-middle income countries in Southeast Asia, mangroves are generally undervalued in decision-making relating to their use, conservation and restoration. The absence of a comprehensive database of mangrove ESEV in Southeast Asia has contributed to this undervaluation of EV of mangrove ES. To fill this knowledge gap, Brander *et al.* (2012) constructed a database containing

130 valuation estimates, for mangrove ESEV in Southeast Asia, standardized to USD per hectare per year in 2007 prices. The mean and median values of ESEV for mangrove ecosystem services are 4,185 and 239 USD/ha/year, respectively. The high variability in ESEV across the study sites stems from the vast variations in the biological-physical features of the site and the social-economic characteristics of the beneficiaries of ES. A meta-analytic value function is used to estimate the changes in ESEV of mangrove ecosystem services in Southeast Asia for the period 2000 to 2050. The estimated foregone annual benefits by 2050 are USD 2.2 billion, with an estimation range of USD 1.6 to 2.8 billion. These estimates of mangrove ecosystem services are particularly important to Indonesia, which has the largest stock of mangroves in 2000, but faces the largest losses of 1.7 million hectares or 38% over the period 2000 to 2050.

6.8.4   *Mangrove ESEV in the Caribbean*

The Wider Caribbean Basin (WCB) economy is heavily dependent on tourism, fisheries and coastal ecotourism. This dependency is driven by coastal ecosystem services provided by mangroves and coral reefs. Overcoming the daunting challenges, coastal ESEV in the WCB has helped justify fishing regulations, establish marine protected areas, award or settle damage claims and identify sustainable sources of finance for conservation, toward long-term viability (Waite *et al.*, 2015). However, the overall economic valuation of ecosystem services in the WCB remains largely unquantified (Schuhmann *et al.*, 2015)). Current ESEV valuation efforts have been piecemeal, fragmented and largely unconnected to policy needs ((Schuhmann *et al.*, 2015)). Integration across disciplines, across institutions and across nations within the WCB can benefit a comprehensive compilation of valuation data and supporting data collection on marine ecosystems within the WCB. Most marine resources within the WCB are transboundary and require international collaboration for sustainable use. ESEV can be mainstreamed into regional policy programs such as the Caribbean Large Marine Ecosystem Strategic Action Programme (CLME-SAP) adopted in 2013 by the countries of the WCB (Mahon *et al.*, 2014).

6.8.5 *Beyond mangroves*

An integrated coastal zone management that links the continent to the ocean would be beneficial, given that these coastal and ocean environments are functionally linked through the physical, biological, chemical and hydrogeological processes at the land–ocean interface (Lundberg and Moberg, 2003; Sheaves, 2009). For example, legislation in Brazil does not currently integrate buffer habitats such as coral reefs and sea grasses to mangrove habitats for mutual protection. However, an example of success in integrated river basin management is the Brazilian government's campaign for a shared co-management of river basins through the participation of different social sectors and diverse stakeholders. This model of the "Basin Committees" could be extended to "Coastal Committees" to encompass the coastal-ocean zones within their frameworks. However, this extension is difficult as the national level of governance suffers from inadequate specialists, insufficient funding and from conflicts within different governance agencies at various levels (Ferreira and Lacerda, 2016).

6.9 Role of Mangrove in Coastal Protection

With extensive leaves, exposed roots and stems, mangrove forests can reduce tsunami and hurricane impacts by attenuating the destructive energy of waves passing through them (Mazda *et al.*, 1997a,b). Tsunami wave energy, heights and velocities may be significantly reduced as the wave propagates through a mangrove forest (Harada and Kawata, 2004). Hence, dense and contiguous mangrove forest covers can provide a certain degree of protection against coastal storm surges such as hurricanes and tsunamis. The 2004 Andaman Tsunami devastated many coastal areas of Asian countries bordering the Andaman Indian Ocean. Affected countries include Indonesia, Sri Lanka, India, Thailand and Malaysia. The tsunami's impacts were the most devastating in locations such as Banda Aceh of Indonesia, where high-density human settlements were inundated by large tsunami waves. By contrast, the communities living behind thick mangroves and other natural coastal forests were less affected (Danielsen *et al.*, 2005; Kathiresan and Rajendran, 2005; NAHRIM,

2006; Tanaka *et al.*, 2007). Consequently, coastal green belts receive increasing attention as potential coastal defense measures that are environmentally friendly and cost-effective, as compared to other high-cost artificial coastal barriers. In addition to reducing tsunami wave heights and energy, mangrove forests also help to retard debris drift, thereby helping to save lives and properties. The effectiveness of mangrove forests as tsunami mitigation measure depends on the forest density and the width of the forest cover. The characteristics and dimension of the mangrove root systems, and the diameters of tree trunks, also play an important role in reducing the impact of tsunamis. An increase in forest width and density can further decrease wave heights and inundation depths and reduce current velocity and hydraulic forces behind the coastal forests. Hence, restoration and rehabilitation of degraded mangroves will improve the effectiveness of mangrove forests to resist the onslaught of tsunamis. However, some controversy regarding the mitigation effects of wave impacts by mangrove forest has been noted in the literature (Chatenoux and Peduzzi, 2007). Mangrove forests could be uprooted by large tsunamis, if the wave heights exceed 4 m, in which case the resulting debris may cause secondary damage (Harada and Imamura, 2005).

6.9.1 *Mangroves can reduce wave heights*

Tsunami waves exceeding 3 m in height can potentially kill. However, dense and deep mangrove forests may significantly reduce tsunami wave heights to safer levels if the wave heights do not exceed 3 m. The effectiveness of mangrove forests in reducing tsunami wave heights has therefore received increasing scientific interest after the 2004 Andaman tsunami. The 2004 tsunami killed a total of 250,000 persons living in affected coastal zones worldwide, particularly in Indonesia, Thailand and India. The tsunami inundated coastal regions in Banda Aceh of Indonesia to depths exceeding 10 m, killing more than 100,000 persons along the beaches. The tsunami waves arriving in Penang and Kedah beaches in northwest Peninsular Malaysia exceeded heights of 3 m at some locations, killing 68 people along the beaches. The most damaging effect of tsunamis occurs when the waves enter the shallow water region near the shore. As the waves propagate into shallow beaches, they amplify in height and they increase in velocity to potentially dangerous levels. The wave heights may reach 3 to 20 m and the velocity may exceed 10 m/s. As observed in Banda Aceh in 2004, tsunami run-up waves inundated thousands of meters inland with depths of 10 m or higher. Recognizing the contribution of mangrove in mitigating tsunami impacts, concerted efforts have been devoted to the rehabilitation and conservation of mangrove forests along these beaches, since the 2004 Andaman tsunami. In this section, we present model simulation of tsunami run-up along the shallow beaches in Penang, Malaysia, to demonstrate the degree of protection against tsunami impacts offered by mangrove forests. Both analytical and numerical models by means of the nonlinear shallow water equation are used. Grid sizes vary from 50 m to 1 m, where a smaller grid size of 1 m is needed to capture the complicated wave structures of the tsunami wave as it propagates up the shallow beaches. The effects of mangrove forest in reducing tsunami wave energy, heights and velocities along beaches will be assessed by incorporating the Morison Equation as a friction term into the momentum equation.

6.9.2   *Morison equation*

Simulations of tsunami wave damping by mangrove forest are typically performed by the inclusion of a friction term into the momentum equation to mimic the drag resistance provided by the mangrove in reducing wave energy. Harada and Imamura (2005) used the Morison Eq. (6.1) to assess the frictional drag C_D exerted by mangrove forests on reducing wave heights and energy of a tsunami. The definitions and units for each of the symbols are given in Table 6.1. Equation (6.1) is discretized by the finite difference method, the details of which are available in Teh *et al.* (2009). The degree of protection and wave reduction offered by mangrove forests against tsunami wave attacks depends primarily by the frictional drag C_D.

$$\frac{\partial \eta}{\partial t} + \frac{\partial M}{\partial x} = 0$$

$$\frac{\partial M}{\partial t} + \frac{\partial}{\partial x}\left(\frac{M^2}{D}\right) + gD\frac{\partial \eta}{\partial x} + \frac{gn^2 M|M|}{D^{7/3}} + \frac{C_D}{2}A_0\frac{M|M|}{D^2} = 0$$

$$(6.1)$$

Table 6.1:   Definition of symbols used in Eqs. (6.1) and (6.2).

Symbol	Unit	Definition
t	s	Time
x	m	Space coordinate along x-direction
M	m^2/s	Discharge fluxes in the x-direction
η	m	Free surface elevation measured from a fixed datum (mean sea level)
d	m	Water depth below fixed datum
D	m	Total water depth, $D = \eta + d$
g	m/s^2	Gravitational acceleration
n	$s/m^{1/3}$	Manning's relative roughness coefficient
C_D	–	Drag coefficient
A_0	per $100\,m^2$	Projected area of trees under water surface
V_0	m^3	Total volume of tree under water surface
V	m^3	Control volume

6.9.2.1   *Mangrove drag coefficient C_D*

Based upon data collected on coastal pine trees forest in Japan, Harada and Imamura (2005) proposed Eq. (6.2) for the calculation of mangrove drag coefficient C_D. The drag coefficient C_D depends on the volume of obstacles presented by the mangrove leaves, stems and root systems to resist tsunami waves. The larger the volume of obstacles, the higher the drag resistance and the better the protection provided by the mangrove forests. Hence, thick and dense mangroves provide the best protection.

$$C_D = 8.4\frac{V_0}{V} + 0.66 \quad \left(0.01 \le \frac{V_0}{V} \le 0.07\right) \qquad (6.2)$$

Based upon Eq. (6.2), C_D varies between 0.744 and 1.248, for values of V_0/V between 0.01 and 0.07 for the Japanese pine forests surveyed. This range of values is small compared to the range of C_D between 0.4 and 10, derived by Mazda *et al.* (1997b) for mangroves. This discrepancy arises because pine forests have a relatively small value of V_0/V compared to mangroves. Based upon mangrove tree characteristics surveyed in Penang, appropriate values of A_0/V and V_0/V are estimated and used to simulate the effect of mangroves in Penang on reducing the wave energy and heights of a tsunami. A reasonable

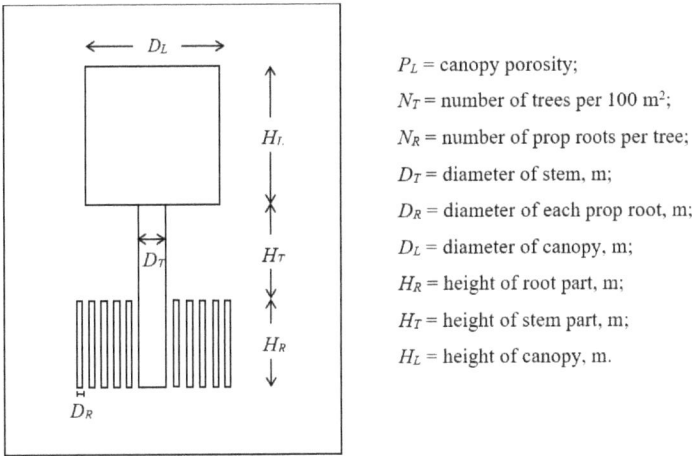

Figure 6.1: Model mangrove tree and parameters.

range of V_0/V values adopted for Penang varies between 0.04 and 0.16, resulting in values of C_D varying between 0.7 and 2.0. The estimation of A_0/V and V_0/V is based upon the concept of a mangrove tree model as described as follows.

6.9.2.2 *Mangrove tree model*

The model mangrove tree as shown in Figure 6.1 is conceptualized as a structure consisting of three parts: root system stem (trunk) and leaf (canopy), modeled as cylinders of different diameters and heights. The parameters on the right of Figure 6.1 refer to the features of the mangrove tree model such as canopy porosity P_L (dimensionless) and height of canopy H_L (m). These parameters are prescribed values based upon a site survey in Penang and Eqs. (6.3) and (6.4). Table 6.2 provides mangrove forest characterization used to represent a typical mangrove in Penang, in conjunction with reported values in the literature (Mazda *et al.*, 1997b).

$$A_0 = A_R + A_T + A_L$$

$$A_R = H_R \left(D_R \frac{N_R N_T}{100} \right), \quad A_T = H_T \left(D_T \frac{N_T}{100} \right)$$

$$\text{and} \quad A_L = P_L \times H_L \left(D_L \frac{N_T}{100} \right) \tag{6.3}$$

Table 6.2:   Parameters of the control forest determined from a field survey (Teh *et al.*, 2009).

Parameter	Unit	Value
Forest width, W	m	1000
Canopy porosity, P_L	—	0.03
Number of roots per tree, N_R	—	300
Number of trees, N_T	—	20
Diameter of each root, D_R	m	0.008
Diameter of trunk, D_T	m	0.18
Diameter of canopy, D_L	m	3.0
Height of root part, H_R	m	0.18
Height of trunk part, H_T	m	5.5
Height of canopy part, H_C	m	3.0

$$V_0 = V_R + V_T + V_L$$

$$V_R = H_R \left(\pi \frac{D_R^2}{4} N_R N_T \right), \quad V_T = H_T \left(\pi \frac{D_T^2}{4} N_T \right)$$

$$\text{and} \quad V_L = P_L \times H_L \left(\pi \frac{D_L^2}{4} N_T \right) \tag{6.4}$$

6.9.3   *Analytical model*

It is useful to have a simple analytical formula to relate wave height
reduction caused by frictional drag presented by the mangrove forest.
The mangrove drag resistance causes the wave height to be reduced
according to the exponential form $\eta = \eta_0 e^{-\beta x}$, where x(m) is the
distance traveled through the mangrove forest and β(m^{-1}) is a decay
constant, indicating how fast the wave height will decay per meter
length of travel. The decay constant β(m^{-1}) depends on Manning's
coefficient n, water depth h, wave period T and maximum particle
velocity v_0. Given the prescribed or computed values of these four
parameters, the value of βm^{-1} can be calculated by Eq. (6.5). The
half distance $\hat{x}$(m) is defined as the distance it takes the wave height
to reduce by half and it is calculated by Eq. (6.6). Further details
regarding the complicated mathematics involved in the derivation of

this analytical model are found in Teh *et al.* (2009).

$$\beta = \frac{n}{h}\sqrt{\frac{2v_0}{Th^{1/3}}}\,\mathrm{m}^{-1} \tag{6.5}$$

$$\hat{x} = -\frac{1}{\beta}\ln(0.5)\,\mathrm{m} \tag{6.6}$$

6.9.4 Wave reduction in Penang mangrove

The values of β depend on four parameters: (a) Manning's coefficient n, (b) wave period T, (c) mean water depth h and (d) averaged incident velocity v_0. For the 2004 tsunami in Penang, the mean water depth is about $h = 2$ m, while the averaged incident velocity v_0 is $2\,\mathrm{m/s}$. Table 6.3 provides the values of decay constant $\beta(\times 10^{-3}\,\mathrm{m}^{-1})$ for the mangrove forest in reducing wave heights for a given set of (a) Manning's coefficients n from 0.1 to $0.7\,\mathrm{sm}^{-1/3}$, and (b) wave periods T from 10 to 60 minutes. Table 6.4 provides the corresponding half distance $\hat{x}(\mathrm{m})$ for this given set of parameters. For example, for the given set of four parameter values consisting of (a) $n = 0.1\,\mathrm{sm}^{-1/3}$, (b) $T = 40$ minutes, (c) $v_0 = 2$ m/s and (d) $h = 2$ m, then β is $1.82 \times 10^{-3}\,\mathrm{m}^{-1}$ from Table 6.3. The corresponding half distance $\hat{x}$ is 381 m from Table 6.4. This means that it will require a mangrove width of 381 m to reduce the wave height by half for the given representative set of values: (a) $n = 0.1\,\mathrm{sm}^{-1/3}$, (b) $T = 40$ minutes, (c) $v_0 = 2\,\mathrm{m/s}$ and (d) $h = 2$ m. Various combinations of this set of four parameters may be used to provide a preliminary assessment of the impact of mangrove forests in reducing tsunami wave heights by means of the analytical model. However, we will also use the numerical model TUNA-RP to simulate the complex wave along the beaches and assess the effect of mangrove drag resistance in reducing tsunami wave heights in a later section.

6.9.5 TUNA simulation for Penang

Mangrove forests fringing the western coast of Penang, mainly at Pantai Acheh, have been credited to have played a role in reducing the 2004 tsunami waves. The 2004 tsunami run-up heights at five locations surveyed in 2005 along the beaches in Penang measured

Table 6.3:   Decay number β ($\times 10^{-3}$ m^{-1}) for prescribed value of $v_0 = 2$ m/s and $h = 2$ m.

T	n (sm$^{-1/3}$)						
(minutes)	0.10	0.20	0.30	0.40	0.50	0.60	0.70
10	3.64	7.27	10.9	14.5	18.2	21.8	25.5
20	2.57	5.14	7.72	10.3	12.9	15.4	18.0
30	2.10	4.20	6.30	8.40	10.5	12.6	14.7
40	1.82	3.64	5.46	7.27	9.09	10.9	12.8
50	1.63	3.25	4.88	6.51	8.13	9.76	11.4
60	1.48	2.97	4.45	5.94	7.42	8.91	10.4

Table 6.4:   Half distance $\hat{x}$ (n, T) given $v_0 = 2$ m/s and $h = 2$ m.

T	n (sm$^{-1/3}$)						
(minutes)	0.1	0.2	0.3	0.4	0.5	0.6	0.7
10	190	95.3	63.5	47.6	38.1	31.8	27.2
20	269	135	89.8	67.4	53.9	44.9	38.5
30	330	165	110	82.5	66.0	55.0	47.2
40	381	190	127	95.3	76.2	63.5	54.4
50	426	213	142	106	85.2	71.0	60.9
60	467	233	156	117	93.4	77.8	66.7

between 2.31 m and 4.0 m (Koh *et al.*, 2009) as shown in Table 6.5. At Pantai Acheh, the measured run-up height was only 2.51 m, despite its low elevation.

On the contrary, the 2004 tsunami simulation by two-dimensional (2D) TUNA-RP without mangrove forests along the beaches in Pantai Acheh estimated a run-up height of 5.17 m. In this 2D simulation, the Manning's n coefficient of the ocean bottom is set to be a constant value of 0.015 (Edward, 2008) to account for the bottom friction effect. The simulated run-up height of 5.17 m is significantly higher than the measured value of 2.51 m by approximately a factor of two. This raises the interest to further investigate the role of the coastal mangrove forest at Pantai Acheh in reducing the tsunami waves by means of TUNA-RP simulation. The dominant mangrove species at Pantai Acheh is *Avicennia marina*

Table 6.5: Comparison of run-up height between survey data (Koh *et al.*, 2009) and TUNA-RP simulated results with and without coastal forest.

Location	Lat (°)	Long (°)	Surveyed (2004 tsunami)	TUNA simulated without forest	TUNA simulated with forest
			Run-up heights (m)		
B. Ferringhi (Teluk Bayu)	5.47	100.24	3.46	3.77	3.78
B. Ferringhi (Miami Beach)	5.48	100.27	4.00	3.38	3.46
Tanjung Tokong	5.46	100.31	2.61	2.63	2.63
Tanjung Bungah	5.47	100.28	2.31	2.13	2.12
Pantai Acheh	5.40	100.18	2.51	5.17	2.96 (*2.58)

Note: *Run-up height using n' value of $0.12\,\mathrm{sm}^{-1/3}$.

(Hamdan *et al.*, 2012; Mansor and Zakaria, 2004); hence, the tree species used in the simulation is *Avicennia marina*. The drag resistance presented by *Avicennia marina* mangrove forest is estimated to have the equivalent Manning's coefficient of $0.1\,\mathrm{sm}^{-1/3}$. The width of the mangrove forest is estimated to be 1,000 m, corresponding to a forest width–wavelength ratio of 0.02. Based upon the one-dimensional (1D) TUNA-RP results shown in Figure 6.2, the mangrove forest in Pantai Acheh with a forest width of 1,000 m and averaged roughness coefficient of vegetation $n' = 0.1\,\mathrm{sm}^{-1/3}$ would provide a wave height reduction ratio of about 0.54. With this reduction ratio of 0.54, the run-up height simulated by the 1D TUNA-RP model of 6.69 m without mangroves would be reduced to 6.69 m × 0.54 ≈ 3.61 m with the presence of mangroves. This estimated run-up height of 3.61 m is nevertheless still higher than the measured run-up height of 2.51 m by 44%. On the contrary, the 2D run-up height simulated by the 2D TUNA-RP model is 5.19 m without mangroves. Hence, with the presence of mangroves, the run-up height would be reduced to 5.19 m × 0.54 = 2.80 m. It is noted

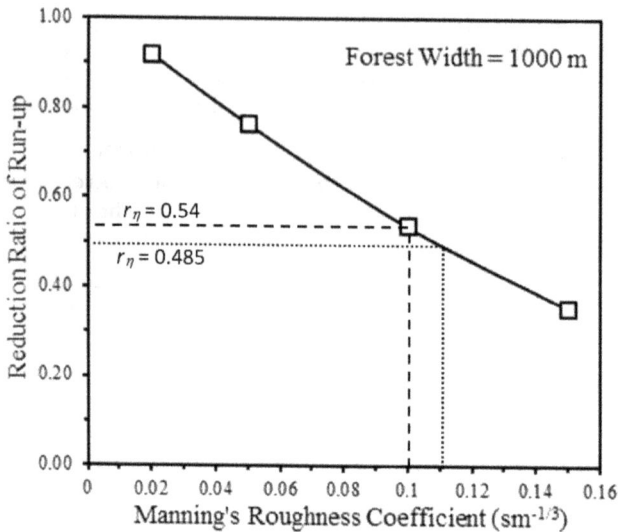

Figure 6.2: Reduction ratio of run-up height r_η as a function of Manning's roughness coefficient n (Chow, 1959).

that simulated waves from the 1D TUNA-RP model tend to focus the wave energy along a designated path. This has the tendency to produce higher waves in 1D models. Propagating waves will typically undergo transformation due to wave interference, refraction and diffraction. Therefore, the waves simulated by 2D TUNA would be slightly lower than those simulated by 1D TUNA. Hence, in the following section, we use 2D TUNA-RP to simulate run-up heights with mangroves and compare the results with the measured run-up heights for the 2004 tsunami as shown in Table 6.5.

6.9.5.1 *TUNA simulation with mangroves*

To assess the impact of mangrove forests in wave reduction, the mangrove forest along the northwestern coast of Penang Island is included in the simulation of TUNA-RP. Table 6.5 shows the surveyed and the simulated run-up heights at five selected beaches in Penang using 2D TUNA-RP. The simulated run-up heights agree well with the observed maximum wave heights reported, with a relative error between 0.76% and 17.9%. This percentage of errors of less than 20% is acceptable when comparing simulation results against field measurements (Horrillo *et al.*, 2015). In the presence of a mangrove forest with a tree density of $0.20 \, \text{trees}/\text{m}^2$, the simulated maximum run-up height at Pantai Acheh by 2D TUNA-RP is now 2.96 m. This appears to be a good match between the measured data and the simulation result obtained by the 2D TUNA-RP model. Further simulation indicates that the simulated maximum run-up height at Pantai Acheh is 2.58 m if the averaged roughness coefficient n' is increased slightly to $0.12 \, \text{sm}^{-1/3}$ to represent the friction provided by the Pantai Acheh mangrove forest. This appears to be a better match between the surveyed data and the 2D model simulation result. Hence, this simulation study suggests that the hydraulic resistance provided by the mangrove forest in Pantai Acheh ranges from 0.1 to 0.12 $\text{sm}^{-1/3}$.

6.9.6 *Controversy over the role of mangroves*

The 2004 Andaman tsunami inflicted a disaster of epic proportions to millions of people worldwide. After the tsunami, much attention has been focused on the role of natural barriers, such as coral reefs, mangroves and sand dunes, in protecting vulnerable coastlines and populations. Traditional knowledge acquired by coastal fishers in Sri Lanka acknowledged the protective role played by mangroves against tsunami impacts in a post-2004-tsunami survey (Venkatachalam *et al.*, 2009). Kathiresan and Rajendran (2005) highlighted the effectiveness of mangrove forests in Tamil Nadu of southern India in lowering human deaths and loss of wealth in areas populated with dense mangrove forests. However, this assertion has been vigorously challenged by many authors over the incorrect statistical analysis they used. A modeling study conducted by Venkatachalam *et al.* (2012) concluded that mangroves were not a significant ameliorative factor in mitigating the impact of tsunamis. The authors had reached this conclusion after considering a combination of several contributing factors, such as the way tsunami impacts were measured and the household damage as well as the human deaths inflicted. The distance of a location from the shoreline played a significant role in reducing the impact of the tsunami, in addition to the mere presence of mangrove forests. In Malaysia, areas with intact and dense mangrove forests, such as Pantai Acheh in Penang Island, suffered much less damage during the 2004 Andaman tsunami.

The dense and intact mangrove forests in Pantai Acheh dissipated wave energy and reduced wave heights to safe levels below 3 m (NAHRIM, 2006; Koh *et al.*, 2009; Hamdan *et al.*, 2012). Mangrove trees can indeed be uprooted to contribute to hazardous debris flows that could inflict even more damage (Shuto, 1987). The devastating damage inflicted during the 2010 Mentawai tsunami with 10 m waves and the 2011 Tohoku tsunami with 20 m waves provided further evidence of the inability of coastal forests to withstand the impact of extreme tsunamis. In both tsunami events, all trees in the paths of these extreme waves were destroyed (Hill *et al.*, 2012; Suppasri *et al.*, 2011). Hence, we should not harbor any false sense of security that bio-shields such as mangroves and other greenbelt vegetations are able to protect against extreme coastal hazards (Wolanski, 2007; Yanagisawa *et al.*, 2009). Further, the presence of forest gaps, such as roads that run perpendicular to the shore, can intensify the force of tsunami waves by channeling the waves into narrow constrictions (Tanaka, 2009; Thuy *et al.*, 2009). In Tamil Nadu of India, during the 2004 Andaman tsunami, the maximum velocity at the exit end of the forest gap was 1.7 times more in comparison with the case of a no-vegetation belt (Mascarenhas and Jayakumar, 2008). Along the gaps, the wave heights increased monotonically with the distance traveled, posing increasing risks further inside the forest gaps. Hence, coastal vegetation intended for wave attenuation should not have significant and continuous gaps in the direction of tsunami flow, which is perpendicular to the shoreline.

6.10　Threats to Mangrove

The future of mangroves is by no means an assured certainty. Mangrove forests worldwide have been threatened by many forms of disturbances such as conversion to agriculture and aquaculture and unsustainable harvest for timber, food, fuel and medicine (Saenger, 2002). Indonesia had lost more than 200,000 ha of its mangroves by 1960s, mostly in Java and Sumatra. In the following three decades up to the 1990s, another 800,000 ha was lost, mainly in Kalimantan and Sulawesi. Shrimp farming (tambak) and the timber industry cleared 600,000 ha more over the following two decades. However, the net loss of mangroves in the coming two decades is expected to be reduced to around 23,000 ha (Ilman *et al.*, 2016). Serious environmental stress, anthropogenic impact, metal pollution, heavy influx of sewage and industrial effluents affect the health of the mangrove ecosystem (Satheeshkumar *et al.*, 2012a,b). Sea level rise coupled with vast urban development accompanied by fast human population growth along coastal areas poses the most serious threats to mangroves long-term survival. Conversion of mangrove habitats to agriculture and aquaculture was a major contributing factor to mangrove loss and degradation in the past. This threat remains unless we put a stop to this conversion. Human populations and development, besides displacing native coastal vegetation, also alter hydrological regimes that govern mangrove habitats. The effects caused by changes in upland freshwater hydrology and alterations in tidal hydrodynamics following SLR may reduce the viability of mangroves. Estuarine and coastal pollution from oil exploration and from wastes in runoff degrades mangrove health, potentially leading to disease and pest impacts that further compound the adverse impacts of anthropogenic activities (Kathiresan, 2002). Other adverse effects are clear-cutting and many forms of overexploitation of mangroves for timber and firewood. Geological disasters such as tsunamis and earthquakes have caused long-term change to mangrove distribution (Roy and Krishnan, 2005). Frequent hurricanes and cyclones cause periodic heavy damage to mangroves (Doyle *et al.*, 1995), the cumulative impacts of which can lead to habitat change and coastal vegetation regime shift detrimental to mangroves (Smith *et al.*, 2009).

6.10.1 *Climate change impact*

The impact of sea level change on mangroves is not yet fully understood. Predicted global climate change (GCC) and SLR could exert multiple impacts with varying intensity on mangroves throughout the world (Ward *et al.*, 2016). Despite inadequate data, some predict that mangroves may face collapse if confronted with SLR of the order of 1.0 to 1.2 mm per year over an extended period of the order of decades. SLR may pose the greatest threat to mangroves where the mangrove sediment surface levels are not keeping pace with SLR. Mangroves in the Key West of Florida, for example, have shifted inland by 1.5 km since the mid-1940s under a regime of SLR of 2.3 to 2.7 mm per year (Ross *et al.*, 2000). The greatest impact will be on those mangroves where there is limited area for landward migration, being constrained by human settlements or local topography (Gilman *et al.*, 2008). Several studies suggested a possible regime shift in community composition involving entire mangrove forests, in response to prolonged SLR. Most vulnerable to the impact of SLR are those mangroves occupying low-relief islands or those located in carbonate settings, where the rates of sediment supply and where available upland space for migration are low, such as small islands and atolls in the Pacific. Also, most vulnerable are coastal forests where the rivers are lacking in sediment supply or where the landform is subsiding. However, globally, mangroves appear to keep pace with SLR, because average sedimentation rates are in equilibrium with the rate of mean SLR. The least vulnerable are those mangroves located in macro-tidal estuaries, in tropical coastal wetlands or in shores adjacent to rivers that provide sufficient flows of sediments. Mangroves that are likely to be resilient to SLR are those that occupy high-relief islands, located in remote areas with minimal human interruptions to inhibit landward migration (Alongi, 2008). Mangroves will move inland if the pace of SLR is moderate and if other environmental, topographic and hydrological conditions are suitable for mangroves to migrate upland. Because of their topographic position in the intertidal zone, mangroves are directly affected by tidal flows and upland river flows. The rate of SLR in relation to local topography, sediment input from upland rivers, coastal slope and tidal elevation is a major factor that exerts the most impact on mangrove resilience. The nature, sources and amount of sediment delivered from upland rivers to the estuary, together with area available for mangrove landward migration, will determine the migration paths of mangroves (Woodroffe, 1990; Gilman *et al.*, 2008). Vertical sediment accumulation within the mangrove prop root systems may allow mangroves to keep up with the rate of SLR in areas of higher elevation coupled with relatively low tidal range. In these areas, mangroves will progressively move inland to areas where suitable conditions are available as sea levels rise (Ellison, 2000). On the contrary, in areas of lower elevation coupled with greater tidal range, sedimentation may not be able to keep pace with SLR (Ellison, 2000; Ward *et al.*, 2016). In these areas, mangroves will be lost without any replacement.

Mangroves in the tropics may not be much affected by rising global temperature. But, in the temperate higher latitudes, increase in temperature under GCC may allow mangroves to migrate and colonize farther poleward. Increasing carbon dioxide and nitrogen enrichment can augment the growth of Spartina to suppress growth of mangrove seedlings, and thus change the competitive relationship between the vegetation types (McKee and Rooth, 2008; Zhang *et al.*, 2012b). Snedaker and Araujo (1998) have shown that increasing carbon dioxide concentrations can alter the competitiveness of mangroves, particularly that of the species *Laguncularia racemosa*. These impacts pose challenges for management and conservation of mangroves in the future. As current approaches may not be adequate or suitable, innovative adaptation is required to confront increasingly more challenging and constantly changing environments.

6.10.2 *Regime shift induced by salinity*

In the past half century, mangroves have migrated about 3.3 km inland in the southern Everglades, largely at the expense of freshwater marshes and swamp forest (Gaiser *et al.*, 2006). Sea level rise and upstream hydrological manipulation have been identified as the main reasons driving this mangrove inland migration. It is speculated that strong hurricane storm surges could accelerate the spread of mangroves into the areas previously occupied by freshwater vegetation like hardwood hammocks. The possibility of vegetation regime shift induced by salinity alteration due to SLR and repeated hurricane storm surges has stimulated research to understand what specific perturbations might trigger and sustain such vegetation changes (Teh *et al.*, 2015). GCC and its associated SLR and increases in hurricane occurrence frequency and intensity are perturbations that might induce the existing vegetation to shift rapidly from one type to the other. Rising global temperature might lead to more severe hurricanes as warmer temperatures ultimately heat up the ocean water, which in turn would supply more thermal energy for storms to intensify as the storms swirl through the ocean (Walsh, 2004; Webster *et al.*, 2005; Emanuel, 1987, 2005). Along most of the US Atlantic and Gulf of Mexico coasts, the sea level has been rising by about 2 to 3 mm per year and sea level is likely to rise by 55 to 60 cm over the next century (Titus and Richman, 2000). These SLRs will likely increase salinity levels in the soil along the affected coastal zones. An increase in salinity of the vadose zone (unsaturated zone of soil) induced by these events might eradicate the salinity-intolerant hardwood hammocks that previously thrived at higher elevations with low salinity (Teh *et al.*, 2009). This will promote landward migration of salinity-tolerant mangroves to replace the demised hammocks. Inland expansion and invasion of mangroves at the expense of freshwater vegetation has been the subject of considerable interest among scientists working on coastal ecosystems (Alexander and Crook, 1974; Gleason *et al.*, 1974; Willard *et al.*, 1999; Williams *et al.*, 1999; Lara *et al.*, 2002).

6.10.3   *Failure of conservation approaches*

Many mangrove conservation programs worldwide are plagued with numerous problems. Successful management, conservation and restoration of mangrove require the commitment of local, state and national-level governments as well as local communities. As trends of human settlement along coastal areas continue around the world, people are increasingly socially and culturally distancing themselves from nature, from mangroves and from nature conservation (Miller, 2005; Zaradic *et al.*, 2009). Additionally, benefits and services derived from mangroves and their conservation might be indirect or poorly understood by local residents. Therefore, they may perceive that they are being excluded from access to mangrove resources (Shackelton *et al.*, 2002).

6.11   Summary

Coastal mangrove forests provide invaluable and irreplaceable services to people and wildlife, making their protection critical. As loss of mangrove habitat continues throughout much of the world, solutions to reverse this downward trend will need to involve commitments from many parties. Policies enacted and enforced by national governments are one part of that protection, but are not sufficient in isolation. An examination of studies throughout the world can help guide the implementation of successful conservation and restoration practices. These restoration efforts, when put into practice, will require the buy-in and commitment of local people with the will to conserve these habitats that could otherwise serve limited personal gain through timber harvest, for example. Citizen science can also be employed to engage people and benefit from their local knowledge. Information and data provided by locals have been used for conservation initiatives for protected area planning (e.g., Scholz *et al.*, 2004), natural resource monitoring (e.g., Giordano *et al.*, 2010) and disease threats (e.g., Dhondt *et al.*, 1998). Data collected can then be used to assist in monitoring, modeling and restoration-conservation planning. With the engagement of local people living within these resources, the current rates of decline can be slowed and ultimately reversed.

There remains a paucity of academic literature examining the utilization of ESEV by decision-makers. Overall, ESEV is used more for communication and advocacy than for decision-making or for setting up economic and financial instruments (Marre *et al.*, 2015). In many nations, ESEV has not been perceived as critical in the design, adoption or implementation of policy measures regarding coastal and marine ecosystems (Marre *et al.*, 2015). Ideally, collaborations between researchers and decision-makers should be enhanced to facilitate research and knowledge transfer, as well as to promote uptake and enhance impact within policy contexts. ESEV studies can generate information on the costs associated with species and habitat loss, the benefits of conservation and restoration efforts, and economic dependence on natural ecosystems. Appropriate and timely ESEV of ecological services that contribute to better informed decision-making can benefit the protection, restoration and development of coastal resources in a sustainable manner.

Climate change and SLR will have varying impacts on mangrove habitats throughout the world. In the face of SLR, some regions, such as Florida, will have the advantage of having more room for mangroves to expand inland. In more populated coastal regions without a protected coastal zone, there is more likely to be a problem that humans occupy areas inland, thus potentially restricting inland migration of mangrove. Higher temperatures might not impact mangroves in tropical regions such as Malaysia. But, away from the equator, warming temperatures may promote northward expansion. Proactive engagement with governments and the public in these potential regions of mangrove expansion may help with conservation efforts. Identifying and implementing approaches to the conservation and restoration of mangrove habitats that can be applied across the globe may accelerate learning and protection efforts. Here, we have provided insights from many countries around the world on various successes and failures in mangrove conservation and restoration. These insights can provide a platform to identify successful solutions. Such global learning and action may secure the future of mangrove forests and the services they provide worldwide.

References

Affandi, N.M., Babak, K., Rozainah, M.Z., Noraini, M.T. and Hashim, R. (2010). Early growth and survival of Avicennia alba seedlings under excessive sedimentation. *Scientific Research and Essays*, 5, 2801–2805.

Aheto, D.W., Kankam, S., Okyere, I., Mensah, E., Osman, A., Jonah, F.E. and Mensah, J.C. (2016). Community-based mangrove forest management: Implications for local livelihoods and coastal resource conservation along the Volta estuary catchment area of Ghana. *Ocean and Coastal Management*, 127, 43–54.

Alexander, T.J. and Crook, A.G. (1974). Recent vegetational changes in southern Florida. In: Gleason, P.J. (ed.), *Environments of Southern Florida: Present and Past*. Memoir 2, Miami Geological Society, Miami, Florida, 61–72.

Allen, J. (1998). Mangroves as alien species: The case of Hawaii. *Global Ecology and Biogeography Letters*, 7, 61–71.

Alongi, D.M. (2008). Mangrove forests: Resilience, protection from tsunamis and responses to global climate change, *Estuarine, Coastal and Shelf Science*, 76, 1–13. doi: 10.1016/j.ecss.2007.08.024.

Alongi, D.M. (2002). Present state and future of world's mangrove forests. *Environmental Conservation*, 29, 331–349.

Alongi, D.M. (2012). Carbon sequestration in mangrove forests. *Carbon Mgnt.*, 3, 313–322.

Anneboina, L.R. and Kumar, K.S.K. (2017). Economic analysis of mangrove and marine fishery linkages in India. *Ecosystem Services*, 24, 114–123.

Armitage, A.R., Highfield, W.E., Brody, S.D. and Louchouarn, L. (2015). The contribution of mangrove expansion to salt marsh loss on the Texas Gulf Coast. *PLoS ONE*, 10, e0125404.

Baldwin, A., Egnotovich, M., Ford, M. and Platt, W. (2001). Regeneration in fringe mangrove forests damaged by Hurricane Andrew. *Plant Ecology*, 157(2), 149–162.

Barbier, E.B. (2016). The protective service of mangrove ecosystems: A review of valuation methods. *Marine Pollution Bulletin*, 109(2), 676–681.

Barbier, E.B., Koch, E.W., Silliman, B.R., Hacker, S.D., Wolanski, E., Primavera, J., Granek, E.F., Polasky, S., Aswani, S., Cramer, L.A. and Stoms, D.M. (2008). Coastal ecosystem-based management with nonlinear ecological functions and values. *Science*, 319, 321–323.

Berkes, F. (2004). Rethinking community-based conservation. *Conservation Biology*, 18, 621–630.

Borrini-Feyerabend, G., Dudley, N., Jaeger, T., Lassen, B., Broome, N.P. and Phillips, A. (2012). Governance of protected areas: from understanding to action. The World Conservation Union (IUCN), Gland, Switzerland.

Brander, L.M., Wagtendonk, A.J., Hussain, S.S. *et al.* (2012). Ecosystem service values for mangroves in Southeast Asia: A meta-analysis and value transfer application. *Ecosystem Services*, 1, 62–69, doi: 10.1016/j.ecoser.2012.06.003.

Buckingham, K. and Hanson, C. (2015). The restoration diagnostic. Case example: Restoration of mangrove forests in Vietnam. World Resources Institute, Washington DC.

Bush, S.R., van Zwieten, P.A.M., Visser, L., van Dijk, H., Bosma, R., de Boer, W.F. and Verdegem, M. (2010). Scenarios for resilient shrimp aquaculture in tropical coastal areas. *Ecology and Society*, 15(2), 15.

Cat, N.N., Tien, P.H., Sam, D. and Binh N. (2006). Status of coastal erosion of Viet Nam and proposed measures for protection. *Food and Agriculture Organization of the United Nations*, Rome, http://www.fao.org/forestry/11286-08d0cd86bc02ef85da8f5b6249401b52f.pdf.

Chatenoux, B. and Peduzzi, P. (2007). Impacts from the 2004 Indian Ocean tsunami: Analysing the potential protecting role of environmental features. *Natural Hazards*, 40(2), 289–304.

Chen, L.Z., Wang, W.Q., Zhang, Y.H. and Lin, G.H. (2009). Recent progresses in mangrove conservation, restoration and research in China. *Plant Ecology*, 2, 45–54.

Chen, L.Z., Zeng, X.Q., Tam, N.F.Y. *et al.* (2012). Comparing carbon sequestration and stand structure of monoculture and mixed mangrove plantations of Sonneratia caseolaris and S. apetala in Southern China. *Forest Ecology and Management*, 284, 222–229, doi: 10.1016/j.foreco.2012.06.058.

Chow, V.T. (1959). *Open Channel Hydraulics*. New York: McGraw-Hill Book Company.

Danielsen, F., Sørensen, M.K., Olwig, M.F., Selvam, V., Parish, F., Burgess, N.D., Hiraishi, T., Karunagaran, V.M., Rasmussen, M.S., Hansen, L.B., Quarto, A. and Suryadiputra, N. (2005). Asian tsunami: A protective role for coastal vegetation. *Science*, 310(5748), 643.

Datta, D., Chattopadhyay, R.N. and Guha., P. (2012). Community based mangrove management: A review on status and sustainability. *Journal of Environmental Management*, 107, 84–95.

Davis, S.M., Childers, D.L., Lorenz, J.J., Wanless, H.R. and Hopkins, T.E. (2005). A conceptual model of ecological interactions in the mangrove estuaries of the Florida Everglades. *Wetlands*, 25, 832–842.

De Graaf, G. and Xuan, T. (1998). Extensive shrimp farming, mangrove clearance and marine fisheries in the southern provinces of Vietnam. *Mangrove Salt Marshes*, 2, 159–166.

Dharmawan, B., Böcher, M. and Krott, M. (2016). The failure of the mangrove conservation plan in Indonesia: Weak research and an ignorance of grassroots politics. *Ocean and Coastal Management*, 130, 250–259.

Dhondt, A.A., Tessaglia, D.L. and Slothower, R.L. (1998). Epidemic mycoplasmal conjunctivitis in house finches from eastern North America. *Journal of Wildlife Diseases*, 34(2), 265–280.

Donato, D.C., Kauffman, J.B., Murdiyarso, D. *et al.* (2011). Mangroves among the most carbon rich forests in the tropics. *Nature Geoscience*, 4, 293–297, doi: 10.1038/ngeo1123.

Doughty, C.L., Langley, J.A., Walker, W.S., Feller, I.C., Schaub, R. and Chapman, S.K. (2016). Mangrove range expansion rapidly increases coastal wetland carbon storage. *Estuaries and Coasts*, 39, 385–396.

Doyle, T.W., Smith III, T.J. and Robblee, M.B. (1995). Wind damage effects of Hurricane Andrew on mangrove communities along the southwest coast of Florida, USA. *Journal of Coastal Research*, 21, 159–168.

Duke, N.C., Pinzon, Z.S. and Prada, M.C. (1997). Large-scale damage to mangrove forests following two large oil spills in Panama. *Biotropica*, 29, 2–14.

Edward, B. (2008). *Tsunami the Underrated Hazard*. Chichester: Praxis Publishing.

Ellison, J.C. (2000). How South Pacific mangroves may respond to predicted climate change and sea-level rise, in: Gillespie, A., Burns, W.C.G. (eds.), Climate change in the South Pacific: Impacts and Responses in Australia, New Zealand, and Small Island States. Springer Netherlands, pp. 289–300.

Emanuel, K. (1987). The dependence of hurricane intensity on climate. *Nature*, 326, 483–485.

Emanuel, K. (2005). Increasing destructiveness of tropical cyclones over the past 30 years. *Nature*, 436, 686–688.

Ewel, K.C., Twilley, R.R. and Ong, J.-E. (1998). Different kinds of mangrove forests provide different goods and services. *Global Ecology and Biogeography Letters*, 7, 83–94.

Fan, H., He, B. and Pernetta, J.C. (2013). Mangrove ecofarming in Guangxi Province China: An innovative approach to sustainable mangrove use. *Ocean and Coastal Management*, 85, 201–208.

Fan, H.Q., Liu, W.A., Zhong, C.R. and Ni, X. (2014). Analytic study on the damages of wood-boring isopod, Sphaeroma, to China mangroves. *Guangxi Science*, 21(2), 140–146, 152.

Ferreira, A.C. and Lacerda, L.D. (2016). Degradation and conservation of Brazilian mangroves, status and perspectives. *Ocean and Coastal Management*, 125, 38–46.

FSI (2011). India State of Forest Report 1987–2011. Forest Survey of India, Ministry of Environment and Forests, Government of India.

Gaiser, E.E., Zafiris, A., Ruiz, P.L., Tobias, F.A.C. and Ross, M.S. (2006). Tracking rates of ecotone migration due to salt-water encroachment using fossil mollusks in coastal South Florida. *Hydrobiologia*, 569, 237–257.

Gilman, E.L., Ellison, J., Duke, N.C. and Field, C. (2008). Threats to mangroves from climate change and adaptation options: a review. *Aquatic Botany*, 89, 237–250.

Giordano, R., Liersch, S., Vurro, M. and Hirsch, D. (2010). Integrating local and technical knowledge to support soil salinity monitoring in the Amudarya river basin. *Journal of Environmental Management*, 91, 1718–1729.

Giri, C., Ochieng, E., Tieszin, L.L., Zhu, Z., Singh, A., Loveland, T., Masek, J. and Duke, N. (2011). Status and distribution of mangrove forests of the world using earth observation satellite data. *Global Ecology and Biogeography*, 20, 154–159.

Gleason, P.J., Cohen, A.D., Brooks, H.K., Stone, P., Smith, W.G. and Spackman, Jr., W. (1974). The environmental significance of Holocene sediments from the Everglades and saline tidal plain. In: P.J. Gleason, (ed.), Environments of Southern Florida: Present and Past. Memoir 2: Miami Geological Society, Miami, Florida, 61–72.

Ha, T.T.T., van Dijk, H. and Bush, S.R. (2012). Mangrove conservation or shrimp farmer's livelihood? The devolution of forest management and benefit sharing in the Mekong Delta, Vietnam. *Ocean and Coastal Management*, 69, 185–193.

Hai, N., Dell, B., Phuong, V. *et al.* (2020). Towards a more robust approach for the restoration of mangroves in Vietnam. *Annals of Forest Science*, 77, 18, doi: 10.1007/s13595-020-0921-0.

Hamdan, O., Khali Aziz, H. and Shamsudin, I. (2010). Kajian Peruba-han Hutan Paya Laut Negeri Selangor, Kedah dan Kelantan (Study on the Changes of Mangrove Forest in Selangor, Kedah and Kelantan). Forest Research Institute Malaysia (FRIM) Reports No. 91. https://info.frim.gov.my/infocenter/Korporat/2003Publicat ions/Links/FRIM11.htm.

Hamdan, O., Khali Aziz, H., Shamsudin, I. and Raja Barizan, R.S. (2012). Status of Mangrove in Peninsular Malaysia. Forest Research Institute Malaysia (FRIM), Kepong, Selangor Darul Ehsan, Malaysia.

Hamilton, S.E. and Casey, D. (2016). Creation of a high spatio-temporal resolution global database of continuous mangrove forest cover for the 21st century (CGMFC-21). *Global Ecology and Biogeography*, 25(6), 729–738.

Harada, K. and Imamura, F. (2005). Effects of coastal forest on tsunami hazard mitigation-a preliminary investigation. In: K. Satake, ed. 2005. *Tsunamis: Case Studies and Recent Developments*. The Netherlands: Springer, pp. 279–292.

Harada, K. and Kawata, Y. (2004). Study on the effect of coastal forest to tsunami reduction. Annuals of Disaster Prevention Institute, Kyoto University 47(C).

Hawkins, S., Robertson, S., Thu Thuy, P., Xuan To, P., McNally, R., Van Cuong, C., Dart, P., Xuan Phuong, P., Brown, S. and Vu, N. (2010). Roots in the water: Legal frameworks for mangrove PES in Viet-nam. Katoomba Group's Legal Initiative Country Study Series. Forest Trends, Washington, DC.

He, Y. and Zhang, M.-X. (2001). Study on wetland loss and its reasons in China. *Chinese Geographical Science*, 11, 241–245.

Hill, E.M., Borrero, J.C., Huang, Z., *et al.* (2012). The 2010 Mw 7.8 Mentawai earthquake: Very shallow source of a rare tsunami earthquake determined from tsunami field survey and near-field GPS data. *Journal of Geophysical Research*, 117, B06402, doi: 10.1029/2012JB009159.

Horrillo, J., Grilli, S.T., Nicolsky, D., Roeber, V. and Zhang, J. (2015). Performance benchmarking tsunami models for NTHMP's inundation mapping activities. *Pure and Applied Geophysics* 172(3), 869–884.

Ilman, M., Dargusch, P., Dart, P. and Onrizal (2016). A historical analysis of the drivers of loss and degradation of Indonesia's mangroves. *Land Use Policy*, 54, 448–459, doi: 10.1016/j.landusepol.2016.03.010.

Kathiresan, K. (2002). Why are mangroves degrading? *Current Science*, 83, 1246–1249.

Kathiresan, K. and Rajendran, N. (2005). Coastal mangrove forests miti-gated tsunami. *Estuarine Coastal and Shelf Science*, 65, 601–606.

Kimball, M. and Teas, H.J. (1975). Nitrogen fixation in mangrove areas of South Florida, in: Walsh, G.E., Snedaker, S.C., Teas, H.J. (eds.), *Proceedings of the International Symposium on Biology and Management of Mangroves*, Vol. 2. University of Florida, Gainesville, USA, pp. 654–661.

Koch, M.S. and Snedaker, S.C. (1997). Factors influencing Rhizophora mangle L. seedling development in Everglades carbonate soils. *Aquatic Botany*, 59, 87–98.

Koh, H.L., Teh, S.Y., Kew, L.M. and Zakaria, N.A. (2009). Simulation of future Andaman Tsunami into Straits of Malacca by TUNA. *Journal of Earthquake and Tsunami*, 3(2), 89–100, doi: 10.1142/S1793431109000470.

Kruczynski, W.L. and McManus, F. (2002). Water quality concerns in the Florida Keys: Sources, effects, and solutions, in: Porter, J.W., Porter, K.G. (Eds.), The Everglades, Florida Bay, and Coral Reefs of the Florida Keys: An Ecosystem Sourcebook. CRC Press, Boca Raton, Florida, USA, pp. 827–881.

Laegdsgaard, P. and Johnson, C. (2001). Why do juvenile fish utilize mangrove habitats? *Journal of Experimental Marine Biology and Ecology*, 257, 229–253.

Lara, R., Szlafsztein, C., Cohen, M., Berger, U. and Glaser, M. (2002). Implications of mangrove dynamics for private land use in Bragança, North Brazil: A case study. *Journal of Coastal Conservation*, 8, 97–102.

Le, H.D., Smith, C., Herbohn, J. and Harrison, S. (2012). More than just trees: Assessing reforestation success in tropical developing countries. *Journal of Rural Studies*, 28, 5–19.

Lebel, L., Tri, N.H., Saengnoree, A., Pasong, S., Buatama, U. and Thoa, L.K. (2002). Industrial transformation and shrimp aquaculture in Thailand and Vietnam: Pathways to ecological, social, and economic sustainability? *AMBIO: Journal of Human Environment*, 31, 311–323.

Lewis III, R.R., Milbrandt, E.C., Brown *et al.* (2016). Stress in mangrove forests: Early detection and preemptive rehabilitation are essential for future successful worldwide mangrove forest management. *Marine Pollution Bulletin*, 109, 764–771.

Li, M.S. and Lee, S.Y. (1997). Mangroves in China: A brief review. *Forest Ecology and Management*, 96, 241–259.

Li, Y.H. (2012a). *Ecology and Conservation of Estuarine Wetland in Quanzhou Bay*. Environmental Science Press, Beijing, China.

Li, Y.H. (2012b). Effect and countermeasure of the global climate change on the mangrove eco-system in Quanzhou Bay estuary. *Straits Science*, 2, 10–12.

Liu, H., Zhang, K., Li, Y. and Xie, L. (2013). Numerical study of the sensitivity of mangroves in reducing storm surge and flooding to hurricane characteristics in southern Florida. *Continental Shelf Research*, 64, 51–65.

Lundberg, J. and Moberg, F. (2003). Mobile link organisms and ecosystem functioning: Implications for ecosystem resilience and management. *Ecosystems*, 6, 87–98.

MA, Millennium Ecosystem Assessment (2003). *Ecosystems and Human Well-being: A Framework for Assessment*, Island Press, Washington DC.

MA, Millennium Ecosystem Assessment (2005). *Ecosystems and Human Well-being: Synthesis*, Island Press, Washington, DC.

Mahon, R., Fanning, L. and McConney, P. (2014). Assessing and facilitating emerging regional ocean governance arrangements in the Wider Caribbean Region. *Ocean Yearbook*, 28, 631–671.

Máñez, K.S., Krause, G., Ring, I. and Glaser, M. (2014). The Gordian knot of mangrove conservation: Disentangling the role of scale, services and benefits. *Global Environmental Change*, 28, 120–128.

Mansor, M. and Zakaria, M.Y. (2004). Ecological survey on mangrove forests: A case study of Balik Pulau and Pantai Acheh. Pulau Pinang: Universiti Sains Malaysia.

MARD (2014). Coastal forest protection and development plan to respond to climate change for the period 2015–2020 (approved at decision 120/QD-TTg dated 22 January 2015 of Vietnamese Prime Minister). Ministry of Agriculture and Rural Development, Hanoi.

MARD (2015). Synthesis report: Master plan on shrimp cultivation in Mekong River Delta of Vietnam to 2020, vision to 2030. Ministry of Agriculture and Rural Development, Hanoi.

Marre, J., Thebaud, O., Pascoe, S., Jennings, S., Boncoeur, J., Coglan, L. (2015). The use of ecosystem services valuation in Australian coastal zone management. *Marine Policy*, 56, 117–124.

Mascarenhas, A. and Jayakumar, S. (2008). An environmental perspective of the post-tsunami scenario along the coast of Tamil Nadu, India: Role of sand dunes and forests. *Journal of Environmental Management*, 89(1), 24–34.

Mazda, Y., Magi, M., Kogo, M., and Hong, P.N. (1997a). Mangrove as a coastal protection from waves in the Tong King delta, Vietnam. *Journal of Mangrove Salt Marshes*, 1, 127–135.

Mazda, Y., Wolanski, E., King, B., Sase, A., Ohtsuka, D. and Magi, M. (1997b). Drag force due to vegetation in mangrove swamps. *Mangroves and Salt Marshes*, 1(3), 193–199.

McIvor, C.C., Ley, J.A. and Bjork, R.D. (1994). Changes in freshwater inflow from the Everglades to Florida Bay including effects on biota and biotic processes: A review, in: Davis, S.M., Ogden, J.C. (Eds.), *Everglades: The Ecosystem and Its Restoration*. St. Lucie Press, Delray Beach, Florida, USA, pp. 117–146.

McKee, K.L. and Rooth, J.E. (2008). Where temperate meets tropical: Multi-factorial effects of elevated CO_2, nitrogen enrichment, and competition on a mangrove-salt marsh community. *Global Change Biology*, 14, 971–984.

McLeod, E., Chmura, G., Bouillon, S., Salm, R., Bjork, M., Duarte, C.M., Lovelock, C.E., Schlesinger, W.H. and Silliman, B.R. (2011). A blueprint for blue carbon: Toward an improved understanding of the role of vegetated coastal habitats is sequestering CO_2. *Frontiers in Ecology and the Environment*, 9, 552–560.

McNally, R., McEwin, A. and Holland, T. (2011). *The Potential for Mangrove Carbon Projects in Vietnam. Netherlands Development Organization* (SNV). The Hague, Netherlands, http://www.bibalex.org/search4dev/files/419125/442099.pdf.

Melana, D.M., Atchue III, J., Yao, C.E., Edwards, R., Melana, E.E. and Gonzales, H.I. (2000). Mangrove Management Handbook. Department of Environment and Natural Resources, Manila, Philippines Through the Coastal Resource Management Project, Cebu City, Philippines.

Michot, T.C., Day, R.H. and Wells, C.J. (2010). Increase in black mangrove abundance in coastal Louisiana. Louisiana natural resource news. *Newsl. Louisiana Association Professor Biology*, January, 4–5.

Miller, J.R. (2005). Biodiversity conservation and the extinction of experience. *Trends in Ecology and Evolution*, 20(8), 430–434.

Nagelkerken, I., van der Velde, G., Gorissen, M.W., Meijer, G.J., Van't Hof, T. and den Hartog, C. (2000). Importance of mangroves, seagrass beds and the shallow reef as a nursery for important coral reef fishes, using a visual census technique. *Estuarine, Coastal and Shelf S.*, 51, 31–44.

NAHRIM (2006). Study of the impact of climate change on the hydrologic regime and water resources of Peninsular Malaysia. Kuala Lumpur: National Hydraulic Research Institute of Malaysia, Kuala Lumpur.

Nelson, G. (1994). *The Trees of Florida: A Reference and Field Guide* (Reference and Field Guides). Pineapple Press, Saratosa, Florida, USA.

Nguyen, C.H. (2012). Investing in coastal ecosystems: A guiding document for journalists about the role and importance of coastal ecosystems. IUCN: Gland, Switzerland, https://www.iucn.org/sites/dev/files/content/documents/iucn_sach_dung_13_12_2012_en_final.pdf.

Nguyen, T.P. and Parnell, K.E. (2017b). Gradual expansion of mangrove areas as an ecological solution for stabilizing a severely eroded mangrove dominated muddy coast. *Ecological Engineering*, 107, 239–243.

Nguyen, T.P., Luom, T.T. and Parnell, K.E. (2017a). Existing strategies for managing mangrove dominated muddy coasts: Knowledge gaps and recommendations. *Ocean and Coastal Management*, 138, 93–100.

Odum, W.E., McIvor, C.C. and Smith III, T.J. (1982). The Ecology of the Mangroves of South Florida: A Community Profile. United States Fish and Wildlife Service, Washington, DC, USA. FWS/OS–81/24.

Osland, M.J., Day, R.H., From, A.S., McLemore, M.L., McCoy, J. and Kelleway, J. (2015). Life stage influences the resistance and resilience of black mangrove forests to winter climate extremes. *Ecosphere*, 6, 1–15.

Osland, M.J., Enwright, N., Day, R.H. and Doyle, T.W. (2013). Winter climate change and coastal wetland foundation species: Salt marshes vs. mangrove forests in the southeastern United States. *Global Change Biology*, 19(5), 1482–1494.

Pelegri, S.P. and Twilley, R.R. (1998). Heterotrophic nitrogen fixation (acetylene reduction) during leaf-litter decomposition of two mangrove species from south Florida, USA. *Marine Biology*, 131, 53–61.

Peng, Y., Zheng, M., Zheng, Z. *et al.* (2016). Virtual increase or latent loss? A reassessment of mangrove populations and their conservation in Guangdong, southern China. *Marine Pollution Bulletin*, 109, 691–699, doi: 10.1016/j.marpolbul.2016.06.083.

Peng, Y.S., Chen, G.Z., Li, S.Y., Liu, Y. and Pernetta, J. (2013). Use of degraded coastal wetland in an integrated mangrove aquaculture system: A case study from the South China Sea. *Ocean and Coastal Management*, 85, 209–213, doi: 10.1016/j.ocecoaman.2013.04.008.

Powell, N., Gerger Swartling, Å. and Hoang, M.H. (2011b). Stakeholder agency and rural development policy: Articulating co-governance in Vietnam. ICRAF World Agroforestry Centre, Hanoi, p. 166.

Powell, N., Osbeck, M., Tan, S.B. and Toan, V.C. (2011a). *Mangrove Restoration and Rehabilitation for Climate Change Adaptation in Vietnam*. World Resources Report Case Study World Resources Report, Washington DC.

Primavera, J.H. and Esteban, J.M.A. (2008). A review of mangrove rehabilitation in the Philippines: Successes, failures and future prospects. *Wetlands Ecology and Management*, 16, 345–358, doi: 10.1007/s11273-008-9101-y.

Proctor, J., Anderson, J.M., Chai, P. and Vallack, H.W. (1983). Ecological studies in four contrasting lowland rain forests in Gunung

Mulu National Park, Sarawak. *Journal of Ecology*, 71, 237–260, doi: 10.2307/2259975.

Queiroz, L.S., Rossi, S., Calvet-Mir, L., Ruiz-Mallén, I., García-Betorz, S., Salvà-Prat, J. and Meireles, A.J.A. (2017). Neglected ecosystem services: Highlighting the socio-cultural perception of mangroves in decision-making processes. *Ecosystem services*, 26, 137–145.

Ren, H., Lu, H., Shen, W., Huang, C., Guo, Q., Li, Z.A. and Jian, S. (2009). Sonneratia apetala Buch. Ham in the mangrove ecosystems of China: An invasive species or restoration species? *Ecological Engineering*, 35(8), 1243–1248.

Rey, J.R., Crossman, R.A. and Kain, T.R. (1990). Vegetation dynamics in impounded marshes along the Indian River lagoon, Florida, USA. *Environmental Management*, 14, 397–409.

Richards, D.R. and Friess, D.A. (2016). Rates and drivers of mangrove deforestation in Southeast Asia, 2000–2012. *Proceedings of the National Academy of Sciences*, 113, 344–349.

Ricklefs, R.E. and Latham, R.E. (1993). Global patterns of diversity in mangrove floras. In: Ricklefs, R.E. and Schulter, D. (eds.) *Species Diversity in Ecological Communities*. University of Chicago Press, Chicago, pp. 215–229.

Ross, M.S., Meeder, J.F., Sah, J.P., Ruiz, P.I and Telesnicki, G.J. (2000). The Southwest Saline Everglades revisited: 50 years of coastal vegetation change. *Journal of Vegetable Science*, 11, 101–112, doi: 10.2307/3236781.

Roy, S.D. and Krishnan, P. (2005). Mangrove stands of Andamans vis-à-vis tsunami. *Current Science*, 89, 1800–1804.

Sabah Biodiversity Centre (2011). Lower Kinabatangan-Segama Wetlands Ramsar Site Management Plan. Volume. I: Sabah State Government, Malaysia, p. 265.

Saenger, P. (2002). *Mangrove Ecology, Silviculture and Conservation*. Kluwer Academic Publishers, Dordrecht.

Sahu, S.C., Suresh, H.S., Murthy, I.K. and Ravindranath, N.H. (2015). Mangrove area assessment in India: Implications of loss of mangroves. *Journal of Earth Science and Climatic Change*, 6(5), 280.

Saintilan, N., Wilson, N.C., Rogers, K., Rajkaran, A. and Krauss, K.W. (2014). Mangrove expansion and salt marsh decline at mangrove poleward limits. *Global Change Biology*, 20, 147–157.

Satheeshkumar, P., Manjusha, U., Pillai, N.G.K. and Kumar, D.S. (2012a). Puducherry mangroves under sewage pollution threat need conservation. *Current Science*, 102, 13–14.

Satheeshkumar, P., Siva Sankar, R, Senthil Kumar, D. and Athithan, A. (2012b). Did mangroves offer an effective barrier to the Thane cyclone surges? *Current Science*, 103, 981–982.

Schmitt, K. and Duke, N.C. (2014). Mangrove management, assessment and monitoring. In: Kohl, M., Pancel, L. (eds.) *Tropical forestry handbook*. Springer, Berlin, pp. 1–29.

Scholz, A., Bonzon, K., Fujita, R., Benjamin, N., Woodling, N., Black, P. and Steinback, C. (2004). Participatory socioeconomic analysis: Drawing on fishermen's knowledge for marine protected area planning in California. *Marine Policy*, 28(4), 335–349.

Schuhmann, P.W. and Mahon, R. (2015). The valuation of marine ecosystem goods and services in the Caribbean: A literature review and framework for future valuation efforts. *Ecosystem Services*, 11, 56–66.

Shackelton, S., Campbell, B., Wollenberg, E. and Edmunds, D. (2002). *Devolution and Community-based Natural Resource Management: Creating Space for Local People to Participate and Benefit?* Overseas Development Institute, London, United Kingdom.

Sheaves, M. (2009). Consequences of ecological connectivity: The coastal ecosystem mosaic. *Marine Ecology Progress Series*, 391, 107–115.

Shuto, N. (1987). The effectiveness and limit of tsunami control forests. *Coastal Engineering in Japan*, 30, 143–153.

Smith, T.J. III, Anderson, G.H., Balentine, K., Tiling, G., Ward, G.A. and Whelan, K.R.T. (2009). Cumulative impacts of hurricanes on Florida mangrove ecosystems: Sediment deposition, storm surges, and vegetation. *Wetlands*, 29, 24–34.

Snedaker, S.C. and Araujo, R.J. (1998). Stomatal conductance and gas exchange in four species of Caribbean mangroves exposed to ambient and increased CO_2. *Marine and Freshwater Research*, 49, 325–327.

Sorensen, J. and McCreary, S. (1990). Institutional arrangements for managing coastal resources and environments, second edition, in: COAST, Renewable Resources Information Series. National Park Service, US Department of the Interior and USAID, Coastal Management Publication No. 1.

Spalding, M., Blasco, F. and Field, C. (1997). World Mangrove Atlas. The International Society for Mangrove Ecosystems, Okinawa, Japan.

Stojanovic, T., Ballinger, R.C. and Lalwani, C.S. (2004). Successful integrated coastal management: Measuring it with research and contributing to wise practice. *Ocean and Coastal Management*, 47, 273–298.

Strong, A.M. and Bancroft, G.T. (1994). Patterns of deforestation and fragmentation of mangrove and deciduous seasonal forests in the Upper Florida Keys. *Bulletin of Marine Science*, 54, 795–804.

Suppasri, A., Shuto, N., Immamura, F., Koshimura, S., Mas, E. and Yalciner, A.C. (2011). Lessons learned from the 2011 Great East Japan tsunami: Performance of tsunami countermeasures, coastal buildings and tsunami evacuation in Japan. *Pure and Applied Geophysics*, 170(6–8), 993–1018.

Swiadek, J.W. (1997). The impacts of hurricane andrew on mangrove coasts in Southern Florida: A review. *Journal of Coastal Research*, 13(1), 242–245.

Takagi, H., Thao, N.D. and Esteban, M. (2014). Tropical cyclones and storm surges in southern Vietnam. In: Thao, N.D., Takagi, H. and Esteban, M. (eds.) *Coastal Disasters and Climate Change in Vietnam: Engineering and Planning Perspectives*. Elsevier, Berlin, pp. 3–16.

Tamin, N.M., Zakaria, R., Hashim, R. and Yin, Y. (2011). Establishment of Avicennia marina mangroves on accreting coastline at Sungai Haji Dorani, Selangor, Malaysia. *Estuarine, Coastal and Shelf Science*, 94, 334–342.

Tanaka, N. (2009). Vegetation bioshields for tsunami mitigation: Review of effectiveness, limitations, construction and sustainable management. *Landscape and Ecological Engineering*, 5(1), 71–79.

Tanaka, N., Sasaki, Y., Mowjood, M.I.M., Jinadasa, K.B.S.N. and Homchuen, S. (2007). Coastal vegetation structures and their functions in tsunami protection experience of the recent Indian Ocean tsunami. *Landscape and Ecological Engineering*, 3(1), 33–45.

Teh, S.Y., Koh, H.L., Liu, P.L.-F., Izani, A.M.I. and Lee, H.L. (2009). Analytical and Numerical Simulation of Tsunami Mitigation by Mangroves in Penang, Malaysia. *Journal of Asian Earth Sciences*, 36(1), 38–46, doi: 10.1016/j.jseaes.2008.09.007.

Teh, S.Y., Turtora, M., DeAngelis, D.L., Jiang, J., Pearlstine, L., Smith, T.J. and Koh, H.L. (2015). Application of a coupled vegetation competition and groundwater simulation model to study effects of sea level rise and storm surges on coastal vegetation. *Journal of Marine Science and Engineering*, 3, 1149–1177, doi: 10.3390/jmse3041149.

Teh, S.Y., Koh, H.L., DeAngelis, D.L., Voss, C.I. and Sternberg, L. (2019). Modeling $\delta^{18}O$ as an early indicator of regime shift arising from salinity stress in coastal vegetation. *Hydrogeology Journal*, 27(4), 1257–1276, doi: 10.1007/s10040-019-01930-3.

Thu, P.M. and Populus, J. (2007). Status and changes of mangrove forest in Mekong Delta: Case study in Tra Vinh, Vietnam. *Estuarine, Coastal and Shelf Science*, 71, 98–109.

Thuc, T., Thang, N., Huong, H., Kien, M., Hien, N., Phong, D. (2016). Climate change and sea level rise scenarios for Vietnam. Ministry of Natural Resources and Environment (MONRE), Hanoi. https://www.researchgate.net/profile/Thuc_Tran/publication/318875854_Climate_Change_and_Sea_Level_Rise_Scenarios_for_Viet_Nam_-_Summary_for_Policymakers/links/5a3cadbaa6fdcc21d878b1

67/Climate-Change-and-Sea-Level-Rise-Scenarios-for-Viet-Nam-Su mmary-for-Policymakers.pdf.

Thuy, N.B., Tanimoto, K., Tanaka, N., Harada, K. and Iimura, K. (2009). Effect of open gap in coastal forest on tsunami run-up-investigations by experiment and numerical simulation. *Ocean Engineering*, 36, 1258–1269.

Titus, J. and Richman, C. (2000). Maps of lands vulnerable to sea level rise: Modeled elevations along the U.S. Atlantic and Gulf Coasts. *Climate Research*, 18, 205–228.

Tobey, J., Clay, J. and Vergne, P. (1998). Maintaining in balance: The economic, environmental and social impacts of shrimp farming in Latin America. Coastal Management Report #2202. University of Rhode Island, Coastal Resources Centre, Narragansett, Rhode Island USA, p. 62.

Tomlinson, P.B. (1986). *The Botany of Mangroves*. Cambridge University Press, New York, New York, USA.

Torres, C. and Hanley, N. (2017). Communicating research on the economic valuation of coastal and marine ecosystem services. *Marine Policy*, 75, 99–107.

Tuan, T.H. and Tinh, B.D. (2013). Cost-benefit analysis of mangrove restoration in Thi Nai Lagoon, Quy Nhon City, Vietnam, Asian Cities Climate Resilience. Working Paper Series, 4: 2013. IIED, London p. 50.

Twilley, R.R. (1998). Mangrove wetlands, in: Messina, M.G., Conner, W.H. (eds.), Southern Forested Wetlands: Ecology and Management. Lewis Publishers, Boca Raton, Florida, USA, pp. 445–473.

UN (2015). Transforming our world: The 2030 Agenda for Sustainable Development, United Nations General Assembly, 21 October 2015, A/RES/70/1, available at: http://www.refworld.org/docid/57b6e3e 44.html (accessed 1 November 2017).

Valiela, I., Bowen, J.L. and York, J.K. (2001). Mangrove forests: One of the world's threatened major tropical environments. *BioScience*, 51, 807–815.

Venkatachalam, A.J., Kaler, J. and Price, A.R.G. (2012). Modelling ecological and other risk factors influencing the outcome of the 2004 tsunami in Sri Lanka. *Ecosphere*, 3, 18.

Venkatachalam, A.J., Price, A.R.G., Chandrasekara, S. and Senaratna Sellamuttu, S. (2009). Risk factors in relation to human deaths and other tsunami (2004) impacts in Sri Lanka: The fishers'-eye view. *Aquatic Conservation: Marine and Freshwater Ecosystems*, 19, 57–66.

Wagner, F.H. (2001). Freeing agency research from policy pressures: A need and an approach. *Bioscience*, 51, 445–450.

Waite, R., Kushner, B., Jungwiwattanaporn, M., Gray, E. and Burke, L. (2015). Use of coastal economic valuation in decision making in the Caribbean: Enabling conditions and lessons learned. *Ecosystem Services*, 11, 45–55.

Walsh, K. (2004). Tropical cyclones and climate change: Unresolved issues. *Climate Research*, 27, 77–83.

Wang, L., Mu, M., Li, X., Lin, P. and Wang, W. (2011). Differentiation between true mangroves and mangrove associates based on leaf traits and salt contents. *Journal of Plant Ecology*, 4, 292–301.

Wang, W.Q. (2007). *Chinese Mangrove*. Science Press, Beijing, China.

Ward, R.D., Friess, D.A., Day, R.H. and MacKenzie, R.A. (2016). Impacts of climate change on mangrove ecosystems: A region by region overview. *Ecosystem Health and Sustainability*, 2(4), e01211.

Webster, P.J., Holland, G.J., Curry, J.A. and Chang, H.R. (2005). Changes in tropical cyclone number, duration, and intensity in a warming environment. *Science*, 309, 1844–1846.

Westing, A.H. (1983). The environmental aftermath of warfare in Viet Nam. *Natural Resources Journal*, 23, 365–390.

Willard, D.A., Holmes, C.W., Orem, W.H. and Weimer, L.M. (1999). Plant communities of the Everglades: A histo of the last two millenia. In: S. Gerould, and A. Higer, (compilers). U.S. Geological Survey Program on the South Florida Ecosystem—Proceedings of South Florida Restoration Science Forum, May 17–19, 1999, Boca Raton, Florida. U.S. Geological Survey Open-File Report, 99–181, Tallahassee, Florida, 118–119.

Williams, K., Pinzon, Z.S., Stumpf, R.P. and Raabe, E.A. (1999). Sea level rise and coastal forests on the Gulf of Mexico. Open-File Report, 99–441. U.S. Geological Survey, Center for Coastal Geology, St. Petersburg, Florida.

Wolanski, E. (2007). Thematic paper: Synthesis of the protective functions of coastal forests and trees against natural hazards, in: Coastal Protection in the Aftermath of the Indian Ocean Tsunami: What Role for Forests and Trees. RAP Publication (FAO), pp. 157–159.

Woodroffe, C.D. (1990). The impact of sea-level rise on mangrove shorelines. *Progress in Physical Geography*, 14, 483–520.

Yanagisawa, H., Koshimura, S., Goto, K., Miyagi, T., Imamura, F., Ruangrassamee, A. and Tanavud, C. (2009). The reduction effects of mangrove forest on a tsunami based on field surveys at Pakarang Cape, Thailand and numerical analysis. *Estuarine Coastal Shelf Science*, 81(1), 27–37.

Zaradic, P.A., Pergams, O.R. and Kareiva, P. (2009). The impact of nature experience on willingness to support conservation. *PLoS ONE*, 4(10), e7367.

Zhang, K., Liu, H., Li, Y., Xu, H., Shen, J., Rhome, J. and Smith III, T.J. (2012a). The role of mangroves in attenuating storm surges. *Estuarine, Coastal and Shelf Science*, 102–103, 11–23.

Zhang, Y., Huang, G., Wang, W., Chen, L. and Lin, G. (2012b). Interactions between mangroves and exotic Spartina in an anthropogenically-disturbed estuary in southern China. *Ecology*, 93, 588–597.

Zuang, X.L., Lin, J. and Li, Y.H. (2011). The present situation and the major control measures of mangroves in southeast of China. *Straits Science*, 7, 19–22.

Chapter 7

Tsunami: Sendai Disaster Risk Reduction

7.1 Introduction

7.1.1 *How tsunamis are generated*

A tsunami is created when the sea surface is abruptly raised or lowered vertically. An uplift of the seabed, caused by an undersea earthquake or by a submarine volcanic eruption can create a tsunami. A tsunami may also be created by a violent horizontal displacement of a large volume of water. A submarine mass failure (SMF) sliding down a large lake or down the ocean can create a tsunami too. A meteorite impact or human activities such as nuclear explosions in the deep ocean may also generate a tsunami. A tsunami begins with an abrupt vertical displacement of massive volumes of water triggered usually by seismic activities. The vertically displaced water column then radiates outward, thus creating a tsunami that travels with high speeds and with potentially high waves. These waves can travel long distances over thousands on kilometers in the deep ocean. For example, the tsunami generated by an earthquake in Chile in 1960 travelled across the Pacific Ocean and struck the Japanese coastlines after one day of long ocean travel. Most natural tsunamis are created by large marine earthquakes that can lift extensive areas of the seabed by several meters. All oceanic regions of the earth are potentially subjected to the threat of marine earthquakes and subsequent tsunamis. However, tsunamis are more concentrated in the

297

Pacific Ocean and its marginal seas, an area popularly known as the Pacific Ring of Fire. The most destructive tsunamis are formed by the occurrence of large marine earthquakes, with epicenters located on the deep ocean floor near populated shorelines. These usually occur in regions of the earth, such as the Pacific Ring of Fire, characterized by high seismic activities. The collision of two plates along tectonic boundaries causes the earthquakes, which in turn create the tsunami.

7.1.2   *Earthquake-generated tsunamis*

On 26 December 2004, a megathrust earthquake with a large magnitude of 9.3 on the Richter scale uplifted extensive areas of the seabed by some 11 m. A volume of water amounting to 200 trillion tons (or 200 trillion m^3) was uplifted vertically in a split second at the source of the submarine earthquake. This massive volume of seawater uplifted by the earthquake thus created the mega tsunami that devastated the shorelines of eleven countries fringing the Indian Ocean and Andaman Sea. That earthquake was one of the most powerful and destructive earthquakes in the past 100 years. The initial waves created at the epicenter then travel radially outward. As tsunami waves travel toward the shallow shore, their celerity (wave profile speed) and wavelength progressively reduce, while the wave heights and water velocities correspondingly increase. For example, the initial Andaman tsunami wave height generated by the earthquake at the source was 11 m, but these waves increased to exceed 30 m as they arrived at the Banda Aceh beaches. Megathrust earthquakes are potentially very destructive and are created when two giant tectonic plates in the earth's crust slip under one another, as is the case of the Andaman 2004 earthquake and tsunami. Most tsunamis that caused severe damage were generated by tsunami sources located at short distances of less than 300 km from the shore, known as near-field tsunamis. When the waves reach the shallow coastal areas, they slow down, and increase in elevations and velocities. Eventually, the waves break along the beaches, creating an environment that can be highly dangerous to lives and properties. This breaking part of a tsunami's evolution creates a dispersion of the waves, giving rise to waves with different frequencies or spectra and with different propagation speeds. Tsunamis generated by SMF can be catastrophic if the SMF occurs near populated coastal regions.

7.1.3 *Submarine mass failure*

Tsunamis generated by abrupt vertical uplifts of water column by the uplift of the seabed due to earthquake fault motion are better understood than those tsunamis caused by submarine landslides, also known as SMF. The higher frequency of tsunamis generated by seismic fault motion such as submarine earthquakes has provided valuable data and insights for the understanding of earthquake-generated tsunamis. For the infrequent tsunamis generated by SMF, the initial waveforms created at the source are more difficult to derive theoretically. This lack of understanding regarding SMF-generated tsunamis has hampered efforts aimed at reducing the risks and impacts of SMF tsunamis. The risks are intensified because tsunamis generated by SMF can be potentially more dangerous than those generated by vertical uplifts of water. A simple reason is that in deep water, the slide depth may be high, of the order of hundreds or even thousands of meters. The enormous potential energy released by an SMF sliding down thousands of meters is transferred into kinetic energy of the SMF. This enormous kinetic energy can create enormous tsunamis with wave heights exceeding 10 m at the source. For vertical uplift of the ocean floor, the initial wave height at the source of tsunami creation is capped by the uplift of the seafloor, which is generally limited to merely a couple of meters. The potential energy released, and the subsequent kinetic energy created, by SMF can therefore generate tsunamis that are much higher and more focused than that created by uplift of the ocean floor. Hence, near-field tsunamis triggered by SMF can be devastating. Yet, most catastrophic tsunamis that occurred in the past had not been predicted before their occurrences. Tsunamis are indeed difficult to predict in advance.

7.1.4 *Tsunamis are difficult to predict*

It should be noted that it is extremely difficult to issue timely and correct tsunami warnings for coastal communities because of two reasons. First, seismic data often translate to tsunami data imprecisely and with a delay. Second, local coastal wave amplification factors vary significantly from coast to coast. This combination renders prediction of tsunami inundation along the affected coasts difficult and inaccurate. Hence, tsunami warning and inundation modeling is a science that is rather imprecise. Out of 20 tsunami warnings issued in Hawaii since 1946, 15 were false alarms. While correct tsunami warnings can save lives, false alarms are expensive and embarrassing. While it is generally accepted that an earthquake of magnitude 7 and above on the Richter scale may be capable of triggering a life-threatening tsunami, a smaller 5.2 earthquake in 1930 generated a tsunami of 6 m in California, a wave height that is far above what is deemed safe (below 3 m). The difficulty in early and accurate prediction of tsunami mandates the development of holistic tsunami mitigation measures, for preparing and protecting coastal communities, including tsunami resilience education.

7.1.5　*Tsunami resilience education*

Since the 2004 Andaman tsunami, several tsunamis of similar scales have occurred, including the tsunami that destroyed the Fukushima power plant on 11 March 2011. There is scientific evidence indicating that significant earthquakes and tsunamis may continue to pose great hazards and risks to these coastal regions. Hence, developing community tsunami disaster risk reduction (DRR) capability is essential to protect coastal communities and to mitigate the impact when a tsunami occurs in the future. In view of the inability to accurately predict tsunami occurrences, it becomes mandatory to develop tsunami resilience education as a primary tool for tsunami DRR. The following features are deemed essential in tsunami education and awareness campaigns: (1) Understand the local history of tsunami occurrences; (2) Recognize the physical signs of an impending tsunami; (3) Understand what areas are at risk; (4) Know when, how and where to evacuate; and (5) Participate in regular training sessions on tsunami RDD measures. Tsunami resilience education should be integrated with other holistic approaches to achieve the goals of tsunami-resilient communities. Toward these goals of building tsunami-resilient communities, it is essential to develop a good understanding of tsunami characteristics, its hazards, vulnerability and risk.

7.1.6　*Tsunami characteristics and hazards*

A tsunami is a long-period wave that is generated by abrupt disturbances in the ocean. It is initiated by a large-scale instantaneous vertical displacement of a massive volume of water, usually triggered by either internal or external sources. Internal sources include a submarine earthquake that occurs beneath the ocean floor, or an SMF or a volcanic eruption in the ocean floor. External sources may be caused by a large meteorite impact on the ocean or large lake. The vertically uplifted water column then propagates outward across the ocean with high speeds exceeding hundreds of kilometers per hours and with potentially high amplitude waves. As the waves subsequently propagate up the shallow beaches or coastal areas, the wave amplitudes amplify to potentially dangerous levels that can inflict significant damage to properties and lives. At this stage of tsunami propagation, along swallow beaches, the waves slow down due to the shallowness of the water, and the wave heights amplify to reach maximum vertical heights onshore. This may create a highly dangerous environment that may cause tremendous damage to properties and lives. The Andaman tsunami that occurred off the coast of Banda Aceh on 26 December 2004 is an example of a catastrophic disaster that inflicted catastrophic loss and pain to coastal communities, especially for those living in the impacted regions. This Andaman tsunami killed around 250,000 people along the affected coastal regions worldwide. Hence, it is important to understand the hazards, vulnerability and risk associated with a tsunami in a hazard zone. The hazards of a tsunami at a specific location may be estimated by a thorough assessment of past tsunamis in the region, undertaken in combination of tsunami numerical simulation models or scaled-down physical simulations. These model simulation results can provide useful insights into developing effective tsunami mitigation measures to reduce the potential impacts and risks to local communities.

7.2   Tsunami Simulation Models

Tsunami simulation typically consists of three phases. Phase (1) begins with creating a credible tsunami initial wave at the source of generation, by means of tsunami generation models such as TUNA-GE, based upon the Okada model concept (Okada, 1985). With this initial tsunami wave as the initial condition, Phase (2) then simulates tsunami propagation in the deep ocean by propagation models such as TUNA-M2 (Koh *et al.*, 2005, 2007). Finally, Phase (3) simulates tsunami wave run-up and inundation along the swallow beaches by means of models such as TUNA-RP (Tan *et al.*, 2014; Koh *et al.*, 2017). The following subsection describes the three models used to simulate these three distinct phases of tsunami evolution consisting of (a) tsunami source generation, (b) tsunami propagation in the deep ocean and (c) tsunami run-up and inundation along beaches.

7.2.1   *Okada tsunami source model*

The initial tsunami waves generated by a marine earthquake can be formulated by the Okada model (Okada, 1985). An earthquake is caused by a sudden dip-slip of a seismic fault. When an earthquake occurs near the seabed, it triggers a vertical displacement of the sea floor. The characteristic time scale involved in this co-seismic surface deformation is much faster than the tsunami wave propagation time scale. Hence, it is normal to assume that the vertical displacement of the sea floor triggers an instantaneous "co-seismic" vertical uplift of the ocean water, thereby generating a tsunami, with dimensions that mimic those of the fault. The length scale of this sea floor deformation (of the order of 100 km) is much longer than the length scale of water depth (of the order of 1 km). Further, the water depth (of the order of 1,000 m) is much larger than the scale of vertical uplift of the seabed (of the order of 1 m). Hence, tsunamis are typically characterized as long waves, with wavelength much longer than the water depth, and the water depth is much larger than the tsunami wave height.

The focal geometry of earthquakes depends on the orientation and characteristics of the source. Dip-slip faults are inclined fractures where the blocks have shifted vertically. If the rock mass above an inclined fault moves down, then the fault is termed normal (Figure 7.1(a)). On the contrary, if the rock mass above the fault moves up, the fault is termed reverse (Figure 7.1(b)). Strike-slip faults are vertical (or nearly vertical) fractures where the blocks have mostly moved horizontally. If the block opposite an observer looking across the fault moves to the left, the slip style is termed left lateral (Figure 7.1(c)). If the block moves to the right, the motion is termed right lateral (Figure 7.1(d)). Figure 7.1 illustrates dip-slip faults and strike-slip faults by Stein and Wysession (2002). To describe the focal mechanism of an earthquake, three angles are necessary, which are the dip angle, slip angle and strike angle.

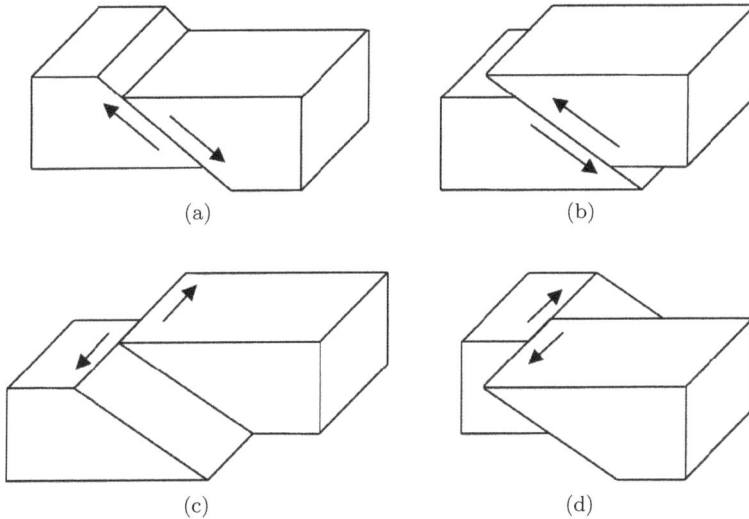

(a)

(b)

(c)

(d)

Figure 7.1: (a) Normal dip-slip fault, (b) Reverse dip-slip fault, (c) Left-lateral strike-slip fault and (d) Right-lateral strike-slip fault.

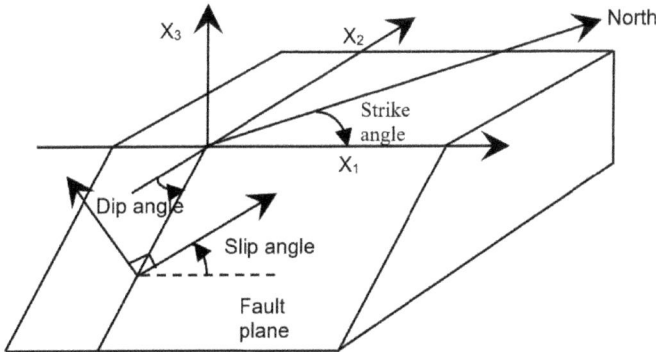

Figure 7.2: Geometry of the earthquake source.

Figure 7.2 shows the geometry of the earthquake source parameters (Stein and Wysession, 2002). The dip angle indicates how steeply the fault penetrates the earth. The slip angle describes the relative displacement of formerly adjacent points on opposite sides of a fault, measured on the fault plan. Finally, the strike angle is the trend or bearing, relative to north, of the line defined by the intersection of a fault and a horizontal surface. Focal depth is the depth from the

Table 7.1:   Definition of symbols used in Eqs. (7.1) to (7.3).

Symbol	Unit	Definition
t	s	Time
x	m	Space coordinate along x-direction
y	m	Space coordinate along y-direction
η	m	Free surface elevation measured from a fixed datum (mean sea level)
d	m	Water depth below fixed datum
D	m	Total water depth, $D = \eta + d$
g	m/s^2	Gravitational acceleration
n	s/m$^{1/3}$	Manning's relative roughness coefficient
u	m/s	Velocity in the x-direction
v	m/s	Velocity in the y-direction
M	m^2/s	Discharge fluxes in the x-direction, $M = u(h + \eta) = uD$
N	m^2/s	Discharge fluxes in the y-direction, $N = v(h + \eta) = vD$

sea floor to the epicenter. Once a tsunami is generated, it will propagate through the deep ocean, a process that can be simulated by a tsunami propagation model.

7.2.2   *TUNA-M2 propagation model*

The propagation of tsunami waves through the deep ocean is simulated by the depth-averaged two-dimensional nonlinear shallow water equation (NSWE). The continuity and momentum equations of the NSWE (IOC, 1997; Hérbert *et al.*, 2005) are described in Eqs. (7.1) to (7.3). The definition and unit of the symbols used in Eqs. (7.1)–(7.3) are listed in Table 7.1. In this chapter, the governing equations are solved using the explicit leap-frog finite difference scheme with an upwind algorithm for the nonlinear advection terms (IOC, 1997). The NSWE equation can also be solved by several other methods, such as the finite element method. The explicit leap-frog finite difference method is employed in this chapter, as it is known to perform well, provided that the time step Δt fulfills the Courant stability criterion. The finite difference method is also employed by many well-known models, such as COMCOT (Liu *et al.*, 1998), TUNAMI-N2 (Imamura *et al.*, 1988) and MOST (Titov and Synolakis, 1998).

The complicated discretization of the explicit leap-frog finite difference method used in this chapter is referred to in Koh *et al.* (2009). The simulation results of TUNA have been verified by comparing them with known analytical solutions in rectangular domains (Koh *et al.*, 2005, 2007). Further verification is performed by comparing TUNA simulation results with those simulated from the COMCOT model (Cornell Multi-grid Coupled Tsunami Model), indicating satisfactory performance of TUNA. TUNA will be used to simulate the 2004 Andaman tsunami generation, propagation and run-up along the impacted northwest coast of Peninsular Malaysia, including Penang and Langkawi.

$$\frac{\partial \eta}{\partial t} + \frac{\partial M}{\partial x} + \frac{\partial N}{\partial y} = 0 \tag{7.1}$$

$$\frac{\partial M}{\partial t} + \frac{\partial}{\partial x}\left(\frac{M^2}{D}\right) + \frac{\partial}{\partial y}\left(\frac{MN}{D}\right)$$
$$+ gD\frac{\partial \eta}{\partial x} + \frac{gn^2}{D^{7/3}}M\sqrt{M^2 + N^2} = 0 \tag{7.2}$$

$$\frac{\partial N}{\partial t} + \frac{\partial}{\partial x}\left(\frac{MN}{D}\right) + \frac{\partial}{\partial y}\left(\frac{N^2}{D}\right)$$
$$+ gD\frac{\partial \eta}{\partial y} + \frac{gn^2}{D^{7/3}}N\sqrt{M^2 + N^2} = 0 \tag{7.3}$$

7.2.3 *TUNA-RP inundation model*

When the tsunami waves reach the beaches, the NSWE is not suitable as the water moves through dry–wet cycles during this phase of the tsunami evolution. Moving boundary algorithm (MBA) is applied to track the movement of the shoreline as the tsunami rises or recedes along the beach. MBA is used to simulate the inundation distances and maximum run-up heights of the tsunami along the coast. Technical details of the mathematical formulation and MBA applied in TUNA are available elsewhere in (Tan, 2017; Tan *et al.*, 2017). Before a tsunami simulation model is applied, the model must be calibrated and verified against tsunami wave data surveyed along the affected beaches soon after the occurrence of the tsunami.

7.3   Post-Tsunami Field Surveys

An earthquake with a magnitude of $M_w = 9.3$ erupted off the western coast of Banda Aceh, North Sumatra, at 08:58:53 Malaysian time on 26 December 2004. The earthquake occurred on the tectonic boundaries of the subduction zones between the Indian plate and the Sunda plate. The instantaneous uplift of the sea floor caused the sea level to be lifted upward by 12 meters at the source, triggering a mega tsunami that killed 250,000 people worldwide. The waves took between 3 and 4 hours to reach the beaches in Peninsular Malaysia. Two field surveys were conducted by the authors, in collaboration with international experts, along the affected beaches in Penang and Kedah to assess the run-up heights and inundation distances as well as to document the damage to properties. The survey data were collated for calibrating and validating the tsunami simulation model TUNA developed in-house in USM (Koh *et al.*, 2009, 2010).

7.3.1   *Beach run-up and inundation*

Highest run-up heights and maximum inundation distances were measured at 14 locations along affected areas in Penang and Kedah. Figure 7.3 shows the map of Peninsular Malaysia (bottom left), and the four most impacted areas, namely, Langkawi, Kedah, Penang and Perak. The location numbers indicated on the maps refer to those listed in Table 7.2, which provides a record of run-up heights and inundation distances. The surveyed tsunami run-up heights along the beaches in Penang vary between 2.3 m and 4.0 m, while those in Langkawi are between 2.2 m and 3.7 m. For the state of Kedah excluding Langkawi, the run-up heights were observed to vary between 0.38 m and 3.8 m exhibiting significant scattering. These scatterings in run-up wave heights along other beaches have also been reported in the literature. Local features such as bathymetry, land curvature and sea–land nonlinear interactions are the significant causes of these variations in run-up wave heights.

7.3.2   *Leading depression wave*

It has been reported that the 2004 tsunami waves that arrived at the northwest peninsular Malaysia were known as leading depression

Figure 7.3: Survey locations along the coast of northwest Peninsular Malaysia.

N-waves. A leading depression N-wave (Figure 7.4(a)) is a wave that travels with a leading depression, i.e., a leading negative wave (with height below sea level). This is followed by a subsequent elevation wave that has a positive wave height (above sea level). A leading elevation N-wave (Figure 7.4(b)), on the contrary, is just the reverse that travels with a leading elevation, i.e., a leading positive wave (with height above sea level). This is followed by a subsequent depression wave that has a negative wave height (below sea level). A survey team was assembled by Komoo and Othman (2006) to record the extent of damage inflicted by this tsunami, immediately after the occurrence of the 2004 Andaman tsunami. It was observed during this survey that the sea level abruptly receded into the sea

Table 7.2:   Survey run-up heights and inundation distances for December 26, 2004 tsunami.

No.	Date (2005)	Location	Latitude (N) Deg.	Min.	Longitude (E) Deg.	Min.	Run-up Height (m)	Inundation Distance (m)
1	20 Apr.	B. Ferringhi (Teluk Bayu)	5	28.26	100	14.63	3.46	19.20
		B. Ferringhi (Miami Beach)	5	28.67	100	16.07	4.000	25.60
2	20 Apr.	Tanjung Tokong	5	27.62	100	18.48	3.650	35.80
		Tanjung Tokong	5	27.57	100	18.41	N/A	190.00
		Tanjung Tokong	5	27.70	100	18.50	2.61	18.30
3	21 Apr.	Tanjung Bungah	5	28.21	100	16.66	2.31	18.38
		Tanjung Bungah	5	28.20	100	16.65	2.94	36.20
4	22 Aug.	Kuala Kedah	6	6.00	100	26.00	0.90	N/A
5	22 Aug.	Yan (Kg. K.S. Limau)	5	53.00	100	21.00	1.23	12.90
6	22 Aug.	Sg udang	5	48.00	100	22.00	1.50	N/A
7	22 Aug.	Tanjung Dawai	5	40.00	100	21.00	0.39	75.32
8	22 Aug.	Kota K. Muda	5	34.00	100	20.00	3.80	100.52
9	23 Aug.	Kuala Kurau	5	0.00	100	25.00	1.93	N/A
10	23 Aug.	Pantai Acheh	5	24.00	100	11.00	2.51	13.40
11	24 Aug.	Pantai Tengah (Lanai Hotel)	6	15.00	99	43.00	3.66	44.50
12	24 Aug.	Pantai Chenang (Pelangi Hotel)	6	17.00	99	43.00	3.75	54.72
13	24 Aug.	Kuala Teriang	6	21.00	99	42.00	3.09	27.04
14	24 Aug.	Pantai Kok (Mutiara Beach Resort)	6	21.00	99	40.00	2.25	50.84
		Pantai Kok (Berjaya Hotel)	6	21.00	99	40.00	2.98	34.88

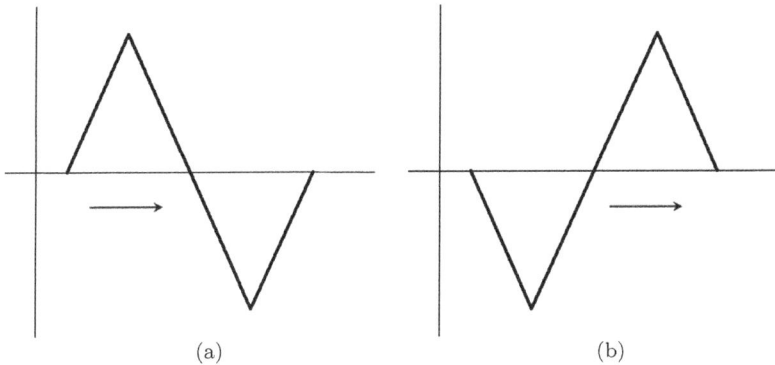

Figure 7.4: *N*-waves of (a) leading depression followed by an elevation and (b) leading elevation followed by a depression.

for distances exceeding 100 meters before the next elevation waves arrived. This means that the tsunami waves were leading depression N-waves. The tsunami waves arrived at Langkawi, Penang and Kota Kuala Muda in Kedah at 12:40 pm, 1:15 pm and 1:40 pm, respectively. This translates to an arrival time of about 3 hours and 40 minutes for Langkawi. It was reported by reliable eyewitnesses in Penang and Langkawi that there were three distinct tsunami waves. The first two waves arrived Penang at approximately 1:15 pm and 1:30 pm. However, the arrival time of the third wave could not be estimated accurately from anecdotes. Similarly, Kota Kuala Muda in Kedah was also reportedly struck by three distinct tsunami waves, at approximately 1:40 pm, 1:50 pm and 2:00 pm. In all these three locations, it was observed that the second tsunami waves were the biggest and the most destructive, claiming lives, causing grave injuries and devastation to property. Along the shores, wave heights exceeding 2 to 4 meters were commonly observed in all three locations. Waves exceeding 3 meters are known to be capable of causing significant damage and inflicting human fatality. Human casualties were often the result of injury caused by debris flows or death by drowning, as the current speeds on shore can exceed 12 m/s. The inundation distances in some places reached 350 m inland. Table 7.3 provides the averaged arrival times, tsunami wave heights and inundation distances surveyed in these three locations. A total of 68 deaths were reported in this tsunami, with 52 deaths in Penang, 12 occurring in Kota Kuala Muda, one in Langkawi and three in Perak.

Table 7.3:   Averaged arrival times, wave heights and inundation distances.

Location	Wave arrival times		Run-up (m)	Inun. Dist. (m)
	First	Second		
Kota Kuala Muda	13:45	14:00	2.0–3.4	200–350
Penang	13:15	13:20	2.5–4.0	150–300
Langkawi	12:35	12:50	2.5–3.0	100–250

7.4   Simulation Results for 2004 Tsunami

Tsunami simulation consists of three phases: (a) Tsunami source generation, (b) tsunami propagation through the deep ocean and (c) tsunami run-up and inundation along the beaches. Tsunami simulation begins with creating a creditable tsunami source based upon reliable earthquake source characteristics such as fault parameterization. The Okada model is used in this chapter to generate the tsunami source, based upon earthquake fault parameters with either one segment or five segments. A simple tsunami source based upon the Gaussian hump shape is also used as a tutorial session. The generated tsunami source based upon earthquake fault characteristics for the 2004 Andaman earthquake is then used as the initial conditions for the propagation simulation. Wave propagation over the deep ocean is simulated by the NSWE. The simulated wave heights offshore at depths of about 50 m are then used to drive the run-up and inundation process. A moving boundary algorithm is employed to track the wave evolution as it runs up and down the shoreline. Simulation of the three phases of tsunami evolution consisting of (1) source, (2) propagation and (3) run-up and inundation is performed by the in-house tsunami simulation model known as TUNA. This section ends with a simulation of tsunami run-up and inundation along four impacted beaches in Penang Island during the 2004 Andaman tsunami.

7.4.1   *Tsunami initial source*

7.4.1.1   *Earthquake source dimension*

There are several estimates reported in the literature (Lay *et al.*, 2005) for the earthquake source dimensions for the 26 December

2004 tsunami. Stein and Okal (2005) had earlier estimated an earthquake source dimension of 1200 km × 200 km × 11 m (length × width × slip). It is generally accepted that the run-up wave height in the near field would not exceed twice the fault slip or height. Measured tsunami run-up wave heights of between 25 m and 30 m were reported in the near field around Sumatra (Borrero, 2005). This would imply that the fault slip might be more than 11 m, possibly between 12 and 15 m. Hence, Okal and Stein (2005) subsequently suggested an alternative earthquake source dimension of 1200 km × 200 km × 13 m. However, Disaster Control Research Center (2005) preferred an earthquake source dimension of 800 km × 85 km × 11 m. This set of earthquake source dimensions provides a basis for generating tsunami sources in the following subsections.

Two types of tsunami source generation models are normally used to generate the initial tsunami waves at the source of generation. The first model is a simple Gaussian hump in the shape of an elongated ellipse (Yoon, 2002), while the second model is one developed by Okada (1985). Three scenarios of tsunami initial wave source generated by the 2004 Andaman earthquake at the earthquake source will be demonstrated in this chapter: (A) A simple Gaussian hump model, (B) a one-segment Okada Model and (C) a five-segment Okada Model. The simple Gaussian hump model provides an easy introductory lesson to learn about tsunami source generation as the model is based upon a simple Gaussian function. However, this Gaussian hump source wave is an elevation wave with positive heights above sea level at all locations. For the tsunami source generated by the 2004 Andaman tsunami, however, the initial wave is a leading depression N-wave. Hence, the Okada tsunami source models are used to reflect the reality on the ground.

7.4.1.2 *Simple Gaussian hump source*

A simple Gaussian hump model was proposed by Yoon (2002) for producing the initial waves generated at the source. In two dimensions, the wave height η is given by $\eta = ae^{-(x/\sigma_x)^2}e^{-(y/\sigma_y)^2}$, where $a = 12.0$ m, $\sigma_x = 60$ km and $\sigma_y = 450$ km. Figure 7.5a shows the map of Southeast Asia where the 2004 earthquake and tsunami originated, and Figure 7.5b depicts the initial tsunami source generated

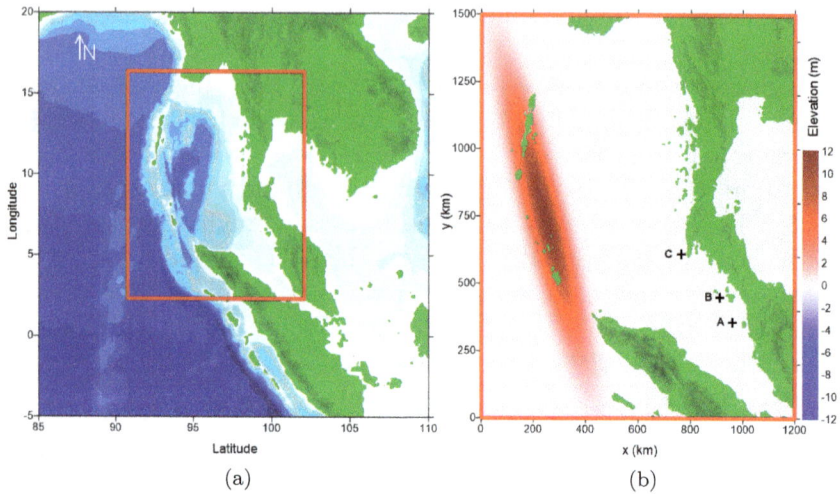

Figure 7.5:   (a) Map of study area with computational domain shown in rectangle and (b) initial Gaussian hump tsunami source.

Table 7.4:   Parameters for one-fault model.

Parameters for fault model	Value
Length of source area (km)	1,000
Width of source area (km)	10
Displacement (m)	20
Focal depth (km)	30
Dip Angle (°)	8
Slip Angle (°)	110
Strike Angle (°)	350

by the Gaussian hump model. The earthquake source stretches over 1200 km by 200 km with a maximum height of 12 m.

7.4.1.3   *Okada one-segment source*

Table 7.4 shows the seven earthquake fault parameters used for generating the Okada one-segment model. Figure 7.6 demonstrates the initial tsunami source generated by this one-segment earthquake fault. The earthquake fault has a length of 1,000 km, a width of 100 km and a vertical displacement of 20 m. The initial waves consist of two distinct portions: a leading elevated wave indicated in red and a

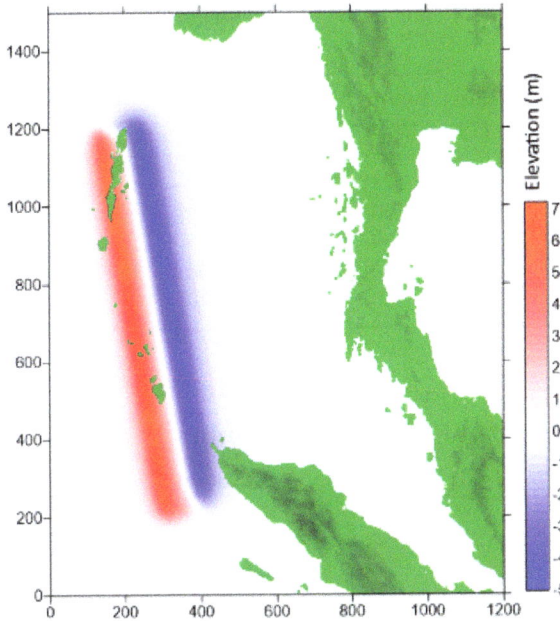

Figure 7.6: One-segment Okada tsunami source generated using parameter values in Table 7.4.

leading depression wave indicated in blue. Elevation waves have positive heights above mean sea level. Depression waves have negative heights below mean sea level. These two distinct waves subsequently travel in two opposing directions. The blue leading depression wave travels eastward toward Thailand and Malaysia, while the red leading elevation wave travels westward toward the Indian continent.

7.4.1.4 *Okada five-segment source*

Table 7.5 shows the earthquake fault parameters used in a five-segment earthquake fault model of Grilli *et al.* (2007), while Figure 7.7 depicts the geographical arrangement of the five fault segments. The generated initial tsunami source waves are shown in Figure 7.8. The tsunami source consists of two distinct waves: (A) A leading depression wave (blue) travelling eastward toward Thailand and Malaysia and (B) a leading elevation wave (red) travelling westward toward the Indian Continent.

Table 7.5:   Source parameters used in a five-segment fault model of Grilli *et al.* (2007).

Segment	Long. (°)	Lat. (°)	Length (km)	Width (km)	Strike (°)	Dip (°)	Slip (°)	Displacement (m)
S1	94.57	3.83	220	130	323	12	90	18
S2	93.90	5.22	150	130	348	12	90	23
S3	93.21	7.41	390	120	338	12	90	12
S4	92.60	9.70	150	95	356	12	90	12
S5	92.87	11.70	350	95	10	12	90	12

7.4.2   *Propagation of 2004 tsunami*

7.4.2.1   *Tutorial session*

We begin with a tutorial session to demonstrate tsunami propagation simulated by TUNA M2 in an ocean within a square computational domain of 10 km by 10 km. Three scenarios are demonstrated. First, we choose a uniform depth of $H = 100$ m (Figure 7.9, top row), indicating a uniform wave propagation speed of $\sqrt{gH} = \sqrt{9.81\,\mathrm{ms}^{-2} \times 100\,\mathrm{m}} = 31.32$ m/s. The wave propagates radially outward, in the form of perfect circles, with the uniform speed of 31.32 m/s. Second, the water depth is increased linearly from 100 m at the western boundary to 1,000 m at the eastern boundary. The waves propagate faster in the deeper eastern half and arrive at the eastern boundary earlier after 40 s. But, across the shallower western half, the waves propagate slower and arrive at the western boundary later at 80 s (Figure 7.9, middle row). By now (at time 80 s), the waves have already moved out of the eastern boundary. Travelling with different speeds, the wave profiles are no longer circular in shape. Third, the depth is increased linearly from 100 m at the eastern boundary to 1,000 m at the western boundary. The wave propagation speed pattern is now reversed. The waves travel faster and arrive at the deeper western boundary earlier after 40 s. The waves travel slower across the shallower eastern half and arrive at the shallower eastern boundary later at 80 s (Figure 7.9, bottom row). By now (at time 80 s), the waves have already moved out of the western boundary. Travelling with different speeds, the wave profiles are no longer circular in shape.

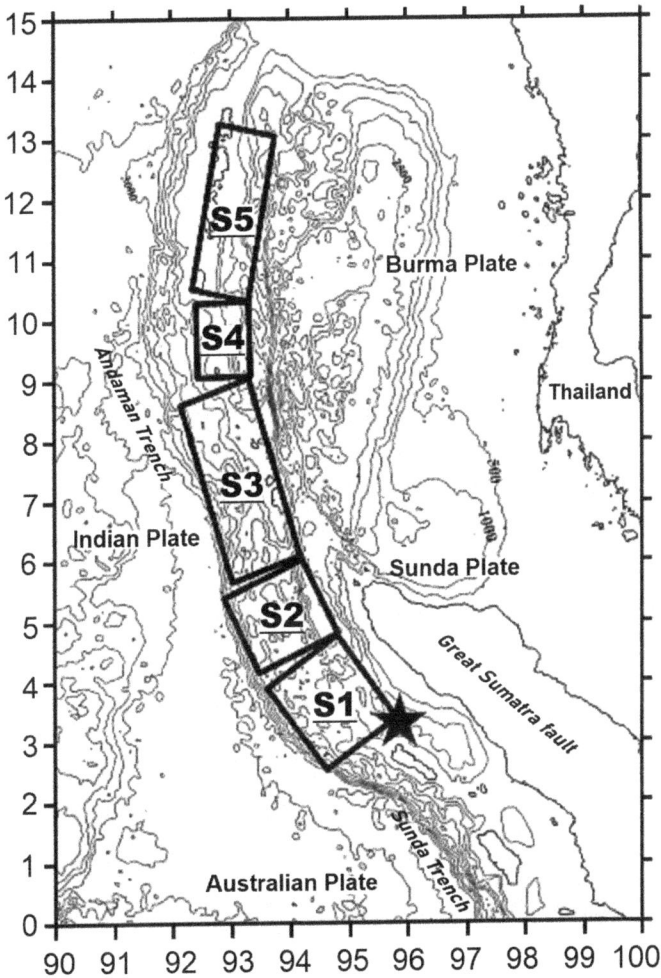

Figure 7.7: Rectangles S1-S5 represent Okada (1985) dislocation model fault segments. (⋆) represents the position of the epicenter for the Sumatra-Andaman earthquake (Ioualalen *et al.*, 2007).

7.4.2.2 *Gaussian hump source*

Figure 7.10 shows four snapshots of the propagation of tsunami waves at intervals of 2,000 s. At $t = 400$ s after the initialization of the tsunami, the wave splits into two elevation waves (red, first frame). The wave propagates through the ocean (red, second frame, $t = 2,400$ s) and arrives at Thailand beach around $t = 4,400$ s, as a

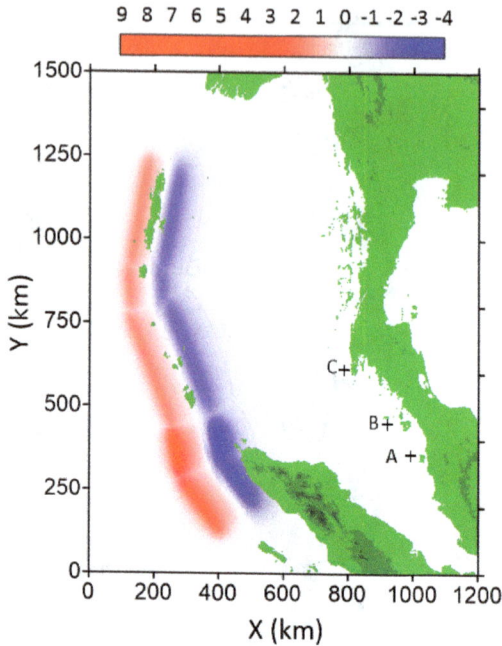

Figure 7.8:   Computational domain with observation points (A, B, C) and initial source wave heights generated by earthquake source with five-segment fault.

leading elevation wave (red, third frame). The leading elevation wave is reflected back into the ocean as a depression wave (blue, final frame) at $t = 6,400$ s.

7.4.2.3   One-segment Okada source

Figure 7.11 shows four snapshots of tsunami propagation from the one-segment Okada source (Figure 7.6) eastward toward Thailand and Malaysia at intervals of 1,800 s. The waves at $t = 1,800$ s (first frame) clearly demonstrate that the tsunami is a leading depression N-wave (blue in front, red at the back). The tsunami begins to arrive at Thailand as a leading depression wave at $t = 7,200$ s (final frame), after some two hours of travel.

7.4.2.4   Five-segment Okada source

Figure 7.12 shows four hourly snapshots of tsunami propagation from the five-segment Okada source (Figure 7.8) eastward toward

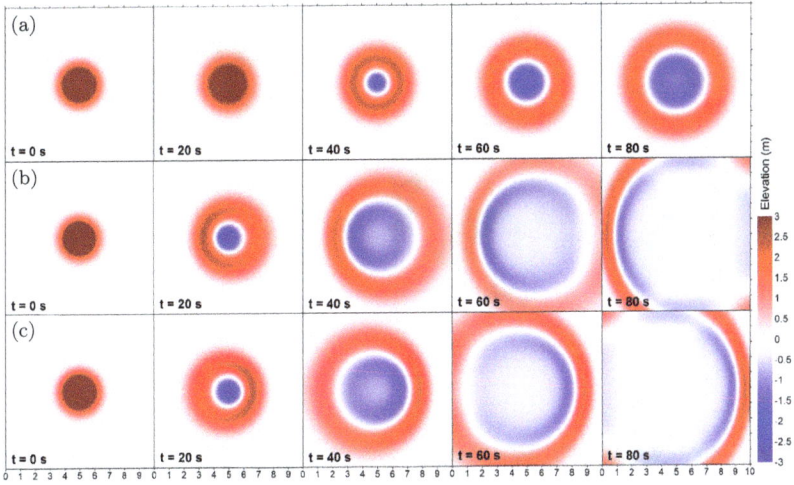

Figure 7.9: Snapshots of tsunami propagation in a square domain with (a) constant depth, (b) eastward linearly increasing depth and (c) westward linearly increasing depth.

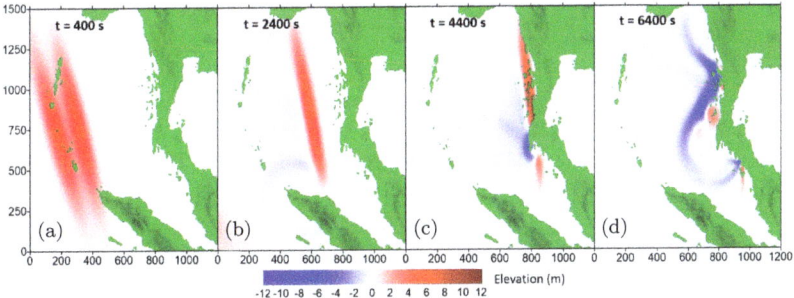

Figure 7.10: Propagation of tsunami waves generated by Gaussian hump source (Figure 7.5b). The snapshots are given here at the interval of 2,000 s, starting at $t = 400$ s.

Thailand and Malaysia. The second frame at $t = 1.0$ hr clearly demonstrates that the tsunami is a leading depression N-wave (blue in front, red at the back). The waves begin to penetrate the Straits of Malacca and begin to arrive at Thailand with a leading depression N-wave after two hours of travel (third frame). The tsunami continues into the straits of Malacca as a leading depression N-wave after three hours of travel (final frame) before reaching Penang an hour later (not shown).

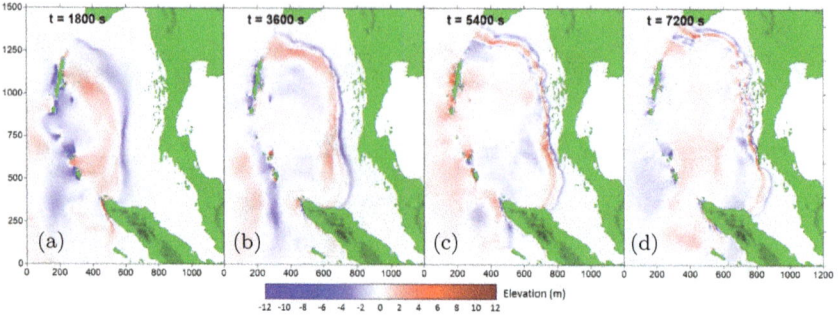

Figure 7.11:   Propagation of tsunami waves generated by one-segment Okada source (Figure 7.6).

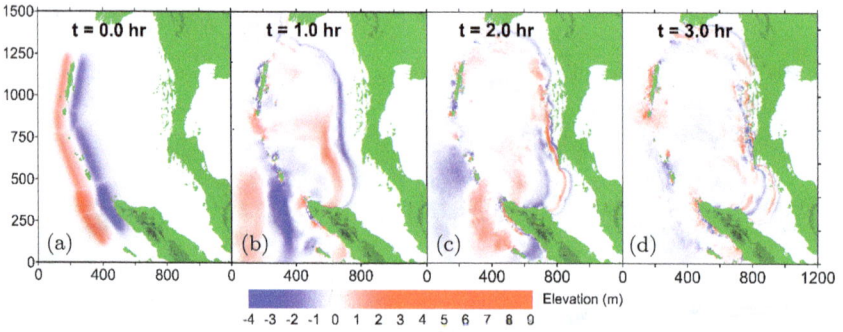

Figure 7.12:   Propagation of tsunami waves generated by five-segment Okada source.

Table 7.6 shows the simulated maximum elevations and arrival times at three offshore locations in Penang, Langkawi and Phuket for the one-segment and five-segment Okada source. Figure 7.13 shows the time series of wave heights at the three locations for the one-segment Okada model. Figure 7.14 shows the time series of wave heights at the three locations for the five-segment Okada model.

7.4.3   *Run-up and inundation of 2004 tsunami*

7.4.3.1   *General observation*

Run-up and inundation simulation is performed by TUNA-RP. To adequately incorporate local topographic features, a refined numerical scheme using moving boundary conditions is used in TUNA-RP,

Table 7.6: Comparison between simulated elevations and arrival times at three selected observation points A, B and C for the Andaman tsunami generated by a one-segment fault and by a five-segment fault.

	One-segment fault		Five-segment fault	
Location	Elevation (m)	Arrival time (h)	Elevation (m)	Arrival time (h)
A: Penang	1.2	3.6	1.2	3.6
B: Langkawi	1.5	3.0	1.0	2.9
C: Phuket	3.6	1.8	2.4	1.7

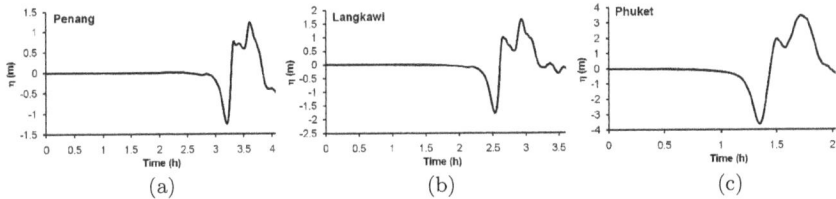

Figure 7.13: Time series of wave heights at three locations due to Andaman tsunami generated by a one-segment fault.

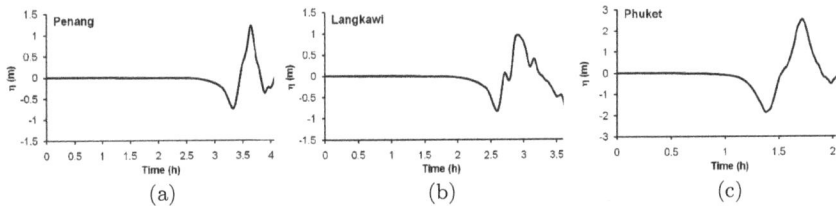

Figure 7.14: Time series of wave heights at three locations due to Andaman tsunami generated by a five-segment fault.

with small grid size. Generally, a tsunami wave arriving offshore at a depth of about 50 m in Penang, simulated by TUNA-M2, is about 1.0 m. The tsunami run-up waves along the beaches may be amplified by a factor of 2 to 3, to reach heights of 2.4 m to 3.6 m, depending on beach roughness, slope and length. These TUNA-simulated run-up wave heights along the beaches agree qualitatively with observed wave heights surveyed. However, for Langkawi and Pantai Acheh in Penang, simulation performed by TUNA resulted in maximum

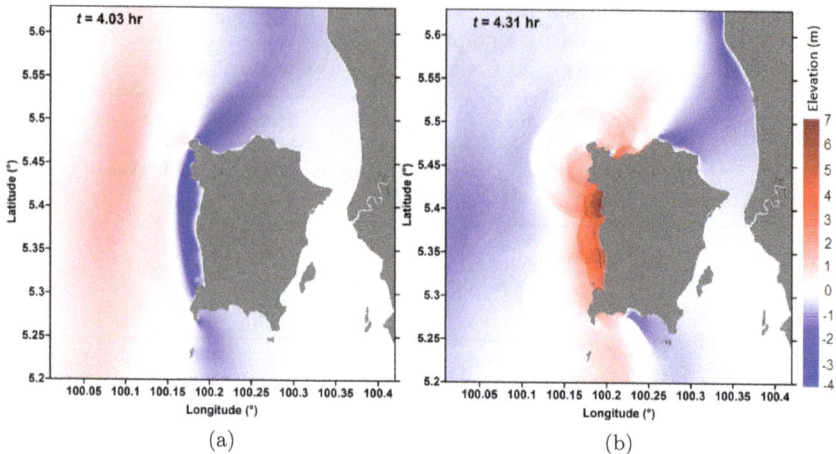

Figure 7.15:   (a) Leading depression tsunami waves (blue) arriving at west coast of Penang Island and (b) Elevation waves (red) reached and run up the coast.

beach run-up heights that over predicted surveyed maximum run-up heights. It has been suggested that the wave breakers located offshore of Langkawi and the presence of mangroves in Pantai Acheh might have provided wave reduction effects by reducing incoming tsunami wave energy and heights. This observation provides the incentive to investigate the role of coastal structures and mangroves on tsunami wave reduction. This investigation will be performed in Section 7.5.

7.4.3.2   *Leading depression wave*

Figure 7.15 (left) shows the arrival of the leading depression (blue) N waves at the west coast of Penang Island at $t = 4.03$ hours after the initialization of the tsunami. Soon after at $t = 4.31$ hours, the elevation waves (red) inundate the west coast of Penang (Figure 7.15, right). Figure 7.16 shows the four inundated areas (A, B, C and D) in Penang Island. The most inundated area D is located along the west coast, while Persiaran Gurney A is the second most inundated. Table 7.7 shows the comparison of run-up height and inundation distance between TUNA-RP and survey data obtained from National Centers for Environmental Information (NCEI) and Koh *et al.* (2009).

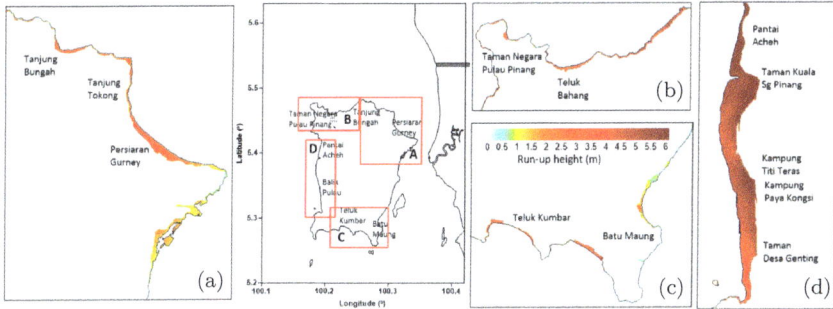

Figure 7.16: Map of simulated run-up height at four selected areas A–D in Penang Island.

The waves arrive at the northwest coast of Penang (Area B) first. Because Area B has cliffy shorelines and steep beach slopes, it experiences relatively low inundation compared to other areas in Penang. Five minutes after Area B is inundated, the tsunami begins to inundate the west coast of Penang (Area D). With the velocity field pointing directly in the onshore direction perpendicular to the shoreline, the waves inundate a very wide area, as may be seen from Figure 7.15 (right panel). Because of its low-lying elevation and mild slope, Area D is severely inundated with the maximum inundation distance extending 1,880 m inland. Area A is also severely inundated by the trapped wave from shoal, particularly in Persiaran Gurney, because it is a concave bay that traps and intensifies waves. A similar tsunami inundation and run-up process is also noted in the south part of Penang (Area C). At the southwest and southeast coasts of Penang Island, trapped waves propagate in alongshore direction toward Area C, inundating low-lying land. Any land reclamation project considered for this area C should incorporate ample provisions for mitigation of tsunami risks.

7.5 Role of Mangrove

Mangrove forests have been observed to have played a role in reducing the impact of tsunamis by reducing tsunami wave heights and velocities after travelling through the forests. Reducing moderate tsunami wave heights to less than 3 m can save lives as tsunamis with wave heights exceeding 3 m have the potential to kill humans.

Table 7.7: Comparison of run-up height and inundation distance between TUNA-RP and survey data obtained from NCEI and Koh *et al.* (2009).

No	Locations	Lat (°N)	Long (°N)	Surveyed		TUNA-RP simulated	
				Height (m)	Distance (m)	Height (m)	Distance (m)
1	Persiaran Gurney	5.4390	100.3080	2.5	200.0	2.8	423.0
2	Tanjung Bungah	5.4670	100.2770	2.5	—	2.62	220.0
		5.4702	100.2776	2.31	18.38	2.46	53.0
4	Tanjung Tokong	5.4603	100.3080	3.65	35.8	3.28	67.0
		5.4617	100.3083	2.61	18.3	3.17	58.0
5	Pasir Panjang	5.2950	100.1830	2.0	70.0	2.94	96.0
		5.3380	100.1950	2.6	3000.0	3.31	1584.0
7	Sg. Batu	5.2830	100.2360	1.6	90.0	1.81	164.0
		5.2800	100.2390	1.8	60.0	2.02	137.0
9	Sg. Pulau Betong	5.3060	100.1920	2.7	100.0	3.0	87.0
		5.2800	100.2390	2.7	100.0	2.75	104.0
11	Pantai Miami	5.4760	100.2660	3.0	10.0	1.42	18.0
		5.4778	100.2678	4.0	25.6	1.47	21.0

However, large tsunamis exceeding 4 m may uproot mangrove plants, and the resulting debris can inflict more damage. Hence, numerical analysis on the role of mangroves in reducing the adverse impact of tsunamis is part of tsunami resilience studies. For this purpose, a numerical simulation model has been developed. The 1D continuity and momentum equations in flux forms in the flow x-direction are expressed as Eqs. (7.4) and (7.5), respectively. The last term in Eq. (7.5) is the resistance force used to model the effects of coastal vegetation such as mangroves (Mazda *et al.*, 1997; Massel *et al.*, 1999). The drag coefficient C_D used in the model is estimated by using Eq. (7.6), while the inertia coefficient $C_M = 1.7$ is used (Harada and Imamura, 2000; Hiraishi and Harada, 2003; Harada and Kawata, 2004). There are other estimates of the drag coefficient C_D used (Quartel *et al.*, 2007). Eqs. (7.4) and (7.5) are then solved by the finite difference approximation in the run-up model TUNA-RP, with the inclusion of the mangrove friction term. The mangroves provide frictional drag to reduce tsunami wave energy and wave heights as will be demonstrated next.

$$\frac{\partial \eta}{\partial t} + \frac{\partial M}{\partial x} = 0 \tag{7.4}$$

$$\frac{\partial M}{\partial t} + \frac{\partial}{\partial x}\left(\frac{M^2}{D}\right) + gD\frac{\partial \eta}{\partial x} + \frac{gn^2 M |M|}{D^{7/3}}$$

$$+ \frac{C_D}{2}A_0\frac{M|M|}{D^2} + C_M\frac{V_0}{V}\frac{\partial M}{\partial t} = 0 \tag{7.5}$$

$$C_D = 8.4\frac{V_0}{V} + 0.66 \quad \left(0.01 \leq \frac{V_0}{V} \leq 0.07\right) \tag{7.6}$$

7.5.1 *Mangroves reduce wave heights*

The role of mangrove forests in reducing the impacts of tsunami waves may be demonstrated in Figure 7.17. The incident solitary wave is a positive half sine curve with a wavelength $L = 12,000$ m, period T $= 0.15$ hour and amplitude $a = 1$ m. This wave enters the computational domain from the left at time 0.0 hour and distance 0.0 m. The wave travels 10,000 m to the right along the flat seabed with constant water depth $h = 50$ m. It then climbs up the beach with a slope of 1:40 for a horizontal distance of 2000 m, climbing

Table 7.8:   Definition of symbols used in Eqs. (7.4) to (7.6).

Symbol	Unit	Definition
t	s	Time
x	m	Space coordinate along x-direction
M	m^2/s	Discharge fluxes in the x-direction
η	m	Free surface elevation measured from a fixed datum (mean sea level)
d	m	Water depth below fixed datum
D	m	Total water depth, $D = \eta + d$
g	m/s^2	Gravitational acceleration
n	$s/m^{1/3}$	Manning's relative roughness coefficient
C_D	–	Drag coefficient
A_0	per $100\,m^2$	Projected area of trees under water surface
V_0	m^3	Total volume of tree under water surface
V	m^3	Control volume
C_M	–	Inertia coefficient

49 m along this stretch from 10,000 m to 12,000 m. A healthy and vibrant mangrove forest is located between 12,000 m and 13,000 m. Figure 7.17 shows snapshots of wave heights η at intervals of 0.05 hours, starting at $t = 0.15$ hour. The wave propagates through the flat computational domain for the first 0.15 hours without amplification as shown in Figure 7.17(a). The resistance of the mangrove causes the wave speed to slow down and the wave heights to amplify to 1.8 m in front of the mangroves as shown in Figure 7.17(b) at $t = 0.20$ hour. More slowing down of the wave caused by dense mangroves creates a very sharp wave gradient and further amplifies the wave heights to 2.5 m right in front of the mangrove at $t = 0.25$ hour (Figure 7.17(c)). The wave amplification of waves *in front* of the mangrove is caused by the reduction of wave velocity in the presence of the dense mangroves. Travelling through a dense mangrove forest over a stretch of 1000 m, the wave energy and heights are much reduced behind the forest (Figure 7.17(c)). A sharp wave gradient in front of the mangroves causes wave breaking and strong back flows that reduce the height to merely 1.2 m in front of the forest, with a small wave behind the mangroves (Figure 7.17(d)). Although the wave heights behind the mangroves are much reduced, the wave heights *in front* of the mangroves are amplified. Hence, mangrove

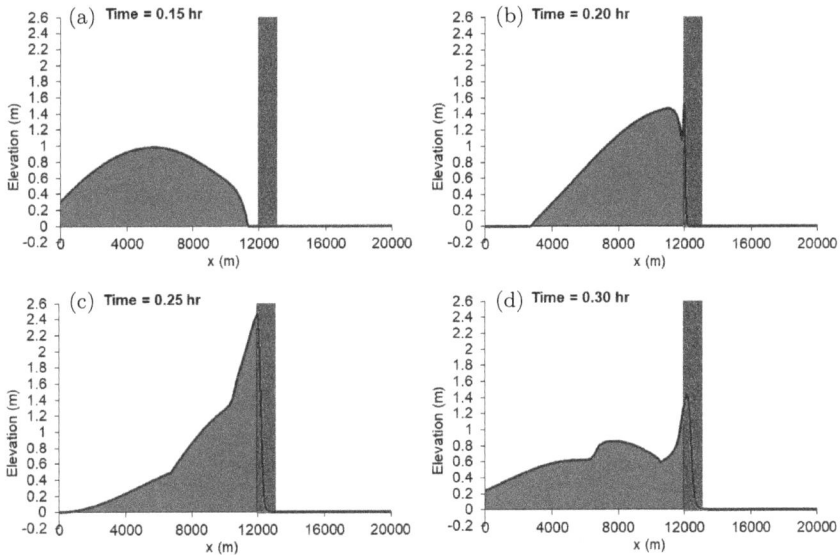

Figure 7.17: Waveforms at interval of 0.05 hour with dense mangrove forest between 12 and 13 km.

forests pose hazards to the zone in front of the forests. This result is consistent with other findings reported in the literature.

7.6 Submarine Mass Failure

Another source of tsunami is submarine mass failure or submarine landslide. Seismic-tectonic activity coupled with sedimentary instability along continental slopes can trigger submarine landslides, also known as SMF. SMFs are major natural marine disasters that could critically damage coastal facilities such as nuclear power plants and oil and gas platforms, as well as destroy submarine pipelines, cables and other submarine facilities. It is therefore essential to investigate SMF for potential tsunami hazard assessment and mitigation. Three-dimensional seismic data from offshore Brunei have revealed a giant seabed mass, with a total seabed volume of 1,200 km^3, an area of 5,300 km^2 and a thickness of 240 m, deposited by a previous SMF sliding along a relatively steep continental slope from 2° to 4° (Gee *et al.*, 2007). The seabed mass extends over 120 km from the continental slope of the Baram Canyon at 200 m water depth to the deep basin floor of the Northwest Borneo Trough (NWNT). The presence of such seabed SMF debris off the coast of Brunei, in the delta of the Baram River, which is characterized as an environment with abundant sediment flux, is a major concern. The potential of an SMF landslide in the North West Borneo Trench (NWBT) cannot be ruled out. SMF can generate large destructive tsunamis if the mass volume exceeds $1\,\mathrm{km}^3$. Large tsunamis generated by SMF exceeding $1\,\mathrm{km}^3$ can cause extensive damage to seafloor infrastructures. The impact on the shoreline can be catastrophic if the shoreline is near the tsunami source. Tsunamis generated by SMF that occur in or near the coast are a major coastal hazard even for moderate earthquakes (Tappin *et al.*, 2001). Large SMFs are often triggered on the continental slope, where such an SMF slide can move vertically downward for several thousand meters. The large potential energy released from this downward movement is then transferred to large kinetic energy, thereby generating large tsunamis. If the tsunami occurs near the coastal region, catastrophic events may occur as there is little time for community preparedness and safe evacuation. At some locations along the U.S. East Coast, nearshore SMF tsunamis are indeed coastal hazards that need to be mitigated. Estimated areas and volumes of potential nearshore SMF are sufficiently large to potentially cause a devastating tsunami for earthquakes of M_w 5.5 or higher if the epicenter is located at the base of the upper slope and if the entire area and volume indeed fail (Brink *et al.*, 2009). The minimum SMF size that can cause a devastating tsunami can be estimated from tsunami run-up models for selected SMFs of various sizes. This information is helpful in the design of infrastructure facilities capable of withstanding the effects of SMF tsunamis. Improvement in the seismic monitoring of continental margins along the U.S. East Coast is required for this approach. It is essential to understand the dynamics of SMF tsunami.

7.6.1 *Dynamics of SMF tsunami*

Fast-moving SMF slides with high kinetic energy are known to produce greater tsunamis. For this rapidly sliding mass, the large kinetic energy of the mass cannot be totally absorbed by the water instantly. This "surplus" kinetic energy results in large volumes of seawater been "pushed" or displaced horizontally and vertically, thereby creating large SMF tsunamis. Conversely, very slow-moving SMF slides produce little or no tsunamis. This is because the small kinetic energy of the SMF mass can be efficiently absorbed by or transferred to the adjacent seawater, resulting in no "surplus" kinetic energy left to lift seawater upward. SMF can generate tsunami with amplitude that far exceeds that generated by co-seismic uplift of the seafloor of the same volume. For earthquake-generated tsunami, the height of the initial tsunami waves is limited to the vertical uplift of the sea floor, typically limited to the order of a few meters. The initial height of the water vertically lifted by an SMF can be of the order of tens of meter. The initial height of an SMF tsunami is highly dependent on the speed at which the mass moves down the slope and across the sea floor. This speed is in turn dependent on the density of the SMF mass and the steepness of the slope. SMF is hazardous to seafloor infrastructures because of its ability to entrain surrounding seafloor materials and cause large substrate erosion of the sea floor. The enlarged SMF can be several times larger than the original SMF. This enlarged total mass and volume would ultimately intensify the ensuing tsunami. Once set in motion, the SMF mass can entrain ambient seawater and seafloor substrates to form longer and larger run-out sediment flows, known as turbidity currents. These turbidity currents can travel long distances down the continental slopes and over hundreds of kilometers on the sea floor, with speeds of up to 20 m/s. SMF and its associated debris flows, and turbidity currents, can pose significant geohazards, and can damage seafloor and shoreline infrastructure. This is particularly so for SMF with high density sliding down steep continental slopes.

7.6.2 *Mass density and slope steepness*

The mass density of SMF and the slope steepness are the dominant parameters governing the energy dynamics of the slide motion and hence the associated tsunami waves' energy characteristics. The sliding SMF with higher mass density moving down a steeper slope will move faster and therefore will generate larger tsunami waves due to the larger kinetic energy in the SMF mass. Conversely, the SMF with lower mass density moving down a milder slope will move slower and therefore generate smaller tsunami waves because of the smaller kinetic energy in the SMF mass. Larger SMF masses generate larger tsunamis than those generated by smaller SMF masses.

7.6.3   *A hypothetical SMF tsunami*

A numerical model known as HySEA has been developed to simulate the entire process of SMF motion sliding down a slope and its subsequent tsunami generation, propagation and beach inundation. A distinguishing feature of the HySEA model is that both the dynamics of the sedimentary fluidized material movements and the upper seawater layer flows are coupled. Further, these two-phase flows interact with each other dynamically, and are computed simultaneously. This numerical model has been used to study the characteristics of an SMF tsunami generated by a hypothetical 1 km^3 SMF sliding down the Alboran Sea Basin (SAB). In this simulation analysis, the HySEA SMF-tsunami model is used to simulate SMF slide dynamics and the subsequent tsunami generation, propagation, run-up and inundation of the shoreline in a single coupled model (Macías *et al.*, 2015). The HySEA-simulated height of the initial wave at the point of generation can exceed a 14-m crust elevation and 25 m of trough depression. A tsunami with heights exceeding 14 m can cause damage of catastrophic proportions. This hypothetical catastrophic SMF tsunami has been observed many times in history.

7.6.4   *Historical cases of large SMF tsunamis*

It has been demonstrated numerically that an SMF of volume 1 km^3 can indeed generate a large tsunami with the potential to cause a catastrophe. Hence, SMFs generated by a large marine earthquake or a large marine volcano lateral collapse can create large and destructive tsunamis. For example, a magnitude M_W 7.1 earthquake in 1946 triggered a giant ($\approx$200 km^3) SMF along the Aleutian Trench, generating the Unimak, Alaska mega tsunami (Fryer *et al.*, 2004). The SMF mass was located within the shallow continental shelf at a 150 m water depth. The SMF mass moved down a 4° slope to the 4,000 m deep Aleutian Trench. The huge potential energy released and transferred to kinetic energy after a drop of 4,000 m created a huge SMF tsunami. Recorded run-up heights for this mega tsunami event reached 35 m above mean sea level at the Scotch Cap lighthouse (Lander, 1996). The tsunami waves propagated across the Pacific Ocean, sustaining a narrow beam of large waves over long distances. Another example of a large SMF tsunami was created by the March 13th, 1888 lateral collapse of the Ritter Island of Papua New Guinea (PNG) into the ocean, which is the largest marine volcano lateral collapse recorded. The collapse that occurred around 6:00 am removed most of the island. This lateral collapse of the volcano was preceded by explosions accompanied by earthquakes. A volume of 4.2 km^3 of mass slid down during the initial collapse, yielding a total volume of 6.4 km^3 of substrate erosion materials consisting of debris flows and turbidity deposits. This demonstrated the efficiency of initial SMF in creating large substrate erosion during the later stage of the SMF movement under the sea (Day *et al.*, 2015). The combination of the initial SMF and subsequent substrate erosion materials can generate a large combined SMF and produce a regionally destructive tsunami. The PNG tsunami's impact was witnessed by literate observers who timed the waves with watches and provided detailed scientific accounts. A third example of an SMF tsunami can be traced to historical records documented three centuries ago. These documents have alluded to a disastrous tsunami on the southwestern Taiwan coast some time between 1781 and 1782, with a reported death toll of more than 40,000. This tsunami was most probably generated by seismically triggered SMF initiated on the upper portion of the continental slope. Further, recent research sponsored by oil and gas exploration and development in China, for example, reveals significant potential tsunami risks arising from SMF in the Northern South China Sea. The Baiyun Depression in this area has been observed to have suffered eleven SMF landslides of different scales in the past. The infamous 2011 Great East Japan Earthquake and Tsunami (GEJET) that crippled the Fukushima Dai-Ichi Nuclear Power Plant is known to be generated by a Mw 9.0 earthquake. However, recorded observation of a concentrated narrow band of large tsunami waves in the Sanriku district combined with marine geophysical data analysis has suggested that an SMF was also involved in the Sanriku area (Tappin *et al.*, 2014). Compiled scientific data have been key to understanding and modeling SMF-generated tsunami and its subsequent propagation (Ward and Day, 2003). Understanding the characteristics of SMF tsunamis is important toward the development of tsunami resilience capability.

7.6.5   *Rigid and deformable SMF*

The dynamics of an SMF mass sliding down a marine slope is complicated. It can be characterized in two distinct approaches, (a) rigid slide or (b) deformable slide, with significantly different kinematics. For the waves generated by a rigid SMF slide, which is the approach used in this chapter, the wave energy is mostly concentrated in a rigid mass with a narrow band moving in the dominant slide direction. As the rigid SMF accelerates down the slope, it can generate very large waves because of its concentration. The ultimate wave heights and energy depend on the mass density and slope steepness and distance travelled. The rigid SMF slide would lead to the worst-case scenario for tsunami hazard assessment. On the contrary, a deformable SMF marine slide would result in more diffusive tsunamis with lesser wave heights. For the waves generated by a deformable SMF slide, wave energy spatial spreading is more significant. The waves generated by a deformable SMF slide would reach their maximum values earlier than those generated by a rigid SMF slide (Ma *et al.*, 2013).

7.6.6   *Depression and elevation waves*

An SMF generates both depression and elevation waves. When an SMF mass slides down a marine slope, the potential energy of the mass released is converted to kinetic energy of the mass, travelling forward in the direction of the slide. These moving masses push the water in front vertically up and generate positive waves, moving in the same direction of the mass sliding. These positive waves are known as elevation waves because their wave heights are above the local mean sea level. The water being pushed away to the front creates a "void or vacuum" behind, resulting in the water behind being drawn vertically downward, generating negative waves with wave heights below the local mean sea level. These negative waves are known as depression waves. Therefore, an SMF generates two waves, a depression wave behind and an elevation wave in the front. The elevation wave crest propagates in the same direction of the slide, with speed faster than that of the sliding mass motion. The depression wave trough, which is tied to the sliding mass motion, propagates with a slower speed, in the direction opposite to the SMF slide direction.

7.7 Simulation of SMF

An SMF-generated tsunami was co-triggered during the 2011 Tohoku submarine earthquake. Detailed analysis performed after the 2011 Tohoku tsunami reveals that the horizontal acceleration from the Tohoku earthquake was sufficiently large to trigger an SMF. The much-elevated tsunami run-ups and extensive inundations measured along the central Sanriku coast exhibited characteristics of an SMF-generated tsunami. The impacts are more directionally focused than would be expected from an earthquake-only source. The conclusion that a significant part of the Tohoku tsunami was generated by an SMF source has important implications for assessment of tsunami hazards in regions with similar geohazards, such as the North West Borneo Trench (Figure 7.18).

7.7.1 *North West Borneo Trench*

The Brunei SMF slide, within the North West Borneo Trench, is an example of a major SMF located on a tectonically active marine

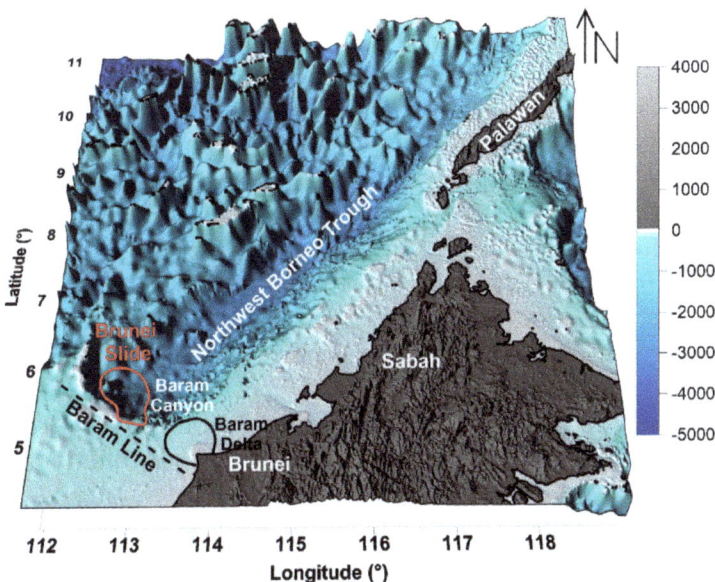

Figure 7.18:   Location of Brunei Slide and the North West Borneo Trough.

margin adjacent to a large river and canyon system within the active Baram Delta. Analysis of three-dimensional seismic data from offshore Brunei has revealed giant seabed mass debris deposited by a previous SMF tsunami, triggered by the Brunei SMF. Initiated on a relatively steep slope varying between $2°$ and $4°$, this giant debris has a total seabed volume of $1,200 \, \text{km}^3$, an area of $5,300 \, \text{km}^2$ and a thickness of 240 m (Gee *et al.*, 2007). A recurrence of the Brunei SMF landslide of a similar magnitude cannot be ruled out. Therefore, hazard and risk assessment on the Brunei SMF tsunami is essential to develop mitigation measures and programs for the affected community. The risk assessment begins with the simulation of an SMF tsunami triggered by a submarine mass failure in the North West Borneo Trench to be performed in this Section 7.7.

7.7.2   *Rigid SMF formulation*

There are two types of SMF landslide models used: (1) Rigid mass and (2) Deformable mass. For the waves generated by rigid SMF landslides, the energy of the waves is mostly concentrated in a narrow band of the dominant slide direction. As the rigid landslide continuously accelerates downslope, it can gain higher velocity, generate more kinetic energy, and hence can eventually produce larger waves. On the contrary, for waves generated by deformable SMF landslides, the spatial spread of the energy is more significant. The spatial energy spread results in smaller tsunami waves that spread over a larger spatial extent. To project the worst-case scenario for hazard assessment of an SMF tsunami in this section, we therefore adopt the rigid SMF empirical formulation proposed by Watts *et al.* (2003) to generate the initial source waves caused by the Brunei SMF Slide. These empirical equations used for generating the SMF source tsunami have been developed from 2D and 3D model studies. These numerical studies systematically vary critical SMF parameters such as slide properties, slope angle and length, basin geometry and water depth (Heller and Spinneken, 2015). Volume of the mass, initial acceleration and maximum velocity are important parameters that have impacts on the initial source tsunami. These numerical studies provided the foundation for generating SMF landslide tsunami sources. A commonly used formulation is based upon the concept of center of mass motion, given by an empirical formulation proposed by Watts *et al.* (2003).

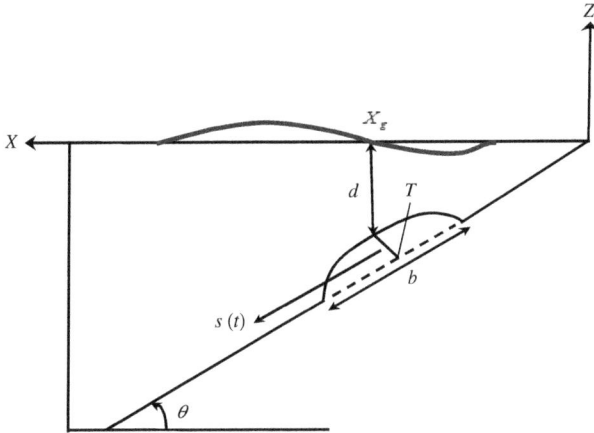

Figure 7.19: Schematic figure for SMF slide (Watts *et al.*, 2003).

7.7.2.1 *Center of mass motion*

The Borneo SMF slide is therefore modeled as a rigid body moving along a planar slope with the center of mass motion $s(t)$ moving parallel to the planar slope. The slide motion is subject to external forces due to mass, gravity and frictional dissipation. The SMF slide geometry is idealized as a mound with semi-elliptical cross section as displayed in Figure 7.19. The mound has a maximum thickness T at the center, a total length b along the planar slope axis and a total width w along the cross-sectional axis. The mound has an initial submergence depth d measured at the center of the mound. This SMF mass slides along the planar slope with an angle θ from the horizontal. Then, following Watts *et al.* (2003), the center of mass motion $s(t)$ is described by Eqs. (7.7) and (7.8), in which the units and definitions are given in Table 7.9. The Borneo SMF tsunami source is generated by this Watts formulation as described in the next subsection.

$$s(t) = s_0 \ln\left[\cosh(t/t_0)\right] \tag{7.7a}$$

$$a_0 \cong 0.30g \sin\theta \tag{7.7b}$$

$$u_t \cong 1.16\sqrt{bg \sin\theta} \tag{7.7c}$$

$$s_0 \equiv u_t^2/a_0 \cong 4.48b \tag{7.7d}$$

$$t_0 \equiv u_t/a_0 \cong 3.87\sqrt{b/g \sin\theta} \tag{7.7e}$$

Table 7.9:   Definition of symbols used in Eqs. (7.7) and (7.8).

Symbol	Unit	Definition
a_0	m/s^2	Initial acceleration
u_t	m/s	Theoretical terminal velocity
s_0	m	Characteristic distance of motion
t_0	s	Characteristic time of motion
g	m/s^2	gravitational acceleration
λ_0	m	Characteristic wavelength
θ	°	Angle of the planar slope
d	m	Initial submergence depth

$$\lambda_0 = t_0 \sqrt{gd} \qquad (7.8)$$

7.7.3   *Borneo SMF tsunami source*

The center of mass motion $s(t)$ for the Borneo SMF tsunami source is described by Eqs. (7.7) and (7.8). The Brunei SMF mound is assumed to have a specific density $\gamma \cong 1.85$, as it consists of a mixture of water and solids. A Coulomb friction coefficient $C_n \cong 0$ and a drag coefficient $C_d \cong 1$ are adopted. Finally, an added mass coefficient C_m is set $\cong 1$. Further details are available in Watts *et al.* (2003). The following subsections present numerical simulations to assess the impacts of a potential tsunami generated by an SMF off the coast of Brunei in an area known as the NWBT. Henceforth, all tsunamis are assumed to be generated by an SMF.

7.7.3.1   *Two slope scenarios*

We consider two scenarios of slope steepness for the Borneo SMF tsunami simulation: (a) 2° slope and (b) 4° slopes. Table 7.10 summarizes the Brunei SMF parameters used to compute the initial slide motions and initial tsunami characteristics at the source of generation as displayed in Table 7.11. The two-dimensional source tsunamis generated by the Brunei Slide are presented in Figure 7.20. The center of the Brunei Slide is initially located at latitude of 5.85° and longitude of 113.61° (Gee *et al.*, 2007). It then slides downward to the seafloor in the northeast direction for about 211 km. This submarine landslide motion then creates an elevation N-wave,

Table 7.10:   Brunei Slide parameters.

Parameters	2° slope	4° slope
Length, b (km)	120	120
Width, w (km)	44	44
Thickness, T (m)	240	240
Initial submergence depth, D (m)	2450	2450

Table 7.11:   Slide motions and initial tsunami characteristics.

Parameters	2° slope	4° slope
a_0 (m/s^2)	0.10	0.20
u_t (m/s)	234	331
s_0 (km)	537	537
t_0 (s)	2290	1620
λ_0 (km)	355	251

moving in the northeast direction, the direction of slide. The elevation N-wave has a velocity field pointing in the northeast direction, parallel to the sliding direction. The SMF moves faster in a steeper slope of 4°. The higher kinetic energy creates greater energy to push more water column forward and upward, generating higher crests and deeper troughs. As observed in Figure 7.20(c), the wave heights of the N-wave generated by the steeper 4° slope are 3 times those generated by the milder 2° slope. As observed in Figure 7.21, the initial tsunami source waves propagate in two distinct directions. The primary dominant paths of tsunami propagation follow the sliding direction, i.e., due northeast. The secondary and less dominant waves propagate in two directions perpendicular to the sliding direction, as leading depression waves (blue) and as leading elevation waves (red). The SMF source tsunami waves are used to simulate the subsequent tsunami propagation through the ocean and ultimate run-up and inundation along the beaches.

7.7.4   *Beach run-up heights*

The generated SMF tsunami source is then used as the initial condition to simulate tsunami propagation and eventual run-up and

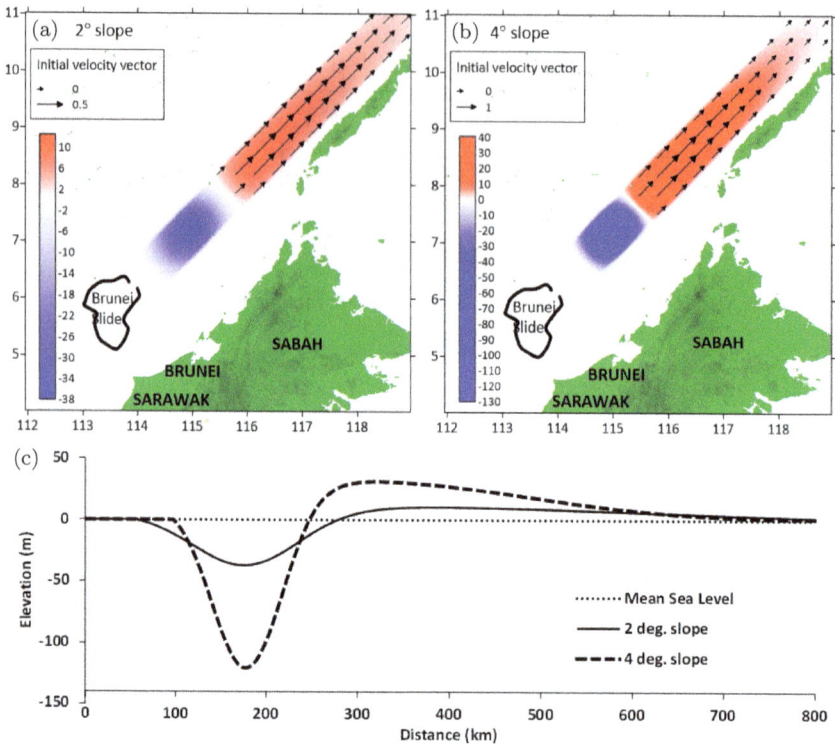

Figure 7.20:   Initial tsunamis generated by Brunei Slide: (a) 2.0° slope, (b) 4.0° slope and (c) cross-sectional view.

inundation by means of TUNA. Figure 7.22 shows the inundation map for Kudat for two slopes of 2° and 4°. The wave may arrive at the nearest beaches in Sabah and Sarawak within an hour, leaving little time for communities to evacuate safely. This Brunei SMF tsunami is potentially destructive, as the source trough may have a maximum wave height reaching 120 m. The waves in the front propagate forward in the primary dominant slide direction due northeast. Fortunately, Sabah, Brunei and Sarawak are positioned in the direction perpendicular to this dominant propagation direction. However, the secondary and less dominant waves travel southeast toward Sabah, and travel southwest toward Brunei and Sarawak. The maximum inundated depths may reach 20 m at Kudat, 26 m at Kota Kinabalu and 15 m at Miri. The maximum inundation distance of 4.86 km may be expected at Miri due to its low-lying coast. The maximum

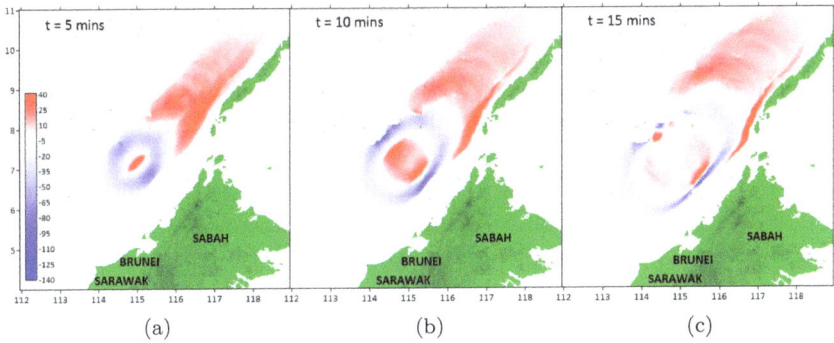

Figure 7.21: Tsunami propagation snapshots for 4° slope at intervals of 5 mins; *x*-axis and *y*-axis are presented in longitude and latitude, respectively.

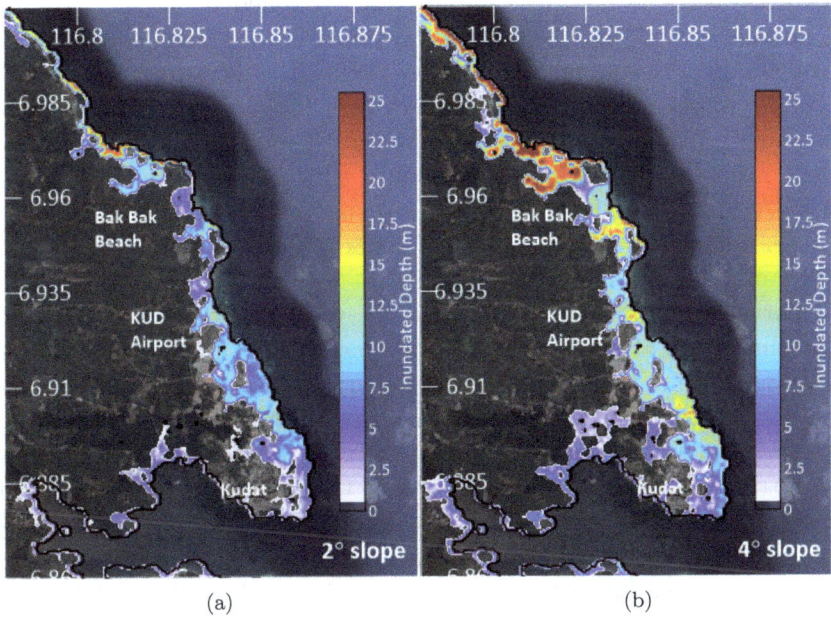

Figure 7.22: Inundation map of Kudat.

inundation distances for Kudat and Kota Kinabalu are, respectively, 1.64 km and 2.64 km. In view of the vulnerability of these locations to the impact of the Brunei SMF tsunami, it is important to develop and implement community resilience programs to mitigate the potential risks.

7.8   Tsunami Disaster Risk Reduction

Tsunamis that were generated close to populated coastal zones can be catastrophic. Hence, tsunami disaster risk reduction programs must be put in place in towns located along these affected coastal zones. An earthquake with a magnitude of $M_w = 9.3$ occurred off the western coast of Banda Aceh, North Sumatra, at 00:58:53 UTC (08:58:53 Malaysian time) on 26 December 2004. It was the largest earthquake ever recorded in the Indian Ocean and the fourth largest in the world. This undersea earthquake instantly lifted two hundred trillion tons of water vertically upward and generated a mega tsunami that killed around 250,000 people worldwide, mostly in Banda Aceh of Indonesia and Phuket of Thailand. Malaysia was not spared the atrocity, with 68 deaths, of which 52 happened in Penang Island. Two field surveys were conducted by the authors along the affected beaches in Penang and Kedah to assess the run-up heights and inundation distances. These surveys provide comprehensive knowledge regarding areas that were badly affected, and the scientific data needed to calibrate and validate a tsunami simulation model TUNA. Universiti Kebangsaan Malaysia (UKM) formed a team to conduct research on the social, economic and cultural impact of this tsunami (Komoo and Othman, 2006). The Malaysian Meteorological Department (MMD) established early warning systems to provide frontline mitigation measures. The Academy of Sciences Malaysia (ASM) coordinated several research projects focusing on earthquake and tsunamis mitigation science. A series of international workshops, conferences and roundtables were held in Malaysia and other affected countries in Asia including Australia, China, India, Indonesia, the Philippines, Singapore, Sri Lanka, Taiwan and Thailand to coordinate research collaboration and to establish effective early warning systems. Other international organizations such as UNESCO-IOC and USGS actively participated in such research and community engagement and other advocacy activities to formulate guidelines for policymakers (Koh *et al.*, 2007, 2009). Since the 2004 mega tsunami, a series of eight destructive tsunamis have occurred, including the catastrophic Fukushima Triple Disaster on 11 March 2011. This 2011 Fukushima tsunami killed more than 20,000 people and caused major radiation leaks. Large tsunami inundations can indeed cause great loss of life and inflict severe damage to properties. The next mega tsunami is long overdue (Sieh *et al.*, 2015). Hence, it is critical to develop disaster risk reduction (DRR) capability to mitigate tsunami impacts in the future. To guide DRR, the United Nations has formulated the Sendai Framework for Disaster Risk Reduction (SFDRR).

7.8.1 Sendai Framework for DRR

The Sendai Framework for Disaster Risk Reduction (SFDRR): 2015–2030 (UNISDR, 2015) calls for an integrated approach to DRR, incorporating multisectoral and transdisciplinary international collaboration and partnerships (ICPs) across borders. The SFDRR highlights the critical role of hazard awareness and education (HAE) in disaster reduction, recovery, rehabilitation and reconstruction. The SFDRR provides a mechanism to identify social, economic, cultural and political factors and processes that require urgent transformation. This transformation is needed to provide practical solutions for DRR and to foster equitable, resilient and sustainable development through research (Munene *et al.*, 2018). SFDRR enhances ICP for more coherent approaches on DRR, including climate change risk and adaptation. ICP helps to enhance integrated risk management capabilities by bridging the gap between science, policy and practice, by developing efficient management of transboundary crises and by assisting developing countries in securing resources to improve their research and innovation capacities in DRR.

7.8.1.1 DRR capability

DRR entails the capability to understand hazard, identify vulnerability, anticipate risk, mitigate the impact and bounce back rapidly through survival and adaptation. Catastrophic disasters such as massive tsunami inundations would invariably result in poverty, create disruptions to food supply and cause unemployment. Hence, DRR should formulate policy and programs that contribute to the United Nations Sustainable Development Goals (UNSDGs), such as poverty elimination (SDG1), hunger reduction (SDG2) and restoration of clean water and sanitation (SDG6). Achieving DRR goals requires close ICP among disaster management organizations. Trust and control are core elements for building confidence among collaboration partners. In all disaster response and recovery cycles, trust and control are essential and are deemed complementary and mutually reinforcing (Kalkman and de Waard, 2017). It is mandatory to establish an authoritative agency to develop best practices for delivery of trusted hazard advice and services, supported by cutting-edge science and hazard impact research (Hemingway and Gunawan, 2018). In Malaysia, a consortium consisting of key government agencies and several higher education institutes (HEI), including USM and Sunway University, provides tsunami hazard advice and services. Workshops and training sessions were organized to develop tsunami resilience consisting of the following five criteria (Bernard, 2005): (a) Understand the nature of the tsunami hazards, (b) Develop the tools to mitigate the hazards, (c) Disseminate information about tsunami hazards, (d) Exchange information with other at-risk areas and (e) Institutionalize planning for tsunami disaster. At the community and local institutional level, it is essential to strengthen HAE.

7.8.2 *Hazard awareness education*

HAE has two primary functions: (a) To enrich awareness and education on hazard, vulnerability and risk and (b) to enable translating HAE to inform decision and actions for community protection during a disaster (Frankenberg *et al.*, 2013). HAE provides the community with the necessary tools such as information, skills and knowledge to cope with disaster (Johnson *et al.*, 2014). Schools and universities have a responsibility of delivering HAE development toward building community resilience against disaster risk (Shaw *et al.*, 2011). As educational institutions having close links with the community, schools and universities have a prime role in building community resilience against disasters through their mandate for education, information sharing and their broad stakeholder networks (Oktari *et al.*, 2015, 2017). Tsunami disaster can be mitigated by appropriate tsunami DRR programs, including school and public HAE programs. Integrated disaster risk management comprises a combination of components that are undertaken before, during and after a tsunami disaster event. Central to the disaster risk management cycle is HAE and effective communication with the public about tsunami hazard, vulnerability, risk and mitigation (Kurita *et al.*, 2006). However, HAE does not necessarily translate into effective preparatory attitude and behavior for disaster mitigation. Concerted efforts are required to persuade and motivate people at risk to convert HAE knowledge into effective preparation and action for disaster mitigation (Abunyewah *et al.*, 2018). Awareness is location specific and depends on a variety of factors such as education, culture and the policies of local administration and national governments (Anh *et al.*, 2017). Awareness alone, however, does not ensure disaster risk preparedness. While local inhabitants may have a high level of awareness about coastal hazards in general, many tend to underestimate their risk severity. This failure could lead to increased risk during a disaster, particularly in vulnerable and impoverished neighborhoods. Hazard and risk communication must be undertaken to persuade people to adopt self-protective behaviors and practices. Residents exposed to tsunami hazard need to be well informed about the various types of risk management and mitigation procedures available to them. HAE through the school curriculum can be effectively integrated with home-based disaster preparedness as they are closely linked (Johnston *et al.*, 2005).

Public HAE is essential in preparing the public against disasters (Esteban *et al.*, 2017) and to take immediate actions for community protection during a disaster (Frankenberg *et al.*, 2013). Research conducted by Ronan *et al.* (2001) on school children revealed that those who had been involved in HAE programs displayed more stable risk perceptions, showed reduced hazard-related fears and demonstrated an increased awareness of critical hazard-related protective behaviors compared to children who are not exposed to HAE. More importantly, by following HAE in schools, children are more likely to interact with and educate their parents and community on HAE issues, which in turn would increase home-based and community-based preparedness for disaster mitigation (Johnston *et al.*, 2005). Outside of the school system, increasing HAE may be achieved through media activities such as television, radio and internet campaigns, and via the distribution of information leaflets, brochures, posters and videos, and public information meetings. Providing HAE risk information to the public has a positive result in an increase in public trust on risk mitigation and planning competence (Schütz and Wiedemann, 2000). However, HAE and risk information and community engagement should be provided to a community at least once every 3 years, to ensure risk knowledge does not decrease with time. This is important because public ignorance and complacency can increase over time due to long intervals between successive hazard events. To render HAE effective, an integrated tsunami risk management system is essential.

7.8.3   *Integrated tsunami risk management system*

An integrated tsunami risk management system consists of the following components: (A) Identify hazard zones, (B) Develop inundation maps, (C) Disseminate evacuation maps, (D) Turn vague concerns and abstract issues into clarified and concrete action plans and (E) Implement the plan during a tsunami event. To achieve these goals, the following infrastructure components are needed: (a) Adequate seismic network, (b) Robust numerical models, (c) Sufficient deep-seabed pressure sensors, (d) Dense coastal tide gauges and (e) Effective school and public HAE Program (Koh *et al.*, 2007). For Malaysia, tsunami inundation maps for affected coastal areas are developed from numerical simulation of the TUNA-RP model (Koh *et al.*, 2017). These inundation maps are useful tools in tsunami HAE programs for schools and the public and for risk managers. An inter-government consortium of Malaysian government agencies and universities led by the Malaysian Meteorological Department and the National Disaster Management Agency has conducted several tsunami HAE and survey programs in three coastal communities identified as tsunami hazard zones: Penang, Langkawi and Kudat. A mixed methodology approach was utilized, using key informant interviews, site surveys and questionnaire surveys to understand the HAE level in the population (Arce *et al.*, 2017). These programs followed closely the SFDRR Framework for DRR. Interviews with teachers, school principals and community leaders were conducted, followed by focus group discussions involving schools, NGOs, local government agencies and universities. Workshops, ground thrusting and tabletop exercises were conducted to train and obtain feedback from participants, leading to informed decisions and actions. Expected strategic outcomes include development of substantial strategic, operational and tactical capability on community DRR for tsunami risk mitigation. Technical outputs include early warning systems at schools and beaches, and training modules including tsunami inundation maps, tsunami evacuation routes and safe assembly points (Koh and Teh, 2019). These technical outputs are useful for conducting regular tsunami evacuation drills, and for identifying safe evacuation routes and assembly points. Figure 7.23 shows an example of safe evacuation routes and assembly points for Zone 1 in Kudat town.

7.8.4 *International collaboration and partnership*

International collaboration and partnership (ICP) can vastly enhance these SFDRR programs and research for tsunami disaster risk reduction (DRR). Natural hazardous events such as tsunamis, earthquakes, storm surges and landslides are natural geo-meteorological occurrences that are impossible to prevent by any conceivable human and technological interventions. However, such natural hazardous events can turn into disasters only if the communities become vulnerable because of unmitigated risks. It is essential to mitigate the physical, social, economic and environmental vulnerability. Disaster risk can be mitigated by reducing the vulnerability via DRR programs including HAE and integrated tsunami risk management. Integrated tsunami risk management programs must subscribe to internationally acknowledged best practices as mandated by the SFDRR. Inspired by the aspiration of ICP, a series of ten tsunami mitigation workshops known as the "South China Sea Tsunami Workshops, SCSTWs" have been held around countries surrounding the South China Sea over the past decade. The authors have been given the task to organize SCSTW3 on 3 to 5 November 2009, held in the campus of Universiti Sains Malaysia in Penang, Malaysia. More than 150 participants from 20 counties took part in SCSTW3. Two workshop proceedings have been published to provide guidelines on tsunami risk reduction measures. The first book focuses on the first objective of cultivating tsunami HAE and preparedness capability (Koh *et al.*, 2011a). The second book covers the second objective of promoting tsunami technical and scientific ICP (Koh *et al.*, 2011b). The ICP network established through the ten SCSTWs has helped to enhance ICP in promoting DRR programs. Higher education institutes can play a vital role in promoting and enhancing tsunami DRR via research, outreach, HAE and ICP programs. Tsunami HAE in schools and HEIs are particularly effective in achieving tsunami DRR as schools and HEIs are an integral component of the community.

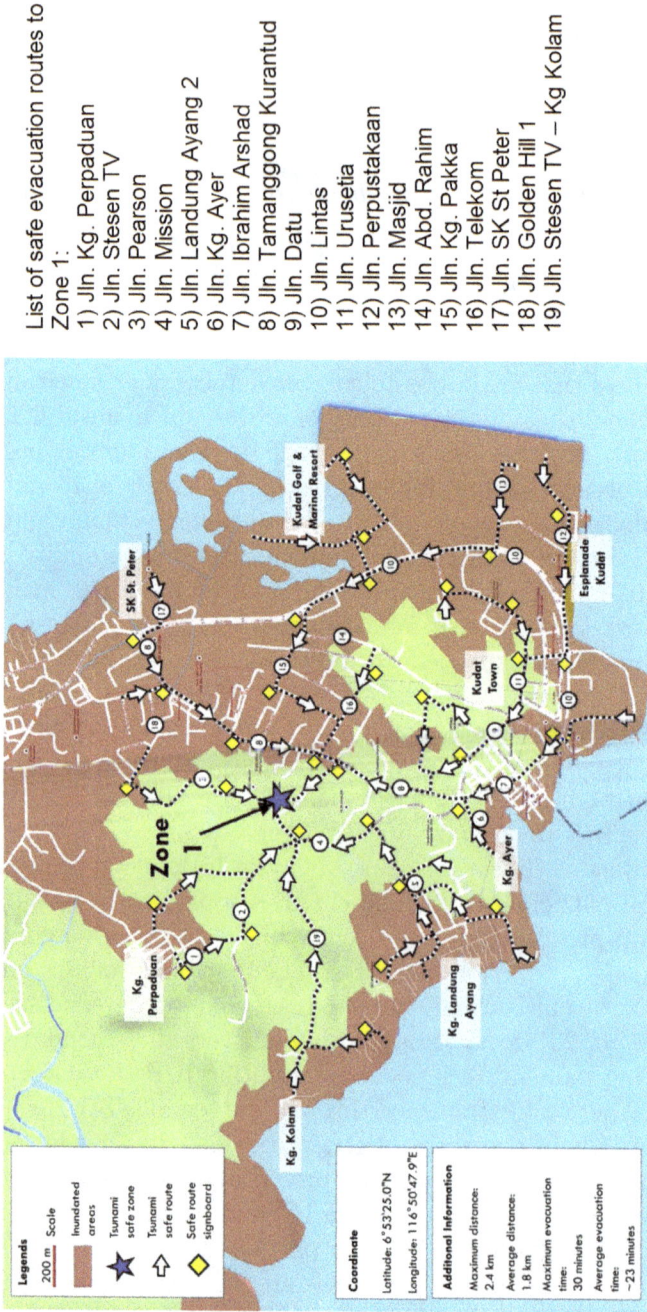

Figure 7.23:   Map showing tsunami safe evacuation zones and routes for Zone 1 in Kudat, Sabah.

7.8.5 *Save life and save livelihood*

Saving lives and saving livelihood is critical during and after a disaster event such as a mega tsunami. A large tsunami inundation can inflict enormous loss of lives and properties to densely populated coastal communities such as Banda Aceh. It is a big challenge to devise an effective DRR strategy to protect Banda Aceh because of the size of the population that needs to be evacuated or relocated to limited higher ground further inland. This daunting challenge is further compounded by the proximity of Banda Aceh to the tsunami source. This remark is equally pertinent to many big coastal cities in developing countries. However, many lives can be saved if the community is well prepared in advance and is equipped with knowledge learned from HAE programs for self-protection. Besides these mitigation measures, other desirable DRR measures, such as adequate infrastructures, must be put in place for safe and orderly evacuation during a tsunami event. Saving lives and saving livelihood is paramount (UNISDR, 2015). Life and livelihood can be saved if the community can receive adequate and regular tsunami DRR training through HAE programs that are well planned and implemented. Pre-existing social-cultural vulnerability should be identified and rectified as such vulnerability can be intensified in times of disaster, thereby creating a condition of multiple vulnerabilities and intensifying the losses and damages (Pongponrat and Ishii, 2017).

References

Abunyewah, M., Gajendran, T. and Maund, K. (2018). Conceptual framework for motivating actions towards disaster preparedness through risk communication. *Procedia Engineering*, 212, 246–253, doi: 10.1016/j.proeng.2018.01.032.

Anh, L.T., Takagi, H., Thao, N.D. and Esteban, M. (2017). Investigation of awareness of typhoon and storm surge in the Mekong Delta – Recollection of 1997 Typhoon Linda. *Japan Society of Civil Engineers, Ser. B3 (Ocean Engineering)* 7(2), 168–173.

Arce, R.S.C., Onuki, M., Esteban, M. and Shibayama, T. (2017). Risk awareness and intended tsunami evacuation behaviour of international tourists in Kamakura City, Japan. *International Journal of Disaster Risk Reduction*, 23, 178–192, doi: 10.1016/j.ijdrr.2017.04.005.

Bernard, E.N. (2005). The U.S. National Tsunami Hazard Mitigation Program: A Successful State-Federal Partnership. Developing Tsunami-Resilient Communities: The National Tsunami Hazard Mitigation Program, Eddie Bernard (ed.). *Natural Hazards*, 35, 5–24.

Borrero, J.C. (2005). Field survey of northern Sumatra and Banda Aceh, Indonesia after the tsunami and earthquake of 26 December 2004, Seismol. *Research Letters*, 75(3), 312–320.

Brink, U.S., Lee, H.J., Geist, E.L. and Twichell, D. (2009). Assessment of tsunami hazard to the U.S. East Coast using relationships between submarine landslides and earthquakes. *Marine Geology*, 264, 65–73.

Day, S., Llanes, P., Silver, E., Hoffmann, G., Ward, S. and Driscoll, N. (2015). Submarine landslide deposits of the historical lateral collapse of Ritter Island, Papua New Island. *Marine and Petroleum Geology*, 67, 419–438.

Disaster Control Research Center (DCRC) (2005). Modeling a tsunami generated by Northern Sumatra Earthquake [12/26/2004], Tsunami Engineering Laboratory, Disaster Control Research Center, Tohoku University, Japan. http://www.tsunami.civil.tohoku.ac.jp/hokusai3/E/events/events.html.

Esteban, M., Takagi, H., Mikami, T., Aprilia, A., Fujii, D., Kurobe, S. and Utamae, N.A. (2017). Awareness of coastal floods in impoverished subsiding coastal communities in Jakarta: Tsunamis, typhoon storm surges and dyke-induced tsunamis. *International Journal of Disaster Risk Reduction*, 23, 70–79, doi: 10.1016/j.ijdrr.2017.04.007.

Frankenberg, E., Sikoki, B., Sumantri, C., Suriastini, W. and Thomas, D. (2013). Education, vulnerability, and resilience after a natural disaster. *Ecology and Society*, 18(2), 16, doi: 10.5751/es-05377-180216.

Fryer, G.J., Watts, P. and Pratson, L.F. (2004). Source of the great tsunami of 1 April 1946: A landslide in the upper Aleutian forearc. *Marine Geology*, 203, 201–218.

Gee, M.J.R., Uy, H.S., Warren, J., Morley, C.K. and Lambiase, J.J. (2007). The Brunei slide: A giant submarine landslide on the North West Borneo Margin revealed by 3D seismic data. *Marine Geology*, 246(1), 9–3.

Grilli, S. T., Ioualalen, M., Asavanant, J., Shi, F., Kirby, J., and Watts, P. (2007). Source constraints and model simulation of the December 26, 2004 Indian Ocean Tsunami. *Journal of Waterway, Port, Coastal, and Ocean Engineering*, 133(6), 414–428.

Harada, K. and Imamura, F. (2000). Experimental study on the resistance by mangrove under the unsteady flow. *Proceeding of the 1st Congress of APACE*, 975–984.

Harada, K. and Kawata, Y. (2004). Study on the effect of coastal forest to tsunami reduction. *Annuals Disaster Prevention Research Institute, Kyoto University*, No. 47 C.

Heller, V. and Spinneken, J. (2015). On the effect of the water body geometry on landslide-tsunamis: Physical insight from laboratory tests and 2D to 3D wave parameter transformation. *Coastal Engineering*, 104, 113–134.

Hemingway, R. and Gunawan, O. (2018). The natural hazards partnership: A public-sector collaboration across the UK for natural hazard disaster risk reduction. *International Journal of Disaster Risk Reduction*, 27, 499–511, doi: 10.1016/j.ijdrr.2017.11.014.

Hérbert, H. Schindelé, F., Altinok, Y., Alpar, B. and Gazioglu, C. (2005). Tsunami in the Marmara sea (Turkey): A numerical approach to discuss active faulting and impact on the Istanbul coastal areas. *Marine Geology*, 215, 23–43.

Hiraishi, T. and Harada, K. (2003). Greenbelt tsunami prevention in South-Pacific region. *Report of the Port and Airport Research Institute*, 42(2), 3–25.

Imamura, F., Shuto, N. and Goto, C. (1988). Numerical simulation of the transoceanic propagation of tsunamis. Proceeding of the Sixth Congress of the Asian and Pacific Regional Division, Int. Assoc. Hydraul. Res., Kyoto, Japan, pp. 265–272.

IOC (1997). Numerical Method of Tsunami Simulation with the Leap Frog Scheme, 1, Shallow Water Theory and Its Difference Scheme. In Manuals and Guides of the IOC, pp. 12–19, Intergovernmental Oceanographic Commission, UNESCO, Paris.

Ioualalen, M., Asavanant, J. Kaewbanjak, N., Grilli, S.T., Kirby, J.T., and Watts, P. (2007). Modeling the 26th December 2004 Indian Ocean tsunami: Case study of impact in Thailand. *Journal of Geophysical Research*, 112, C07024.

Johnston, D., Paton, D., Crawford, G.L., Ronan, K., Houghton, B. and Bürgelt, P. (2005). Measuring tsunami preparedness in coastal Washington, United States. *Natural Hazards*, 35, 173–184, doi: 10.1007/s11069-004-2419-8.

Johnson, V.A., Ronan, K.R., Johnston, D.M. and Peace, R. (2014). Evaluations of disaster education programs for children: A methodological review. *International Journal of Disaster Risk Reduction*, 9, 107–123, doi: 10.1016/j.ijdrr.2014.04.001.

Kalkman, J.P. and de Waard, E.J. (2017). Inter-organizational disaster management projects: Finding the middle way between trust and control. *International Journal of Project Management*, 3(5), 889–899, doi: 10.1016/j.ijproman.2016.09.013.

Koh, H.L. and Teh, S.Y. (2019). Disaster Risk Reduction and Resilience through Partnership and Collaboration. In: Leal Filho, W., Azul, A., Brandli L., Özuyar, P., Wall, T. (eds.) *Partnerships for the Goals. Encyclopedia of the UN Sustainable Development Goals*. Springer, Cham, doi: 10.1007/978-3-319-71067-9_49-1.

Koh, H.L., Teh, S.Y. and Izani, A.M.I. (2005). Meso Scale Simulation of December 26 2004 Tsunami With Reference to Malaysia and Thailand. Proceedings of the Third International Symposium on Southeast Asian Water Environment, 6–8 December 2005, The University of Tokyo and Asian Institute of Technology, Bangkok, Thailand, pp. 89–96.

Koh, H.L., Teh, S.Y. and Izani, A.M.I. (2007). Tsunami Mitigation Management. Special Feature: Natural disaster management technologies. The United Nations Asian and Pacific Centre for Transfer of Technology (UN-APCTT) Nov–Dec 2007. *Asia Pacific Tech Monitor*, 24(6), 47–54.

Koh, H.L., Teh, S.Y., Liu, P.L.-F., Izani, A.M.I. and Lee, H.L. (2009). Simulation of Andaman 2004 Tsunami for assessing impact on Malaysia. *Journal of Asian Earth Sciences*, 36(1), 74–83.

Koh, H.L., Teh, S.Y., Liu, P.L.-F. and Che Abas, M.R. (2010). Tsunami Simulation Research and Mitigation Programs in Malaysia Post 2004 Andaman Tsunami. Chapter 2 in Tsunamis: Causes, Characteristics and Warnings, and Protection. Neil Veitch and Gordon Jaffray (eds.). Nova Science Publishers, Inc., Hauppauge, New York, pp. 29–56.

Koh, H.L., Liu, P.L.-F. and Teh, S.Y. (2011a). *Tsunami Education, Protection and Preparedness*. Penerbit Universiti Sains Malaysia, Pulau Pinang, p. 247.

Koh, H.L., Liu, P.L.-F. and Teh, S.Y. (2011b). *Tsunami Simulation for Impact Assessment*. Penerbit Universiti Sains Malaysia, Pulau Pinang, p. 249.

Koh, H.L., Teh, S.Y., Tan, W.K. and Kh'ng, X.Y. (2017). Validation of tsunami inundation model TUNA-RP using OAR-PMEL-135 benchmark problem set. *IOP Conference Series: Earth and Environmental Science*, 67, 012030, doi: 10.1088/1755-1315/67/1/012030.

Komoo, I. and Othman, M. (2006). The 26.12.04 tsunami disaster in Malaysia: An environmental, socio-economic and community well-being impact study. Akademi Sains Malaysia, Kuala Lumpur and LESTARI, Bangi, Selangor Darul Ehsan, p. 168.

Kurita, T., Nakamura, A., Kodama, M. and Colombage, S. (2006). Tsunami public awareness and the disaster management system of Sri Lanka. *Disaster Prevent Manage*, 15, 92–110, doi: 10.1108/09653560610654266.

Lander, J.F. (1996). Tsunami affecting Alaska 1737–1996. Publication 31, Nat. Geophysical Data Ctr., Nat. Envir. Satellite, Data, and Info. Service, Nat. Oceanic and Atmospheric Admin., U.S., Dept. of Commerce, Boulder, CO, p. 205.

Lay, T., Kanamori, H., Ammon, C.J., Nettles, M., Ward, S.N., Aster, R.C., Beck, S.L., Bilek, S.L., Brudzinski, M.R., Butler, R., DeShon, H.R., Ekström, E, Satake, E. and Sipkin, S. (2005). The Great Sumatra-Andaman Earthquake of 26 December 2004. *Science*, 308, 1127–1133.

Liu, P.L.-F., Woo, S.B. and Cho, Y.S. (1998). Computer programs for tsunami propagation and inundation, Cornell University, Sponsored by National Science Foundation, 1998, p. 104.

Ma, G., Kirby, J.T. and Shi, F. (2013). Numerical simulation of tsunami waves generated by deformable submarine landslides. *Ocean Modelling*, 69, 146–165.

Macías, J., Vázquez, J.T., Fernández-Salas, L.M., González-Vida, J.M., Bárcenas, P., Castro, M.J., Díaz-del-Río, V. and Alonso, B. (2015). The Al-Borani submarine landslide and associated tsunami: A modeling approach. *Marine Geology*, 361, 79–95.

Massel, S.R., Furukawa, K. and Brinkman, R.M. (1999). Surface wave propagation in mangrove forests. *Fluid Dynamics Research*, 24(4), 219–249.

Mazda, Y., Wolanski, E., King, B., Sase, A., Ohtsuka, D. and Magi, M. (1997). Drag force due to vegetation in mangrove swamps. *Mangroves and Salt Marshes*, 1, 193–199.

Munene, M.B., Swartling, Å.G. and Thomalla, F. (2018). Adaptive governance as a catalyst for transforming the relationship between development and disaster risk through the Sendai Framework? *International Journal of Disaster Risk Reduction*, 28, 653–663, doi: 10.1016/j.ijdrr.2018.01.021.

Okada, Y. (1985). Surface deformation due to shear and tensile faults in a half-space. *Bulletin of the Seismological Society of America*, 75(4), 1135–1154.

Okal, E.A. and Stein, S. (2005). Ultra-long period seismic moment of the Sumatra earthquake: Implications for the slip process and tsunami generation, Eos Trans. American Geophysical Union (AGU) 86 (18), Jt. Assem. Suppl., Abstract U43A-02, Meetings 2005 Joint Assembly. 23–27 May 2005, New Orleans , Louisiana, USA.

Oktari, R.S., Shiwaku, K., Syamsidik, Munadi, K. and Shaw, R.A. (2015). A conceptual model of a school–community collaborative network in enhancing coastal community resilience in Banda Aceh, Indonesia. *International Journal of Disaster Risk Reduction*, 12, 300–310, doi: 10.1016/j.ijdrr.2015.02.006.

Oktari, R.S., Shiwaku, K., Munadi, K., Syamsidik and Shaw, R. (2017). Enhancing community resilience towards disaster: The contributing factors of school-community collaborative network in the tsunami affected area in Aceh. 2017. *International Journal of Disaster Risk Reduction*, 29, 3–12, doi: 10.1016/j.ijdrr.2017.07.009.

Pongponrat, K. and Ishii, K. (2017). Social vulnerability of marginalized people in times of disaster: Case of Thai women in Japan tsunami 2011, *International Journal of Disaster Risk*, 27, 133–141, doi: 10.1016/j.ijdrr.2017.09.047.

Quartel, S., Kroon, A., Augustinus, P.G.E.F., Van Santen, P. and Tri, N.H. (2007). Wave attenuation in coastal mangroves in the Red River Delta, Vietnam. *Journal of Asian Earth Sciences*, 29(4), 576–584.

Ronan, K.R., Johnston, D.M., Daly, M. and Fairley, R. (2001). School children's risk perception and preparedness: A hazards education survey. *Australasian Journal of Disaster and Trauma Studies*, 1, 32.

Schütz, H. and Wiedemann, P.M. (2000). Hazardous incident information for the public: Is it useful? *Australasian Journal of Disaster and Trauma Studies*, 2.

Shaw, R., Takeuchi, Y., Gwee, Q.R. and Shiwaku, K. (2011). Disaster education: An introduction. In: Shaw, R., Shiwaku, K. and Takeuchi, Y. (eds), Disaster Education. Community, Environment and Disaster Risk Management, Vol. 7. Emerald Group Publishing Limited, Bingley, pp. 1–22.

Sieh, K., Daly, P., Mckinnon, E.E., Pilarczyk, J.E., Chiang, H.-W., Horton, B., Rubin, C.M., Shen, C.-C., Ismail, N, Vane, C.H. and Feener, R.M. (2015). Penultimate predecessors of the 2004 Indian Ocean tsunami in Aceh, Sumatra: Stratigraphic, archaeological, and historical evidence. *Journal of Geophysical Research*, 120, 308–325, doi: 10.1002/2014JB011538.

Stein, S. and Wysession, M. (2002). *An Introduction to Seismology, Earthquakes and Earth Structure*, 1st Edition. Wiley-Blackwell, p. 498.

Stein, S. and Okal, E. (2005). Long period seismic moment of the 2004 Sumatra Earthquake and implications for the slip process and Tsunami Generation, Northwestern University.

Tan, W.K. (2017). Developing tsunami run-up models and inundation maps for Malaysia. PhD Thesis (Penang: Universiti Sains Malaysia).

Tan, W.K., Teh, S.Y., Koh, H.L. and Che Abas, M.R. (2014). Simulating Tsunami Run-up on a Planar Beach by TUNA-RP. Proceedings of the 3rd International Conference on Fundamental and Applied Sciences (ICFAS2014), 3–5 June 2014, Kuala Lumpur, Malaysia. *AIP Conference Proceedings*, 1621, 388–393. doi: 10.1063/1.4898496.

Tan, W.K., Teh, S.Y. and Koh, H.L. (2017). Tsunami Runup and Inundation Along the Coast of Sabah and Sarawak, Malaysia Due to a Potential Brunei Submarine Mass Failure. *Environmental Science and Pollution Research*, 24(19), 15976–15994, doi: 10.1007/s11356-017-8698-x.

Tappin, D.R., Watts, P., McMurtry, G.M., Lafoy, Y. and Masumoto, T. (2001). The Sissano, Papua New Guinea tsunami of July 1998 – offshore evidence on the source mechanism. *Marine Geology*, 175, 1–23.

Tappin, D.R., Grillin, S.T., Harris, J.C. *et al.* (2014). Did a submarine landslide contribute to the 2011 Tohoku tsunami? *Marine Geology*, 357, 344–361.

Titov, V.V., and Synolakis, C.E. (1998). Numerical modeling of tidal wave runup. *Journal of Waterway, Port, Coastal, and Ocean Engineering*, 124(4), 157–171.

UNISDR (2015). Sendai framework on disaster risk reduction 2015–2030. United Nations Office for Disaster Risk Reduction, Geneva, Switzerland. Available via UNISDR. https://www.unisdr.org/files/43291_se ndaiframeworkfordrren.pdf. (accessed 2 January 2019).

Ward, S.N. and Day, S.J. (2003). Ritter Island volcano lateral collapse and tsunami of 1888. *Geophysical Journal International*, 154, 891–902.

Watts, P., Grilli, S.T., Kirby, J.T., Fryer, G.J. and Tappin, D.R. (2003). Landslide tsunami case studies using a Boussinesq model and a fully nonlinear tsunami generation model. *Natural Hazards and Earth System Sciences*, 3, 391–402.

Yoon, S.B. (2002). Propagation of distant tsunamis over slowly varying topography. *Journal of Geophysical Research*, 107 (C10), American Geophysical Union, 4.1–4.11.

Chapter 8

Integrated Coastal Zone Management

8.1 Introduction

Coastal zones are highly productive in providing valuable ecosystem services and in supporting vibrant economic development (Romañach *et al.*, 2018). However, they are constantly threatened by anthropogenic activities, environmental degradation and climatic shocks (Koh *et al.*, 2018). These hazards confronting coastal zones can result in deterioration of water quality and impairment of ecosystem services, amounting to billions of Malaysian Ringgit annually (Koh and Teh, 2019). The hazards of climate change and anthropogenic activities on coastal resources must be adequately mitigated to allow future generations to have similar access to these resources, the fundamental principle of the UNSDGs. STEM (Science, Technology, Engineering and Mathematics) plays an important role in providing deep knowledge and core technical skills to support such research (Teh and Koh, 2020) and to devise effective mitigation and adaptation measures to reduce the risks to life below water (SDG14) and life on land (SDG15), potentially impacted by climate change. However, STEM must be integrated with government, industry and community for effective governance and implementation of viable solutions for reducing risks via partnership and collaboration (SDG17) (Koh and Teh, 2019). Research and community outreach are essential for the preservation of coastal resources at local, regional and international levels. Integrated with strategic capability empowered by research and community engagement, STEM had played a crucial role in providing deep knowledge, insights and convictions in achieving the UNSDGs (Teh and Koh, 2020), in particular in promoting an integrated approach to coastal zone management.

Integrated Coastal Zone Management (ICZM) entails diverse scientific and humanity disciples and several SDGs and involves multiple stakeholders. Hence, an integrated approach is ideal to adapt to the complexity in coastal concept and best practices. This ICZM practice offers a golden opportunity for universities to provide quality education (SDG4), to prepare talents to face the complexity and uncertainty in managing future sustainable cities (SDG11), in adapting to climate change (SDG13) and in protecting valuable coastal resources below the sea (SDG14). Coastal modeling is an integral part of the planning and implementation of ICZM. This chapter will discuss the application of model simulations to address the major hazards confronting Malaysian and other national coastal zones arising from anthropogenic activities along the coastal regions. A typical example is the coastal reclamation and development around Pulau Tekong near Singapore requiring Environmental Impact Assessment (EIA), for which impact modeling is a mandatory component. Coastal reclamation is practiced in many countries to create more land along the coast by extending the coastline into the offshore area through reclamation. Changing an existing natural coastline by reclamation may have adverse environmental impacts on coastal hydraulics, water quality and other adverse ecological impacts. In 2003, Malaysia instituted arbitral proceedings against the Singapore government for its proposal to reclaim the coastal sea around Pulau Tekong and Changi Bay to cater to the demand for more land in Singapore. The scientific and technical issues relating to the proposed reclamation will be discussed in this chapter via model simulations to evaluate the impacts. The international legal and institutional issues and procedural matters will be discussed in Chapter 10 on governance over water security.

8.2 Tribunal for the Law of the Sea

Malaysia instituted arbitral proceedings against Singapore under Annex VII to the Law of the Sea Convention (The Convention) on 4 July 2003, regarding a dispute concerning land reclamation by Singapore in and around the Straits of Johor. A request was filed on 5 September 2003 by Malaysia against Singapore with the International Tribunal for the Law of the Sea (ITLOS or The Tribunal) for the prescription of provisional measures under article 290, paragraph 5, of the Convention. The request was made in relation to the dispute concerning land reclamation by Singapore in and around the Straits of Johor. The Tribunal may prescribe measures to preserve the respective rights of the parties to the dispute and to prevent serious harm to the marine environment and its ecosystems. The Convention takes an integrated approach to the issues it covers including various uses of the sea, such as navigation, fishing and other forms of resource exploitation. Protection and preservation of the marine environment and its ecosystems requirements have to be integrated into all uses of the seas. Apparent conflicts between economic, social and environmental considerations (the three pillars of sustainable development) have to be resolved in an integrated manner. The substantive right of a coastal state to have unimpeded maritime access to its ports and resources is accompanied by its procedural right to be consulted on plans for artificial alterations which may impact common navigation channels and safety of navigation. The Articles 194 and 204 of the Convention in effect require the application of the precautionary principle. It is widely agreed that the core of the precautionary principle is well reflected in Principle 15 of the Rio Declaration on environment and development. Principle 15 provides the following: "In order to protect the environment, the precautionary approach shall be widely applied by states according to their capabilities. Where there are threats of serious or irreversible harm, lack of full scientific certainty shall not be used as a reason to postponing cost-effective measures to prevent environmental degradation." Article 206 of the Convention elaborates on this by specifying the requirement of EIA. Article 206 provides the following: "When States have reasonable grounds for believing that planned activities under their jurisdiction may cause substantial pollution of or significant and harmful changes to the marine environment, they shall assess the potential effects of such activities on the marine environment and shall communicate reports of the results of such assessments". The duty and onus to conduct EIA rests on the Singapore government. However, the authors took on the voluntary task of performing modeling simulations to assess the potential impact of reclamation by Singapore on the marine environment in the Straits of Johor. The Tribunal public sitting was held on Thursday, 25 September 2003, at 3.00 p.m., at the International Tribunal for the Law of the Sea, Hamburg, Germany, with President L. Dolliver M. Nelson presiding (ITLOS, 2003).

8.2.1   *Impact of reclamation on marine environment*

Construction related to coastal reclamation has various environmental and ecological consequences (Ghaffari *et al.*, 2017). The onsite and offsite extraction and material transfers of the building materials can cause many environmental issues (Padmalal *et al.*, 2008), leading to potential ecological and geomorphological consequences at the source and at the destination. Geo-engineering construction techniques may entail the conversion of coastal wetlands into tidal mudflats and sub-tidal zones (Zhu *et al.*, 2016), leading to biodiversity loss as wetlands are rich and diverse ecosystems. Loss of biodiversity in the Chinese coastal region due to building artificial land has been widely reported (Duan *et al.*, 2016). Since 1949, the loss of coastal wetlands in China to reclamation is estimated to amount to 22,000 km^2 (Tian *et al.*, 2016). Dredging of the ocean floor for the construction of the island chain in the South China Sea has severely damaged coral reefs and the associated marine ecosystem, particularly fishery (Hutchings and Wu, 1987). In Hong Kong, sea water has been seriously polluted due to the inflow of construction wastes and effluents released during the process of building artificial land (Chan *et al.*, 2017). The seaward land expansion may result in irreversible damage to coastal wetland ecosystems, and may impair the sustainability of coastal resources, which may be exacerbated by sea level rise (UN-Habitat, 2016). Sun *et al.* (2017a) integrated the ecosystem service valuation into the assessment, management, restoration and compensations of coastal ecological infrastructures for Lianyungang City, a rapidly developing and agriculture-dominated coastal city in Eastern China. Their results showed that 323 km^2 ha of coastal wetlands had been lost to reclamation, resulting in USD806 million per year loss in coastal agriculture-related ecosystem service, consisting of food production, raw material supply, disturbance regulation and water purification. Land reclamation permanently alters the natural properties of coastal resources and the environment and has caused significant damage to marine ecosystem services. Four categories of marine ecosystems services are involved: fishery resources (provisioning), water quality (regulating), natural seascapes (cultural) and marine biodiversity (supporting). Marine biodiversity and water quality were the two main ecosystem service losses associated with land reclamation in Jiaozhou Bay in China (Shan and Li, 2020). The annual ecological damage in Jiaozhou Bay caused by land reclamation was 12.46 billion CNY.

8.2.2 *Impacts of Singapore reclamation*

The coastal reclamation by Singapore is located around Pulau Tekong and Changi, at the confluence of Sungai Johor and Selat Johor, between Malaysia and Singapore. Model simulation results presented to the Tribunal public sitting on 25 September 2003 indicated that the tidal velocity is likely to increase in the order of 70% at some locations in the Straits of Johor (ITLOS, 2003). Professor Falconer offered his view to the Tribunal that the corresponding increase in energy, turbulence and sediment transport is higher than 70% because of nonlinearity in the relationship. Our simulation indicated that tidal velocity is likely to increase by 56% in the vicinity of Pulau Tekong, where the maximum vertically integrated tidal velocity may reach 0.8 m/s (Syamsidik and Koh, 2003). The tidal current velocity near the surface is likely to reach 1.2 m/s (Koh *et al.*, 2004). These levels of tidal currents are likely to pose hazards to safe passage for ships passing through the narrow channels in the reclaimed sites and may also increase coastal erosion. Reclamation activity will create large quantities of sediments, pollute the waters at and around the sites, and pose serious adverse ecological impacts. Our simulation shows that the levels of SS around the reclamation sites in Pulau Tekong may exceed 640 mg/l around Changi Bay, posing serious hazards to the marine environment and its ecology, particularly sea grass, which forms the primary food of the dugong. Hence, reclamation around Pulau Tekong had to be conducted with extreme care (Koh *et al.*, 2004). Following the precautionary principle, an EIA study was conducted by the authors to evaluate the impact of reclamation at Tekong on the environment and its ecosystems in the Straits of Johor.

8.3　EIA Simulation Study on Singapore Reclamation

The reclamation is expected to lead to several adverse impacts to the marine environment and the ecosystems. It will reduce the width of the navigational channels, increase the tidal velocity and pose hazards to the safe navigation of ships. Suspended sediment (SS) concentrations may increase significantly, particularly during the reclamation phase, harming sensitive marine ecosystems in the area. Concerned over these environmental pollutions and their impacts, Malaysia instituted arbitral proceedings under the International Tribunal for the Law of the Sea (ITLOS) for resolution of the conflicts. The authors performed model simulations to address the issues raised. It articulates an approach conducive to bilateral cooperation between the two countries, following a model similar to the Joint International Commission between the US and Canada. Before this cooperation can proceed, a thorough scientific study must be conducted as detailed in Section 8.3.

8.3.1　*Study site around Tekong Island*

The study area is located around Pulau Tekong and Changi, at the border shared between Malaysia and Singapore (Figure 8.1). The coastal area around here has been undergoing regular reclamation in the past (Chia *et al.*, 1988) to cater to the development needs of Singapore. Further information relevant to the study site is available from Koh *et al.* (1991), Koh *et al.* (1995), Koh and Lee (2001) and Murray-North (1994). The reclamation projects were divided into two phases: (a) Reclamation I will reclaim land eastward of Changi Bay, involving 200 million m^3 of landfill materials and (b) Reclamation II around Pulau Tekong, covering an additional 180 million m^3 of landfill. The total coastal area to be reclaimed is expected to reach 100 km^2. The reclamation of this immense dimension is expected to raise undesirable environmental and ecological impacts. This section presents simulation results to evaluate these potential impacts by means of the model known as AQUASEA.

8.3.2　*Simulation results on tidal currents*

The simulation model AQUASEA was developed by Vatnaskil Consulting Engineers (VCE, 1998). It consists of two components

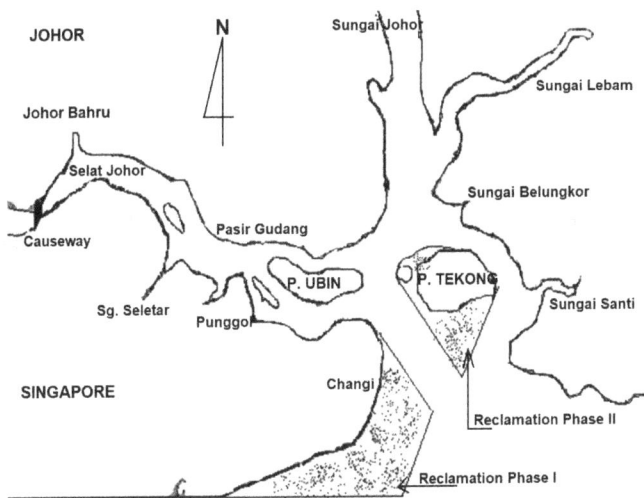

Figure 8.1: Location of the study area around Pulau Tekong and Changi, at the border shared between Malaysia and Singapore.

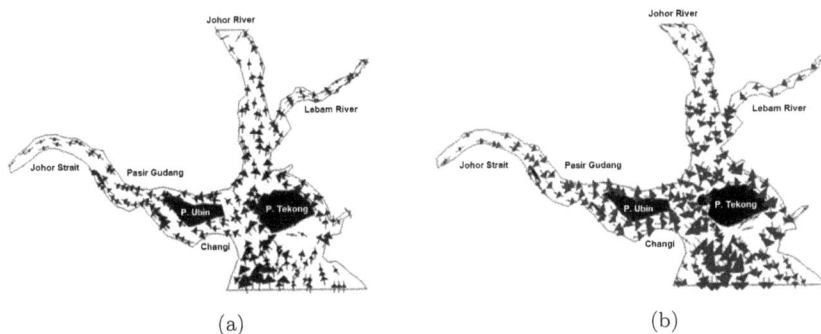

(a)

(b)

Figure 8.2: (a) Flood and (b) Ebb tidal currents during a spring tide under the preexisting condition before the reclamation.

for simulating: (a) tidal hydrodynamic regimes, and (b) transport and transformation of pollutants. For simulation of tidal hydro-dynamic regimes, the spring amplitude of 1.1 m (refer Admiralty Chart No. 2585) is used, with a semidiurnal tide's tidal period of 12.42 hours. A spring tide here has maximum amplitude of 1.25 m, while a neap tide has maximum amplitude of 0.5 m. Details regard-ing the simulation model AQUASEA are available from Koh *et al.* (2004). Figure 8.2 shows the tidal currents during a spring tide under

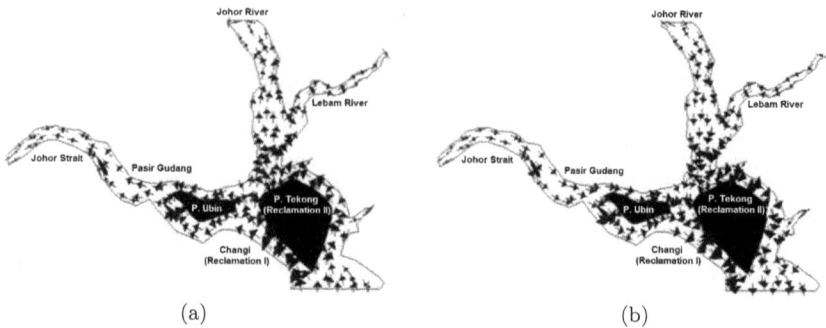

Figure 8.3:   (a) Flood and (b) Ebb tidal currents during a spring tide after the completion of Reclamation II.

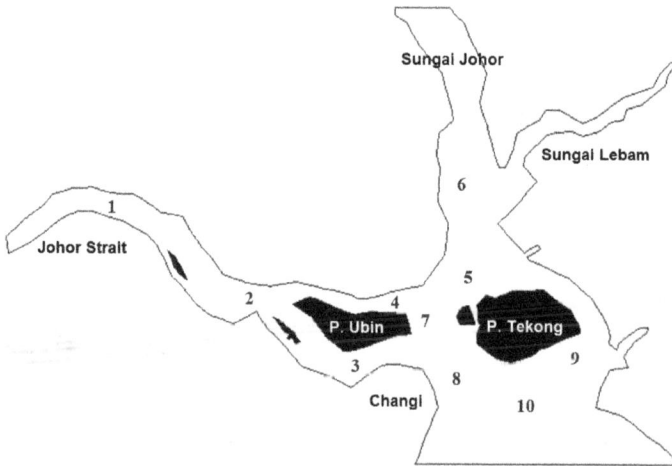

Figure 8.4:   Ten observation points, at which the current velocity is observed.

the preexisting condition before the reclamation, while Figure 8.3 illustrates the resulting tidal currents after the completion of Reclamation II. The currents after Reclamation II will increase significantly around the sea between Pulau Tekong and Changi. Figure 8.4 shows the ten observation points, at which the current velocity will be observed. Figure 8.5 shows the simulated current velocity over a tidal cycle during a spring tide at four locations: Point 8 and Point 9, both are near the reclamation sites, and points 7 and 10 further away.

Table 8.1 shows the comparison of tidal currents at ten observation points between preexisting conditions and after Reclamation II.

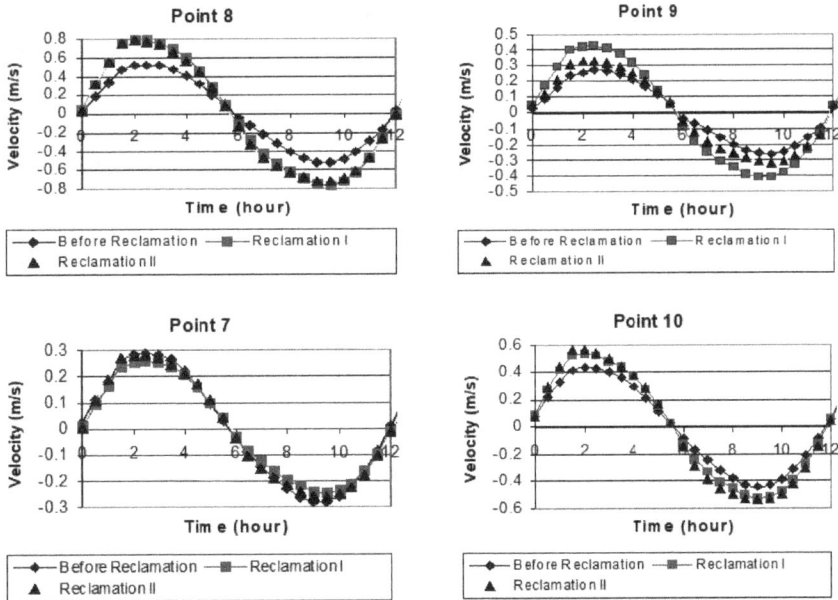

Figure 8.5: Simulated current velocity over a tidal cycle during a spring tide at Points 7 to 10 as indicated in Figure 8.4.

Table 8.1: Comparison of tidal currents at ten observation points. Negative sign means the current reverses direction from flood tide.

Location	Existing condition (m/s)		Reclamation I (m/s)		Reclamation II (m/s)	
	Flood tide	Ebb tide	Flood tide	Ebb tide	Flood tide	Ebb tide
1	0.09	−0.09	0.09	−0.09	0.09	−0.09
2	0.22	−0.23	0.23	−0.22	0.24	−0.22
3	0.40	−0.41	0.41	−0.4	0.43	−0.39
4	0.29	−0.3	0.31	−0.3	0.32	−0.29
5	0.24	−0.26	0.22	−0.22	0.2	−0.19
6	0.23	−0.23	0.23	−0.22	0.24	−0.22
7	0.29	−0.29	0.26	−0.25	0.28	−0.26
8	0.52	−0.52	0.80	−0.76	0.80	−0.73
9	0.27	−0.26	0.43	−0.41	0.33	−0.32
10	0.43	−0.44	0.54	−0.52	0.57	−0.54

After Reclamation II, the channel between Pulau Tekong and Changi Bay will be significantly narrower, leading to a bottleneck effect that would increase tidal velocity. The vertically averaged velocity amplitude at locations 8 and 9 before reclamation is 0.52 m/s and 0.27 m/s, respectively. However, after Reclamation II, the values are increased to 0.80 m/s and 0.33 m/s, respectively. The velocity at location 8 will increase significantly by 54% after reclamation II. The tidal energy, tidal turbulence and erosion will increase by much more than 54%, since the relationship is nonlinear. It is generally accepted that a tidal current change of more than 25% is considered as highly significant. As such, the Singapore government needs to perform further investigations regarding the impacts of reclamation on tidal ecology, erosion and safe navigation. For example, a coupled hydrodynamic and ecological simulation for prognosticating land reclamation impacts in river estuaries has been performed. The results indicate that a mere 16.7% increase in water velocity may decrease the growth suitability index (GSI) of phytoplankton by up to 12%. This reduction in GSI may cut down the stability of phytoplankton species (Xu *et al.*, 2018), which in turn may lead to adverse impact on organisms higher than phytoplankton in the food chain. Further, the velocity at point 8 near the water surface, being higher than the vertically averaged velocity by some 50%, is likely to increase to 1.2 m/s. This high tidal velocity is hazardous to safe navigation and coastal erosion. At locations relatively far from the reclamation sites, e.g., Location 6 in Johor River, and Locations 1 to 4 in the straits of Johor, the increases in velocity are minor. During a spring tide with amplitude of 1.1 m, the velocity amplitude at location 1 before reclamation and after Reclamations I and II remains relatively small at around 0.09 m/s. The velocity at this Point 1, being far away from the reclamation site, is not altered much by the two reclamation schemes, around Changi Bay and Pulau Tekong. The changes in tidal velocity at the 10 observation locations due to the reclamations are tabulated in Table 8.1. Figure 8.5 shows simulated vertically averaged current velocity at Points 7 to 10 over a semidiurnal tidal cycle with period of 12.42 hours during a spring tide with amplitude 1.1 m. Large increases in tidal velocity occur at Points 8 and 9, being near the reclamation sites. The velocity increase at Point 10 is relatively modest, as compared to those

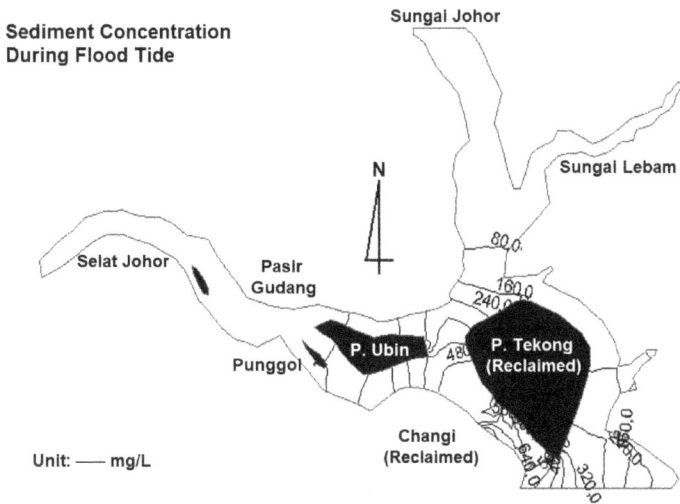

Figure 8.6: Simulated SS levels around the reclamation sites.

increases at Points 8 and 9. For Point 7, the tidal velocity remains virtually unchanged.

8.3.3 *Simulation results on suspended solids*

Suspended sediment (SS) is simulated by the AQUASEA transport component. The source of SS loadings is assumed to come only from the reclamation at Pulau Tekong site. We assume a spill rate of 5%, that is, 5% of landfill materials will contribute as SS in the water column. We assume a landfill material density of 2650 kg/m^3. The reclamations are expected to be completed over a period of 8 years and will consume a total of 380 million m^3 of landfill materials. The simulation SS around the reclamation sites are plotted in contour lines in Figure 8.6, indicating significant levels of SS concentrations in the water system around the reclamation sites. The SS concentrations at location 8 (shown in Figure 8.4), near the reclamation site, may exceed 600 mg/l during reclamation. Figure 8.7(b) shows SS concentrations at 72 selected observation points shown in Figure 8.7(a). Near Changi, at observation points 70 to 72, SS concentrations exceed 600 ml/L. At observation points 25 and 26,

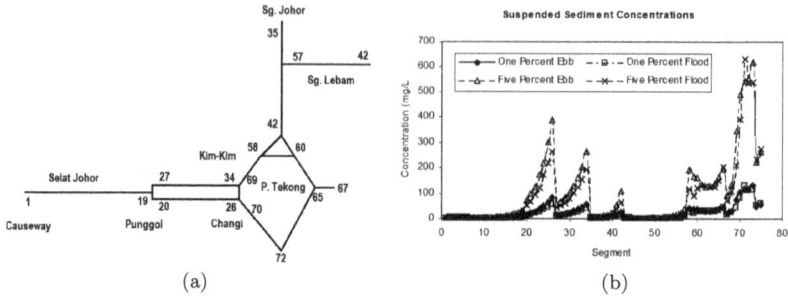

Figure 8.7:   (a) Seventy-two selected observation points and (b) SS concentration at these 72 selected observation points.

the SS is almost 400 mg/L. At observation point 35 and 36, in Johor River, the SS exceeds 60 mg/l, even though it is farther away. Compared to the natural background SS of 20 mg/l before reclamation, these heightened SS levels arouse grave concern. The spill rate may be higher if the control measures are not effective or if there is an increase in the rate of reclamation at certain times. This projected sharp increase in SS levels around the reclamation sites would have adverse impacts on marine organisms living in the area (Dollar and Grigg, 1981; Johnston, 1981), which include seagrass, fish, mussels and corals. It may also cause other adverse impacts such as an increased rate of sedimentation. Thus, care should be exercised by the government of Singapore to ensure sustainable coastal development in the area (Griffith and Ashe, 1993). This research has identified significant adverse impacts on the environment and the marine ecosystems around the reclamation sites. Hence, it is prudent to have a standing working group composed of relevant experts from Malaysia and Singapore to regularly and constantly collaborate to ensure that such unfavorable scenarios identified would not prevail, in line with the temporary provisions stipulated by the ITLOS order. This recommendation is also consistent with the spirit of cooperation over transboundary water issues that has been in operation between the US and Canada, working under the auspices of the International Joint Commission between the two countries. It is hoped that this spirit of cooperation would be harnessed for the good of both countries in all future transboundary projects.

8.4   Medini Iskandar Development

8.4.1   *Introduction*

The Medini Iskandar Development is a 2300 acre multipurpose urban development comprising residential, commercial, educational, business and recreational areas. The developer submitted the EIA report for the Medini Iskandar development to the Department of Environment (DOE) Johor in July 2008 for approval. A conditional approval of the EIA report was granted by the DOE subject to some conditions. The conditional approval stipulated that sewage effluent from the centralized sewage treatment plant (STP) is not permitted to be discharged into Sg Pendas. A suitable location for the discharge of the STP effluent into the Selat Johor is to be identified, based on a hydraulic and water quality modeling investigation. This modeling simulation study, as detailed in Section 8.4, aims to evaluate the impact of the discharge of STP effluent on the marine water quality and on the aquatic resources in Selat Johor. The supplementary EIA report detailing the findings of the model simulation study was submitted to the DOE. Approval was granted in December 2011 for the construction and operations of the marine sewage outfall in Selat Johor. This section presents the sampling and simulation results for key hydraulic and environmental parameters suitable for sustaining acceptable water quality criteria in Selat Johor. The simulation models used include USEPA WASP7, developed by the US Environmental Protection Agency (USEPA, 2013), and AQUASEA, developed by Vatnaskil Consulting Engineers (VCE, 1998). Water quality modeling for the proposed development project in Medini Iskandar was performed to (a) evaluate the impact of discharging secondary treated sewage on water quality and to (b) identify an appropriate discharge location that can provide adequate dilution and dispersion to ensure that the existing fish cage culture and aquaculture in or around Sg Pendas and Selat Johor is not adversely impacted. The USEPA stipulates that the maximum FC level suitable for human contact activity such as swimming and fish cage culture is 200 FC/100 ml. The results of the EIA study are described in Section 8.4. A collection of sewage water quality parameter and their values are provided in Tables A.1–A.5 in the Appendix. Receiving water quality criteria for Malaysia is compared to those in USA and China in Table A.6 while the parameter limits of sewage effluent standard in Malaysia are listed in Table A.7.

8.4.1.1 *Sg Pendas not suitable for Sewage Discharge*

Simulation results indicated that the options of discharging treated sewage into Sg Pendas are not acceptable, as it would violate the USEPA standard on *faecal coliform*. This is because the limited dilution and dispersion provided by the small tidal flows in Sg Pendas are not able to reduce the primary pollutant concerned, i.e., *faecal coliform* (FC), to an acceptable level, below 200 FC/100 ml. The option of discharging treated sewage into the Selat Johor, with much larger tidal flows, at an appropriate location, is promising. The much larger tidal flows in Selat Johor will dilute and disperse the effluent to low levels that would not adversely downgrade the existing water quality in Selat Johor or Sg Pendas. Existing fish cage culture or aquaculture activities in or around Sg Pendas or in Selat Johor will not be adversely impacted. Based upon simulation results, the FC levels in or around the river mouth of Sg Pendas and around the four fish cage areas in Selat Johor are less than 20 FC/100 ml (tidally averaged), if the effluent is treated to have an FC concentration of 4×10^5 FC/100 ml or equivalent to 99% removal of FC from raw sewage. Hence, the water quality around the four fish cage areas will be suitable for human contact activity and fish health.

8.4.2 *Faecal coliform characteristics*

Sewage contains high concentrations of faecal bacteria, which are the most common cause of gastrointestinal infections like diarrhea, cholera and typhoid (Reder *et al.*, 2015). FC bacteria are frequently used as an indicator for these group of pathogen organisms, e.g., by (a) the European Community Bathing Water Directive (76/160/EEC) (EU, 1986), by (b) the United States Environmental Protection Agency (USEPA) for recreation waters (USEPA, 1986, 2012) and (c) by the World Health Organization (WHO) in the guidelines on the use of wastewaters in agriculture and aquaculture (WHO, 1989). The threshold limit for permissible FC concentration is 200 MPN/100 mL for recreational water with physical body contact such as swimming and fish cage cultivation activities. In this Section, we focus on simulations of FC in Selat Johor and in Sg Pendas arising from the discharge of treated sewage from the sewage treatment plant (STP 6) into the waterways in the project site. The project will accommodate a total of 400,000 persons equivalent (PE), contributing an estimated sewage flow rate of 1.04 m^3/s. Based upon water usage of 225 liters per person per day (225 L/p $\cdot$ d), the sewage flow is equal to 225 L $\times$ 400,000 per day, or 9×10^4 m^3/day, or equivalent to 1.04 m^3/s. The sewage is anticipated to be treated to remove 99% of FC or an FC concentration of 4×10^5 FC/100 mL in the sewage effluent.

8.4.2.1 *Effluent FC concentration*

Raw sewage has been reported in the literature to have FC concentrations in the range of 1×10^7 to 4×10^7 FC/100 mL (Koh *et al.*, 1995; Westcot, 1997). Sewage may be treated to achieve 99% removal (Pescod, 1992; Ludwig, 1988). Hence, if the treated sewage has an FC concentration of 4×10^5 FC/100 mL, then the loading rate of FC into Selat Johor is 4×10^5 FC/100 mL $\times$ (1,000 mL/1 L) $\times$ 225 L $\times$ 400,000 or 3.6×10^{14} FC per day. Based upon FC decay rates reported in the literature (Thomann and Mueller, 1987) and the salinity in Selat Johor (2%), we estimate the FC decay rate to be around $T_{90} = 4$ hours. This means that FC takes 4 hours to achieve 90% kill-off. Water quality simulations are then performed by means of WASP7 and AQUASEA, to estimate the levels of FC in the waterways.

8.4.3 *Project site descriptions*

Figure 8.8 shows the project site contained within the dashed lines, indicating the locations of the sewage treatment plant (STP 6) and six sampling points A (downstream), B, C, D, E and F (upstream) along Sg Pendas. The STP 6 effluent was to be discharged into two locations: (a) location C along Sg Pendas and (b) a location within the Malaysian boundary in the Selat Johor. Figure 8.9(a) shows the project study area consisting of the western section of Selat Johor, stretching from the Johor Causeway at the northeast to the river

Figure 8.8:   Project site showing six sampling points, STP 6 and project boundary within dash lines.

mouth near Sg Tengeh at the southwest, with Sg Pendas enclosed in a red rectangle. Figure 8.9(b) shows the locations of aquaculture ponds and four fish cage areas indicated in triangles along Sg Pendas (Area 1) and in Selat Johor (Areas 2, 3 and 4). The treated sewage is conveyed by sewage pipes from STP 6 for discharge into Selat Johor via a marine outfall, located between fish cage areas 2 and 3. The sewage discharge is located about 400 m into the Selat Johor, well within the Malaysian boundary, and hence will not encroach into Singapore territorial rights.

8.4.4   *Methodology*

Sewage discharged into Selat Johor undergoes various processes such as dilution, advection and dispersion as well as kinetic

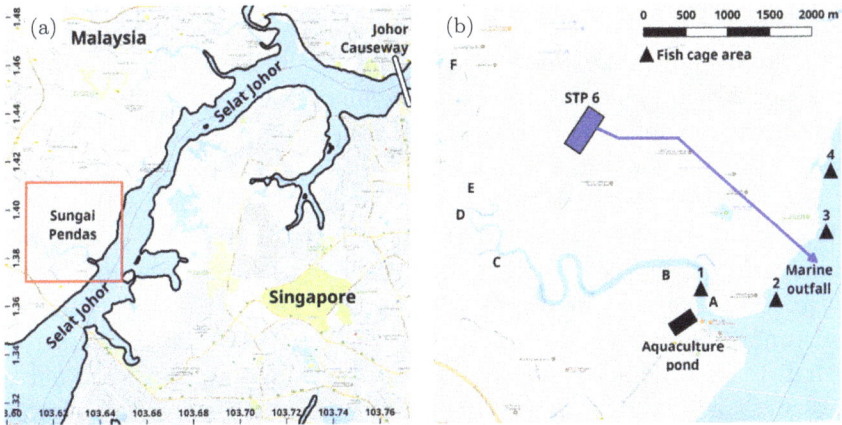

Figure 8.9: (a) Study area showing Selat Johor and Sungai Pendas (Sg Pendas) and (b) locations of six field sampling points A to F along Sg Pendas, STP 6, marine outfall and fish cage areas.

transformations such as decay or stabilization. Analytical and numerical models attempt to represent these transformations and physical processes to simulate FC concentrations in space and time. A combination of analytical and numerical simulation models is used in this study to predict and evaluate potential scenarios of FC levels due to the discharge of treated sewage from the project into Selat Johor and Sg Pendas. Analytical models are ideal for capturing essential summary features of complex situations, and can help to properly conceptualize the various scenarios. Analytical solutions, however, may encounter the problem of inadequate representation and insufficient resolution due to simplification used. Numerical models are then used to incorporate higher levels of resolution and refined representation in order to simulate the various scenarios as realistically as possible. The well-documented and popular numerical model known as WASP7 used in this study for simulating one-dimensional river transport of pollutants in the small Sg Pendas was developed by the USEPA (2013). To simulate pollution transport in two-dimensional coastal seas such as Selat Johor, the model AQUASEA (VCE, 1998) is used. These models have been successfully used by the study team in several studies in Malaysia (Koh *et al.*, 1991, 1995, 1997; Koh and Lee, 2006), which provide some environmental parameter inputs used in this study. We begin with a simple conceptual box model to derive at preliminary solutions.

8.4.4.1   *Simple conceptual models*

Simple conceptual box models are frequently utilized to provide quick back-of-envelop calculations to derive at preliminary solutions. Two illustrative examples are provided. First, a simple box model analysis indicates that Sg Pendas's upstream tributaries near the sampling locations E and F have low flow rates of less than 0.03 m^3/s. This low flow rate contributes negligibly to the total flows in the main and downstream reaches of Sg Pendas. Hence, these small tributaries are ignored in the WASP7 simulations, without sacrificing accuracy. Similarly, this small river flow rate of 0.03 m^3/s is too small to provide adequate dilution to the much larger sewage effluent with a flow rate of 1.04 m^3/s. Hence, based upon this simple model analysis, this stretch of Sg Pendas is not suitable to receive sewage effluent from STP 6. This conclusion will be confirmed in Section 8.4.5 by WASP7 simulation. Second, the western Selat Johor can be simplified to a simple rectangular box with equivalent width W m, length L m and mean depth H m. The semidiurnal tidal period is given by $P = 12.42 \times 3{,}600$ s, and R m is the tidal range. Then, the tidal flow Q m^3/s is given by $Q = \pi WLRP^{-1} \sin(2\pi t/P)$, where t is time in seconds. The depth-integrated velocity V m/s is given by $V = Q/WH$. These simple box models give reasonable accuracy without heavy computational cost (Koh *et al.*, 2004). This simple analytical model provides quick and reasonably accurate estimates for the tidal current V m/s and tidal flows Q m^3/s in Selat Johor. Near the proposed sewage discharge location, the tidal flow is estimated by this analytical model to be about 1,350 m^3/s, which is about 1350 times the sewage discharge rate. This high level of dilution coupled with strong dispersion will bring the level of FC down to acceptable levels near the fish cage cultivation sites. This conclusion will be confirmed in Section 8.4.6 by numerical simulations to be performed by AQUASEA. It is worth noting that similar simple box models have been successfully developed to realistically represent riverine freshwater inputs into the ocean components of the Earth System Models (ESM). A physically based Estuary Box Model (EBM) is coupled to the ESM to parametrize the mixing processes in the Columbia River Estuaries systems with good results, and without significantly increasing computational time (Sun *et al.*, 2017b).

8.4.4.2 *Model WASP7*

WASP is one of the most widely used water quality models in the United States and throughout the world. Because of the model's capabilities of handling multiple pollutant types, it has been widely applied in the development of Total Maximum Daily Loads in the US and worldwide. WASP has the capabilities of linking with other hydrodynamic and watershed models, which allows for multiyear analysis subject to varying meteorological and environmental conditions. WASP has been applied to all of the major estuaries in Florida where it was linked with a hydrodynamic and watershed model simulating 12 continuous years to aid USEPA in the development of numeric nutrient criteria. WASP7 is a dynamic compartment water quality model that can be linked with hydrodynamic and watershed models. The model is designed to provide a broad framework applicable to many environmental problems and to allow the user to match the model complexity with the requirements of the problem. The main objective of the modeling study is to predict the water quality in the river located downstream of the project site due to pollutants such as FC discharged from the project site.

8.4.4.3 *Model AQUASEA*

The simulation model AQUASEA is based upon the two-dimensional depth-averaged hydrodynamic and water quality shallow water equations. The water quality model simulates the fate of a substance, such as FC, in the water environment under the influence of tidal flows and dispersion processes and subject to the FC die-off. Eqs. (8.1) to (8.4) show the two-dimensional equations for the transport and decay of FC. The explicit staggered finite difference method is used in solving the shallow water equations (Koh *et al.*, 2004). The hydrodynamic model provides flow information to the transport model. The definitions and units of the symbols used in Eqs. (8.1) to (8.4) are listed in Table 8.2.

$$\frac{\partial \eta}{\partial t} + H\frac{\partial U}{\partial x} + H\frac{\partial V}{\partial y} = 0 \tag{8.1}$$

Table 8.2:   Definition of symbols used in Eqs. (8.1) to (8.4).

Symbol	Unit	Definition
t	s	Time
x	m	Space coordinate along x-direction
y	m	Space coordinate along y-direction
η	m	Water elevation above mean sea level
H	m	Mean water depth
g	m/s^2	Gravitational acceleration
n	s/m$^{1/3}$	Manning's relative roughness coefficient
U	m/s	Velocity in the x-direction
V	m/s	Velocity in the y-direction
l	MPN/100 mL	FC concentration
W	MPN/s	FC loading rate
α	s^{-1}	FC decay rate
E_x	m^2/s	Dispersion along x-direction
E_y	m^2/s	Dispersion along y-direction

$$\frac{\partial U}{\partial t} + H\frac{\partial (UV)}{\partial y} + H\frac{\partial (U^2)}{\partial x} + g\frac{\partial \eta}{\partial x} + \frac{gn^2}{H^{4/3}}U\sqrt{U^2 + V^2} = 0$$

$$(8.2)$$

$$\frac{\partial V}{\partial t} + H\frac{\partial (UV)}{\partial x} + H\frac{\partial (V^2)}{\partial y} + g\frac{\partial \eta}{\partial y} + \frac{gn^2}{H^{4/3}}V\sqrt{U^2 + V^2} = 0$$

$$(8.3)$$

$$\frac{\partial l}{\partial t} = E_x\frac{\partial^2 l}{\partial x^2} + E_y\frac{\partial^2 l}{\partial y^2} - U\frac{\partial l}{\partial x} - V\frac{\partial l}{\partial y} - \alpha l + W \qquad (8.4)$$

We use a combination of appropriate modeling approaches to simulate water quality in Sg Pendas and in Selat Johor and to evaluate the impact of discharging treated sewage on FC levels in the receiving waterways. The goal is to select an appropriate sewage outfall location with the assimilative capacity to adequately dilute and disperse the discharged effluent. This is to ensure that the existing fish cage and aquaculture areas in Sg Pendas and in Selat Johor are not adversely impacted. Two options of discharging treated sewage are evaluated in this study. One option consists of discharging treated sewage into Sg Pendas, at several locations some 3 km, 4.5 km and 6 km from the headwater. The second option is to discharge treated

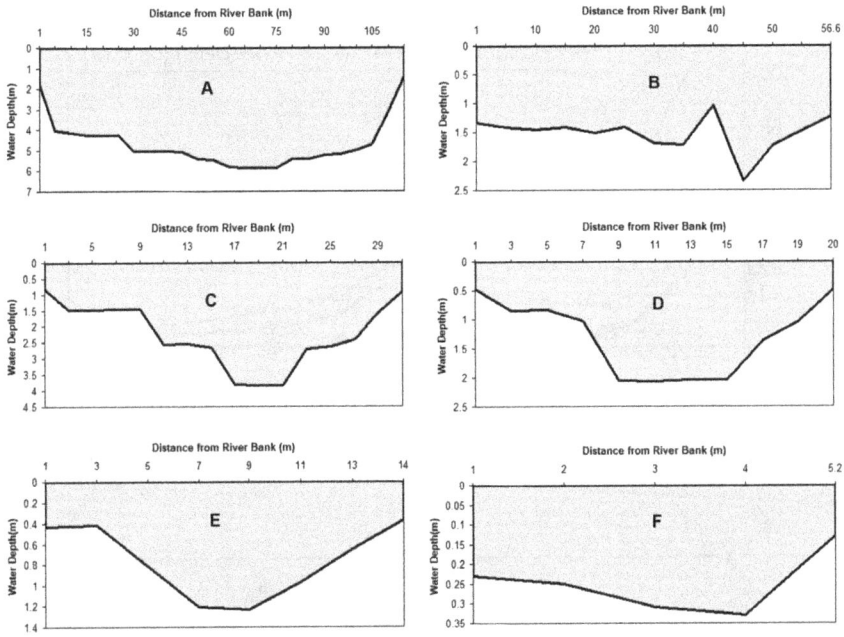

Figure 8.10: Water depths along six river transects along Sg Pendas.

sewage into Selat Johor at a suitably selected location. The evaluations of these two options are performed by simulations of the impact of discharging treated sewage from STP 6 into the two waterways: (a) Sg Pendas and (b) Selat Johor.

8.4.5 *Simulation of FC in Sg Pendas*

8.4.5.1 *Sg Pendas Hydrology*

Figure 8.10 shows the measured water depths at six transects along Sg Pendas, from downstream location A1 to upstream location F. The river width is about 115 m, near the Sg Pendas river mouth (A1), at the confluence with Selat Johor. The river width decreases to 56.6 m at transect B1, and decreases progressively further upstream, where at transect F the width is only 5.2 m. The river length is about 9.5 km. Figure 8.11 shows average measured water flow (m^3/s), width (m) and depth (times 10^{-1}m) at six transects in Sg Pendas. Figure 8.12 shows tidal elevations in the study area, indicating a

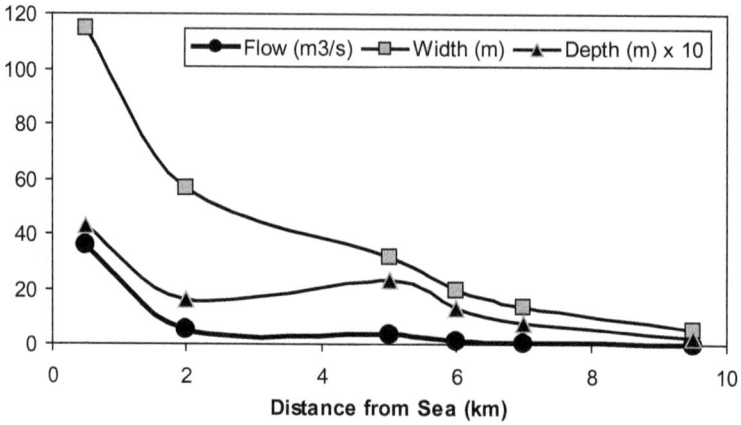

Figure 8.11:   Average measured flow, width and depth (10^{-1}) at six transects in Sg Pendas.

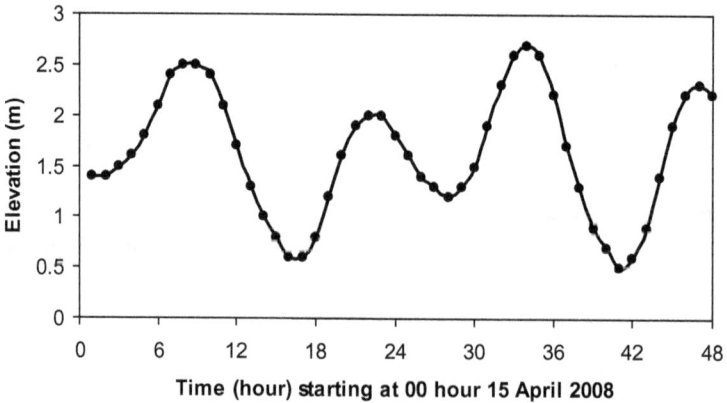

Figure 8.12:   Tidal Elevations in the project study area indicating a semidiurnal tide with amplitude of 1.0 m.

semidiurnal tide with amplitude of 1.0 m. Figure 8.13 shows the comparison between measured flows and WASP7 simulated flows along Sg Pendas, indicating good agreement.

8.4.5.2   *Simulated FC in Sg Pendas*

It is assumed that treated sewage has an FC concentration of 4×10^5 FC/100 ml and the FC die-off rate is $T_{90} = 6$ hours for Sg Pendas, where the water is less saline. This means that it takes

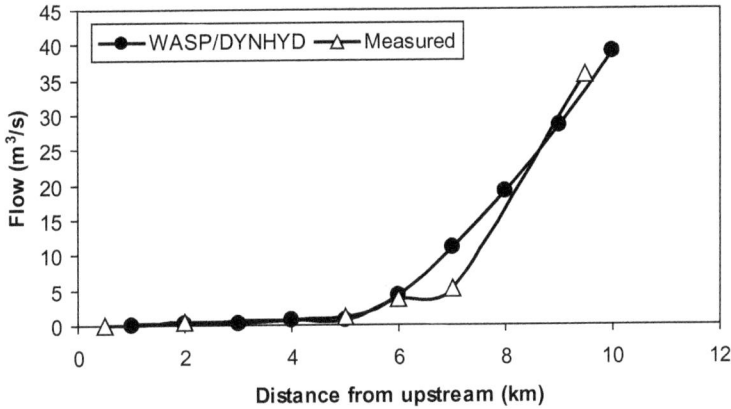

Figure 8.13: Comparison between Measured flows and WASP7 simulated flows.

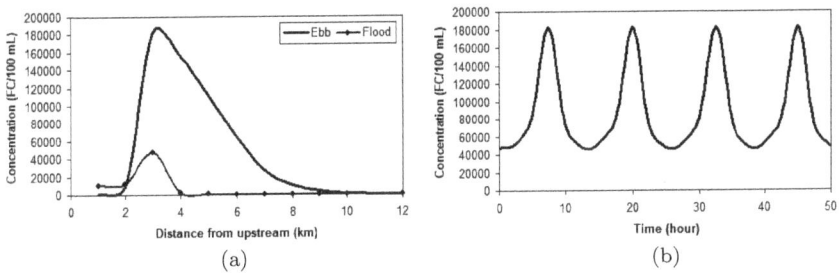

(a)   (b)

Figure 8.14: (a) Simulated FC concentration spatial distribution along Sg Pendas, with sewage effluent concentration of 4×10^5 FC/100 ml and $T_{90} = 6$ hours, and (b) corresponding simulated time series of FC at discharge location.

6 hours for FC to decrease its concentration by 90% to 10% of its original values. Figure 8.14(a) shows the FC spatial distributions along Sg Pendas at two distinct stages of a tide, i.e., during the highest flood tide and during the lowest ebb tide. The highest FC concentration occurs during the lowest ebb tide because of its low volume of water and small flow rates. The concentration can reach a peak of 180,000 FC/100 ml at the release point C. However, the value near the river mouth is lower at about 5,000 FC/100 ml because of decay and dilution as the river flows downstream. During the maximum flood tide, the peak FC concentration may reach 50,000 FC/100 ml near the discharge point C. This lower value is because of a high volume of water and large flow rates during flood tide to provide for

better dilution and dispersion. Figure 8.14(b) shows the time series of FC concentrations at the release location C, 3 km from headwater or 6.5 km from river mouth, with sewage effluent concentration of 4×10^5 FC/100 ml and $T_{90} = 6$ hours. The peak FC level is 180,000 FC/100 ml, while the trough FC level is 5,000 FC/100 ml. This set of high FC levels in Sg Pendas means that the option of discharging treated sewage from STP 6 into Sg Pendas is not acceptable, as it does not meet the standards for FC of less than 200 FC/100 ml, stipulated by the USEPA for contact activities. Hence, the remaining option left is to explore and evaluate the possibility of discharging treated sewage into Selat Johor.

8.4.6   *Simulation of FC in Selat Johor*

8.4.6.1   *Simulation mesh*

AQUASEA, a two-dimensional depth-averaged shallow water equation simulation model, is used to simulate hydrodynamic and water quality in Selat Johor. Figure 8.15(a) shows the finite element mesh used in this simulation study for the Selat Johor, with higher resolution meshes (Figure 8.15(b)) being employed around the sewage outfall site to allow better simulation results. The sewage outfall location is roughly indicated by a red dot (between Sg Pendas and Tg Bunga).

8.4.6.2   *Simulated tidal currents in Selat Johor*

Figure 8.12 depicts hourly spring tidal elevations at the nearby tidal station Pelabuhan Tanjung Pelepas beginning at 00-hour 15 April 2008 for two days, indicating a spring tidal amplitude of 1.0 m. The tidal boundary elevation η at the mouth of Selat Johor near Tengeh is defined as $\eta = 1.0\sin(\sigma t)$m, with $\sigma = 2\pi/(12.42 \times 3,600 \text{ s})$, where time t has the unit of s (seconds). The tidal velocity is set to 0.0 m/s at all times at the Johor Causeway, as it is a solid and impermeable wall. Figure 8.16 shows the simulated current flows at three distinct phases of the semidiurnal tidal cycle. Figure 8.16(a) displays the tidal currents during the ebb tide. During an ebb tide, tidal currents flow southwest out of Selat Johor, for six hours, beginning from Maximum High Water (MHW) to Minimum Low Water (MLW). Figure 8.16(b) shows the slack tidal currents when the currents are weak. Soon after

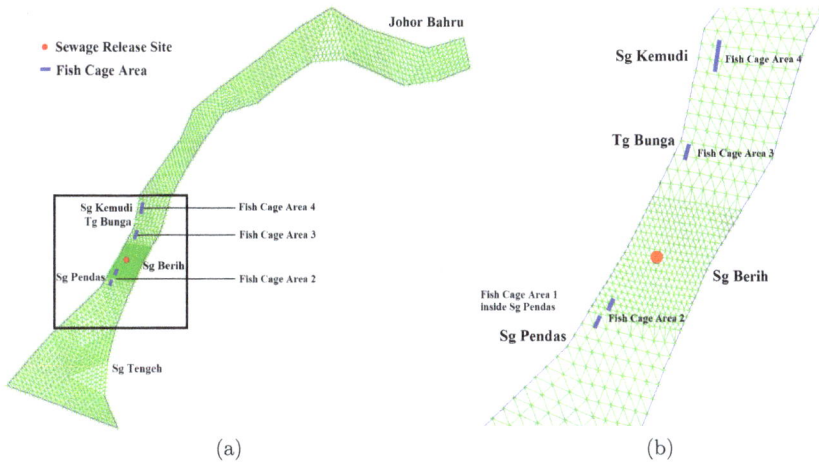

(a)   (b)

Figure 8.15:   (a) Simulation Mesh in Selat Johor, and (b) refined mesh around discharge location.

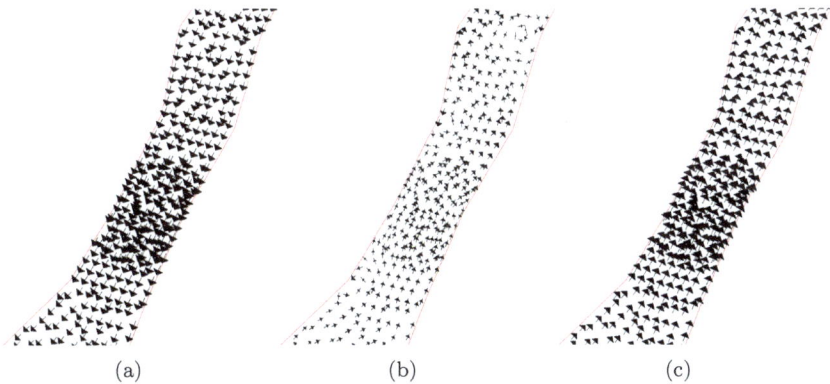

(a)   (b)   (c)

Figure 8.16:   Tidal current flows during (a) ebb tide, (b) slack tide and (c) flood tide.

the slick tide, the flood tidal currents begin to reverse direction to flow due northeast into Selat Johor (Figure 8.16(c)), beginning from (MLW) to (MHW) for six hours.

8.4.6.3   *Simulated FC in Selat Johor*

The simulated FC concentrations (MPN/100 mL) are displayed in contours in Figure 8.17 for three distinct phases of the tidal cycle:

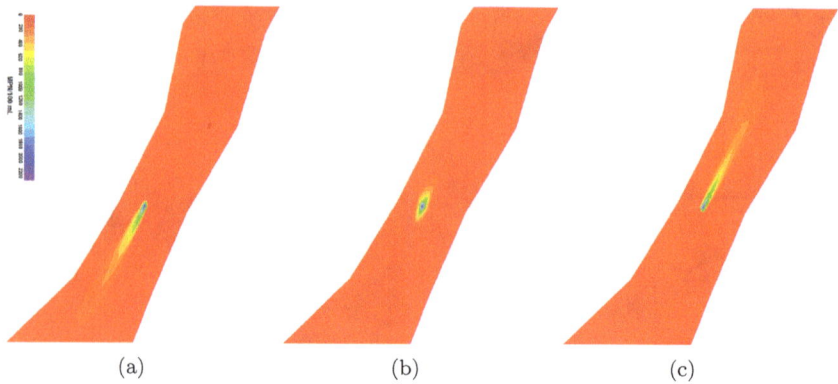

(a)   (b)   (c)

Figure 8.17:   FC contours for three distinct phases of the tidal cycle: (a) ebb tide, (b) slack tide and (c) flood tide.

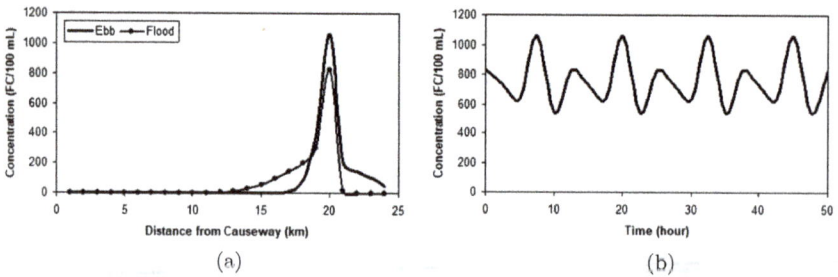

(a)   (b)

Figure 8.18:   (a) Spatial distribution of peak FC at each location along the FC plume center line and (b) Temporal distribution of FC at the discharge location.

(a) ebb tide, (b) slack tide and (c) flood tide. The peak FC concentrations occur during slack water when the tidal currents are weak. This stage of weak tidal currents occurs every six hours, right in between MHW and MLW. Figure 8.18(a) shows the spatial distribution of peak FC at each location along the FC plume center line. Figure 8.18(b) shows the temporal distribution of FC at the discharge location. Averaged over the tidal cycle, the mean FC concentration at the discharge location is about 800 FC/100 mL. Coastal dilution, dispersion and decay will further reduce the FC concentration to lower values 1 km away from the discharge site. For example, at the fish cage areas 2 and 3, the peak FC concentration is less than 50 MPN/100 mL, as may be seen from Figure 8.19. Tidally averaged FC values remain around 10 MPN/100 mL near the fish cage

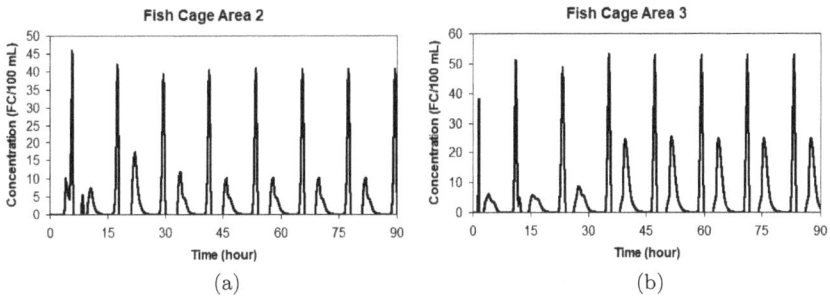

Figure 8.19: FC concentration at the fish cage areas 2 and 3.

areas 2 and 3. At fish cage culture areas 1 and 4, the FC levels are even lower as the two sites are further away from the discharge location than areas 2 and 3. These levels of FC are well below the permissible threshold limit of 200 MPN/100 mL stipulated for recreational waters and contact activities such as fish cultivation (EU, 1986; USEPA, 1986, 2012). Hence, the treated sewage can be discharged into Selat Johor at the indicated location. The FC will not pose any health hazards to fish and humans working in the aquaculture farms. However, it should be noted that the absence of coliforms does not ensure that the water is free of other pathogens (NRC, 2004; Wu *et al.*, 2011). In conclusion, an acceptable option for the discharge of treated sewage from STP 6 is to discharge the effluent into Selat Johor at the location indicated by a red dot shown in Figure 8.15.

8.5 Challenges Confronting Coastal Reclamation

Highly dynamic in its socio-economic development, the coastal zone is facing increasingly complex land-use conflicts among the stakeholders over the limited resources. Increasing coastal risks caused by future climate change have made coastal zone management a complex task. To cope with the context of global climate change, integrated coastal zone management is important in building adaptation capability of human systems and in ensuring sustainable development along the coasts. Thus, integrated coastal zone management has become a focus of many studies. Research and management of complex coastal systems is underpinned by natural sciences, engineering and social sciences. Hence, effective management of coastal systems should integrate and leverage the strength of both sciences and engineering. This (ICZM approach would support the sustainable use of coastal resources (Gallagher, 2010; Pickaver *et al.*, 2004), by holistically managing coastal zones (Wheeler *et al.*, 2011), rather than by applying a sector-by-sector approach. By integrating and harmonizing various policies and decision-making processes in ICZM, one would develop an integrated process conducive to the protection of coastal zones and the sustainable utilization of their resources (Tissier *et al.*, 2010). This ICZM approach will cultivate community awareness regarding the importance of coastal zones to society and to the overall marine ecosystem (Tiller *et al.*, 2012). Best practices in ICZM in various countries will be discussed in the remaining subsections.

The aim of these subsections is to provide a brief overview of current status and future trends of ICZM in general, and coastal reclamation in particular, in Asia and to reveal potential problems to be faced today and in the future. Coastal reclamation takes place in a sensitive marine zone that has hydrogeological and environmental linkages with two other distinctive zones: (a) the upland terrestrial habitats on which all cities are built and (b) the environmentally sensitive marine habitats of mangroves, seagrass, corals and fisheries. It is clear that managing coastal reclamation in a sensitive marine zone is a complex task that is dependent on good principles and best practices in ICZM techniques. Global climate challenges and sea level rise implications would compound the complexity of this task by the requirement of climate action planning (SDG13). The task mandates the requirement of marine conservation to protect fishery and life below water (SDG14). Future cities are likely to be built along the coasts, some on reclaimed land. ICZM principles and best practices would help to enable future cities to be sustainable (SDG11), the achievement of which depends critically on partnership and collaboration (SDG17). Higher education institutes must task themselves with the core responsibility to prepare future talents with core sustainability competences by providing quality education (SDG4).

8.5.1   *ICZM in Penang*

Based upon the integration concept discussed above, it may be surmised that ICZM efforts in Penang had been hampered by three deficiencies: (a) the absence of an authoritative government agency having overall responsibility and jurisdiction to coordinate efforts from various government agencies, (b) poor institutional capacity for understanding the sciences of ICZM, the inherent complexity and uncertainty involved and (c) a lack of integration between the different sectors, diverse stakeholders and various administrative levels (Albotoush and Tan, 2019). Regulations and acts concerning the coastal zones are implemented by different agencies at the State and Federal levels, without good coordination. Many differences exist (a) between coastal zones with diverse characteristics, (b) between local, regional and national governing agencies, and (c) between the governance hierarchy and local communities regarding societal engagement in ICZM. ICZM in Penang, however, is gradually becoming an effective tool for managing coastal zones for long-term sustainability of their resources (Albotoush and Tan, 2019). An interface between upland and the sea, coastal wetlands, such as those in northern China, are vital in their provisions of important ecosystem services, the protection and provision of which is central to the concepts of sustainability.

8.5.2 *Protecting coastal wetlands in China*

An essential component of coastal zones, coastal wetlands provide human beings with many irreplaceable ecosystem services. They play an important role in regulating climate, improving water quality and maintaining biodiversity (Costanza *et al.*, 1997), including that of migratory waterbirds. But, they are under stress from sea level rise, from human activities and from landscape development (Ivajnsic and Kaligaric, 2014). The rapid changes in coastal landscapes and socio-economic development have impaired the health of coastal wetlands and have truncated their provision of ecological services. This loss of coastal ecosystem services is duly acknowledged by many governments, including China. Since 2000, China started to implement a range of measures to reduce landscape development stress in coastal wetlands, including strict control of sea reclamation and marine development (Guo and Zhang, 2019). A number of coastal protection areas, including 33 coastal national nature reserves (CNNR) and 67 national special marine reserves (NSMR), have been established to protect some 17% of the coastline, and to promote the conservation of biodiversity and natural ecosystems.

These protection areas are mostly in the northern part of the Chinese coastline because of higher concentrations of coastal wetlands and migratory waterbirds in the north (Lei *et al.*, 2017). The coastal zones of the Huanghai Sea and the Bohai Sea are the focus of coastal wetland conservation. These coastal wetlands are critical to the survival of many migratory waterbirds using the East Asian-Australasian Flyway, by provision of critical staging and over-wintering habitats for more than 200 migratory bird species (Xia *et al.*, 2017). Over 70% of the globally threatened waterbird species on this flyway critically depend on essential provisions rendered by China's coastal wetlands (Bai *et al.*, 2015). The health of these migratory waterbirds may be impacted by unsustainable marine reclamation in a hurry to create land for development. Other critical issues related to hydraulic alterations, physical damage, water quality degradation and ecosystem impairments resulting from coastal reclamations abound, particularly in Asia.

8.5.3   *Critical issues confronting reclamation*

Coastal land reclamation has emerged as a popular solution to meet the high demand of land along the coasts, especially in Asia. However, the sustainability of these newly reclaimed lands under the combined onslaught of increasing population, SLR, higher frequency of extreme meteorological events and land subsidence is largely unknown. Low-lying coastal areas, particularly reclaimed lands, are highly vulnerable to both anthropogenic and natural disturbances such as storm surges and coastal floods. Given current rates of sea level rise, physical and environmental changes at the coast can induce dramatic impacts. It is generally accepted that land reclamation of salt marshes, tidal flats and mangroves represents a major risk to the sustainability of coastal environment and marine ecosystems. This risk will intensify by global climate change and sea level rise impacts. Reclamation in such environmentally sensitive areas has led to significant losses of natural biota, especially in the intertidal zone. Disaster and ecosystem risks resulting from reclamation activities at the coastal frontier remain a controversial concern, as they might damage coastal ecosystems and increase exposure to other disaster risks (Asdak *et al.*, 2018) as well as degrading the water quality of coastal groundwater (Kim *et al.*, 2019). These critical issues, including trade-off, economic compensation for ecosystem damage and land subsidence, will be discussed in the following subsections under Section 8.5.3 by way of several case studies in Asia.

8.5.3.1 *Reclamation in Yalu River Estuary*

The impacts of a coastal reclamation activity on estuarine sedimentation in the Yalu River Estuary (YE), China, have been a focus of research. The estuarine sedimentary environment had experienced dramatic changes after the land reclamation activity completed in 1975. Model simulations show that the estuary became better mixed after the land reclamation, due to the stronger tidal current and enhanced turbulence (Cheng *et al.*, 2020). As a result, suspended sediment concentrations (SSCs), especially those in the bottom layer, had increased due to increased bed erosion caused by higher turbulence. Because of enhanced bed sediment resuspension induced by stronger currents, the YE may turn into a sediment source, instead of a sink, if more land is reclaimed in the future. Model simulations on the impact of land reclamation indicated that the reclamation may have two significant impacts causing (1) the increase in magnitude of the tidal currents during both the spring and neap tides and (2) the intensification of turbulent mixing. The strong hydrodynamics would lead to an increase in sediment loads post reclamation, detrimental to marine ecosystems. The consequences of reclamation on the marine environment are threefold: (a) significant increase in SSC in the main branch of YE by 140 mg/L; (b) significant vertical gradient between surface and bottom SSC; and (c) higher horizontal gradient between the inner and outer estuary SCCs. These changes in sediment transport would lead to increased erosion in the estuary, and may change the estuary from a sediment sink to a sediment source if the reclamation continues. The environmental impacts of such changes have not been adequately quantified.

8.5.3.2 *Reclamation in the Tianjin Harbor Industrial Zone*

Large-scale reclamation projects in the Tianjin Harbor Industrial Zone have already caused a series of negative impacts on the marine ecological environment (Nie and Tao, 2008). Reclamation projects have been observed to have significantly affected the growth, reproduction and distribution of phytoplankton, and the zooplankton that feed on the phytoplankton. Dredging work on the sea bottom had directly affected the living environment of benthos (Liu *et al.*, 2010) and had led to the extinction of benthos organisms. Land-sourced pollutants and industrial sewage had led to negative impacts on marine life. To evaluate the impacts of reclamation on the marine ecosystems, research was conducted three times between 2006 and 2008 in the Tianjin Harbor Industrial Zone. The objective was to evaluate the living environment of marine planktons and benthos that may have been affected by reclamation projects covering 80 square kilometers. The Shannon-Weaver diversity index (SWDI) of benthos dropped from 1.28 to 0.0 as the benthos were completely destroyed by dredging work on the sea bottom. The SWDI of phytoplankton and zooplankton decreased sharply from 3.01 to 1.71 and from 1.7 to 0.58, respectively. The results showed that reclamation projects had adversely changed the living environment of marine organisms, had decreased the biodiversity and had altered the structure of the marine community (Li *et al.*, 2010). The huge social-economic benefits gained from the reclamation projects had not been accompanied by ecological conservation, casting in doubt the sustainability of the coastal ecological environment.

8.5.3.3   *Reclamation in Wenzhou Shoal*

To fulfill demand for land in Wenzhou in Zhejiang Province, China, coastal reclamation was proposed to create an area of 88 km^2. A study was conducted to examine the feasibility of the proposed land reclamation in Wenzhou Shoal, located between Lingkun Island and Niyu in the Oujiang Estuary. An analysis of the natural hydrodynamic regimes, sediment transport and the seabed evolution was performed. A numerical model of tidal current, sediment movement and seabed deformation under the combined action of tidal current and wave is applied in the feasibility study. The concerned area has weak tidal currents, is constantly aggrading and is not a tidal passage, conditions that are in favor of reclamation. The study concluded that the proposed reclamation in the Wenzhou Shoal will have little impact on changing (a) the hydrodynamic and sediment environment in the area, (b) the flood discharge and drainage of the Oujiang River, and (c) the marine resources of harbors and navigational channels. Therefore, the proposed reclamation is deemed feasible from the environmental and ecological perspectives, as the impacts are negligible.

Indeed, the reclamation is considered beneficial for the maintenance and stability of the navigational channels and would bring about huge and favorable socio-economic impacts (Li, 2010).

8.5.3.4 *Reclamation in Penang*

Comparisons of historical and current topographic maps for Penang Island in Peninsular Malaysia revealed that land formerly consisting of coastal swamps and mangrove forests have been lost to large-scale land reclamations. Between 1960 and 2015, reclaimed land and artificial shoreline have been created to cater to the needs of development. The total coverage of mangrove forests is relatively stable, as significant losses on the east coast are balanced with similar increases on the west coast. Coastal reclamation on the island is still ongoing with plans for the construction of five artificial islands. Another two coastal reclamation projects are scheduled for the near future. In total, these projects will result in reclaiming 32.3 km^2 of coastal land out of 321.8 km^2 of the island, equivalent to a 10% increase (Chee *et al.*, 2017). The associated negative impacts on the island's natural coastal habitats are inevitable. Sections of the coast of Penang Island are in dire need of monitoring, conservation and management. With appropriate planning and long-term management, eco-engineering represents a valuable adaptive management tool to mitigate the impact of harmful coastal development (Mayer-Pinto *et al.*, 2017) for the island of Penang. However, ICZM best practices had rarely been observed in the past.

8.5.3.5 *Reclamation in Johor*

The Johor River is the main source of freshwater in Johor Bahru and Singapore as well as a vital maritime passage in the Straits of Johor. The Johor River Estuary has been experiencing environmental degradation due to the combined impact of reclamation and anthropogenic activities. The area of reclaimed land increased more than 10-fold from 13 km^2 (1973) to 135 km^2 (2017). Each reclamation showed an increasing tendency to migrate toward the mouth of the Johor River (Wang *et al.*, 2019). Over the last four decades, reclamation around the Tekong Island contributed an additional land area of 40 km^2. To better protect the coastal environment, reclamation activities should be planned properly by the government to protect mangroves in the area. On both sides of the Johor River, mangrove reserve areas should be designated to prohibit illegal logging, and to protect the flood storage capacity and water purification functions of mangroves. It should be noted that Johor harbors three Ramsar sites, namely, Pulau Kukup (647 ha), Sungai Pulai (9,126 ha) and Tanjung Piai (526 ha), all of which must be protected.

8.5.3.6   *Trade-off in Dapeng Bay, Taiwan*

A study was conducted to examine the environmental and biogeochemical changes in Dapeng Bay in southwestern Taiwan, following two major reclamation works performed between 1999 and 2010. The semi-enclosed coastal lagoon was largely occupied by oyster culture racks and fish farming cages before December, 2002. Substantial external inputs of nutrients and organic carbon coupled with the long water residence time γ (10 days) caused the lagoon to be eutrophic, particularly in the inner lagoon that directly received nutrient inputs. Up to 2002, the lagoon had very low DO concentrations and high nutrient concentrations in the bottom layer of the inner lagoon, as a result of excessive external and internal inputs of nutrients and organic carbon. The estimated net ecosystem production (NEP) in 2002 during the first stage was 5.8 mol $C/m^2/yr$. After January 2003, the aquaculture structures were completely removed, tidal flushing had increased and residence time γ had decreased to 6 days. The NEP increased to 7.7 mol $C/m^2/yr$ after the aquaculture structure removal, as a result of improved light availability, better water quality and enhanced biological community structures caused by improved habitat environment. The annual mean concentrations of dissolved oxygen increased, and nutrients decreased substantially, due to increased flushing, the absence of aquaculture feeding and increased biological utilization. The second reclamation work began in July 2006, focusing on establishing artificial wetlands for wastewater treatment and on dredging bottom sediment to remove nutrients and to increase storage. The water residence time γ increased slightly to 8 days due to increase in storage. Substantial decreases in nutrient concentrations and dissolved organic matter were noted, while the NEP improved to 13.4 mol $C/m^2/yr$.

reclamation has caused loss of large tracks of coastal wetlands and natural shorelines, leading to deterioration of marine ecological environment. China has attached great importance to strengthening reclamation management for the protection of marine environment. In the future, China will continue to implement strict reclamation management based on the principles of ecological system recovery and public participation (Yue *et al.*, 2016). Strict reclamation management has provided a guarantee for the sustainable development and utilization of China's marine resources and environment. Reclamation demand for industrial construction and township expansion has decreased in the recent past, while demand for coastal tourism and public leisure space close to the sea has gradually increased. Recently, China has advanced the developmental goal of building up its marine ecological sites and of establishing itself as a marine global power. China has established and implemented strict guidelines for the total amount of reclamation allowed as well as for an intensive management system for reclamation. The Chinese central government has increased the protection of coastal ecosystems and environment and has tightened the approval process for coastal reclamation proposals (Wang *et al.*, 2014). For example, Shandong province was allowed to reclaim only 345 km^2 of its original request of 520 km^2, while only 506 km^2 of the Zhejiang request of 1747 km^2 was approved. A reclamation management system based on the ecology of reclaimed coastal systems will be developed according to reclamation demand.

8.5.4.2 *Future reclamation in Singapore*

Habitat loss associated with land reclamation and shoreline development is prevalent as coastal cities expand. The majority of Singapore's mangrove forests, coral reefs and sand-mudflats disappeared between the 1920s and 1990s (Lai *et al.*, 2015). The total cover of intertidal coral reef flats has reduced from 17.0 km^2 to 9.5 km^2, and sand-mudflats have decreased from 8.0 km^2 to 5.0 km^2, largely because of intensive land reclamation. Conversely, mangrove forests have increased from 4.8 km^2 to 6.4 km^2 due to restoration efforts and greater regulatory protection. All coastal habitats are predicted to shrink further as new reclamations are scheduled to be completed. Such coastal habitat decline may be counteracted, at least in part, if ecological engineering is used to help conserve biodiversity. The most efficacious way to achieve implementation is via a top-down approach to ensure smooth coordination among the various government agencies that are involved in coastal planning. The loss of natural habitats and their associated biodiversity is likely to accelerate going forward. Finding a balance between conservation and development of its shores can help inform urban marine sustainability planning in Singapore as in other coastal cities facing similar challenges.

8.5.4.3   *Future reclamation in Korea*

Large-scale reclamation of tidal wetlands continues unabated in Korea. The argument that public waters such as tidal wetlands must be conserved and be used in a sustainable way is nominally accepted within relevant public policies and acts, thereby creating a discrepancy within the legal system. Large-scale reclamations of public waters in Korea will continue, unless the Public Waters Management and Reclamation Act (PWMRA) is retrofitted toward a conservation-oriented legislative policy. The PWMRA has basically evolved to respond to newly created reclamation needs. Existing conservation-orientated acts that have the potential to regulate reclamation were developed to be not in conflict with PWMRA. The legal system should accept the precautionary principle as the legal doctrine in protecting public waters from unsustainable use. To safeguard the sustainable use of tidal wetlands, reclamation that would irreversibly change the natural features of tidal wetlands should be prohibited. The reclamation deterrence principle should be stipulated as the underlying principle. The approval or rejection of a reclamation plan should be determined in a way that favors, prima facie, the conservation value over the use value (Park, 2014). A reclamation plan may only be authorized when it is able to withstand a four-tier test as follows: (1) the reclamation plan is suitable for coast-related plans established by other acts, and will not adversely impact protected areas or other zones of importance designated by other acts; (2) there are no other feasible alternatives that have a lower impact on the coastal and marine ecosystems; (3) appropriate mitigation measures that are feasible in practice and that will not cause significant damage to the coastal and marine ecosystem must be prepared in advance; and (4) the "water dependency" requirement must be met if special aquatic ecosystems, such as wetlands, are subject to reclamation (Houck, 1989). This four-tier test indeed poses significant challenges for coastal reclamations in the future. The associated ecological, financial and social conflicts remain a formidable challenge.

8.5.5 *Formidable challenges ahead*

Almost 40% of the global population resides within 100 km of the coast and around 704 million people live contiguous to seacoasts at elevations less than 10 m above mean sea level. The demands for coastal land to support human settlements and the associated anthropogenic activities continue to drive coastal reclamation. The spatial distributions and temporal trends in recent reclamation projects worldwide are the focus of a study, by mapping and tabulating the annual magnitude of change in coastal land gained from 1988 to 2018 for eight major Asian coastal cities (Sengupta *et al.*, 2020). Subsidence is observed over many recently reclaimed coastal lands at alarming rates, e.g., the Incheon's international airport (28.5 cm/year), followed by Singapore's Pasir Panjang Terminal (14.7 cm/year) and Shanghai's Pudong international airport (10 cm/year). In coastal cities such as Shenzhen, Jakarta and Shanghai, land subsidence, triggered principally by excessive groundwater extraction, has emerged as a major environmental challenge. The degree of land subsidence relates principally to the characteristics of the underlying geological substrate. These coastal cities experience an increasing frequency of heavy rainfall and tropical cyclones that will pose severe threats if reclaimed land continues to subside. Across these cities, both the spatial extent and rate of reclamation are remarkable; some 700 km^2 has been reclaimed in just three decades. More than 35% of this new coastal land has been reclaimed in Shanghai alone (562 km^2), while Singapore and Incheon have also experienced substantial land gains through reclamation. These three cities alone account for almost 10% of all the land gained globally over the last three decades.

More than 95% of reclaimed land in Tokyo, Singapore, Jakarta, Karachi and Manila is used for various categories under "built-up", including airports and harbors. In both Shanghai and Singapore, substantial proportions (58% and 29%, respectively) of the newly reclaimed land remains vegetated. An analysis of the spatial-temporal patterns reveals that recently reclaimed areas are predominantly characterized by construction, including ports, airports, commercial and residential uses, driven primarily by economic development. Shanghai, however, represents a significant departure from this trend, whereby more than 50% of the new coastal land gained during the recent past has not been devoted to construction projects and is vegetated or has been constructed as artificial wetland parks, suggesting a different policy context. This may be attributed to the ecological policy enacted by the Chinese government in 2015. Further, in recent times, the smart and eco-cities have been constructed on reclaimed coastal land along the congested coasts. The notion of sustainable floating cities has been proposed recently by UN-Habitat (UN-Habitat, 2016). In Shanghai, the Jinshan coastal tourism resort is projected to become the "Gold Coast" of China for economic, tourism and Eco-city development. However, the associated ecological, financial and social conflicts remain a formidable challenge. Coastal reclamation at this alarming rate carries the implication that potential risks may not have been thoroughly evaluated in the context of climate action, from the perspectives of financial feasibility verses ecological impairment, and from the perspective of economic returns verses social equity.

References

Albotoush, R. and Tan, A.S.H. (2019). Evaluating integrated coastal zone management efforts in Penang Malaysia. *Ocean and Coastal Management*, 181, 104899.

Asdak, C., Supian, S. and Subiyanto, A. (2018). Watershed management strategies for flood mitigation:a case study of Jakarta's flooding. *Weather and Climate Extremes*, 21, 117–122. doi: 10.1016/j.wace.2018.08.002.

Bai, Q.Q., Chen, J.Z., Chen, Z.H., Dong, G.T., Dong, J.T., Dong, W.X., Fu, V.W.K., Han, Y.X., Lu, G. and Li, J. (2015). Identification of coastal wetlands of international importance for waterbirds: A review of China Coastal Waterbird Surveys, 2005–2013. *Avian Research*, 6, 1–16.

Chan, J.T.K., Leung, H.M., Yue, P.Y.K., Au, C.K., Wong, Y.K., Cheung, K.C., *et al.* (2017). Combined effects of land reclamation, channel dredging upon the bioavailable concentration of polycyclic aromatic hydrocarbons (pahs) in Victoria Harbour sediment, Hong Kong. *Marine Pollution Bulletin*, 114, 587–591. doi: 10.1016/j.marpolbul.2016.09.017.

Chee, S.Y., Othman, A.G., Sim, Y.K., Mat Adam, A.N. and Firth, L.B. (2017). Land reclamation and artificial islands: Walking the tightrope between development and conservation. *Global Ecology and Conservation*, 12, 80–95.

Cheng, Z., Jalon-Rojas, I., Wang, X.H. and Liu, Y. (2020). Impacts of land reclamation on sediment transport and sedimentary environment in a macro-tidal estuary. *Estuarine, Coastal and Shelf Science*, 242, 106861.

Chia, L.S., Khan, H., and Ming, C.L. (1988). The Coastal Environmental Profile of Singapore, ICLARM, Manila-Philippines.

Costanza, R., D'Arge, R., Groot, R.D., Farber, S., Grasso, M., Hannon, B., Limburg, K., Naeem, S., Nelll, R.V.O., Paruelo, J., Raskin, R.G., Sutton, P. and Belt, M.V.D. (1997). The value of the world's ecosystem services and natural capital. *Nature*, 25, 3–15.

Dollar, S.J. and Grigg, R.W. (1981). Impact of a Kaolin clay spill on a coral reef in Hawaii. *Marine Biology*, 65, 269–276.

Duan, H., Zhang, H., Huang, Q., Zhang, Y., Hu, M., Niu, Y. and Zhu, J. (2016). Characterization and environmental impact analysis of sea land reclamation activities in China. *Ocean and Coastal Management*, 130, 128–137. doi: 10.1016/j.ocecoaman.2016.06.006.

Erkens, G., Bucx, T., Dam, R., Lange, G. De and Lambert, J. (2015). Sinking coastal cities. *Proc. IAHS*, 372, 189–198. doi: 10.5194/piahs-372-189-2015.

EU, European Commission (1986). Bathing water directive. *Official Journal of the European Communities*, 76/160/EEC, 5.2.76, No L 31/1–No L 31/7, 1975.

Gallagher, A. (2010). The coastal sustainability standard: A management systems approach to ICZM. *Ocean and Coastal Management*, 53(7), 336–349.

Ghaffari, K., Habibzadeh, T., Mortaza, A.N. and Mousazadeh, R. (2017). Construction of artificial islands in southern coast of the Persian Gulf from the viewpoint of international environmental law. *Journal of Politics and Law*, 10(2), 263–275. http://dx.doi.org/10.5539/jpl.v10n2p264.

Griffith, M.D. and Ashe, J. (1993). Sustainable development of coastal and marine areas in small Island developing states: A basis for integrated coastal management. *Ocean and Coastal Management*, 21, 269–284.

Guo, Z. and Zhang, M. (2019). The conservation efficacy of coastal wetlands in China based on landscape development and stress. *Ocean and Coastal Management*, 175, 70–78.

Houck, O.A. (1989). Hard choices: The analysis of alternatives under section 404 of the Clean Water Act and similar environmental laws. *University of Colorado Law Review*, 60, 778.

Hung, J.-J., Huang, W.-C. and Yu, C.-S. (2013). Environmental and biogeochemical changes following a decade's reclamation in the Dapeng (Tapong) Bay, southwestern Taiwan. *Estuarine, Coastal and Shelf Science*, 130, 9–20.

Hutchings, P.A. and Wu, B.L. (1987). Coral reefs of Hainan island, South China sea. *Marine Pollution Bulletin*, 18(1), 25–26. http://dx.doi.org/10.1016/0025-326X(87)90652-7.

Ishenda, D.K. and Shi, G. (2019). Determinants in relocation of capital cities. *Journal of Public Administration and Governance*, 9(4), 200–220. doi: 10.5296/jpag.v9i4.15983.

ITLOS (2003). Case concerning Land Reclamation by Singapore in and around the Straits of Johor. Request for provisional measures, Malaysia v. Singapore. Verbatim Record. International Tribunal for the Law of the Sea, Hamburg, Germany. TLOS/PV.03/02/Corr.1

Ivajnsic, D. and Kaligaric, M. (2014). How to preserve coastal wetlands, threatened by climate change-driven rises in sea level. *Environmental Management*, 54, 671–684.

Johnston, S.A. (1981). Estuarine dredge and fill activities: A review of impacts. *Environmental Management*, 5(5), 427–440.

Kim, R.H., Kim, J.H., Ryu, J.S. and Koh, D.C. (2019). Hydrogeochemical characteristics of groundwater influenced by reclamation, seawater intrusion, and land use in the coastal area of Yeonggwang, Korea. *Journal of Geosciences*, 23, 603–619. https://doi.org/10.1007/s12303-018-0065-5.

Koh, H.L. (2004). Pemodelan Ekosistem dan Alam Sekitar (Ecosystem and Environmental Modelling). USM Press, Pulau Pinang, Malaysia.

Koh, H.L. and Lee, H.L. (2001). Modeling Water Quality in Sg. Lebum, Johor, Malaysia. In: Recent Advances in Marine Science and Technology, 2000. Ed. Narendra Saxena, pp. 199–208. PACON International, USA, June 2001.

Koh, H.L. and Lee, H.L. (2006). Catchment management modeling: From headwater to the coasts. *Aquatic Ecosystem Health and Management*, 9(2), 261–268.

Koh, H.L. and Teh, S.Y. (2019). Climate change mitigation and adaptation: Role of mangroves in Southeast Asia. In: Leal Filho, W., Azul, A., Brandli, L., Özuyar, P., Wall, T. (eds.) Climate action. Encyclopedia of the UN sustainable development goals. Springer, Cham. doi: 10.1007/978-3-319-71063-1_107-1.

Koh, H.L. and Teh S.Y. (2019). Disaster Risk Reduction and Resilience through Partnership and Collaboration. In: Leal Filho, W., Azul, A.,

Brandli, L., Özuyar, P., Wall, T. (eds.) *Partnerships for the Goals. Encyclopedia of the UN Sustainable Development Goals.* Springer, Cham. doi: 10.1007/978-3-319-71067-9_49-1.

Koh, H.L., Lim, P.E. and Midun, Z. (1991). Management and control of pollution in Inner Johore strait. *Environmental Monitoring and Assessment*, 19, 349–359.

Koh, H.L., Lim, P.E. and Lee, H.L. (1995). Water Quality Modeling for an Estuary in Johore. *Water Quality Research Journal of Canada*, 30(1), 45–52.

Koh, H.L., Lim, P.E. and Lee, H.L. (1997). Impact modeling of sewage discharge from Georgetown of Penang, Malaysia on coastal water quality. *Journal Environmental Monitoring and Assessment*, 44, 199–209.

Koh, H.L., Syamsidik, Lee, H. and Zubir, D. (2004). Pulau Tekong and Changi Reclamation: Modeling and Institutional Perspectives, *Proceedings Second International Symposium on Southeast Asian Water Environment*, Editors: Shinichiro Ohgaki, Kensuke Fukushi, Hiroyuki Katayama and Satoshi Takizawa, pp. 271–278, University of Tokyo.

Koh, H.L., Teh, S.Y., Kh'ng, X.Y. and Barizan, R.S.R. (2018). Mangrove forests: Protection against and resilience to coastal disturbances. *Journal of Tropical Forest Science*, 30, 446–460.

Lai, S., Loke, L.H.L., Hilton, M.J., Bouma, T.J. and Todd P.A. (2015). The effects of urbanisation on coastal habitats and the potential for ecological engineering: A Singapore case study. *Ocean and Coastal Management*, 103, 78–85.

Lei, G.C., Zhang, Z.W., Yu, X.B. and Zhang, M.X. (2017). *Strategic Research on Coastal Wetland Conservation and Management in China.* Higher Education Press, Beijing.

Li, K., Liu, X., Zhao, X. and Guo, W. (2010). Effects of reclamation projects on marine ecological environment in Tianjin Harbor industrial zone. *Procedia Environmental Sciences*, 2, 792–799.

Li, M.-G. (2010). The effect of reclamation in areas between islands in a complex tidal estuary on the hydrodynamic sediment environment. *Journal of Hydrodynamics*, 22(3), 338–350. doi: 10.1016/S1001-6058(09)60063-9.

Liu, X.B., Zhang, W.L. and Tian, S.Y. (2010). Characteristics of macrobenthos in Tianjin Intertidal zone. *Journal of Salt and Chemical Industry*, 39(1), 31–35.

Ludwig, R.G. (1988). Environmental Impact Assessment: Siting and Design of Submarine Outfalls. The Monitoring and Assessment Research Centre (MARC), University of London, London.

Mayer-Pinto, M., Johnston, E.L., Bugnot, A.B., Glasby, T.M., Airoldi, L., Mitchell, A. and Dafforn, K.A. (2017). Building 'blue':

An eco-engineering framework for foreshore developments. *Journal of Environmental Management*, 189, 109–114. doi: 10.1016/j.jenvman.2016.12.039.

Murray-North (SEA) Pte. Ltd., Scott & Furphy Pte.Ltd., Lawson and Treloar Pte. Ltd., Europasia Engineering Services Sdn. Bhd., Asian Wetlands Bureau Malaysia and University of Malaysia. (1994). Hydraulic and Water Quality Study of the Strait of Johor, Final Report, Ministry of the Environment, Republic of Singapore and Ministry of Science, Technology and Environment of Malaysia.

Nie, H.T. and Tao, J.H. (2008). Impact of coastal exploitation on the eco-environment of Bohai Bay. *The Ocean Engineering*, 26(3), 44–50.

NRC, National Research Council (2004). Indicators for Waterborne Pathogens. The National Academics Press, Washington.

Padmalal, D., Maya, K., Sreebha, S. and Sreeja, R. (2008). Environmental effects of river sand mining: A case from the river catchments of Vembanad lake, Southwest coast of India. *Environmental Geology*, 54(4), 879–889. doi: 10.1007/s00254-007-0870-z.

Park, T. (2014). Analysis of relevant laws on reclamation of Korean tidal wetlands and court debates observed at the Saemangeum reclamation lawsuit. *Ocean and Coastal Management*, 102, 583–593.

Pescod, M.B. (1992). Wastewater treatment and use in agriculture. Food and Agriculture Organization of the United Nations, Rome.

Pickaver, A.H., Gilbert, C. and Breton, F. (2004). An indicator set to measure the progress in the implementation of integrated coastal zone management in Europe. *Ocean and Coastal Management*, 47(9), 449–462.

Reder, K., Flörke, M. and Alcamo, J. (2015). Modeling historical fecal coliform loadings to large European rivers and resulting in-stream concentrations. *Environmental Modelling and Software*, 63, 251–263.

Romañach, S.S., DeAngelis, D.L., Koh, H.L., Li, Y.H., Teh, S.Y., Barizan, R.S.R. and Zhai, L. (2018). Conservation and restoration of mangroves: Global status, perspectives, and prognosis. *Ocean and Coastal Management*, 154, 72–82.

Scott, M. and Lennon, M. (2020). Climate disruption and planning: Resistance or retreat? *Planning Theory and Practice*, 21, 125–154. doi: 10.1080/14649357.2020.1704130.

Sengupta, D., Chena, R., Meadows, M.E. and Banerjee, A. (2020). Gaining or losing ground? Tracking Asia's hunger for 'new' coastal land in the era of sea level rise. *Science of the Total Environment*, 732, 139290.

Shan, J. and Li, J. (2020). Valuing marine ecosystem service damage caused by land reclamation: Insights from a deliberative choice experiment in Jiaozhou Bay. Marine Policy 122, 104249. https://doi.org/10.1016/j.marpol.2020.104249.

Sun, X., Li, Y., Zhu, X., Cao, K. and Feng, L. (2017a). Integrative assessment and management implications on ecosystem services loss of coastal wetlands due to reclamation. *Journal of Cleaner Production*, 163, S101–S112.

Sun, Q., Whitney, M.M., Bryan, F.O. and Tseng, Y.-H. (2017b). A box model for representing estuarine physical processes in Earth system models. *Ocean Modelling*, 112, 139–153.

Syamsidik and Koh, H.L. (2003). Impact Assessment Modeling on Coastal Reclamation at Pulau Tekong. 8p. *In: Proceedings of Regional Conference on Integrating Technology in the Mathematical Sciences*, 14–15 April 2003, USM, Penang.

Teh, S.Y. and Koh, H.L. (2020). Education for sustainable development: the STEM approach in Universiti Sains Malaysia. In: Leal Filho, W., *et al.* (eds.) Universities as living labs for sustainable development. World sustainability series. Springer, Cham, pp. 567–587. doi: 10.1007/978-3-030-15604-6_35.

Thomann, R.V. and Mueller, J.A. (1987). Principles of surface water quality modeling and control. Harper-Collins, New York.

Tian, B., Wu, W., Yang, Z. and Zhou, Y. (2016). Drivers, trends, and potential impacts of long-term coastal reclamation in China from 1985 to 2010. *Estuarine, Coastal and Shelf Science*, 0272–7714170, 83–90. doi: 10.1016/j.ecss.2016.01.006.

Tiller, R., Brekken, T. and Bailey, J. (2012). Norwegian aquaculture expansion and Integrated Coastal Zone Management (ICZM): Simmering conflicts and competing claims. *Marine Policy*, 36, 1086–1095. doi: 10.1016/j.marpol.2012.02.023.

Tissier, M.D., Le, A.A. and Hills, J.M. (2010). Ocean and coastal management practitioner training for building capacity in ICZM. *Ocean and Coastal Management*, 53, 787–795. doi: 10.1016/j.ocecoaman.2010.10.018

UN-Habitat (2016). Urbanization and development: Emerging futures. UN Habitat World Cities Report 2016 http://wcr.unhabitat.org/main-report/, Accessed date: 15 November 2020.

USEPA (1986). Ambient Water Quality Criteria for Bacteria. United States Environmental Protection Agency, Washington.

USEPA (2012). Recreational Water Quality Criteria. United States Environmental Protection Agency, Washington, 2012.

USEPA (2013). WASP7, a Hydrodynamic and Water Quality Model. United States Environmental Protection Agency, Washington.

VCE (1998). AQUASEA Tidal flow in Estuaries and Coastal Areas Lake Circulation Transport Modelling, Vatnaskil Engineering Consultant, Iceland.

Wang, W., Liu, H., Li, Y. and Su, J. (2014). Development and management of land reclamation in China. *Ocean and Coastal Management*, 102, 415–425.

Wang, X.G., Sua, F.Z., Zhang, J.J., Cheng, F., Hua, W.Q. and Ding, Z. (2019). Construction land sprawl and reclamation in the Johor River Estuary of Malaysia since 1973. *Ocean and Coastal Management*, 171, 87–95.

Westcot, D.W. (1997). Quality control of wastewater for irrigated crop production. Food and Agriculture Organization of the United Nations, Rome.

Wheeler, P.J., Peterson, J.A. and Gordon-Brown, L.N. (2011). Spatial decision support for integrated coastal zone management (ICZM) in Victoria, Australia: Constraints and opportunities. *Journal of Coastal Research*, 27(2), 296–317.

WHO (1989). Health Guidelines for the Use of Wastewater in Agriculture and Aquaculture, Technical Report Series No. 2009/3 778. World Health Organization, Geneva.

Wu, J., Long, S., Das, D. and Dorner, S.M. (2011). Are microbial indicators and pathogens correlated? A statistical analysis of 40 years of research. *Water Health*, 9(2), 265–278.

Xia, S.X., Yu, X.B., Millington, S., Liu, Y., Jia, Y.F., Wang, L.Z., Hou, X.Y. and Jiang, L.G. (2017). Identifying priority sites and gaps for the conservation of migratory waterbirds in China's coastal wetlands. *Biological Conservation*, 210, 72–82.

Xu, Y., Cai, Y., Sun, T., Yang, Z. and Hao, Y. (2018). Coupled hydrodynamic and ecological simulation for prognosticating land reclamation impacts in river estuaries. *Estuarine, Coastal and Shelf Science*, 202, 290–301.

Yue, Q., Zhao, M., Yu, H., Xu, W. and Ou, L. (2016). Total quantity control and intensive management system for reclamation in China. *Ocean and Coastal Management*, 120, 64–69.

Zhu, G., Xie, Z., Xu, X., Ma, Z. and Wu, Y. (2016). The landscape change and theory of orderly reclamation sea based on coastal management in rapid industrialization area in Bohai Bay, China. *Ocean and Coastal Management*, 133, 128–137.

Appendix

Table A.1:   Some Water quality parameters in raw sewage.

Parameters (units)	Mean concentration (range of mean)	Remarks
Faecal coliform	8.3 (0.3–49)	21 U.S. cities
(10^6 num /100 ml)[b]	30 (69)[a]	2 Houston plants – 14 days
	120	Lima, Peru
	40[d]	Penang, Malaysia
Total Coliform	21.9(1.6–47.4)	14 U.S. cities
(10^6 num/100 ml)[b]	412 (1011)[a]	2 Houston plants – 14 days
	(70–733)	Mexico
	200 or 350[e]	Area of Rio de Janeiro, Brazil
	180	Lima, Peru
BOD_5 (mg/l)[c]	112	City of Davis, California
	184	San Diego, California
	200[e]	Area of Rio de Janeiro, Brazil
Suspended Solids	185	City of Davis, California
(mg/l)[c]	200	San Diego, California
	250[e]	Area of Rio de Janeiro, Brazil

Note: [a]Standard deviation.
Source: From [b]Thomann and Mueller (1987); [c]Pescod (1992); [d]Koh *et al.* (1997); [e]Ludwig (1988).

Table A.2:   Estimated per capita contribution of indicator faecal coliforms from human beings and some animals.

Animal	Average indicator number (10^6) per gram of faeces	Average contribution (10^6)/capita / day
Chicken	1.30	240
Cow	0.23	5400
Duck	33.00	11000
Human	13.00	2000
Pig	3.30	8900
Sheep	16.00	18000
Turkey	0.29	130

Source: From Westcot (1997).

Table A.3:   Approximate removal efficiencies of sewage parameters from municipal waste treatment.

Parameters	Treatment	Approximate % removal
Faecal Coliform[a]	Primary (sedimentation)	75
	Secondary (activated sludge & chlorination)	99
Total Coliform[b]	Primary(without chlorination)	25–75
	Secondary (without chlorination)	90–99
	Secondary (with chlorination)	99.9–99.99
BOD_5	Preliminary (using 1.0 mm aperture milliscreens)	20[c]
	Primary	25–50[d]
Suspended Solids	Preliminary (using 1.0 mm aperture milliscreens)	10[c]
	Primary	50–70[d]

Source: From [a]Koh *et al.* (1997); [b]Thomann and Mueller (1987); [c]Ludwig (1988); [d]Pescod (1992).

Table A.4:   Some reported overall decay rates, k for faecal coliforms.

Coliforms	k (day^{-1})	Remarks
Total coliform	1–5.5	Freshwater – summer (or 20°C), seven locations
	0.8	Average freshwater, 20°C
	1.4 (0.7–3.0)	Seawater, 20°C
	48 (8–84)	From 14 ocean outfalls (variable temperature)
Total or faecal	0–2.4	New York Harbor Salinity: 2–18‰ Dark Samples
	2.5–6.1	New York Harbor Salinity: 15‰ Sunlit Samples
Faecal coliform	37–110	Seawater, sunlit
E. coli	0.08–2.0	Seawater, 10–30‰

Source: From Thomann and Mueller (1987).

Table A.5: T_{90} values determined from actual sewage effluent/seawater fields in tropical or semi-tropical waters.

Location	T_{90} values, hours
Honolulu, Hawaii	0.75 or less
Mayaguez Bay, Puerto Rico	0.7
Rio de Janeiro, Brazil	1.0
Nice, France	1.1
Accra, Ghana	1.3
Montevideo, Uruguay	1.5
Santos, Brazil	0.8–1.7
Fortaleza, Brazil	1.3 ± 0.2
Maceio, Brazil	1.35 ± 0.15

Source: From Ludwig (1988).

Table A.6: Water quality criteria for Malaysia, America and China.

Country	MALAYSIA[i]			AMERICA[ii]		CHINA[iv]				
Parameters (units)	Classes[a] I	IIA	III	Fishing	Shell fishing	Classes[h] II	III	Classes[i] I	II	Standard[j]
Faecal Coliform* (count/100ml)	10	100	500 (2000)#	100[b] 1000[c]	14[f,iii]			≤200 (14)†		
Total Coliform (count/100ml)	100	1000	50000	5000[d] 1000[e]	70[g]		≤1000	1000 (70)†		500 (50)†
BOD$_5$ (mg/l)	1	3	6			≤3	≤4	≤1	≤3	≤5
DO (mg/l)	7	5–7	3–5			>6	>5	>6	>5	>5
Suspended Solids (mg/l)	25	50	150					≤10	100	≤10

Notes: [a]DOE Interim National Water Quality Standard:
Class I   Fishery I – very sensitive aquatic species.
Class IIA   Fishery II – sensitive aquatic species.
Class III   Fishery III – common, of economic value, and tolerant species.
[b]North Carolina   [c]South Carolina
[d]Massachusetts   [e]New Hampshire
[f]Venezuela, Mexico and USEPA   [g]United States
[h]Environmental Quality Standards for Surface Water:
Class II is mainly applicable to the protected areas for rare fishes, and the spawning fields of fishes and shrimps.
Class III is mainly applicable to the protected areas for the common fishes.
[i]Marine Water Quality Standard:
Class I is applicable to the marine fishery areas, marine nature reserves and the protected areas for rare fishes.
Class II is applicable to the spawning fields of aquacultures.
[j]Water Quality Standard for Fisheries
* = Geometric mean
# = Maximum not to be exceeded
† = Standard for shellfishing
Source: From [i]UM/UKM/USM/UPM/UTM Joint Water Quality Consultancy Group (1994); [ii]Thomann and Mueller (1987); [iii]USEPA, Koh *et al.* (1997); [iv]Koh, H.L. and Lee, H.L. (2001). Sewage treatment and discharge: A global review. Report submitted to Lyonnaise Des Eaux, Paris, France.

Table A.7: Sewage Effluent Standards Parameter Limits For Standard A and Standard B Effluent.

	Parameter	Unit	Standard A	Standard B
(i)	Temperature	°C	40	40
(ii)	pH	–	6.0–9.0	5.5–9.0
(iii)	BOD$_5$ at 20°C	mg/L	20	50
(iv)	COD	mg/L	50	100
(v)	Suspended Solid	mg/L	50	100
(vi)	Mercury	mg/L	0.005	0.05
(vii)	Cadmium	mg/L	0.01	0.02
(viii)	Cromium Hexavalent	mg/L	0.05	0.05
(ix)	Arsenic	mg/L	0.05	0.1
(x)	Cianide	mg/L	0.05	0.1
(xi)	Plumbum	mg/L	0.10	0.5
(xii)	Cromium, Trivalence*	mg/L	0.20	1.0
(xiii)	Tembaga*	mg/L	0.20	1.0
(xiv)	Mangan*	mg/L	0.20	1.0
(xv)	Nickel*	mg/L	0.20	1.0
(xvi)	Timah*	mg/L	0.20	1.0
(xvii)	Zinc	mg/L	1.0	1.0
(xviii)	Boron	mg/L	1.0	4.0
(xix)	Besi (Fe)	mg/L	1.0	5.0
(xx)	Phenol**	mg/L	0.001	1.0
(xxi)	Free chlorine**	mg/L	1.0	2.0
(xxii)	Sulfide	mg/L	0.50	0.50
(xxiii)	Oil and grease	mg/L	tak boleh dikesan	10.0

Notes: *If two or more of these metals are present in the effluent, the concentration of these metals cannot exceed.

(a) 0.5 mg/L for all metals added up, where Standard A may apply.

(b) 3.0 mg/L for all metals added up, and 1.0 mg/L for all types of solvent forms added, where Standard B can be applied.

** If Standard B is applicable and both phenol and free chlorine are present in the same effluent, the phenol concentration individually should not exceed 0.2 mg/L and the free chlorine concentration should not exceed 1 mg/L.

Chapter 9

Oil and Gas Offshore Disposal of Drilling Wastes

9.1 Introduction

The oil and gas industry is an important source of revenue for Malaysia and several other neighboring countries such as Indonesia. The industry helps achieve the United Nations Sustainable Development Goal (UNSDG) 8: decent work and economic growth by employing tens of thousands of Malaysian workers with decent pay. However, some of its offshore operations may harm the marine environment and its ecosystems, contravening SDG14: life below water. One such offshore operation is the discharge of treated drilling wastes into the open sea, which may have an adverse impact on the marine ecosystems including corals and fish. The UN SDG calls for a balance between the three pillars of sustainable development: economic, social and environmental. Maintaining and enhancing this delicate balance is a daunting challenge. To improve oil and gas supply to match demand, offshore oil and gas exploration and exploitation (OOGEE) has become more commonplace in the recent past. To access oil and gas reserves buried in the seabed, the OOGEE involves drilling of the seabed that generates drill cuttings of various volumes, sizes and characteristics. For Malaysian OOGEE, the drill volume generated in a single drilling operation varies significantly from a low of 141 m^3 (Teh and Koh, 2016) to a high of 8000 m^3 (Koh and Teh, 2011), over a period from several days to slightly more than a month. Most of the cuttings or particles settle onto the seabed near the drilling well,

405

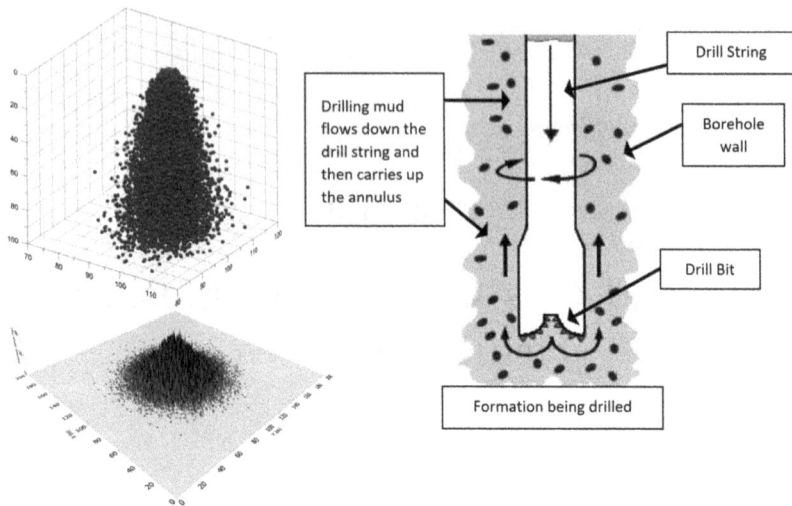

Figure 9.1: Drill cuttings settling onto seabed (left); circulation of drilling fluids (right) (OGP, 2003).

within a radius of about 100 m. The maximum heights of sediment deposited on the seabed due to the discharge of these drill cuttings vary from 0.15 m to 5 m, located usually near the well. However, strong tidal currents may carry the particles farther from the well site to settle on the seabed at distances of more than 100 m from the well. To protect sensitive ecosystems such as corals and fish, the Malaysia Department of Environment (DOE) has protocols known as Environmental Impact Assessment (EIA) put in place to regulate disposal of drill cuttings into the sea to ensure that such disposal will not be harmful to the marine environment. Approvals are required before such disposal can be legally carried out. To save time and cost, these OOGEE drill cuttings are normally discharged into the sea for disposal, upon receiving approval from the DOE. The disposed particles settle downward (Figure 9.1, left top) to form mounds or piles on the seabed (Figure 9.1, left bottom). If the drill well is close to the coral sites, these settled particles (referred to as sediment in the literature) may accumulate to a thickness that may harm the coral communities, as thick sediment layers may smother the corals. The oil and gas industry in Norway has developed a qualitative guideline to recommend that sediment thickness in the vicinity of corals should not exceed the threshold of 6.3 mm (Purser and Thomsen, 2012). However, this threshold guideline is rarely observed.

Drilling uses a rotating drill bit attached to the end of the drill pipe (drill string) to bore into the earth to reach oil and gas deposits. A set of three progressively smaller diameter drill casings from 50 cm to 20 cm is used (Figure 9.1, right). The difference in diameters implies different drilling rates (from 1 to 3 kg/s), which has implications on the rate of drilling waste generation during the drilling operation. The rotating drill bit breaks off small pieces of rocks as it penetrates into rock strata, generating rock cuttings in the process. The rock cuttings typically range in sizes from fine clay to coarse gravel and their composition varies depending on the type of sedimentary rock penetrated by the drill bit. Larger particles would readily settle on the seabed nearer the well, while finer particles will disperse as they settle farther away from the well. Drilling fluids (muds) are pumped down the drill string during drilling to maintain a positive pressure in the well, to cool and lubricate the drill bit, to protect and support the exposed formation in the well and to lift the cuttings from the bottom of the hole to the surface platform. Drilling mud is slurry material consisting of various solids and chemical additives used to control fluid properties such as density. The drillings mud contains toxic chemicals harmful to the marine environment and ecosystems. These rock cuttings and drilling mud (collectively called "cuttings" hereafter) are brought to the surface platform for treatment and eventual disposal. There are three main disposal options: (1) shipping to shore for onshore treatment and disposal, (2) reinjection into existing or new well and (3) discharge offshore overboard after treatment. In most countries that do not have onshore treatment facilities, cuttings are normally discharged into the sea. The reinjection option avoids direct marine and seabed impact and onshore impact, but is dependent on available technology and on suitable seabed geological formation. Reinjecting cuttings into the subsea environment is not considered a viable option technically. Offshore overboard disposal into the sea is a simple and inexpensive option, compared to onshore land disposal. Onshore, the drilling wastes will be classified as hazardous wastes, to be treated with extreme care that will incur high costs. Offshore overboard disposal is the preferred option in Malaysia and elsewhere. When cuttings are released into the sea, cutting piles or mounds (Figure 9.1 left, bottom) will form on the seabed. These pile formations may hinder future undersea operations if the mounds are large and high. For Malaysian OOGEE, this is not a major concern as the volume is not that large. The cuttings, however, may have adverse impacts on corals if the thickness of the pile over the coral beds is persistently high, exceeding 6.3 mm over a long period. This chapter aims to provide the methodology, conduct literature review and develop insights regarding the modeling, assessment and mitigation of potential adverse impacts of released cuttings on the marine and seabed environment. The basic methodology is based upon the numerical simulation of the transport, dispersion and deposition of cuttings to determine the sediment thickness and sedimentation rate on the seabed, particularly near or over the coral beds. The simulated sediment thickness (mm) and the simulated sedimentation rates ($mg/cm^2/day$) are important parameters critical in the assessment of adverse impacts on the corals.

There are four categories of hazards which may be posed to coral reefs due to the discharge of drill cuttings into the marine environment. These hazards are (1) direct mechanical damage; (2) exposure to waste drilling products; (3) exposure to waste production products (produced water); and (4) acute exposure to accidentally released hydrocarbons. To mitigate impacts on the corals due to exposure to waste drilling products, the oil and gas industry has developed risk assessment guidelines stipulating that the maximum cutting depositional depth or thickness should not exceed the threshold of 6.3 mm (Purser and Thomsen, 2012). This chapter is devoted to the modeling, assessment and mitigation of exposure to cuttings, with a focus on the cuttings' adverse impact on the corals. Generally, corals are susceptible to high levels of sediment cover (more than 6.3 mm) over extended periods (weeks). However, many species are able to overcome exposure to low and medium levels of sediment cover over several days. To provide some quantitative perspective, it has been reported that the species *Lophelia pertusa* is able to tolerate moderately high short-term sediment deposition (6.3 mm over 10 days) before coral mortality sets in (Purser and Thomsen, 2012). Mortality is most likely due to oxygen deficiency. Cuttings that settle onto the coral would lead to oxygen deficit or depletion on the coral surface, depending on the thickness deposited and the duration of exposure, as well as on the sea current flow conditions that help to flush off sediment from the coral beds. Upon oxygen depletion, anaerobic microenvironments on the coral surface would form, providing suitable habitats for sulphate-reducing microbial assemblages.

The production of hydrogen sulphate (H_2S) by these assemblages during sulphate reduction would present a potential danger to the coral tissues as sulphide is a cell toxin (Bagarinao, 1992). Sub-lethal effects on *L. pertusa* from exposure to benthic sediments include loss of tissues (Larsson and Purser, 2011), as well as reduced skeletal growth and reduced larval survival (Larsson *et al.*, 2013). Cuttings in the vicinity of OOGEE activities would not cause coral death within 11 days, with sediment thickness of 2 to 3 mm. Although all drilling campaigns are different, the area of thick deposition during a drilling campaign is observed normally within a distance of 100 m. For example, an in-depth survey of drilling campaign showed that visible cutting accumulation seldom exceeded 100 m from the point of discharge (Bell *et al.*, 1998; Gates and Jones, 2012). However, a precautionary principle should prevail for Malaysian OOGEE, mandating a careful study of the potential impacts of released cuttings on the marine environment and ecosystems. This type of study typically would require numerical simulations of the transport, dispersion and deposition of cuttings in the water column and their ultimate settlement onto the seabed. Generally, cuttings may be divided into two main groups according the particle sizes. About 50% of cuttings are "coarse" particles that readily settle onto the seabed to form mounds or piles within a short distance from the well (50 to 70 m at water depth of 100 m). The remaining 50% consists of "fine" particles that would remain in the water column as suspended solids. These suspended solids will eventually settle on the seabed at distances that vary, depending on tidal current and eddy dispersion strength. Strong tidal currents with strong eddy dispersion would carry the particles further away and settle on the seabed far from the well (100 to 200 m, at water depth of 100 m). For simulating the transport, dispersion and eventual deposition of cuttings onto the seabed, an in-house simulation model known as TUNA-PT has been developed (Teh and Koh, 2014). The particle settling process is governed by the cutting characteristics such as settling velocity and coastal conditions such as depth of water, tidal currents and eddy dispersions (Teh and Koh, 2016).

Corals differ greatly in their ability to resist sedimentation, through a variety of mechanisms, often assisted by tidal currents that clean off sediments that settled on coral tissues, albeit at the expense of metabolic energy. Corals can withstand a certain amount of sediment settling over short periods of sedimentation, as this occurs naturally, allowing the coral communities to adapt and evolve (Perry and Smithers, 2010; Rogers, 1977, 1990). Tolerance of coral to sedimentation and water-column suspended solids is related to the intensity, duration and frequency of exposure (Erftemeijer *et al.*, 2012). Scientific literature derived from field-based and laboratory studies has provided some qualitative guidelines regarding the tolerance of coral reef systems to suspended solids and sedimentation. Quantitative regulatory framework regarding critical thresholds is, however, yet to be formulated, as most research on tolerance of corals to suspended solids or sedimentation is rarely quantified. It should be appreciated, however, that over half of coral reef complexes are made up of sediments (Dudley, 2003), suggesting that corals and sediments do coexist in a natural environment due to natural adaptation over time. Long-term maximum sedimentation rates that can be tolerated by different corals have been reported in the literature to have a wide range from 10 mg/cm^2/day to over 400 mg/cm^2/day. This wide range of tolerance provides limited guidelines on quantitative regulatory framework. There is little information in the public literature about the conditions that would prevail in situ during drilling around coral reefs. Little is known about the biological and ecological response and adaptation of corals to sediments created by these operations. Hence, protecting corals from the harmful impacts of released cuttings is a daunting challenge, if the cuttings are discharged near coral beds (within 200 m). Nevertheless, the Malaysian DOE has mandated the protection of the corals and the marine environment and ecosystems, in which numerical simulations are an integral component of the assessment protocol.

Many oil and gas exploration and production wells are situated in the South China Sea (SCS), which is rich in marine life such as corals, mangroves, fishery and other marine resources. Improper treatment and inappropriate disposal of drilling wastes into the SCS marine environment can harm these valuable resources. Hence, offshore disposal of drilling wastes including cuttings is regulated by the Malaysian DOE. The regulatory process for offshore disposal of cuttings from OOGEE wells requires an Environmental Impact Assessments (EIAs) to be undertaken to address and mitigate the potential adverse impacts of cuttings on the marine and benthic environment and ecosystems. This EIA assessment process typically requires environmental modeling as an integral component. Modeling the transport, dispersion and deposition of cuttings in the sea is essential to provide scenario analysis critical to a proper and comprehensive assessment of the impacts of released cuttings on the marine environment and on the seabed marine ecosystems. The cuttings contain particles of various sizes with vastly different settling velocities. About 50% of the cuttings are larger particles with higher settling velocity and they readily settle on the seabed to form a pile within a short horizontal distance from the base of the release point (within 50 to 70 m, at water depth of 100 m). The remaining 50% consist of finer particles with slower settling velocity. These finer particles will settle onto the seabed farther away horizontally from the release point (beyond 100 m, at water depth of 100 m). The ultrafine particles tend to remain in the water column as suspended solids, being carried in the water by tidal currents and eddy dispersion. They settle onto the seabed far away from the base of release point (200 m, at water depth of 100 m). Based upon this general observation, we recommend that drilling wells should be located at least 200 m away from the coral beds, for normal oil and gas cutting discharge with discharge volume of about 1000 m^3. The precise scenario analysis on cuttings potential impacts on the corals should be performed on a case-by-case basis. This scenario analysis is the focus of this chapter.

This chapter is organized as follows: Section 9.2 presents a brief introduction to the cuttings' physical properties such as particle size and settling velocity distributions crucial in the simulation of TUNA-PT. The chemical and toxicological characteristics of the cuttings are briefly summarized to provide indication of potential toxicity impact of cuttings on the marine environment. Section 9.3 presents a regression analysis of a seabed cutting pile's physical dimensions in the North Sea compiled by Bell *et al.* (1998), including pile heights, radii, volumes and water depths. This regression analysis would provide modelers with good insights on conceptualizing relationships between cuttings' physical dimensions, as a first step in the simulation analysis. Based upon this regression, a modeler would be able to formulate a preliminary strategy for simulation of the cuttings' distribution on the seabed. Test simulations would then be performed to verify the preliminary understanding of the seabed distribution of cuttings. A refined strategy would then be conducted to provide good predictions of the cuttings' deposition on the seabed for assessment and mitigation analysis. Cutting pile thickness on the seabed is primarily governed by particle sizes and their associated settling velocities, as well as tidal current velocity and coastal dispersion. Section 9.4 presents the main mathematical formulations of the particle-tracking TUNA-PT model. Section 9.5 performs hypothetical simulations of the cuttings' deposition onto the seabed subject to variations in these three governing parameters. This hypothetical simulation and analysis would provide theoretical guidance on formulating preliminary approaches to performing real simulation study at a chosen site. These site-specific simulations would be the basis for assessment and mitigation of the cuttings' impact on the marine environment, once the site-specific data required have been compiled. This includes tidal flows and eddy dispersion parameters, total mass of cuttings discharged, the discharge schedule and particle size distributions. With the modest experience and expertise gained from Sections 9.5–9.6 proceeds to perform cutting deposition simulations on three selected sites in the North Sea, with data provided by Bell *et al.* (1998) to verify that TUNA-PT can indeed be used to predict cuttings pile formation for the purpose of impact assessment. A major goal of Chapter 9 is to empower the reader to confidently perform simulation, assessment and mitigation of impacts due to cuttings discharged overboard into the sea. In the SCS, particularly off the coast of Sabah, adequate meta-ocean data are readily available from public domain (DHI, 2009). Coupled with site- and project-specific data, Section 9.7 conducts a case study on modeling and assessing the impact of cuttings on the marine environment at a site, known as K-1, off the coast of Sabah. Corals are located within a short distance (ca 100 m) from the proposed drilling site. An in-depth literature review of coral vulnerability and tolerance to suspended solids and sediment depositions is provided in Section 9.8. This literature review provides information and insights on a proper assessment and mitigation suitable for minimizing the adverse impact of released cuttings on the corals, which is presented in Section 9.9.

9.2   Drill Cuttings' Characteristics and Properties

There are two broad categories of drilling fluids used: Water-Based Fluids (WBFs) and Non-Aqueous Drilling Fluids (NADFs). For both

Table 9.1:   NADF classification groups and descriptions (ERM, 2009).

Classification	Base Fluid	Aromatics	Aromatic (%)	PAH (%)
Group I	Diesel and Conventional Mineral Oil	high aromatic content	>5	>0.35
Group II	Low toxicity mineral oil	medium aromatic content	0.5–5.0	0.001–0.35
Group III	Enhanced mineral oil and synthetics (esters, olefins and paraffins)	low to negligible aromatic content	<0.5	<0.001

types, a variety of chemicals are added to modify the properties of the fluids to facilitate drilling and to help bring the drilling mud and cuttings to the surface. NADFs are divided into three groups according to the level of aromatic contents and toxicity (Table 9.1). Group I NADF is largely discontinued now as the low-toxicity Group II fluids become readily available. Recently, Group III fluids, which have low to negligible aromatic contents and toxicity, have been developed to address environmental issues related to overboard discharge and occupational health of drill workers. Water-Based Mud (WBM) is generally considered less harmful compared to NADF as the former contains water, rather than oil as its base fluid. However, WBM may also contain additives (barite) and may include various salts and minerals added to improve performance. A main mitigation measure to reduce toxicity is the use of a solid control system including a dryer to minimize oil on cuttings to less than 5% by weight. During drilling, fluid or mud is pumped into the drilling string to be mixed with the rock cuttings to facilitate drilling. The combination of muds and cuttings (cuttings) is pumped to the surface for treatment and disposal. The cuttings consisting of particles of various sizes and settling velocities are grouped into ten size classes, as shown in Table 9.2. This distribution of ten classes of particles will be used in the modeling of cutting deposition in this chapter. Larger particles with higher settling velocity will settle onto the seabed nearer the base of the discharge location, while smaller particles with slower

Table 9.2:   Cutting particle diameter and settling velocity (Rye *et al.*, 2006).

Size Class	Diameter (mm)	Weight (%)	Density (tonnes/m^3)	Velocity (m/s)	Velocity (m/day)
1	0.007	10	2.4	1.90E-05	1.7
2	0.015	10	2.4	8.80E-05	7.6
3	0.025	10	2.4	2.50E-04	21.2
4	0.035	10	2.4	4.80E-04	41.6
5	0.05	10	2.4	9.80E-04	84.9
6	0.075	10	2.4	2.20E-03	191
7	0.2	10	2.4	1.60E-02	1356.5
8	0.6	10	2.4	5.70E-02	4898.9
9	3	10	2.4	2.10E-01	17988.5
10	7	10	2.4	3.20E-01	27483.8

settling velocity will settle farther away. The settled particles form a mound with an inverted V shape, with higher pile height at the center of the mound, as shown in Figure 9.1 left bottom.

9.3   Analysis of Pile Dimensions

The disposed cuttings will eventually be deposited on the seabed to form a "pile" beneath the platform. Most of the cuttings deposited on the seabed are expected to be concentrated around the base of the release location, with the pile patterns determined by the overall hydrodynamic flows and cutting settling velocities. Pile height or thickness is at maximum directly below or near the release location, gradually decreasing with increasing distance away. Cuttings may be discharged above, at or below the sea surface from the platform to achieve optimal performance. Larger cuttings will settle quickly down the water column and be deposited on the seabed near the release location. Smaller cuttings will be dispersed horizontally before gradually settling down on the seabed at distances farther away. Very fine particles will remain in the water column for a long period as suspended solids. Marine lives such as pelagic fish may become exposed to these suspended solids and attached toxic substances. An oxygen demand may also be exerted in the water column. The concentration of suspended solids in the water column is primarily due to drilling fluids since these smaller particles have lower settling velocity and remain suspended in the water column for longer periods of time. In contrast, rock cuttings settle quickly on the seabed because of their larger particle sizes. Large volume of released cuttings may form cutting piles with high thickness or heights in the vicinity of the well. Persistent cutting heights of more than 6.3-mm thickness may result in the smothering of benthic organisms consisting of mainly sessile species, including corals. Smothering impacts are normally limited to a small area around the drill well. Different fauna groups are tolerant to different degrees of smothering; for example, burrowing organisms are more tolerant compared to surface living bottom feeders.

Various physical dimensions of mapped drill cutting piles on the seabed resulting from discharge of cuttings at various sites in the North Sea over a long period have been reported in the literature (Bell *et al.*, 1998). Regression analysis is performed on this set of data as plotted in Figure 9.2, to provide relationships between any two selected dimensions. For convenience of analysis, pile areas are converted to a circular pile radius. The depths of the seabed vary

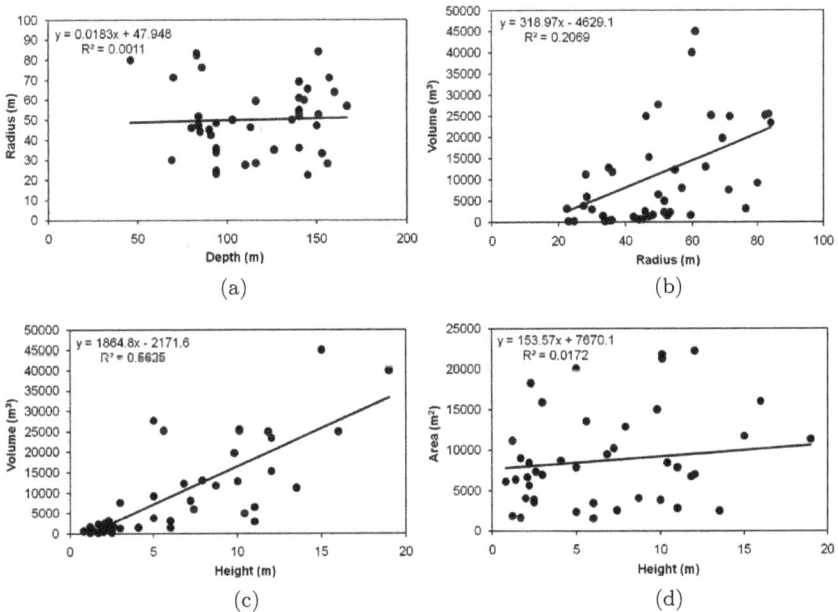

Figure 9.2: Regression between various parameters.
Source: Bell *et al.*, 1998.

from 50 to 170 m, while the pile radius varies between 20 and 80 m. The pile radius does not correlate well with the depth of the seabed ($R^2 = 0.0011$), with a mean radius of 50 m. Intuitively, pile radius should be positively correlated with water depth, as deeper water would allow more time for the cuttings to spread over a larger area. This would be the case for a single campaign of drilling discharge of cuttings, where cutting size distribution, marine hydrodynamic conditions and water depth are similar. However, for a compilation over multiple campaigns of cutting discharges that occurred over a long period, this lack of positive correlation could be the result of a compilation of data over multiple sites with vastly different geological, marine, environmental and operational conditions. For example, drill cuttings from different geological origins have vastly different properties that alter their spread and settling behavior. Pile volume increases linearly with pile radius with $R^2 = 0.2069$, as should be expected. Pile volume exhibits strong linear correlation with pile heights with $R^2 = 0.5625$. As more cuttings are discharged, these cuttings will continue to pile on top of previously deposited cuttings, increasing both heights and volume simultaneously. Finally, pile areas (hence radius) are not well correlated with pile heights with weak correlation $R^2 = 0.0172$. Pile heights are built up from additional deposition of cuttings on top of previously deposited cuttings, which would not increase pile radius unless the heights have exceeded a limit. These observations regarding the correlations will provide valuable insights on simulation preparation and analysis. Further, this set of data from the North Sea will be used to calibrate and validate TUNA-PT.

9.4   Mathematical Model

9.4.1   *TUNA-PT*

Particles of smaller diameters (less than 1 mm) will be transported horizontally by ocean currents and eddy diffusion away from the discharge platform. This advection–diffusion simulation will be performed by TUNA-PT, a conservative model developed in-house to track particles over long distances. The word "conservative" refers to a property of the simulation model that conserves mass, by tracking each and every particle in the flow field. This is a desirable property for a simulation model. Currently, there are several models (Hannah *et al.*, 2003) for calculating particle concentration in the water column and on the seabed, including bblt (Drozdowski *et al.*, 2004), MUDMAP (Spaulding *et al.*, 1994) and ParTrack (Rye *et al.*, 1998, 2006). However, no single fully validated and universally applicable drilling waste transport/deposition model is currently available (Niu *et al.*, 2009). Developed by Bedford Institute of Oceanography, the bblt (Benthic Boundary Layer Transport) model is based upon tracking particles. Similarly, ParTrack is also based upon tracking each individual particle released into the sea from the release platform. Most particle tracking models keep track of these released particles as they move in the ocean and sink to the seabed. Since each particle is tracked and accounted for, conservation of mass is assured. This "conservative" particle-based approach is adopted by TUNA-PT to simulate the impact of drilling wastes on the ocean environment. The models aggregate the spatially distributed particles to calculate concentration by adding the packets of particles that enter a computation box (or grid) in a user-specified volume (in the water column) or area (on the seabed) and dividing the aggregated mass by box volume or box area to produce concentration in the water column (g/m^3) or on the seabed (g/m^2), respectively.

At each time step, a prescribed mass of particles is released, one for each particle class. Particles of lowest diameters (less than 0.05 mm) that remain in suspension for a long time may be merged into one single class. Four groups of larger sized particles are included. They have settling velocity exceeding 1 cm/s or 1 km/day and they settle to the seabed relatively close to the source (within a radius of ~100 m), the distance depending on the depth, settling velocity and current velocity. To provide good resolution for particle settling near the release location, a length scale of the order of 100 to 200 m, with grid size of 1 m, is appropriate. Finer materials will not sink readily and will be transported to a long distance by currents before they settle on the seabed. Generally, particles of diameter less than 0.05 mm, having settling velocity below 10^{-3} m/s, will remain in suspension for a long distance. The transport and deposition of particles are highly sensitive to particle sizes. This implies that several length and time scales will be involved in the calculation. Hence, TUNA-PT will incorporate three or more length scales in order to provide optimal and adequate resolutions. For water column concentration, a depth of plume trapping of about 20 m to 25 m is assumed, based upon theoretical estimates. This far-field plume will travel long distances, for which a length scale of 10 km, with grid sizes of 100 m, is required. In the near-field travel domain, appropriate resolution requires a length scale of the order of 1 km, with grid size of 10 m.

9.4.2   *Settling velocity*

Existing literature clearly indicates that all models presently in use are sensitive to settling velocity (Carles and Bryden, 1999) and ocean environment, as should be surmised from the physics of sedimentation. For example, Niu *et al.* (2009) reported that increasing settling velocity 15 times (from 0.1 cm/s to 1.5 cm/s) may result in a 40 times increase (from 0.5 g/m^2 to 20 g/m^2) in the peak concentration. Three formulae can be used to calculate settling velocity, depending on the buoyancy index (B_i). The sinking velocity (v_s) is calculated as a function of the diameter of the particle (d), the specific gravity of the particle (γ_s), the specific gravity of the ambient fluid (γ), gravitational acceleration (g) and the fluid kinematic viscosity (v). A kinematic viscosity of 1.858×10^{-6} m^2/s for salt water is used. As given in CERC (1984), the fall velocity depends upon B_i, where

$$Bi = [(\gamma_s/\gamma) - 1]gd^3/v^2 \qquad (9.1)$$

The buoyancy index determines the appropriate equation to be used for v_s as follows:

$$v_s = (\gamma_s/\gamma - 1)gd^2/18v \, (B_i < 39) \qquad (9.2)$$

$$v_s = [(\gamma_s/\gamma - 1)g]^{0.7}d^{1.1}/6v^{0.4} \, (39 < B_i < 10^4) \qquad (9.3)$$

$$v_s = [(\gamma_s/\gamma - 1)gd/0.91]^{0.5}d^{1.1} \, (B_i > 10^4) \qquad (9.4)$$

Once released, the motions of particles depend on settling velocity, ambient–current advection and ocean diffusion. Thus, the 3D velocity vector ($\vec{v}$) of a particle is the sum of an advection velocity ($\vec{v}_a$) and a random motion ($\vec{v}_r$) vector (where the three components of diffusive velocity v_r are scaled by the horizontal and vertical diffusion coefficients, respectively):

$$\vec{v} = \vec{v}_a + \vec{v}_r \tag{9.5}$$

The advection velocity vector $(\vec{v}_a)$ contains a horizontal velocity in two dimensions imported from an external database. The vertical component is derived from the particle settling velocity. Horizontal dispersion or diffusion is parametrized through horizontal diffusivity parameters, K_x and K_y. For the particle-based approach, the horizontal diffusivity is related to the time that has elapsed since the release of the particle according to the relation (Bowden, 1983)

$$K_x = K_y = 1.17 \times 10^{-6} t^{1.34} \tag{9.6}$$

which is valid for $K_{x,y}$ in m^2/s and t in seconds. For long distance transport, in order to limit the size of K_x and K_y, an upper bound is chosen between 1 and 10 m^2/s for near-coastal regions. The vertical diffusion coefficient is specified by the user, between 10^{-4} m^2/s (calm conditions) and 10^{-2} m^2/s (rough weather conditions). A particle-based random walk algorithm (Reed, 1980) is used to simulate both horizontal and vertical dispersion in the water column. Particles diffuse with velocities v_i given by the formula

$$v_i = R^* \sqrt{6K_i/\Delta t}. \tag{9.7}$$

In the above, the subscript $i = 1, 2, 3$ corresponds to the horizontal and vertical directions x, y, z, respectively, K_i is the associated diffusivity, Δt is the model time step and R^* is a random variate uniformly distributed over the interval $-1.0 \leq R^* \leq 1.0$. The value of K_z is usually selected between 10^{-4} m^2/s (calm conditions) and 10^{-2} m^2/s (rough weather conditions). The values of K_x and K_y are computed from Eq. (9.6).

9.5 Hypothetical Cuttings' Simulation

The cuttings' particle transport mechanism consists of horizontal movements carried by tidal currents and horizontal diffusion via eddy currents followed by vertical downward deposition onto the seabed (Rye *et al.*, 1998, 2006; Eames *et al.*, 2002). Currently, there are several models for calculating particle concentration in the water column and on the seabed, including bblt (Niu *et al.*, 2009) and Par-Track (Rye *et al.*, 1998, 2006). The bblt (Benthic Boundary Layer Transport) model and the ParTrack model are based upon tracking each individual particle released into the sea from a release platform (Figure 9.1 left). Most particle-tracking models keep track of each of these released particles as they move in the ocean and sink to the seabed (Figure 9.1, left). Since each particle is tracked and accounted for, conservation of mass is assured. We will adopt this "conservative" particle-based approach in TUNA-PT to simulate the impact of cuttings on the marine environment and the ecosystems. The model aggregates the spatially distributed particles to calculate concentration in a specified volume or over a specified area. The model sums up the mass of packets of particles that enter a computation box in a user-specified volume (in the water column) or area (on the seabed) and divide the aggregated mass by the box volume or by the box area to produce concentration in the water column (mg/L) or on the seabed (mg/cm^2), respectively. Existing literature clearly and correctly indicates that all simulation results are sensitive to settling velocity and the marine environmental conditions such as tidal currents, eddy dispersion and water depths (Bell *et al.*, 1998; Carles and Bryden, 1999). In this and subsequent sections, available data on Malaysian OOGEE and coastal water environment (DHI, 2009) will be used in the simulations. We present simulation analyses of particle transport, dispersion and deposition, subject to tides, coastal dispersion and particle settling. The aim is to illustrate the science and art of modeling and assessing the potential impact of drill cuttings on the marine environment and ecosystems, with the goal of protecting them from harm by mitigation measures. It has been correctly reported that simulation results are highly sensitive to settling velocity of the particles and to coastal hydrodynamics and diffusion (Carles and Bryden, 1999; Niu *et al.*, 2009). Hence, we analyze the dependency of simulated pile formation characteristics subject to variations of these two important parameters, namely, settling velocity V cm/s and dispersion coefficients D m/s$^{1/2}$. For clarity of analysis, we momentarily set the net horizontal current velocity to u = 0.0 m/s. We will perform two sets of scenarios, for which we choose the same input: (a) water depth = 100 m and (b) cutting volume = 1000 m^3 with density of 2400 kg/m^3, giving a total cutting mass = 2.4×10^6 kg. Based upon analysis of Malaysian coastal dispersion studies conducted in the past (Koh *et al.*, 1991, 1997), the vertically averaged dispersion coefficient D m/s$^{1/2}$ varies within the range of 0.3 to 0.8 m/s$^{1/2}$. For settling velocity V, we choose a range of values with V = 1, 2 and 4 cm/s, representing the medium to larger particles that would readily settle on the seabed. The first set of demonstrative examples consists of particle settling subject to settling velocity of 1 cm/s, and with three dispersion coefficients D = 0.3 m/s$^{1/2}$, 0.6 m/s$^{1/2}$ and 0.8 m/s$^{1/2}$. The second set consists of a fixed dispersion coefficient D = 0.6 m/s$^{1/2}$, and with three settling velocities of 1, 2 and 4 cm/s. Through these sets of six demonstrative examples, we show how settling velocity and dispersion coefficients influence the particle deposition patterns on the seabed. The simulation results would provide preliminary understanding of particle settling useful in any real simulation analysis.

9.5.1   *Effects of diffusion*

For the first set of demonstrations, settling velocity is chosen as 1 cm/s. Three values of vertically and tidally averaged horizontal diffusion coefficients D m/s$^{1/2}$ used for this simulation are (a) D = 0.3 m/s$^{1/2}$, (b) 0.6 m/s$^{1/2}$ and (c) 0.8 m/s$^{1/2}$, for both x and y directions. Because of the presumed absence of tidal lows, the simulated sediment distributions on the seabed have the appearance of circles, with peak thickness located at the well. Figure 9.3 shows the simulated distributions of particles deposited on the seabed in 3D (top row), 2D (middle row) and 1D (bottom row). We choose water depth = 100 m, volume = 1,000 m^3 and settling velocity = 1 cm/s. Here, the arrangements are D = 0.3 (left column), D = 0.6 (middle column) and D = 0.8 m/s$^{1/2}$ (right column). A small dispersion of D = 0.3 m/s$^{1/2}$ would result in a maximum pile height of 0.52 m at the discharge location, and a radius of 50 m. Increasing D to 0.6 m/s$^{1/2}$ would decrease the maximum pile height to 0.17 m while increasing the pile radius to 75 m. Further increase in D to 0.8 m/s$^{1/2}$ would further decrease the maximum pile height to 0.10 m while further increasing the pile radius to 100 m. This suggests that we should maintain a minimum separation of more than 100 m between the drilling well and the coral beds to protect the corals.

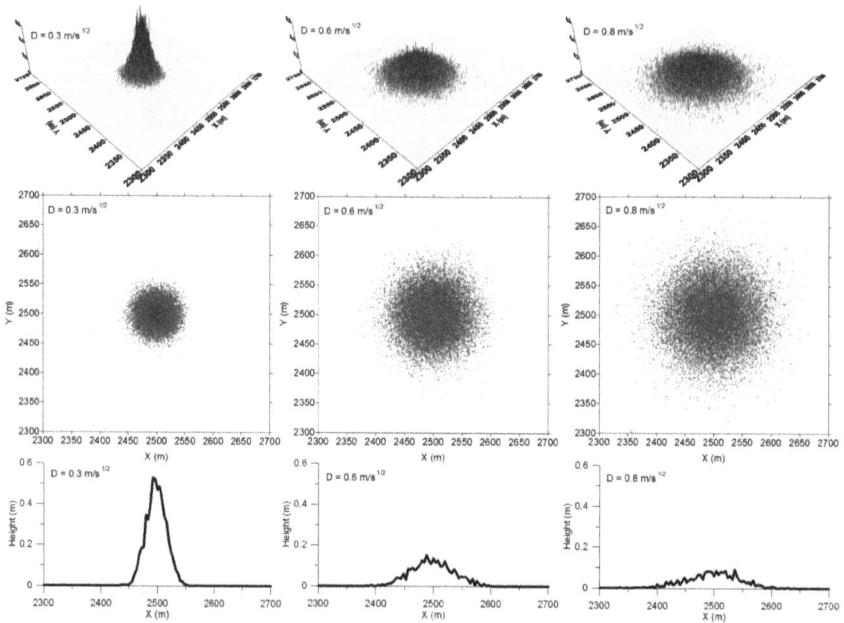

Figure 9.3: Simulated distributions of particles deposited on seabed in 3D (top row), 2D (middle row) and 1D (bottom row), given water depth = 100 m, volume = 1,000 m^3 and settling velocity = 1 cm/s. Here, the arrangements are D = 0.3 (left column), D = 0.6 (middle column), D = 0.8 m/s$^{1/2}$ (right column).

9.5.2   *Effects of settling velocity*

For the second set of demonstrations, the dispersion coefficient is fixed at a moderate value of 0.6 m/s$^{1/2}$, for both x and y directions. Three values of settling velocity are used for this simulation: (a) V = 1 cm, (b) V = 2 cm/s and (c) V = 4 cm/s. Once again, the deposited particles concentrate around the well location and form circular patterns, due to the presumed absence of tidal currents. Figure 9.4 shows the simulated distributions of particles deposited on seabed in 3D (top row), 2D (middle row) and 1D (bottom row), given water depth = 100 m, volume = 1,000 m^3 and dispersion of D = 0.6 m/s$^{1/2}$. Here, the arrangements are V = 1 cm/s (left column), V = 2 cm/s (middle column) and V = 4 cm/s (right column). A settling velocity of V = 1 cm/s would result in a maximum pile height of 0.17 m and a radius of 75 m. Increasing V to 2 cm/s would increase the maximum pile height to 0.28 m while decreasing the

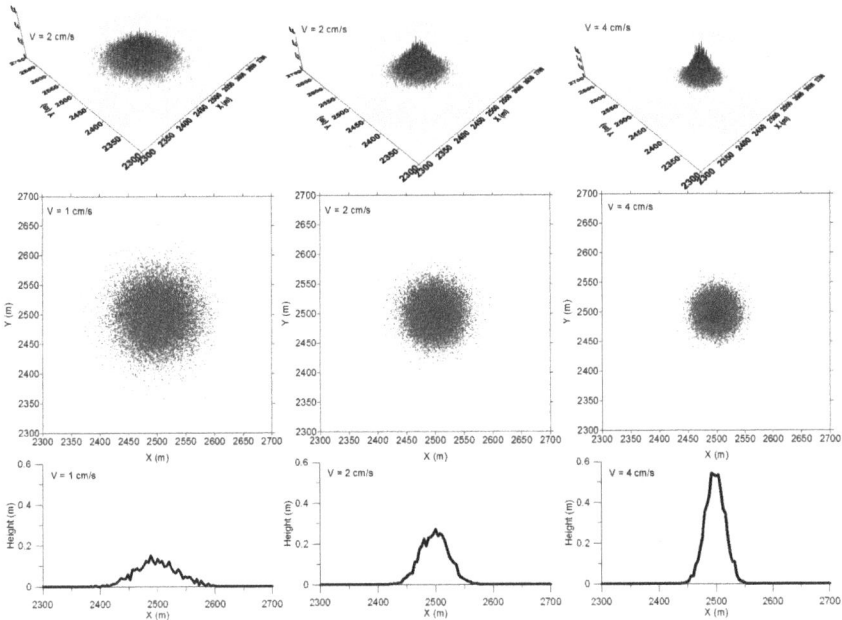

Figure 9.4: Simulated distributions of particles deposited on seabed in 3D (top row), 2D (middle row) and 1D (bottom row), given water depth = 100 m, volume = 1,000 m^3 and dispersion of D = 0.6 m/s$^{1/2}$. Here, the arrangements are V = 1 cm/s (left column), V = 2 cm/s (middle column) and V = 4 cm/s (right column).

pile radius to 60 m. Further increase in V to 4 cm/s would further increase the maximum pile height to 0.55 m while further decreasing the pile radius to 50 m. The simulated radii of the deposited particles appear to fit well with the simulated radii to be presented in the next section for three case studies of the North Sea.

9.6   Two Simulation Case Studies in the North Sea

Three drill cutting disposal sites in the North Sea compiled by Bell *et al.* (1998) are chosen for TUNA-PT simulation analysis. Table 9.3 gives the water depth, pile area, height, volume and radius at each of the three sites called Brae A, N.W. Hutton and Ninian N. For demonstration purposes, we choose one averaged settling velocity = 1 cm/s. The net tidally averaged currents are set to 0.0 m/s for

Table 9.3: Dimensions of mapped drill cutting piles at three selected sites (Bell *et al.*, 1998).

Name of site	Water depth (m)	Pile area (m^2)	Pile height (m)	Pile volume (m^3)	Pile radius (m)
Brae A	113	6746	11.8	25000	46.3
N.W. Hutton	145	13514	5.6	25226	65.6
Ninian N	140	11700	15.0	45000	61.0

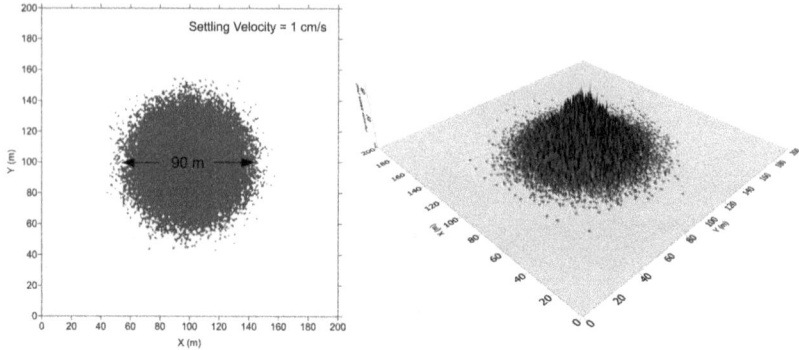

Figure 9.5: Distribution of particles on the seabed at site Brae A in 2D (left) and 3D (right).

all sites. The tidally averaged dispersion D m/s$^{1/2}$ is 0.35 m/s$^{1/2}$ for all sites. Figure 9.5 shows the distribution of particles on the seabed at site Brae A, in 2D (left) and 3D (right) plots, showing a pile radius of about 45 m. Figure 9.6 displays the vertical profile of the particle heights along the centerline of the pile, with simulation grid sizes of 1 m (left) and 4 m (right), respectively, indicating maximum pile height of about 12 m for Brae A. Figure 9.7 shows the distribution of particles on the seabed at site N.W. Hutton, in 2D (left) and 3D (right) plots, showing a pile radius of about 65 m. Figure 9.8 displays the vertical profile of the particle heights along the centerline of the pile, with simulation grid sizes of 1 m (left) and 4 m (right), respectively, indicating maximum pile height of about 5 m for N.W. Hutton. These two simulation results can indeed replicate observed pile dimensions at the two sites selected. Further, the observed pile dimensions recorded at some other sites can also be replicated by TUNA-PT model simulations. Figure 9.9 shows the distributions of

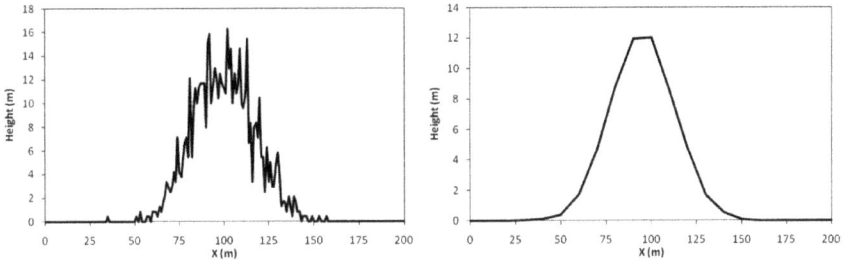

Figure 9.6: Profiles of particle heights at site Brae A, grid sizes of 1 m (left) and 4 m (right).

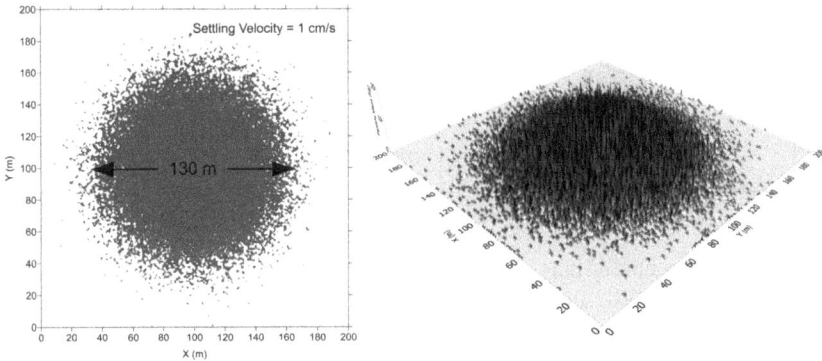

Figure 9.7: Distribution of particles on the seabed at site N.W. Hutton in 2D (left) and 3D (right).

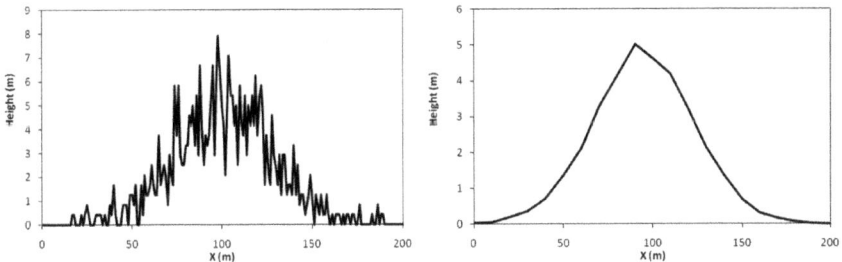

Figure 9.8: Profiles of particles at site N.W. Hutton, grid sizes of 1 m (left) and 4 m (right).

particles deposited on the seabed at the third site Ninian N in 1D, 2D and 3D, indicating good agreement with observed data at site Ninian N shown in Table 9.3. Figure 9.10 shows profiles of particles deposited on the seabed at the site Ninian N with grid sizes of 1 m

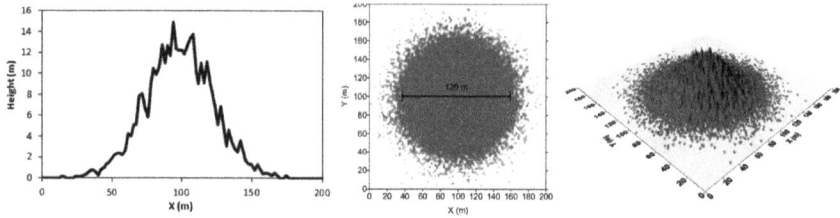

Figure 9.9:   Distributions of particles deposited on seabed at site Ninian N in 1D, 2D and 3D.

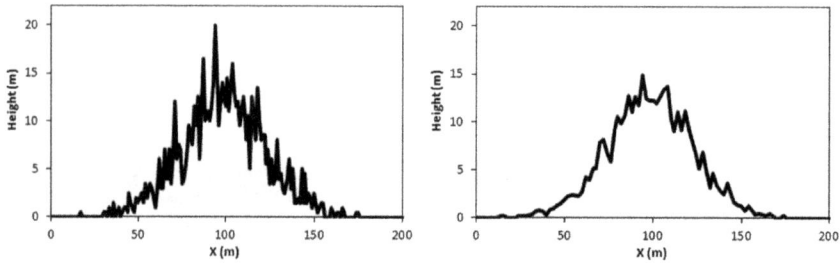

Figure 9.10:   Profiles of particles deposited on seabed at site Ninian N with grid sizes of 1 m (left) and 2 m (right).

(left) and 2 m (right), indicating that a grid size of 2 m is an adequate resolution.

9.7   A Malaysian Case Study at K-1 off Sabah

We proceed to analyze the potential impact of cutting discharge at the drilling site K-1 (5°49'N, 115°8'E) located in the Kimanis Bay in the SCS off the coast of Sabah, for which the coastal hydrodynamics conditions are available (Gao *et al.*, 2015; DHI, 2009; Zu *et al.*, 2008). Drilling generates a volume of 500 m^3 of cuttings, and a total mass of 1.2×10^6 kg, with density of 2400 kg/m^3. The drilling bits consist of three circular sections of decreasing sizes of 17.5, 12.5 and 8.5 inches. The drilling rates over a period of 8 to 9 days for these three sections are 3.5, 2.0 and 1.0 kg/s, respectively, with intermittent breaks of no drillings in between section changes. A coral community is located about 100 m away from the drill site. A conservative and precautionary rule of thumb for cutting impact on corals in the

SCS is that the impact would be negligible if the minimum distance between the drill and coral sites is at least 200 m for cuttings of volume 1,000 m^3 or less. A simulation study was conducted to assess the potential impacts of cutting sedimentation on the environment and ecosystem at the coral site. The purpose is to ensure that the sedimentation on the coral seabed will not have adverse impacts on the coral ecosystem. For this purpose, we perform model simulations via TUNA-PT to determine particle deposition patterns and sediment thickness distributions on the seabed. Based upon the simulation results, we will estimate sediment thickness (mm) and sedimentation rate $(mg/cm^2/day)$ on the coral seabed due to the discharge of 1.2×10^6 kg of cuttings, over a period of 8 to 9 days. These two sediment rates will allow us to assess the potential impacts of sedimentation on coral community. Table 9.4 provides distributions of annually averaged current velocity in the study area, while Table 9.5 provides mean and maximum current speeds at various depths at Kimanis-2 (K2), located just outside the Kimanis Bay entrance. The current velocity at Kimanis-2 is used in this study as the study site K-1 is close to Kimanis-2. The predominant tidal currents flow in the Northeast (NE) and Southwest (SW) directions. This implies that the sediment deposition on the seabed would conform to the flow pattern with predominant orientation in the NE–SW direction. The velocity attains maximum values at the water surface, having mean and maximum values of 0.24 m/s and 0.80 m/s, respectively. The velocity decreases with depth in general, with mean and maximum velocity at depth of −10 m given by 0.1 m/s and 0.16 m/s, respectively. We therefore adopt three simulation scenarios regarding the depth-averaged velocity amplitudes: (A) 0.00 m/s, (B) 0.10 m/s and (C) 0.05 m/s. The temporal depth-averaged velocities are then (A) 0.00 sin (σt) m/s, (B) 0.05 sin (σt) m/s and (C) 0.10 sin (σt) m/s where σ is the frequency of the semidiurnal tide with period of 12.42 hours. The dispersion coefficients used are 0.3 and 0.4 $m/s^{1/2}$.

We consider four scenarios in total to cover a wide spectrum of possibilities regarding the tidal current and dispersion regimes in the study area. The meta-ocean data available from DHI (2009) provides macro-data regarding the tidal conditions in the SCS. The first scenario (Section 9.7.1) represents a sea condition with zero net horizontal velocity as given by 0.0 m/s in all directions at all times. The particles will settle down on the seabed with horizontal dispersion in

Table 9.4:   Percentage distribution of annually averaged current velocities in the study area.

		Direction (45° Sector)							
All Year		**N (%)**	**NE (%)**	**E (%)**	**SE (%)**	**S (%)**	**SW (%)**	**W (%)**	**NW (%)**
Current	0.05	4.37	7.52	6.70	4.56	5.21	5.72	4.13	3.17
Speed	0.15	2.73	12.26	3.99	0.62	2.23	7.89	2.93	1.01
(m/s)	0.25	0.39	10.07	0.74	–	0.17	2.51	0.48	0.01
	0.35	0.03	5.80	0.05	–	0.02	0.24	0.02	–
	0.45	–	2.54	–	–	–	0.07	–	–
	0.55	–	1.20	–	–	–	–	–	–
	0.65	–	0.44	–	–	–	–	–	–
	0.75	–	0.08	–	–	–	–	–	–
	0.85	–	0.07	–	–	–	–	–	–
	0.95	–	0.02	–	–	–	–	–	–

Table 9.5:   Mean and maximum current speeds at various depths at Kimanis-2, at Kimanis Bay mouth.

Depth (m)	**Mean Speed (m/s)**	**Max Speed (m/s)**	**Std Dev (m/s)**
Average	0.16	0.58	0.10
−2	0.24	0.80	0.13
−3	0.23	0.75	0.13
−4	0.21	0.74	0.12
−5	0.19	0.67	0.11
−6	0.18	0.61	0.10
−7	0.16	0.54	0.09
−8	0.14	0.51	0.08
−9	0.12	0.50	0.07
−10	0.10	0.16	0.05

a random manner. This scenario represents a situation where most particles settle down on the seabed near the well, with maximum sediment thickness at the well. This scenario would also represent a best-case scenario for corals, as most particles would have settled on the seabed before they reach the coral site. The random horizontal dispersion is described mathematically by D m/s$^{1/2}$. This would be the case if the tidal currents flow in a random fashion, resulting in no net tidal flows. However, as they descend vertically downward on the seabed, the particles will disperse horizontally by eddy dispersions given by D = 0.3 m/s$^{1/2}$, representing a low vertically averaged dispersion. Since there is no net tidal flow, the particle deposition on the seabed will form a circle, where the radius is dependent on water depth and dispersion. The radius will increase with increase in water depth and increase in dispersion. Deep water allows the particles to take more time to settle on the seabed, thereby permitting the particles to disperse wider horizontally. High dispersion will allow the particles to move over more distance horizontally as the particles settle onto the seabed, resulting in a larger radius. Larger radius implies reduced heights given the same particle volume. The second scenario (Section 9.7.2) reflects the predominant tidal flow regimes in the study area (Table 9.4), given by velocity = 0.10 sin (σt) m/s, flowing in the NE and SW directions, coupled with eddy dispersions of 0.4 m/s$^{1/2}$. As the coral site is oriented in the NE direction from the drill site, the NE–SW flows have the potential of increasing sediment thickness and sedimentation rate at the coral site. This NE flows will bring the particles closer to the coral site, during the NE current flow for half of the tidal cycle of about 6 hours. For the second half of the tidal cycle, the tidal currents flow in the reverse SW direction for about 6 hours. This tidal cycle will result in the sediment deposition on the seabed to have two humps with higher sediment thickness near the well, on the NE and SW sides of the well. The third scenario (Section 9.7.3) reflects the predominant tidal flow regimes in the study area, but given by a smaller velocity = 0.05 sin (σt) m/s, flowing also in the NE and SW directions, coupled with the same eddy dispersions of 0.4 m/s$^{1/2}$. Once again, as the coral site is oriented in the NE direction from the drill site, the NE–SW flows have the potential of bringing more particles to the coral site, thereby increasing sediment thickness and sedimentation rate at the coral site. As it turns out, this third scenario indeed brings more particles closer to the coral sites, resulting in higher sediment thickness at the coral site. The sediment thickness exceeds the recommended threshold of 6.3 mm, adopted by the Norwegian oil and gas industry. However, the duration of the sediment cover is short, of about 6 to 7 days. With no quantitative guideline regarding the sediment thickness deemed harmful to the corals, the potential adverse impact on the corals is subject to interpretations, based upon extensive literature review to be discussed in Section 9.8. Finally, the fourth scenario (Section 9.7.4) reflects sea conditions that incorporated annually averaged tidal flows, with 8-sector velocity profiles as given in Table 9.4. For this scenario, the tidal current flows in eight directions with probability given in Table 9.4 by a long-term study (DHI, 2009). The sediment thickness distribution on the seabed takes the form that reflects the 8-sector velocity flow, with eight uneven sector patterns. The particles are now spread over eight various directions, resulting in lower sediment cover over the coral beds. In short, the most adverse impact on the coral comes from sediments deposited by the NE tidal currents that flow from the drilling well toward the coral site.

9.7.1   *Sediment with velocity amplitude $= 0.0$ m/s, $D = 0.3$ m/s$^{1/2}$*

Where tidal currents flow in a random fashion as given by net flow equal to 0.0 m/s in all directions, the deposited particles will form a circular pattern, with peak sediment thickness at the well and lower thickness at the coral site. Figure 9.11 shows the deposited particles on the seabed in 2D (left) and in 3D (right). The "radius" of the particle formation is about 35 m, with visible particles spreading outside the circle of radius of 35 m. Ninety percent of particles settle within a radius of 35 m. The heavier particles would settle inside this circle, while the finer and lighter particles will spread outside the circle. There are very few particles that settle outside the square of 200 m by 200 m, indicating that the sediment thickness is small at the coral site located outside this square, under the scenario of no net tidal velocity. Hence, the sediment thickness at the coral site is very thin and will not pose any harm to the corals. Figure 9.12 depicts simulated sediment thickness distribution on the seabed along the radial line from the cutting release site (well) to the coral site with current of $0.0\sin(\sigma t)$ m/s. The left figure shows the thickness profile with a grid resolution of $DX = DY = 1$ m, while the right figure shows the thickness profile with a grid resolution of $DX = DY = 4$ m. The maximum sediment thickness is about 0.25 m or 250 mm located directly under the well, 100 m from the corals. At the coral site located at 100 m away from the well, the sediment thickness is 0.05 mm, accumulated over a period of 8 to 9 days. Given that 1 mm $= 240$ mg/cm^2, this

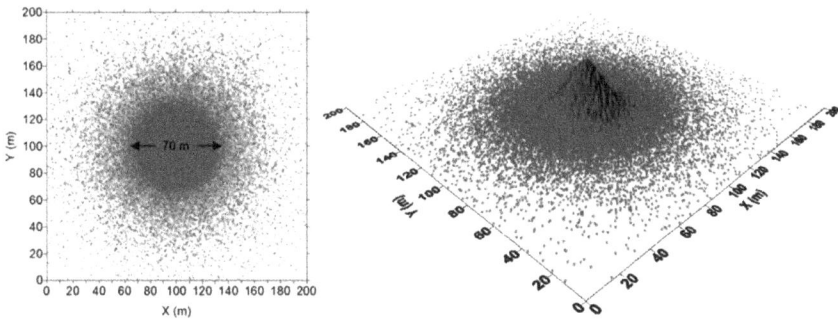

Figure 9.11:   Distribution of seabed particles in 2D (left) and 3D (right) under sea conditions represented by velocity amplitude = 0.0 m/s and D = 0.3 m/s$^{1/2}$.

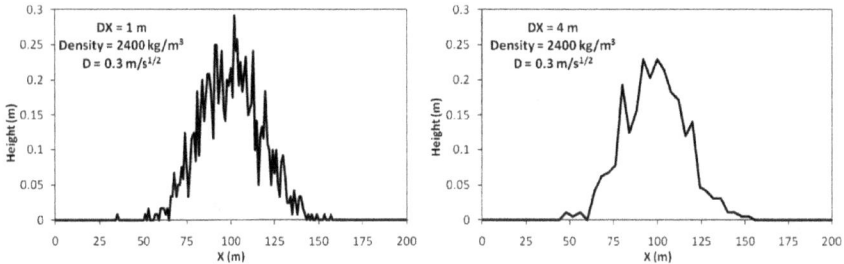

Figure 9.12:   Distribution of sediment thickness on the seabed with sea conditions represented by velocity amplitude = 0. 0 m/s and D = 0.3 m/s$^{1/2}$.

sediment thickness of 0.05 mm over 8 days is equivalent to a sedimentation rate of 0.05 × 240 mg/cm^2/8 days = 1.5 mg/cm^2/day. The small sediment thickness (0.05 mm) and low sedimentation rate (1.5 mg/cm^2/day) are considered insignificant and will not harm the corals, based upon an extensive literature review on corals' resilience to sedimentation to be discussed in Section 9.8. We will deliberate more on this in Section 9.9: Impact Assessment and Mitigation.

9.7.2   *Sediment with velocity amplitude = 0.1 m/s,* $D = 0.4\ m/s^{1/2}$

The second scenario reflects a situation that is probable but not likely to happen as the tidal current is presumed to flow only in the NE–SW direction with velocity of 0.1 sin (σt) m/s for 8 to 9 days consistently. Figure 9.13 shows the sediment deposition thickness in 3D (left) and

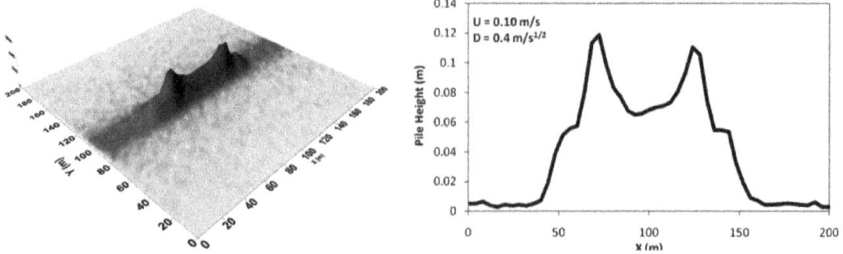

Figure 9.13:   Distribution of sediment thickness in 3D (left), and along plume centerline in the NE–SW direction (right) under sea conditions represented by velocity amplitude = 0.10 m/s, D = 0.4 m/s$^{1/2}$.

in 1D (right), indicating two distinctive peaks of heights 0.12 and 0.11 m. The sediment thickness at the well is 0.065 m. The NE currents bring the particles from the drilling site toward the coral site, resulting in higher sediment thickness at the coral site compared to the previous scenario 9.6.1. The sediment thickness at the coral site is about 0.005 m or 5 mm, which is close to the maximum threshold of 6.3 mm recommended by the Norwegian oil and gas industry. This thickness accumulates gradually over a period of 8 to 9 days, with 0.0 mm thickness at the beginning of the drilling period. The sedimentation rate is 150 mg/cm^2/day. Although this scenario arouses some concern, this maximum sediment thickness of 5 mm that occurs over 1 to 2 days is unlikely to harm the corals. The moderate sedimentation rate of 150 mg/cm^2/day may be reduced by drilling over a longer period, with longer breaks in between change of drilling sections. Lowering the sedimentation rate to 100 mg/cm^2/day would require the drilling operation to be extended to 12 days, with possible increase in cost. A lower sedimentation rate allows (a) tidal currents to help clear the sediment deposited over the coral beds and (b) the corals to self-clean the sediment. This would improve the resilience of corals.

9.7.3   *Sediment with velocity amplitude = 0.05 m/s, D = 0.3 m/s$^{1/2}$*

As with the second Scenario 9.6.2, the third scenario represents a situation that is probable but not likely to happen as the tidal current is presumed to flow only in the NE–SW direction with velocity of 0.05 sin (σt) m/s for 8 to 9 days consistently. Figure 9.14 shows

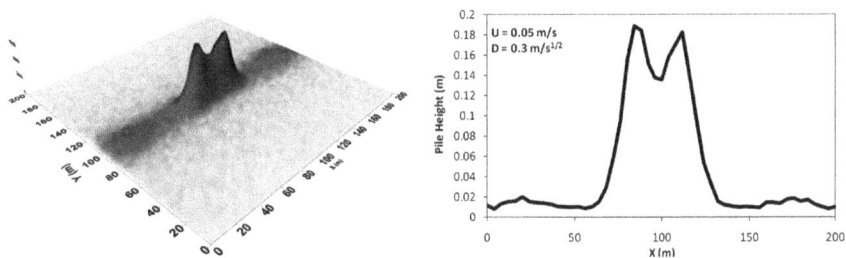

Figure 9.14: Distribution of sediment thickness in 3D (left) and along plume centerline in the Northeast direction (right) under sea conditions represented by velocity amplitude = 0.05 m/s, D = 0.3 m/s$^{1/2}$.

the sediment deposition thickness in 3D (left) and in 1D (right), indicating two distinctive peaks of heights 0.19 and 0.18 m. The sediment thickness at the well is 0.14 m. Compared to scenario 9.6.2, the reduced currents help to focus the sediment deposition close to the well and coral site, resulting in two higher peak sediment thicknesses of 0.19 and 0.18 m. The sediment thickness at the coral site is about 0.018 m or 18 mm, covering an extensive stretch of the coral seabed, accumulated over a period of 8 to 9 days. For about 6 days, the corals are covered with sediment of thickness exceeding 6.3 mm. The corresponding sedimentation rate is 540 mg/cm^2/day, which is considered very high and harmful to corals. The high sediment thickness (18 mm) and high sedimentation rate (540 mg/cm^2/day) are indeed a real concern to coral health. An effective mitigation measure is to move the drill site away from the well by at least 100 m. Further, the well should be positioned so that it is not located in the SW direction from the corals. With this arrangement of the well, the sediment cover over the coral site would not be harmful to the corals as the levels are now low.

9.7.4 Sediment with 8-sector velocity rose, D = 0.4 m/s$^{1/2}$

The fourth scenario represents an annually averaged current flow situation with eight sectors separated by 45°. The tidal currents flow in eight sectors with probability given in Table 9.4. Figure 9.15 shows the sediment deposition thickness in 3D (left) and in 1D (right), indicating a distribution of sediment thickness along eight sectors, following the pattern of current flows, with two distinctive peaks of

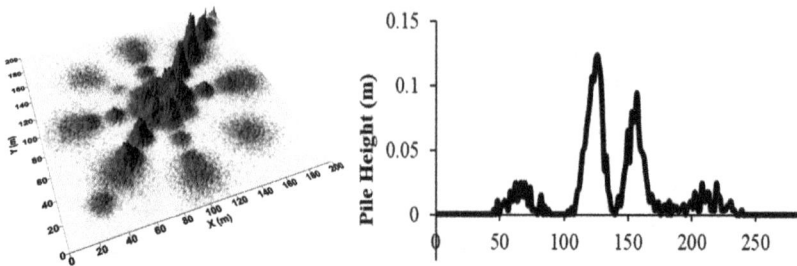

Figure 9.15:   Distribution of sediment thickness in 3D (left) and along plume centerline in the Northeast direction (right) under sea conditions represented by 8-sector velocity rose and D = 0.4 m/s$^{1/2}$.

heights 0.12 and 0.09 m near the well, about 70 m away from the coral site. There are two smaller peaks of thickness of 0.025 m or 25 mm at about 40 m away from the coral site. The sediment thickness around the well is about 0.005 m or 5 mm. Over the coral site at 100 m from the well, the sediment thickness is about 6 mm over an extended stretch. The sediment thickness of 6 mm is close to the recommended threshold of 6.3 mm. The corresponding sedimentation rate is 180 mg/cm^2/day. Although this scenario arouses some concern, this maximum sediment thickness of 6 mm that occurs over 1 to 2 days is unlikely to harm the corals. The moderate sedimentation rate of 180 mg/cm^2/day may be reduced by drilling over a longer period, with longer breaks in between changing of drilling sections. Lowering the sedimentation rate to 100 mg/cm^2/day would require drilling operations to be extended to 14 days, with possible increase in cost. A lower sedimentation rate allows (a) tidal currents to help clear the sediment deposited over the coral beds and (b) the corals to self-clean the sediment. This would improve the resilience of corals. An alternative is to move the well away from the corals by an additional 50 m to reduce the sediment cover over the coral site.

9.8   Vulnerability and Tolerance of Corals to Sedimentation

9.8.1   *Simple calculus of sedimentation*

In most scientific literature discussing the impact of sedimentation on coral, two important parameters are typically used. One parameter measures the sediment thickness or cover (mm), while the second parameter measures the sedimentation rate ($mg/cm^2/day$). The sedimentation rate $mg/cm^2/day$ measures the mass (mg) of sediment that settles over an area of 1 cm^2 in 1 day. For example, a sedimentation rate of 33 $mg/cm^2/day$ means that a total of 33 mg of sediment settles over an area of 1 cm^2 in 1 day. Sediment thickness of 1 mm is equivalent to 220 to 240 mg/cm^2, depending on the sediment density of 2.2 to 2.4 g/cm^2, respectively. If this sediment cover is delivered over 10 days, for example, then the sedimentation rate is 22 to 24 $mg/cm^2/day$ (240 mg/cm^2/10 days = 24 $mg/cm^2/day$). To put this into perspective, Rogers (1990) had suggested the sedimentation lethal limit of 10 $mg/cm^2/day$. This level appears to be too stringent and not enforceable by the OOGEE. Hence, we need to conduct extensive literature review to derive at a more realistic reference sedimentation rate useful for the OOGEE industry. Allers *et al.* (2013) used 66 mg/cm^2 of sediment cover as a nominal reference value, corresponding to a sediment thickness of 0.3 mm with density of 2.2 g/cm^2. If this sediment cover is delivered in 2 days, for example, then the sedimentation rate is 66 mg/cm^2/2 days = 33 $mg/cm^2/day$. This simple conversion calculus is useful in interpretation of results obtained from simulations or experimentation. We discuss the impact of sedimentation on corals in the following sections. Finally, we interpret the simulation results for our study site K-1 in the context of accumulated knowledge, as reported in the literature, on coral resilience and vulnerability to sedimentation.

9.8.2 *Experimental results on sedimentation burden*

As noted earlier, the oil and gas industrial guideline stipulates that maximum sediment thickness should not exceed the threshold of 6.3 mm. In close vicinity of a drill well, this threshold depth of 6.3 mm is rarely enforced and moderate layers of deposited materials, often of coarse grain sizes, may occur (Allers *et al.*, 2013). Hence, if corals are sited close to the drill well, these thick layers of sediment may pose hazards to them. The thresholds that must not be exceeded for sediment thickness and sedimentation rates to ensure coral safety are not quantified in most guidelines. In lieu of this lack of quantitative guideline, we conducted an extensive literature review of sedimentation burden in various sites and its impact on coral health. It has been reported that the sediment cover of 462 mg cm^{-2} (2 mm thickness) resulted in significant accumulation of sediments on top of the coral branches (Allers *et al.*, 2013). The thickness of sediment layers ranged from 1.5 to 15.3 mm (median 3 mm), revealing a large heterogeneity in sediment coverage across and even within branches. However, even this sediment load did not result in complete coverage of coral branches, with most polyps still able to extend at least their tentacles above the sediment horizon. As reported in Allers *et al.* (2013), the gradual local decrease of oxygen at the coral surface where layers of sediments built up did not observably affect *L. pertusa* for the duration of the sedimentation experiment of 11 days. Settling of benthic sediment and cuttings in the vicinity of OOGEE activities would not cause coral death within 11 days, even at concentrations of 3 times (198 mg/cm^2) to 7 times (462 mg/cm^2) the nominal cover of 66 mg/cm^2. The sedimentation experiment demonstrated that a very high sediment load is needed to achieve persistent sediment coverage of living coral branches. However, no concrete data are available to provide quantitative guidelines. The amount of sediment added in the experiments was above natural sedimentation regimes in living cold-water coral ecosystems (Gass and Roberts, 2006). Natural mass fluxes to cold-water coral ecosystems range from 0.5 to 3.7 g m^{-2} d^{-1} (equivalent to 0.05 to 0.37 mg/cm^2/day), as reported by Duineveld *et al.* (2004) for Galicia Bank and Mienis *et al.* (2009) for the Southwest Rockall Trough margin. Note that 1 g/m^2 = 0.1 mg/cm^2. However, both studies also reported very high mass flux pulses, which they attributed to resuspension events resulting from strong currents. In the sedimentation experiment conducted by Allers *et al.* (2013), exposure to sediment cover of 66 mg/cm^2, as well as 3 times to 7 times this value, was delivered to the corals during one depositional event. This experiment did not result in the death of corals. A sediment cover 20 times the nominal value (1320 mg/cm^2 or 6 mm) would completely cover the entire coral structures. The 6 mm sediment cover should therefore be used as a reference thickness. In an authentic drilling situation, however, the discharges occur periodically throughout the drilling period and not all at once (Purser and Thomsen, 2012). Hence, estimates of sediment cover and sedimentation rate should be carefully interpreted.

9.8.3   Coral vulnerability and tolerance to sediments

Many studies conducted to understand coral vulnerability and tolerance to sediments have demonstrated that some corals can indeed survive well in turbid environments laden with sediments (Mapstone *et al.*, 1989; Hopley *et al.*, 1993; Larcombe *et al.*, 1995; Anthony and Larcombe, 2002; Larcombe and Woolfe, 1999; Perry and Larcombe, 2003; Perry, 2005). These and other studies have demonstrated the occurrence of coral reefs, often with high live coral covers, in areas of high and fluctuating SS, turbidity and sedimentation, for example, in the inner shelf of the Great Barrier Reef. In a study conducted in Tanzania over a two-year period from October 2006 to March 2009, the monthly sedimentation rate monitored ranged from 0.2 to 41.5 mg/cm^2/day at the Bawe reefs, and from 0.8 to 65.8 mg/cm^2/day at the Chumbe reefs, with both sites having high live coral covers and high coral genera diversity (Muzuka *et al.*, 2010). This suggests that corals at Chumbe and Bawe have probably adapted to the sedimentation regimes and thus are able to overcome the sediment lethal limit of 10 mg/cm^2/day suggested by Rogers (1990). The current threshold level for sediment thickness adopted for the Norwegian Continental Shelf in environmental risk assessment models by the offshore industry is 6.3 mm (Allers *et al.*, 2013). Mimicking a typical drilling event in an experimental study, Larsson and Purser (2011) exposed *L. pertusa* to doses of fine-grained drill cuttings (less than 63 μm) over 3 weeks to reach total sediment thicknesses of either 6.3 mm (1 time) or 19 mm (3 times), corresponding to average daily doses of ca. 65 and 195 mg/cm^2/day, respectively. The number of fragments observed with smothered tissue in the 6.3 mm treatment was significant. The number of coral fragments with smothered tissues and polyp mortality increased with the higher sediment load of 19 mm (Larsson and Purser, 2011). It can be concluded that a 3-week burial by drill cuttings at the threshold level of 6.3 mm (currently adopted in environmental risk assessment models) may result in damage to *L. pertusa*. In another repeated exposure study, *L. pertusa* was exposed to 33 mg/cm^2 every second day for 45 days, corresponding to a sedimentation rate of 16.5 mg/cm^2/day, without showing any signs of exhaustion of sediment rejection mechanisms, suggesting that sediment rejection in *L. pertusa* is rather efficient. Net deposition rates in the nearshore coral settings on the inner shelf of the central Great Barrier Reef (GBR) averaged between 3 and 7 mg/cm^2/day over the course of a normal year (Browne *et al.*, 2012). Sedimentation rates exceeding 200 mg/cm^2/day for periods of days to weeks are not uncommon on fringing reefs of the GBR. Further, recent studies from nearshore reefs in the GBR provide convincing evidence of spatially relevant and temporally persistent reef-building having occurred over millennial timescales (Larcombe *et al.*, 1995; Anthony and Larcombe, 2002). This suggests that corals are able to adapt to changing sedimentation regimes within some limits.

9.8.4 *More on coral resilience*

Rice and Hunter (1992) noted that long-term exposure to elevated levels of SS (50 to 100 mg/l) and high levels of coral bed sediment (tens of mg/cm^2/day) can cause reduced coral growth and reduced reef development. However, recent studies indicated that the observed adverse impacts are often less severe than what had been previously reported. Perry and Smithers (2010) observed that corals can indeed survive well in turbid environment with high SS and high sedimentation. Five species of gorgonians in the highly sedimented waters of Singapore showed growth rates ranging from 2.3 to 7.9 cm/yr, which are comparable to published growth rates from non-sedimented environments (Goh and Chou, 1995). The reason for the observed good growth rate despite high SS and high sedimentation could be due to flushing by tidal currents that can efficiently remove sediments from corals (Riegl *et al.*, 1996; Anthony and Larcombe, 2002). This efficient removal of sediments from corals by tidal flushing is applicable to the present study site K-1, which is located in an area with good tidal flushing augmented by waves and internal eddies (DHI, 2009). Corals that are naturally exposed and adapted to high and variable background conditions of SS, turbidity and sedimentation (e.g., due to tides, storms and monsoon, as is the case with the study site K-1) will have higher tolerances to short-term pulses in SS, turbidity or sedimentation caused by dredging or drilling operations (DHI, 2009). This remark certainly applies to the study site at K-1.

Sedimentation rate is usually highest on inshore reefs as well as in reef systems that are sheltered from waves, but decreases with distance from shore and with increasing exposure to wave energy (Wolanski *et al.*, 2005). The study site K-1 is in an area that may be termed in between. Hence, the long-term sedimentation rates that can be tolerated by corals in the study site K-1 may well be in the range of about 50 to 100 mg/cm^2/day. Short-term (2 to 3 days) sedimentation rates around 100 mg/cm^2/day may well be tolerated by corals in K-1. Several studies revealed that many coral species and reefs are capable of surviving sedimentation rates of 100 mg/cm^2/day for several days to several weeks without any major negative effects, while some (nearshore) reefs naturally experience sedimentation rates well over 200 mg/cm^2/day. Nearshore fringing reefs in the Great Barrier Reef region that are characterized by high and variable sedimentation rates, from 2 to 900 mg/cm^2/day (short-term rates) with long-term means of 50 to 110 mg/cm^2/day, were found to sustain highly diverse coral growth with a mean coral cover of 40–60% (Ayling and Ayling, 1991). A few coral species, such as *Montastraea cavernosa* and *Astrangia poculata*, can tolerate sedimentation rates as high as 600 to 1,380 mg/cm^2/day (Lasker, 1980; Peters and Pilson, 1985). Field and laboratory experiments in Florida, USA, have shown that some of the tolerant coral species in the Caribbean can survive complete burial with sediment for periods ranging from 7 to 15 days (Rice and Hunter, 1992). Sedimentation at the rate of 200 mg/cm^2/day lasting 45 days has caused no effect on *Acropora cervicornis* (Rogers, 1979, 1990). Further, a sedimentation rate of 200 mg/cm^2/day for 6 continuous weeks has caused only minor tissue damage and bleaching for the species *Sinularia dura*, *Gyrosmilia interrupta*, *Favites pentagona*, *Favia favus* and *Sinularia leptoclados* (Riegl, 1995; Riegl and Bloomer, 1995). This wide range of tolerance limits suggests that different coral species and corals in different geographic regions may respond efficiently to increased thickness and increased rates of sedimentation due to evolutional adaptation. For our study site at K-1, a sediment thickness of 3.3 mm and the corresponding sedimentation rate of 100 mg/cm^2/day over a duration of 8 to 9 days may not pose serious harm to the corals. For other sites with different sediment exposure durations, frequencies and intensities, a case-specific simulation analysis is required.

9.9   Impact Assessment and Mitigation

It is beneficial for impact assessment and mitigation analysis for K-1 to summarize and recap previous conclusions arrived at in the previous section based upon extensive literature review. For our study site at K-1, a sediment thickness of 3.3 mm (the reference thickness) and a sedimentation rate of 100 mg/cm^2/day (the reference rate) over a duration of 8 to 9 days may not pose serious harm to corals. For other sites with different sediment exposure durations, frequencies and intensities, a case-specific simulation analysis is required. The tidal current and dispersion regimes at K-1 as shown in Table 9.4 show significantly high variability. This high variability in the tidal regimes results in equally high variability in the simulated sedimentation regimes over the coral seabed when different tidal and dispersion regimes are used in the simulations. We selected four simulation scenarios to cover the entire spectrum of this variability in tidal and dispersion regimes. The four potential scenarios regarding sediment thickness and sedimentation rate on the coral seabed due to the discharge of 500 m^3 of cuttings over 8 days are summarized as follows:

(a) Scenario 1: sediment thickness = 0.05 mm, sedimentation rate = 1.5 mg/cm^2/day;
(b) Scenario 2: sediment thickness = 5.00 mm, sedimentation rate = 150 mg/cm^2/day;
(c) Scenario 3: sediment thickness = 18.0 mm, sedimentation rate = 540 mg/cm^2/day;
(d) Scenario 4: sediment thickness = 6.00 mm, sedimentation rate = 180 mg/cm^2/day.

Scenario 1 will not pose any adverse impact on the corals as the sediment thickness of 0.05 mm and the sedimentation rate of 1.5 mg/cm^2/day are low compared to the reference safe levels of 3.3 mm and 100 mg/cm^2/day, respectively. Scenarios 2 and 4 have a sediment thickness of 5 to 6 mm, respectively, which is close to the recommended maximum threshold of 6.3 mm and which exceeds our safe reference level of 3.3 mm. The corresponding sedimentation rate is 150 to 180 mg/cm^2/day, respectively, which is moderately high compared to the safe reference level of 100 mg/cm^2/day. This rate may be reduced by drilling over a longer period with longer breaks in between drill section changes. Lowering sedimentation rate to 100 mg/cm^2/day would require drilling operations to be extended to 12 days, with possible increases in cost. A lower sedimentation rate allows (a) tidal currents to help clear the sediment deposited over the coral beds and (b) the corals to self-clean the sediment by sediment rejection mechanism. This would improve the safety of corals. An alternative is to move the drilling well an additional distance of 50 m away from the corals. The real concern is Scenario 3, with sediment thickness of 18 mm and sedimentation rate of 540 mg/cm^2/day, which may harm the corals. However, the likelihood of occurrence of this scenario is low, as this scenario presumes that the current flows in the NE–SW with velocity of 0.05 sin (σt) m/s consistently over 8 to 9 days, which is highly unlikely. Nevertheless, the precautionary principle mandates that the scenario should be ruled out completely. Hence, we recommend that the drill site be moved 100 m further away from the coral site to create a distance of 200 m between the drill and coral sites, preferably not aligned in the NE–SW direction. This will reduce the sediment thickness to less than 2 mm and sedimentation rate to less than 50 mg/cm^2/day, which are well tolerated by corals, based upon the literature review discussed in Section 9.8.

References

Allers, E., Abed, R.M.M., Wehrmann, L.M., Wang, T., Larsson, A.I., Purser, A. and de Beer, D. (2013). Resistance of *Lophelia pertusa* to coverage by sediment and petroleum drill cuttings. *Marine Pollution Bulletin*, 74, 132–140.

Anthony, K.R.N. and Larcombe, P. (2002). Coral reefs in turbid waters: Sediment-induced stresses in corals and likely mechanisms of adaptation. *Proceedings Ninth International Coral Reef Symposium*, 1, 239–244.

Ayling, A.M. and Ayling, A.L. (1991). The effect of sediment run-off on the coral populations of the fringing reefs at Cape Tribulation. Great Barrier Reef Marine Park Authority Research Publication No. 26, Townsville.

Bagarinao, T. (1992). Sulphide as an environmental factor and toxicant: Tolerance and adaptations in aquatic organisms. *Aquatic Toxicology*, 24, 21–62.

Bell, N., Cripps, S.J., Jacobsen, T., Kjeilan, G. and Picken, G.B. (1998). Review of Drill Cuttings Piles in the North Sea. Cordah, UK.

Bowden, K.F. (1983). *Physical Oceanography of Coastal Waters*, Ellis Horwood Ltd, John Wiley & Sons, New York, p. 302.

Browne, N.K., Smithers, S.G., Perry, C.T. and Ridd, P.V. (2012). A field-based technique for measuring sediment flux on coral reefs:

Application to turbid reefs on the Great Barrier Reef. *Journal of Coastal Research*, 284, 1247–1262.

Carles, L.J. and Bryden, I.G. (1999). The sensitivity of a dispersion model to cuttings settling speeds. *Society of Underwater technology Journal*, 24, 19–24.

CERC (1984). Shore Protection Manual, Vol. 1, Chap. 4. Coastal Engineering Research Center, Vicksburg, Mississippi, USA.

DHI (2009). Metocean Criteria for Kebabangan Field Development, Sabah, prepared by Danish Hydraulic Institute for Kebabangan Petroleum Operating Company, Malaysia, 615.

Drozdowski, A., Hannah, C. and Tedford, T. (2004). bblt Version 7 user's manual, Canadian Technical Report of Hydrography and Ocean Science, p. 69.

Dudley, W.C. (2003). Coral Reef Sedimentology, in Mare 461 – Spring 2003 [Online], 23. Available: http://www.kmec.uhh.hawaii.edu/QUESTIn fo/reefsEDM.pdf.

Duineveld, G.C.A., Lavaleye, M.S.S. and Berghuis, E.M. (2004). Particel flux and food supply to a seamount cold-water coral community (Galicia Bank, NW Spain). *Marine Ecological Progress Series*, 277, 12–23.

Eames, I., de Leeuw, B. and Coniff, P. (2002). Formation and remediation of drill cutting piles in the North Sea. *Environmental Geology*, 41, 504–524.

Erftemeijer, P.L.A., Hagedorn, M., Laterveer, M., Craggs J., and Guest, J.R. (2012). Effect of suspended sediment on fertilisation success in the scleractinian coral Pectinialactuca. *Journal of the Marine Biological Association of the United Kingdom*, 92, 741 745.

ERM (2009). Ghana Jubilee Field Phase 1 Development: Environmental Impact Statement. Submitted by Tullow Ghana Limited on 27 November 2009. M. Irvine, A. de Jong and A.K. Armah (Lead Authors). *Environmental Resources Management*. Available online at http://www.erm.com/Public-Information-Sites1/Tullow-Jubilee/.

Gao, X., Wei, Z., Lv, X., Wang. Y and Fang, G. (2015). *Ocean Modelling*, 92, 101–114.

Gass, S.E. and Roberts, J.M. (2006). The occurrence of the cold-water coral *Lophelia pertusa* (Scleractinia) on oil and gas platforms in the North Sea: Colony growth, recruitment and environmental controls on distribution. *Marine Pollution Bulletin*, 52, 549–559.

Gates, A.R. and Jones, D.O.B. (2012). Recovery of benthic megafauna from anthropogenic disturbance at a hydrocarbon drilling well (380 m depth in the Norwegian Sea). *PLoS ONE*, 7(10), e44114. doi: 10.1371/journal.pone.0044114.

Goh, N.K.C. and Chou, L.M. (1995). Growth of five species of gorgonians (Sub-Class Octocorallia) in the sedimented waters of Singapore. *Marine Ecology*, 16, 337–346.

Hannah, C.G., Drozdowski, A., Muschenheim, D.K., Loder, J.W., Belford, S. and MacNeil, M. (2003). Evaluation of drilling mud dispersion models at SOEI Tier I Sites: Part 1 North Triumph, Fall 1999. *Canadian Technical Report Hydrography and Ocean Science*, 232, pp. v+51.

Hopley, D., van Woesik, R., Hoyal, D.C.J.D., Rasmussen, C.E. and Steven, A.D.L. (1993). Sedimentation resulting from road development, Cape Tribulation Area. Great Barrier Reef Marine Park Authority Technical Memorandum 24, Great Barrier Reef Marine Park Authority, Townsville.

Koh, H.L., Lim, P.E. and Midun, Z. (1991). Management and control of pollution in Inner Johore Strait. *Environmental Monitoring and Assessment*, 19(1–3), 349–359, Kluwer Academic Publishers, Netherlands.

Koh, H.L. and Teh, S.Y. (2011). Simulation of Drill Cuttings Dispersion and Deposition in South China Sea. *Proceedings of the International MultiConference of Engineers and Computer Scientists*, 2011 Vol. II, IMECS 2011, 16–18, Hong Kong, International Association of Engineers (IAENG), S.I. Ao, Oscar Castillo, Craig Douglas, David Dagan Feng and Jeong-A Lee (Eds.), pp. 1501–1506.

Koh, H.L., Lim, P.E. and Lee, H.L. (1997). Impact modeling of sewage discharge from Georgetown of Penang, Malaysia on coastal water quality. *Journal Environmental Monitoring and Assessment*, 44, 199–209.

Larcombe, P. and Woolfe, K.J. (1999). Increased sediment supply to the Great Barrier Reef will not increase sediment accumulation at most coral reefs. *Coral Reefs*, 18, 163–169.

Larcombe, P., Ridd, P.V., Prytz, A. and Wilson, B. (1995). Factors controlling suspended sediment on the inner-shelf coral reefs, Townsville, Australia. *Coral Reefs*, 14, 163–171.

Larsson, A.I. and Purser, A. (2011). Sedimentation on the cold-water coral *Lophelia pertusa*: Cleaning efficiency from natural sediments and drill cuttings. *Marine Pollution Bulletin*, 62, 1159–1168.

Larsson, A.I., van Oevelen, D., Purser, A. and Thomsen, L. (2013). Tolerance to long-term exposure of suspended benthic sediments and drill cuttings in the cold-water coral *Lophelia pertusa*. Marine Pollution Bulletin. doi: 10.1016/j.marpolbul.2013.02.033.

Lasker, H.R. (1980). Sediment rejection by reef corals: The roles of behavior and morphology in Montastraea cavernosa (Linnaeus). *Journal of Experimental Marine Biology and Ecology*, 47, 77–87.

Mapstone, B.D., Choat, J.H., Cumming, R.L. and Oxley, W.G. (1989). The fringing reefs of Magnetic Island: Benthic biota and sedimentation.

A baseline study. A report to the Great Barrier Reef Marine Park Authority, Townsville, p. 88.

Mienis, F., de Stitger, H.C., de Haas, H. and van Weering, T.C.E. (2009). Near-bed particle deposition and resuspension in a cold-water coral mound area at the Southwest Rockall Trough margin, NE Atlantic. *Deep Sea Research. Pt.* I, 56, 1026–1038.

Muzuka, A.N.N., Dubi, A.M., Muhando, C.A. and Shaghude, Y.W. (2010). Impact of hydrographic parameters and seasonal variation in sediment fluxes on coral status at Chumbe and Bawe reefs, Zanzibar, Tanzania. *Estuarine, Coastal and Shelf Science*, 89, 137–144.

Niu, H., Drozdowski, A., Husain, T., Veitch, B., Bose, N. and Lee, K. (2009). Modeling the dispersion of drilling muds using the bblt model: The effects of settling velocity. *Environmental Modeling and Assessment*, 14(5), 585–594.

OGP (2003). Environmental aspects of the use and disposal of non aqueous drilling fluids associated with offshore oil & gas operations. Report No. 342. International Association of Oil & Gas Producers, London, UK.

Perry, C.T. (2005). Structure and development of detrital reef deposits in turbid nearshore environments, Inhaca Island, Mozambique. *Marine Geology*, 214, 143–161.

Perry, C.T. and Larcombe, P. (2003). Marginal and non reef building coral environment. *Coral Reefs*, 22, 427–432.

Perry, C.T. and Smithers, S.G. (2010). Evidence for the episodic "turn on" and "turn off" of turbid-zone coral reefs during the late Holocene sea-level highstand. *Geology*, 38, 855–858.

Peters, E.C. and Pilson, M.E.Q. (1985). A comparative study of the effects of sedimentation on symbiotic and asymbiotic colonies of the coral Astrangia danae Milne Edwards and Hime 1849. *Journal of Experimental Marine Biology and Ecology*, 92, 215–230.

Purser, A. and Thomsen, L. (2012). Monitoring strategies for drill cutting discharge in the vicinity of cold-water coral ecosystems. *Marine Pollution Bulletin*, 64, 2309–2316.

Reed, M. (1980). An oil spill fishery interaction model: Development and applications. Ph.D. dissertation, Department of Ocean Engineering. U. Rhode Island, Kingston, R.I., USA, p. 235.

Rice, S.A. and Hunter, C.L. (1992). Effects of suspended sediment and burial on scleractinian corals from West Central Florida patch reefs. *Bulletin of Marine Science*, 51, 429–442.

Riegl, B. (1995). Effects of sand deposition on scleractinian and alcyonacean corals. *Marine Biology*, 121, 517–526.

Riegl, B. and Bloomer, J.P. (1995). Tissue damage in hard and soft corals due to experimental exposure to sedimentation. *Proceedings 1st European Regional Meeting ISKS, Vienna*, 20, 51–63.

Riegl, B., Heine, C. and Branch, G.M. (1996). The function of funnel-shaped coral-growth in a high-sedimentation environment. *Marine Ecology Progress Series*, 145, 87–94.

Rogers, C.S. (1977). The response of a coral reef to sedimentation. Ph.D. dissertation, University of Florida, Gainesville.

Rogers, C.S. (1979). The effect of shading on coral reef structure and function. *Journal of Experimental Marine Biology and Ecology*, 41, 269–288.

Rogers, C.S. (1990). Responses of coral reefs and reef organisms to sedimentation. *Marine Ecology Progress Series*, 62, 185–202.

Rye, H., Reed, M. and Ekrol, N. (1998). The PARTRACK model for calculation of spreading and deposition of drilling mud, chemicals and drill cuttings. *Environmental Modelling and Software*, 13, 431–441.

Rye, H., Reed, M., Frost, T.K. and Utvik, T.I.R. (2006). Comparison of the ParTrack mud/cuttings release model with field data based on use of synthetic-based drilling fluids. *Environmental Modelling and Software*, 21(2), 190–203.

Spaulding, M.L., Isaji, T. and Howlett, E. (1994). MUDMAP: A model to predict the transport and dispersion of drill muds and production water, Applied Science Associates Inc, Narragansett, RI.

Teh, S.Y. and Koh, H.L. (2014). Drill cuttings mound formation study. AIP Conference Proceedings 1605, American Institute of Physics, Melville, NY, pp. 434–439.

Teh, S.Y. and Koh, H.L. (2016). Modelling drill cuttings sedimentation on corals for exploration Wells Z-1 and B-1, Offshore Sabah. *International Journal of Environmental Science and Development*, 7(1), 913–920.

Wolanski, E., Fabricius, K., Spagnol, S. and Brinkman, R. (2005). Fine sediment budget on an inner-shelf coral-fringed island, Great Barrier Reef of Australia. *Estuarine, Coastal and Shelf Science*, 65, 153–158.

Zu, T., Gan, J. and Erofeeva, S.Y. 2008. Numerical study of the tide and tidal dynamics in the South China Sea. *Deep-Sea Research I*, 55, 137–154.

Chapter 10

Governance for Water Security

10.1 Introduction

We conclude this book with some remarks on governance for water security in an era of climate warming and sea level rise, amid an environment of ever-growing population density in cities. Climate change and population growth will continue to exert increasing pressure on the balance between demand and supply of water resources to these densely populated cities. Transboundary conflicts over water issues should be resolved through international laws and partnerships. The United Nations Convention on the Law of the Sea (UNCLOS) and the Coral Triangle Initiative (CTI) are two excellent examples. The dispute between Singapore and Malaysia over coastal reclamations around the Tekong Island by Singapore was amicably resolved through arbitration by the International Tribunal for the Law of the Sea (ITLOS) created by the mandate of UNCLOS. The CTI partnership was created by the six member countries (known as CT6), comprising Indonesia, Malaysia, the Philippines, Papua New Guinea, Solomon Islands and Timor-Leste. The CTI aims to resolve daunting challenges confronting the six countries regarding sustainable exploitation of the coastal and marine resources, shared by 350 million, most of whom live around vulnerable coastal zones fringing the coral triangle (CT). Overexploitation of groundwater to meet the demand in cities had led to severe land subsidence across many Asian cities, one of which is Jakarta. As a consequence, the government of Indonesia has announced a plan to relocate the capital Jakarta to the island of Borneo. The new future capital city must be rebuilt to be sustainable (SDG11), sustainable in the sense of environmental, social and economic viability over the long run. The new city must be resilient to internal and external risks (SDG9), resilient to climate impacts (SDG13), resilient to the hazards of poverty (SDG1), resilient to health concerns (SDG3), resilient to water shortage (SDG6) and resilient to food insecurity (SDG2). Good governance is essential to ensure water security and food security. Protection and conservation of vital coastal resources such as coastal wetlands, mangroves, seagrasses and coral reefs are critical in order to preserve the integrity of their ecosystems and their services, on which the coastal communities depend for their survival. In the context of climate change, the task ahead is challenging, given the vast uncertainty in understanding the drivers, their impacts and responses. For countries that share borders with numerous neighboring countries, such as the CT6, partnership for the goals (SDG17) is essential to reap the benefits of synergy of collaboration and to attain peace and inclusiveness (SDG16). Good governance over transboundary water is crucial to preserving life under water (SDG14) and to sustaining decent work and economic growth (SDG8).

10.2   Transboundary Water Governance

Transboundary waters are defined as aquifers, lakes, river basins and oceans shared by two or more countries. An example is the Great Lakes Basin (GLB) shared between the USA and Canada or the Mekong River Basin (MRB) shared among six countries in Southeast Asia, namely, China, Myanmar, Vietnam, Cambodia, Laos and Thailand. Transboundary water treaties based on legal and institutional frameworks can be effective instruments to deal with challenges confronting transboundary waters. Transboundary water governance is crucial for addressing water-related issues and conflicts caused by growing populations, increasing demands, human intervention and climate change impacts on water resources. However, few treaties are successful and many transboundary water basins are in constant tension because the countries involved do not comply with these treaties, especially in developing regions. However, a successful treaty can effectively resolve the tension and provide mutually supporting social, economic and environmental benefits to ensure long-term sustainability of the regions. An example of this is the transboundary water governance stipulated by the UNCLOS.

10.2.1 *UNCLOS*

The UNCLOS is an international agreement that resulted from the third United Nations Conference on the Law of the Sea (UNCLOS III) that took place between 1973 and 1982. UNCLOS was open for signature on 10 December 1982, and entered into force on 16 November 1994, a year after Guyana became the 60th nation to ratify the treaty. The Convention defines the rights and responsibilities of nations with respect to their use of the world oceans. It establishes guidelines for businesses, for the environment and for the management of marine natural resources. The UNCLOS, concluded in 1982, replaced the quad-treaty 1958 Convention on the High Seas. As of June 2016, 167 countries and the European Union have joined in the Convention. The UN has no direct operational role in the implementation of the Convention. There is, however, a role played by organizations such as the ITLOS, the International Maritime Organization (IMO), the International Whaling Commission and the International Seabed Authority (ISA). The ISA was established by the UN Convention. Part XII of UNCLOS contains special provisions for the protection of the marine environment, obligating all States to collaborate in this matter, as well as placing special obligations on flag States to ensure that ships under their flags adhere to international environmental regulations, often adopted by the IMO. The United Nations Sustainable Development Goal 14 (UNSDG14) has several targets such as Target 14.1 to protect the marine environment from pollution of all kind and Target 14.2 to sustainably manage and protect marine and coastal ecosystems in order to achieve healthy and productive oceans. These targets are consistent with the UNCLOS and ITLOS legal framework to prevent any serious transboundary environmental harm to the ocean. A case in point is ITLOS Case 12, the subject matter for discussion in the following subsections.

10.2.1.1   *ITLOS Case 12*

Land Reclamation in and around the Straits of Johor (Malaysia v. Singapore), Provisional Measures, Order, 8 October 2003 International Tribunal for the Law of the Sea.

On 4 July 2003, Malaysia instituted arbitral proceedings under Annex VII to the Convention against Singapore, in a dispute concerning land reclamation by Singapore in and around the Straits of Johor. On 5 September 2003, a request for the prescription of provisional measures under article 290, paragraph 5, of the Convention was filed with the Tribunal by Malaysia against Singapore (ITLOS, 2003). The request was made in relation to the dispute concerning land reclamation by Singapore in and around the Straits of Johor. The Tribunal may prescribe measures to preserve the respective rights of the parties to the dispute and to prevent serious harm to the marine environment (Target 14.1) and its ecosystems (Target 14.2). The Tribunal delivered its Order on 8 October 2003. In its Order, the Tribunal unanimously prescribes the following provisional measures under article 290, paragraph 5, of the Convention. *First,* Malaysia and Singapore shall cooperate and shall enter into consultations in order to (a) establish promptly a group of independent experts with the mandate to conduct a study to determine, within a period of one year, the effects of Singapore's land reclamation and to propose measures to deal with any adverse effects of such land reclamation; (b) exchange information on, and assess risks of Singapore's land reclamation works; (c) avoid any action incompatible with their effective implementation, and consult to reach a prompt agreement on such temporary measures with respect to Area D at Pulau Tekong, including suspension of reclamation. *Second,* the Tribunal unanimously directs Singapore not to conduct its land reclamation in ways that might cause irreparable prejudice to the rights of Malaysia or cause serious harm to the marine environment, taking especially into account the reports of the group of independent experts. The provisional measure and the tribunal proceeding regarding the protection of the marine environment in the Straits of Johor ordered by the Tribunal specifically addressed Targets 14.1 and 14.2.

10.2.1.2   *USM EIA simulation results*

The authors performed a detailed Environmental Impact Assessment (EIA) study on the impact of reclamation by Singapore.

Our simulation results indicated that tidal velocity is likely to increase by 56% in the vicinity of Pulau Tekong, and that the maximum vertically integrated tidal velocity may reach 0.8 m/s (Syamsidik and Koh, 2003). An increase in tidal velocity by 25% is considered as significant. The tidal current velocity near the surface is likely to reach 1.2 m/s (Koh *et al.*, 2004). These levels of tidal currents are likely to pose hazards to safe passage for ships passing through the narrow channels in the reclaimed sites and may also increase coastal erosion. Reclamation activity may create a large quantity of suspended sediment (SS), thereby polluting the waters around the reclamation sites and posing serious adverse ecological impacts. Our simulation shows that the levels of SS around the reclamation sites in Pulau Tekong may exceed 640 mg/l around Changi Bay, posing serious hazards to the marine environment and its ecology, particularly to seagrass, which forms the primary foods of the dugong. Hence, Singapore's reclamation around Pulau Tekong had to be conducted with extreme care (Koh *et al.*, 2004).

10.2.1.3 *The Tribunal proceedings*

Several related laws were presented during the Tribunal. For example, Oppenheim's International Law states that "No State is allowed to alter the natural condition of its own territory to the disadvantage of the natural conditions of a neighboring state territory. A state cannot build embankments and the like, without a previous agreement with the neighboring state". This cooperative approach is endorsed and embodied in the 1982 Convention. The Law of the Sea Convention takes an integrated approach to the issues it covers including various uses of the sea, such as navigation, fishing and other forms of resource exploitation. Protection and preservation of the marine environment requirements have to be integrated into all uses of the seas. Apparent conflicts between economic, social and environmental considerations (the three pillars of SDGs) have to be resolved in an integrated manner, protecting both substantive and procedural rights.

10.2.1.4 *Substantive and procedural rights*

The Convention incorporates both substantive and procedural rights (ITLOS, 2016). The substantive right of a coastal State to use its territorial waters pursuant to its own developmental and environmental policies and in accordance with its duty to protect the marine environment is qualified by the procedural right of its neighboring State to prior notification, consultation and monitoring should serious transboundary environmental harm be likely to occur. Similarly, the substantive right of a coastal State to have unimpeded maritime access to its ports is accompanied by its right to be consulted on plans for artificial alterations which may impact common navigation channels and safety of navigation. The interrelationship of and interaction between substantive and procedural rights contribute to the integrated approach envisaged by the 1982 Convention. This approach is relevant for the proper management of a sea area such as the Straits of Johor, which is an intensively used economic area with a fragile marine ecosystem. Such a situation calls for close cooperation between Malaysia and Singapore in order to allow for the sustainable use of this sea area. That no State has the right to carry out activities within its jurisdiction which cause damage to other States and their environment is part of the integrated approach. Such a duty to cooperate and the duty of prior notification and consultation receive additional prominence in a case of a semi-enclosed sea such as the Straits of Johor. Article 123 of the UN Convention on the Law of Sea stipulates that "States bordering an enclosed or semi-enclosed sea should cooperate with each other in the exercise of their rights and in the performance of their duties under this Convention". Article 122 defines semi-enclosed seas as "a gulf, basin or sea surrounded by two or more States and connected to another sea or the ocean by a narrow outlet". The Straits of Johor consists of the two territorial seas belonging to Malaysia and Singapore and constitutes an area of the sea surrounded by two States, with a narrow outlet to another sea. This definition of a semi-enclosed sea fits the Straits of Johor well.

10.2.1.5   *Provisions for semi-enclosed seas*

Article 123 specifies this increased duty to cooperate which is incumbent on states bordering a semi-enclosed sea, both in exercising their rights and in performing their duties under the Convention. This includes the coordination in the management, conservation, exploration and exploitation of the living resources of the sea, and coordination in the implementation of their rights and duties with respect to the protection and preservation of the marine environment. In view of the irreparable harm that may result from the land reclamation works by Singapore, the provisional measures as sought by Malaysia with respect to its rights in the Straits Johor as a semi-enclosed sea are fully justified. Further, the precautionary principle applies in this case.

10.2.1.6 *Precautionary principle*

The Articles 194 and 204 of the Convention in effect require the application of the precautionary principle. It is widely agreed that the core of the precautionary principle is well reflected in Principle 15 of the Rio Declaration on environment and development, which provides that "In order to protect the environment, the precautionary approach shall be widely applied by states according to their capabilities. Where there are threats of serious or irreversible harm, lack of full scientific certainty shall not be used as a reason to postponing cost-effective measures to prevent environmental degradation". Article 206 of the Convention elaborates on this by specifying the requirement of an environmental impact assessment. It provides, "When States have reasonable grounds for believing that planned activities under their jurisdiction may cause substantial pollution of or significant and harmful changes to the marine environment, they shall assess the potential effects of such activities on the marine environment and shall communicate reports of the results of such assessments".

Subscribing to the precautionary principle, an EIA study was conducted by the authors to evaluate the impact of coastal reclamation on the environment and its ecosystems in the Straits of Johor. Simulations indicated that tidal velocity is likely to increase by 56% in the vicinity of Pulau Tekong (Syamsidik and Koh, 2003). Further, the tidal current near the surface may reach 1.2 m/s (Koh *et al.*, 2004). These levels of tidal currents pose hazards to safe passage through the narrow channels, and may also increase coastal erosion. Reclamation activities of this scale will create large quantity of sediments, will pollute the waters and will pose serious adverse ecological impacts. Simulations further show that the levels of suspended solids around the reclamation sites may exceed 640 mg/l, posing serious hazards to the marine ecosystems, particularly seagrasses, on which dugongs depend as their primary foods. Hence, the onus is on Singapore to conduct a careful EIA study, in accordance with the precautionary principle (Koh *et al.*, 2004). The precautionary principle requires bilateral cooperation in order to prevent or to contain and solve transboundary problems, in a small sea with a sensitive ecological system such as the Straits of Johor.

10.2.1.7   *Transboundary problems*

A precautionary approach is central to a sustainable use of a territorial sea by committing a state to avoid human activity which may cause significant harm to the natural resources and the ecosystem. At a minimum this mandates notification and prior information on planned activities which may impact the rights of the neighboring state. The precautionary approach requires that, when there are threats of serious or irreversible damage, lack of full scientific certainty shall not be used as a reason for postponing cost-effective measures. An independent environmental impact assessment is a central tool of the international law of the precautionary principle. Malaysia is entitled to protection by the provisions of the 1982 Convention on the Law of the Sea pertaining to the territorial sea in general and the semi-enclosed sea in particular and to the protection and preservation of the marine environment. Malaysia has a right that its coastline not be seriously eroded or be the subject of major sedimentation, that its seagrass areas not be destroyed, that its fisheries not ruined by increased water flows, that its mangrove zones not degraded and that its navigation not impaired. On the basis of this right enshrined in UNCLOS principles, a request for the prescription of provisional measures under article 290, paragraph 5, of the Convention was filed with the Tribunal by Malaysia against Singapore (ITLOS, 2003). This is an excellent example of resolving transboundary conflicts through a tribunal provided by ITLOS, as mandated under UNCLOS. Another excellent example of transboundary partnership is the Coral Triangle Initiative between six countries to protect the fisheries and other marine resources shared among the six countries.

10.2.2   *Coral triangle initiative*

Fisheries and other marine resources are critically dependent on coral reefs and their associated mangroves and seagrass beds.

The extensive CT is situated between the Indian and the Pacific oceans, and consists of some of the lands and seas of six countries (CT6), comprising Indonesia, Malaysia, the Philippines, Papua New Guinea, Solomon Islands and Timor-Leste (Figure 10.1). Endowed with exceptionally high marine biodiversity, the CT harbors 76% of the 798 known coral species (Veron, 2000) and 37% of the 6,000 coral reef fish species (Allen, 2008) in the world. While the Coral Triangle occupies only 1.6% of the world's oceans, it has the largest coral reef coverage of 73,000 km^2 or 29% of the global coral reef area (Burke *et al.*, 2012). Coral reefs in the CT6 countries cover a total area of 98,577 km^2, with Indonesia having the largest coral reef area of 51,000 km^2, followed by the Philippines (26,000 km^2), PNG (13,840 km^2) and Malaysia (3,600 km^2). The CT has the most biodiverse coral reefs in the world, coupled with a high endemism of marine organisms (Veron *et al.*, 2009). However, the CT ecosystem health is declining and needs urgent rehabilitation.

10.2.2.1 *Declining marine resources in CT*

The high biodiversity and extensive habitat coverage help sustain the lives and livelihoods of 120 million people in the CT. The coral coverage in the CT6 countries, however, had been declining since the 1980s. In the entire CT Indo-Pacific region, coral coverage declined from 42% in the early 1980s to 22% by 2003 (Reef Check, 2010). On top of global stressors, significant local and regional anthropogenic pressures have caused degradation to the coral reefs and their associated marine ecosystems and habitats in the CT. The primary threats to coral reef ecosystems are overfishing and destructive fishing, and excessive nutrients and pollution from both watershed and marine sources. Overfishing has greatly reduced reef fish biomass and biodiversity, especially in the Philippines (Nañola *et al.*, 2011). Threatened by global stressors and anthropogenic pressures, the ecological status of the CT will be further exacerbated by climate change impacts and the associated extreme natural disturbances (Burke *et al.*, 2011). In response to these series of threats, the Coral Triangle Initiative was thus born.

10.2.2.2 *The coral triangle initiative*

To reverse this declination trend of coral coverage, the CTI on Coral Reefs, Fisheries, and Food Security (CFF) was launched in 2007, as

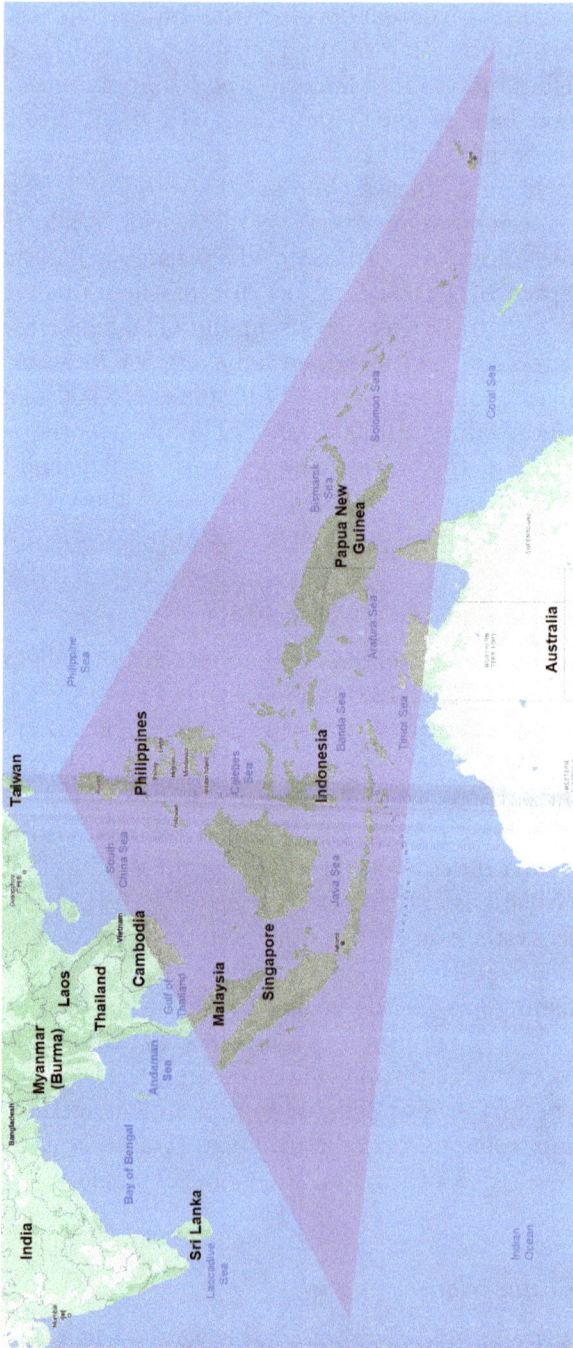

Figure 10.1:   Location of the extensive CT.

a multilateral partnership between the six governments of the CT6. The establishment of the CTI is very timely, given the recognition of the region's importance in global coral reef biodiversity, in abundant fisheries and in food security derived from marine resources. In 2009, the CT6 adopted a 10-year regional plan to improve management and enhance protection of the region's coastal and marine ecosystems and resources, and to ensure food security and sustainable livelihoods for all residents of CT6 (ADB, 2014). The CTI aims to achieve five goals, of which three are most important for CT6 countries: (Goal 1) adopt an ecosystem approach to fisheries management (EAFM), (Goal 2) develop and manage marine protected areas (MPAs) and (Goal 3) adopt climate change adaptation (CCA) measures. The vision of the CTI is to (a) stabilize coral reef ecosystem integrity and services; (b) improve fish stocks and diversity; and (c) improve food security. Reef fish biomass values are not estimated in many parts of the CT6 countries and may vary significantly across the CT6 countries, ranging from less than 10 ton/km^2 in the Philippines (Nañola *et al.*, 2006) to more than 100 ton/km^2 in the Solomon Islands (Green *et al.*, 2006). The total fish catch in the CT may have already exceeded the carrying capacity of their demersal and pelagic fishery resources (Lymer *et al.*, 2010). Indonesia, Malaysia and the Philippines have been fishing down the food web since the 1950s and catching lower trophic-level species, based on marine trophic indexes. The long-term sustainability of fisheries and food security in the CT is cast in doubt. These common threats must be countered by collective responses undertaken by the CTI.

10.2.2.3   *Threats and responses*

Global and regional stressors put severe threats on fisheries and marine resources in the CT. These life-sustaining fisheries are heavily dependent on coral reefs, mangroves and seagrass beds that cover an area of 132,800 km^2 in the CT region. Fish remains a significant source of food, contributing 14% to 19% of dietary protein in the six CT6 countries (FAO, 2010). Capture reef fishery is valued at $9.9 billion, representing 10% of the global market. The annual net benefits, including tourism and other ecosystem services, are estimated to be $12 billion (Burke *et al.*, 2002). Over 120 million people are dependent on ecosystem functions, goods and services that collectively contribute significantly to the regional GDP. However, these myriads of ecosystem services are under immense stress. CTI needs to formulate and implement a concerted series of responses to reduce the stress. Major drivers of these stresses that exert pressures are coastal development, demand for fish and climate change. To achieve the goals of the CTI, CT6 needs to build the capacity of local, national and regional bodies in planning and implementing the CTI national and regional plan of action (NPOA), through science-based learning and research, and through empowerment of equitable partnerships. Working with the Asian Development Bank, the CTI has compiled baseline status of the CT region, and provided the biophysical, governance and socio-economic features of the CT6 countries. This baseline status enables elucidation on the relationships between the ecological and the socio-economic characteristics in the CT6 countries and the identification of the threats, vulnerabilities and emerging issues faced by each member country. The primary concerns are focused on addressing poverty alleviation, population density and governance reforms to achieve sustainable development. In the Philippines, for instance, the high degree of poverty and high population density in coastal communities exacerbate the overexploitation of marine resources and cause widespread degradation of the local environment (White and Cruz-Trinidad, 1998; Green *et al.*, 2003). Climate impacts will intensify the degradation of the marine environment and its precious resources, and will diminish the prospects of sustainable development in CT6. Managing the declining marine resources within CT6 is a challenge, given an environment of diverse national governance systems.

10.2.2.4   *Diverse national governance*

Countries in CT6 have diverse systems of governance in managing their marine resources. The CT-Pacific countries, for example, often have established cultural tenure systems, based upon past experiences and belief systems. A majority of fishing areas in CT-SEA countries are, however, subject to de facto open access and face the dilemma characterized as tragedy of the commons. The tragedy of the commons is a situation in a shared-resource system where individual users, acting independently according to their own self-interest, behave contrary to the common good of all users by depleting or

in the shared basin areas. Hence, understanding the sustainability aspects of bilateral water governance can be very helpful to transboundary water governance for other river basins such as the MRB.

10.2.4 Mekong River Basin

The MRB is shared among six countries in Southeast Asia, namely, China, Myanmar, Vietnam, Cambodia, Laos and Thailand. Aquatic biodiversity in the Mekong River system is the second highest in the world, second only to the Amazon. The Mekong River boasts the most concentrated biodiversity per hectare of any river. Wild-capture fisheries play an important role in supporting livelihoods for the poor rural people who can access the river for food and income, contributing to SDG2 and SDG1. The Mekong Basin has one of the world's largest and most productive inland fisheries. An estimated two million tons of fish are landed each year, in addition to another half a million tons of other aquatic animals, contributing US$4 billion to US$7 billion a year. Aquatic resources make up between 50% and 80% of animal protein in rural diets for people who live in the Lower Mekong Basin. The Mekong River Commission (MRC) was jointly set up in 1995 by the governments of Cambodia, Laos, Thailand and Vietnam to jointly manage the shared water resources and the sustainable development of the Mekong River Basin. Its mission is to promote and coordinate sustainable management and development of water and water-related resources for the countries' mutual benefit and the people's well-being. The emphasis is on the livelihoods of the people in the Mekong region, emphasizing bottom-up solutions. China and Burma became dialogue partners of the MRC since 1996 and steadily escalated their non-binding participations in its various forums. Agricultural expansion, population pressure and dam constructions are the major causes of land-use and landscape changes that have altered the Eco hydrology of the Mekong. Both drought and flood are common hazards in the Mekong Delta, which is sensitive to upstream hydrological changes and GCC. Coordinating large infrastructure projects to cater to hydropower generation, flood control and irrigation needs appears a daunting task for the MRC. The level of success achieved in the GLB is not duplicated in the MRB, perhaps because of the relatively less developed economies in the MRB. The GLB and the MRB are typical examples of transboundary coalitions comprising traditional cities linked by some common dreams or goals. In the context of climate action and adaptation, complicated by population growth and increasing demand for affluence, cities need to evolve to face the challenges of the future. The top priority for future cities is sustainability and resilience.

10.3   Sustainable and Resilient Future Cities

The government of Indonesia has announced its plan to relocate the current capital Jakarta to the island of Borneo. The new capital city of Jakarta should be built to be sustainable and resilient. What are the desirable and priority features of a sustainable and resilient city? Clean water and sanitation (SDG6) and healthy life below water (SDG14) constitute the foundation of a sustainable resilient future city such as the new Jakarta (SDG11). Capable of planning and implementing timely climate actions, cities, like Jakarta, are the optimum strategic scale for sustainability action on climate change mitigation and adaptation (SDG13) to combat the high levels of greenhouse gas emissions in urban cities. Cities have the commanding influence of municipal governments over sustainability goals, covering local land use, carbon control policies and transformation toward a green circular economy (Koh and Teh, 2020a). Endowed with high-quality education such as universities, cities have the moral obligation to provide leadership in education for sustainable development (ESD) and to actively engage with communities toward the SDGs. The sustainability goals generally refer to three areas: (1) zero carbon (green transportation, green building), (2) zero waste (effective waste management and policy) and (3) healthy ecosystems (clean water and sanitation, nature, clean air, local food). Many cities are located along coastal zones, subject to vulnerability of coastal hazards and disasters such as tsunamis, storm surges, coastal erosion, marine pollution and loss of protective mangrove habitats. To be sustainable, future cities must pay close attention to resolving issues related to these coastal vulnerabilities and disturbances. Facing critical and chronic water shortages, future cities must ensure sustainable clean water supply. About 50% of the current world population lives in the cities, i.e., about 3.6 billion people live in urban cities out of the

7.2 billion total world population. The high concentration of people living in the city exposes the urban population to high vulnerability to natural and anthropogenic hazards and disasters, in the forms of acute shocks and chronic stresses. The urban population is projected to increase to 6.3 billion or 64% of the total world population by 2050 (UN DESA, 2012) because of the continuing mass migration of people from rural areas to cities. Hence, sustainable development in the future is primarily determined by its success or failure in the cities. This mass migration would vastly intensify future cities' vulnerability to acute shocks and chronic stresses. Failure to ensure water security, failure to sustain food security and failure to resolve issues connected to severe floods will severely undermine the sustainability and resilience of a city, current and future.

10.3.1 *Concepts of resilience and sustainability*

Future cities must therefore develop resilience to ensure their long-term sustainability. Resilience is the adaptive capacity of a city potentially exposed to hazards and disasters to achieve and maintain an acceptable level of functioning and structure (UNISDR, 2005). Sustainability is the state in which the needs of the present are met without compromising the ability of future generations to meet their needs. A sustainable city would have the ability to maintain and support its activity over the long term by balancing economic viability, environmental integrity and social equity. A resilient city can organize itself to increase this adaptive capacity to learn from past disasters for better future protection and better risk reduction. Future cities, therefore, need to develop resilience to both natural and anthropogenic hazards and disasters in the forms of acute shocks and chronic stresses. Hence, a resilient future city must develop the capability to absorb acute shocks and chronic stresses and to quickly recover from the resulting failures. The most important acute shocks and chronic stresses confronting a future city are intimately connected with the water sectors and their complex and interconnected components (Maruyama, 2016). Urban resilience mandates the capacity to think holistically of the city as an urban being with many interconnected and complex systems (Renner, 2018) and to comprehend how the water systems dominate the web connecting these integrated and complex systems. The need of urban water security presents complicated and convoluted challenges that are difficult to solve. These "wicked" challenges are difficult to solve because they involve convoluted and dynamic processes covering natural science, social science and public policy that span over extensive spatial and temporal scales. For the future cities, sustainability and resilience to these wicked shocks and stresses are keys to their success. These wicked shocks and stresses are dominated by the lack of urban water security.

10.3.2   *Urban water security*

Urban water security is the key to sustainable development in future cities. Water is essential to the protection of crucial ecological services rendered by ecosystems and is imperative to the protection of the biosphere. As a global public good linking food, health, urban and environmental security, water security was declared one of the five priority issues at the 2002 World Summit on Sustainable Development in Johannesburg. The 2002 World Summit calls for a stop to unsustainable exploitation of water resources. It urges wise water reuse and it mandates improvement to the regulation of flow of nutrients to increase crop cultivation per drop of water used and to reduce environmental stress and degradation. Water consumed in food production surpasses the combined amount of water utilized for household, urban and industrial uses by two orders of magnitude. Hence, water security and food security are synonymous because food productions lock in an astronomical proportion of water, known as virtual water.

10.3.3   *Urban virtual water*

Urban dwellers consume far more water per capita compared with their rural counterparts. Water conservation at the city scale is therefore the key to water security for the country. Most of the water consumed in the city is virtual water. Virtual water refers to the water virtually embedded in food products and food productions, amounting to around 1,100 km^3/yr worldwide. Of this, about 90% or 1,000 km^3 of virtual water is traded annually at the global level, suggesting a strong imbalance in water and wealth distribution worldwide. Seven out of ten food-importing countries are water short, while a similar proportion of food-exporting countries are water rich. This accounts for a high proportion of virtual water traded globally, often over long distances incurring high financial and energy costs. Each of us drinks 2 to 5 liters of water and uses 40 to 400 liters for hygiene and sanitation each day. This vast discrepancy in individual water consumption is reflective of uneven wealth distribution and is indicative of wastes and inefficiency in managing water consumption. More than 1.0 billion people today lack satisfactory access to safe drinking water and another 2.0 billion go by without adequate sanitation. This water shortage will increase, given population growth and lifestyle changes that will increase individual consumption of water, particularly in the cities. Water crisis in cities appears inevitable if business as usual continues. Urban rain-fed agriculture may be a way forward by storing and utilizing excess rainwater runoff for crop cultivation within the city. Crops and food plants within cities enhance local food security, create esthetic values, help regulate urban temperature and reduce atmospheric carbon content, while rainwater harvesting moderates urban floods, improves water quality and promotes water security at the local level. In short, more crop per drop.

10.3.3.1 *More crop per drop*

Food production is the major driver of human water use. Future cities can contribute by growing crops and food plants within the cities, using technology and human ingenuity that produce more crop per drop of water used. Agriculture dominates humanity's water needs, consuming 70% to 80% of the world's water. Developing countries receive some 80% of export earnings from the agricultural sector. For each kilogram of grain produced, about 500 to 4,000 liters of water is consumed or lost to evaporation. For beef, it is of the order of 10,000 liters per kilogram of beef produced. With this in perspective, it is not surprising that each of us consumes between 2,000 and 5,000 liters of *virtual* water a day, considering every possible source of water consumption, mainly from food. Uneven geographical distribution of water further renders the efficient utilization of water extremely difficult. The United States, for example, has a storage capacity of about 5,000 m^3 per person, while in part of Nile region, the storage capacity is only about 50 m^3 per person, a difference of two orders of magnitude. Equalizing this vast imbalance of water storage capability is virtually impossible. Feeding the world population on an acceptable nutritional level in 2030 would require an additional water appropriation of 3,800 km^3/yr, the provision of which is by no means assured. Hence, more crop (yield) per drop (of water) consumed is essential to sustain growing human food requirements. Rain-fed agriculture globally consumes approximately 4,500 km^3 of water annually, compared to the 2,500 km^3 from irrigation. Rain-fed agriculture is far more efficient in water usage than irrigation-driven agriculture. This perspective should be kept in clear view. China, ironically, is a water-rich country with 2,200 m^3 of water per person, mostly in the south. By 2030, this figure will decline to 1,700 m^3 per person. For China, the most arable land is in the north, where water is scarce. Hence, there is an urgent need for the south-to-north transfer of water from the Yangtze River, a mega project challenging from the financial, environmental, ecological and engineering perspectives. This mega water transfer scheme underlines the importance of water conservation, including the culture of more crops per drop of water consumed in crop and food productions. Good governance in the water and food sectors is critical to achieving more crop per drop through holistic back-to-nature approaches while preventing environmental pollution and ecological degradation. Legal and institutional framework should be put in place to incentivize urban rainwater harvesting and rain-fed urban agriculture to enhance water security and food security at the city scale.

10.3.4   *Urban rainwater harvesting*

One of the UN Millennium Development Goals (UNMDGs) is to reduce by half the proportion of people without sustainable access to safe drinking water. Another goal is to reduce by half the proportion of people who suffer from hunger. In some countries, both goals are far from being fulfilled by 2015 when the UNMDG agenda ended. One quarter of the global population lives in coastal regions that contain less than 10% of the global renewable water supply. These coastal urban cities are undergoing rapid population growth, further exerting additional stress on water demand. Salinity intrusion due to excessive water withdrawals from coastal aquifers is expected to be exacerbated by the effect of sea level rise, leading to reduction of freshwater availability. Taken together, these potential changes in the quantity, timing and quality of surface water and groundwater will impact the reliability of safe water supplies by various degrees (Alcamo *et al.*, 2007). Water scarcity is a major problem in many countries, particularly in urban cities. Rainwater constitutes a potential source of potable and non-potable water that could mitigate water and food crisis in some of these regions. Rainwater harvesting (RWH) is a technology where surface runoff is effectively collected during rain for later consumption at source. Facing a serious water crisis, Australia has highly promoted the use of rainwater tanks for rainwater storage for both outdoor and indoor uses. Rainwater tanks are used by about 16% of Australian households, out of which about 13% use rainwater tanks as their main source of drinking water (Australian Bureau of Statistics, 2006). To further promote RWH, Australian state governments and local councils have offered cash rebates to support installation of rainwater tanks in households. Cost-effectiveness on the use of rainwater tanks for Australian residential environment is examined in seven cities, Gold Coast, Brisbane, Melbourne, Sydney, Adelaide, Perth and Canberra. Cost for installing and operating rainwater tanks is compared to the cost of alternative water sources including building additional dams and desalination plants. It is found that using rainwater is an economically viable option for households in Gold Coast, Brisbane and Sydney (Tam *et al.*, 2010). Sustainable, resilient cities should therefore explore the option of utilizing RWH techniques to fulfill a substantial part of the water need. The new Jakarta can learn a good lesson from Australia in the governance for water and food security.

10.4   Governance for Water and Food Security

10.4.1   *Transparency and accountability*

Transparency and accountability are the keys to more crop per drop to ensure water security, to ensure ecosystem security and to ensure food security in the cities. To achieve more crop per drop, it is essential to have a clear separation of roles and responsibilities between the policymakers, the regulators and the service providers to prevent water wastage by controlling corrupt practices in the water sectors. Corruption in the water sector is a serious problem that results in less crop per drop, that reduces economic growth and incurs social cost. It has been estimated by the World Bank Institute that more than USD 2 trillion is paid in bribes annually in developed and developing countries in the water sector alone. Corruption is a global phenomenon that affects all societies and that threatens economic growth, political stability and sustainable development. Countries that tackle corruption and improve their rule of law can increase their national incomes by as much as four times, according to one study. Countries that can tackle corruption effectively will have a far better chance of securing water security. Beyond corruption, the approach to water security should integrate transboundary perspectives across the border between natural sciences, engineering, social sciences and the political dimensions to harmonize the intricate web of dependencies and not to succumb to the complexity.

10.4.2   *Transboundary perspectives*

To achieve comprehensive goals, we need to adopt a holistic and transboundary approach that integrates across disciplinary dichotomy, that transcends social-political divides and that connects the urban–rural border. We need to promote transboundary perspectives on all water and environmental issues that involve inter-basin transfer and that can reconcile any major differences that might arise across boundaries. The perceived hazards for the technocrats in stepping outside their comfort zone can be prohibitive and hence must be minimized. Equally, the hazards of failing to engage the political world can lead to misguided and costly policies and practices. The success in water security will depend both on integration and governance of all sectors and actors in the water nexus. Politicians work on their perceptions of the problems, which may not be scientifically sound. The recognition that agricultural water usage exerts a high price in terms of desiccated rivers, gross global pollution and disappearance of wetlands will compel the adoption of this integrative approach. With the resources reallocated after achieving water security, we may have a greater capacity to develop systems to cope with the impact of other major disasters associated with climate change. Specifically, climate change needs to be incorporated in national development strategy and national water policy. Rain-fed agriculture (the use of green water) and irrigation agriculture (the use of blue water) as well as gray water (the use of treated wastewater) must be integrated with the best practices of producing water-wise crops. This holistic integration contributes to more crop per drop, and empowers the essential resources to other major challenges, and to the provision of adequate clean water and satisfactory sanitation.

10.4.3 *Holistic integration*

The world may soon run out of water to grow more food, to provide adequate water supply and satisfactory sanitation, all of which are basic human needs. Reallocation of water from agricultural usage to the cities will further aggravate this water shortage, linking issues of urban security and food security together. Food production needs two orders of magnitude more water than other uses combined; hence, the urgent need is to start looking at the food sector to vastly improve water consumption efficiency. A hard engineering approach relying on large-scale infrastructure to provide the much-needed water may be vital for economic growth and improved living standards, funded by a mix of public and private equity. As energy is needed for water supply and sewage treatment, any disruption in energy supply is a threat to safe water and sanitary services. This further expends and complicates the delicate dependencies and linkages of the various sectors in the water nexus.

Governance in these sectors is weak and ineffective, characterized by a lack of transparency and accountability, particularly in large-scale infrastructure developments. Soft social approaches involving the delicate human dimension of water management are getting more important. A mundane example includes low-flush toilets and low-flow showers to conserve precious urban water. Subsidies, including political aids, are unsustainable in the long run and must be gradually phased out to adequately address the actual cost of water procurement and delivery and to promote water conservation. Improving institutional and governance arrangements in conjunction with building more dams and constructing longer delivery pipes is essential. Institutions per se cannot deliver water, but good governance can reflect the actual cost of water procurement and delivery and to promote active and sustainable conservation. To be effective, governance must embrace a clear separation of policy, regulatory and service delivery functions. Politicians and the public often act from perceived problems, while technical experts work with diagnosis-based problems, giving rise to a dichotomy between the two sectors. This dichotomy is the stumbling block to urban water security. Small storage systems operating onsite can help to vastly improve the efficiency of rain-fed agriculture. Working in coordination with improving water availability at the root zones and proper management of soil, appropriate use of fertilizers and tillage has been proven scientifically to be capable of achieving a 100% increase in crop production in rain-fed farming as compared to merely a 10 to 15% increase achieved in irrigation systems. The future of food security may lie in this simple truth that back-to-natur e approaches are more efficient in conserving precious resources by harmonizing the abundant renewable resources at source through gravity flows.

10.4.4   *Back-to-nature approaches*

Nature organizes the conveyance of water, pollutants and sediments out of the watershed to the sea in the most energy-efficient way possible through gravity. This efficient water cycle sustains the entire biosphere. The more diverse the water cycle system is, the better it can cope with disturbances, which occur naturally at regular intervals. However, large-scale human-induced disturbances have induced the Hurst Phenomenon, with occurrences of extremes floods and prolonged draughts of long durations at increased frequency, far beyond the capability and endurance of mankind. The recent extreme floods in China, India, the USA and elsewhere are a timely reminder of what may become a common occurrence due to global climate change impacts and other human abuses. Anthropogenic abuses and GCC impacts collectively have led to the increase in extreme hazards related to extreme climatic events, such as floods, droughts and related disasters such as epidemics. Of the anthropogenic abuses, wetland removal covering extensive areas to cater to urban growth constitutes the worst abuse, as in the case of the Greater Everglades. Hence, conservation and restoration of wetlands are essential in reversing the impacts of anthropogenic abuses, particularly in urban settings such as Miami. However, such restorations are costly and time consuming, as exemplified by the Greater Everglades Ecosystem Restoration (GEER) and the Comprehensive Everglades Restoration Plan (CERP) that cost more than USD 10 billion over a duration of 30 years.

10.4.5 *Wetland conservation and restoration*

Restoring wetlands is an important part of giving back to nature some of the land that has been robbed, drained, changed, polluted and mistreated. Restoring wetlands is essential to control extreme hazards related to extreme events as wetlands can provide an effective and efficient mechanism to "absorb" the impacts of extreme floods and droughts. Removal of extensive wetlands may aggravate flood or drought hazards. The extensive damage inflicted during the 1993 Great Flood of America along the Mississippi-Missouri basin is a consequence of removal of a large expense of riparian wetlands along the Mississippi-Missouri river basin. Wetlands provide the best wastewater treatment possibilities for poor regions of the world. Protecting the wetlands instead of trying to restore them is the best alternative where possible. Once disturbed, wetland restoration is an extremely complex, very costly and long-term endeavor. Extensive resources are allocated for the ecological restoration of the Greater Everglades in Florida, USA. The need to return cultivated land to its former forest, grassland or wetland status in the Everglades is beyond doubt. Attention should be devoted to the water needed to support riparian wetlands and aquatic ecosystems, as well as to the capacity required to absorb the fury of floods. We need to share the best practices for biodiversity and ecological conservation, and we need to promote better use of water and to avoid deforesting and draining of wetlands. To achieve these goals, we need to formulate criteria and goals for environmental sustainability with clear and objective rules, with full and fair pricing, giving due recognition to life cycle costs. We need to nurture innovation, to set priority, to reward small-scale solutions on site and to share global best practices. We also need to accommodate various interest groups and engage them in deliberations and we also need to beware of those with self-interest. We need to reduce agriculture based upon irrigation and to promote onsite rain-fed agriculture that can better withstand dry spell damage by supplementation based upon water harvesting on-site. To extract every possible crop from every drop, we need to promote effective sequential and cascading reuse of water before it drains into the sea, before it reaches the last phase of the water cycle where it is lost, unable to be utilized for any meaningful purposes. This must be supplemented by good control and regulation of consumptive use in the upstream regions to protect and accommodate downstream users. Water released through saving from more crop per drop approach will benefit aquatic biodiversity and enhance habitat resilience.

10.4.6   *Aquatic biodiversity*

Globally, loss of 50% of aquatic biodiversity has been reported since 1970. Eighty percent of marine water has been degraded due to land-based human activities. Collectively, the biodiversity loss and ecological degradation pose threats to aquatic and marine ecosystem resilience and environmental security. Ever-increasing nitrate and phosphorus concentrations in the aquatic and marine environment, coupled with global warming and inadequate water resources, have caused increasing algal blooms that produce deadly toxins, depress oxygen levels and clog fish gills. The Mekong River basin, that supports 60 million people via rice production and capture fishing, has already been threatened significantly by eutrophication, altered ecohydrology and biodiversity loss. In South Asia, uncontrolled shrimp farms destroy mangrove areas and contribute to the collapse of sensitive coastal ecosystems. Evaporating agriculture green water depletes the freshwater flow and further induces higher concentrations of pollutants and nutrients. A proper system for the management and restoration of river, lake and wetland water quality and quantity is needed. Groundwater pollution is a major environmental peril because of the long time and high cost of remedial measures. Political and public recognition of this problem is very low since the problem is hidden below the ground: out of sight, out of mind. Soil-gully erosion and landslides following a heavy rain have caused major environmental disasters and biodiversity loss around the world that further reduce the fertility and availability of agricultural land. Hazardous substances such as mercury and PCBs in the aquatic environment may further reduce the availability of gray water for agriculture that may lead to less crop per drop. Contaminated soil and polluted water help spread diseases between the cultivated species and wild species of avian stocks that may trigger a pandemic among humans. These are the potential perils induced by a lack of water security that requires integrated and holistic approaches including appropriate climate actions (SDG13). For coastal zones, mangrove conservation and rehabilitation are integral components of climate actions to enhance the myriad of ecosystem services and biodiversity sustained by mangroves, essential to a vibrant and robust city such as Miami. Climate change and other anthropogenic disturbances have increased the magnitude and frequency of extreme events with a concomitant increase in economic and social costs from natural hazards such as hurricanes, droughts and famine. Increasing productivity in crop yield in semi-arid areas via improved dryland agricultural technology constitutes another major contribution to climate action (SDG13).

10.4.7 *Dryland agriculture*

In arid and semi-arid areas, rainfall often varies in a range of 200 to 600 mm/year (Falkenmark *et al.*, 2001) Potential evapotranspiration can vary between 1,500 and 2,300 mm. The ratio of rainfall to evaporation is unsatisfactory, resulting in poor crops. The relation between rainfall and the potential evapotranspiration determines the growing period lasting about 2.5 to 4 months in semi-arid zones. Three case examples are provided to illustrate the basic concept of integrating rainwater harvesting with appropriate agricultural techniques to sustain dryland agriculture for food security in semi-arid regions.

10.4.7.1 *Loess plateau in China*

In semi-arid areas practicing dryland farming, such as the Loess plateau in China, water shortage severely constrains its economic development and agricultural productivity. This semi-arid region has an annual rainfall ranging from 250 mm to 400 mm. With wide annual variability and uneven seasonal distribution, it receives 60% of its rainfall mainly in July to September. The rainy season, however, does not match the stages of crop growth, resulting in soil and water loss and frequent occurrence of droughts, crop failures and food insecurity. Therefore, it is necessary to exploit other water resources such as rainwater harvesting (RWH) to improve crop water use efficiency, the fundamental approach used for the development of dryland farming in semi-arid lands. In several studies, RWH has been shown to boost the growth of crops, ornamental and forest plants, and can mitigate irrigation shortage and reduce irrigation costs. By integrating rainwater with agricultural techniques, the integrated system is capable of increasing grain yields, improving water use efficiencies and helping to use light rain efficiently. Field experiments were conducted to determine the effects of utilizing rainwater furrow ridge (RWFR) systems on improving water use efficiency and on increasing grain yield of spring corn. Rainfall scenarios varied between 230 mm and 440 mm. Spring corn yield in the RWFR systems was 83% higher when the rainfall scenario was 230 mm compared with the control with no RWH. But, the yield was only 43% higher in the 340 mm rainfall scenario, and merely 11% higher in the 440 mm rainfall scenario. Similarly, water use efficiency was 77% higher in the 230 mm rainfall scenario, but was only 43% higher in the 340 mm scenario, and merely 10% higher in the 440 mm scenario compared to the control with no RWH under the corresponding rainfalls. For the scenario with rainfall of 440 mm, the yield increase of spring corn cannot compensate for the additional costs involved in labor and plastic film used for ridge construction. So, the RWFR is economically viable for adoption when the rainfall is below 440 mm during the whole period of corn growth (Ren *et al.*, 2008).

10.4.7.2   *Gansu China*

In the semi-arid region of Gansu in China, economic analysis was performed to examine the feasibility of dryland agriculture using rainwater harvesting coupled with supplemental irrigation. The results positively indicated the financial feasibility of using open-air-hardened surfaces such as highway surfaces, courtyards and roofs to collect rainwater since it is the cheapest way. In addition, unoccupied land should be used to augment rainwater catchment areas to supplement the harvest of rainwater. The adoption of water conservation techniques including seepage control should be utilized to maximize the utility of the harvested rainwater. For optimal investment, it is essential to select crops with a water requirement schedule that coincides with local rainfall events. Potato was found to be the most suitable crop in the studied region (Tian *et al.*, 2003). The economic returns for potato were superior to spring wheat, corn and wheat–corn intercropping. Therefore, potato production using rainwater harvesting and supplemental low-impact irrigation is the best option for cropping systems in the semi-arid region of Gansu, China.

10.4.7.3 *Sub-Saharan Africa*

The semi-arid savannah environment (SASE) of sub-Saharan Africa is characterized by low erratic rainfall with risk of droughts, intra-seasonal dry spells and frequent food insecurity. The subsistence rain-fed agriculture and livestock production compete for the limited water resources. There is an increased interest in improving rain-fed agriculture through adoption of rainwater harvesting (RWH) technologies. Rainwater harvesting has the potential of addressing spatial and temporal water scarcity for domestic consumption, crop production and livestock development. Upscaling RWH techniques may contribute to environmental management and overall water resource management in SASE (Ngigi, 2003). This RWH potential should be exploited to combat persistent low agricultural production and food shortage in sub-Saharan Africa to enhance food security and to maintain hydro-ecological balance in the semi-arid savannahs of Africa. Agricultural innovation coupled with RWH can be a promising tool to achieve water and food security in semi-arid regions. Another crucial innovation in climate action is to conserve mangrove–hammock ecosystems and the associated fresh surface and groundwater resources by concerted efforts of management. An example of this is the GEER and CERP in the Greater Everglades in southern Florida.

10.5 Climate Action for Mangrove Conservation

Current projections of GCC and SLR indicate that SLR could have multiple impacts on mangroves throughout the world (Ward *et al.*, 2016). Because of their landscape position in the intertidal zone, mangroves are directly affected by SLR, but the effects will depend on many factors including local topography, slope, rate of SLR, sources and amount of sediments, and extent of areas available for landward migration (Woodroffe, 1990). Vertical sediment accumulation in mangrove root systems may allow mangroves to keep up with the pace of SLR in areas of higher elevation and in areas with relatively low tidal range. On the contrary, areas located at lower elevation and those areas with greater tidal range may not be able to keep up with SLR (Ellison, 2000; Ward *et al.*, 2016). Given the right conditions, mangroves will continue to progressively move inland as sea levels rise, to areas where suitable conditions prevail (Ellison, 2000). Rising global temperature may have little effect on areas already occupied by mangroves in the tropics, but could allow mangroves to migrate and colonize land farther poleward. Increasing carbon dioxide and nitrogen enrichment can augment the ability of Spartina to suppress growth of mangrove seedlings and change the competitive relationship between the vegetation types (McKee and Rooth, 2008; Zhang *et al.*, 2012). Increasing carbon dioxide concentrations can alter the competitiveness of mangroves, particularly of *Laguncularia racemosa* (Snedaker and Araujo, 1998). Mangroves also serve as sinks for carbon, thereby helping to mitigate the adverse impact of climate change. Carbon sinks are created through the accumulation of living biomass, through litter and deadwood deposition, and via the trapping of sediments delivered from the uplands. Carbon in mangrove sediments does not turn over as quickly as it does in terrestrial soil, but rather builds up vertically in response to SLR (McLeod *et al.*, 2011). In the context of climate action, several issues are pertinent, such as the ecosystem services provided by mangroves that could be vulnerable to GCC and the associated SLR.

10.5.1   *Ecosystem services*

Mangrove forests support a myriad of physical and ecosystem services (Stokstad, 2005) and provide economic values that are just beginning to be recognized (Barbier *et al.*, 2008; Perillo *et al.*, 2009). Mangrove forests live in the tropical and subtropical regions at the interface between marine and terrestrial environments. They grow around the mouths of rivers, in tidal swamps and along coastlines, where they are regularly inundated by saline or brackish water. Changing ecohydrology caused by GCC and SLR can pose hazards and vulnerability to mangrove survival. Numerous studies show that mangroves are the nursery habitats for juvenile coral reef fishes of many species. Mangroves, especially the prop roots of Rhizophora, provide structural heterogeneity that is favorable both to prey attempting to avoid predators and to predatory fish searching for invertebrate prey hiding within the root structure.

Covering about 15 million ha worldwide, mangrove forests are considered among the world's most treasured but also the most severely threatened tropical ecosystems. Found in tropical and subtropical regions, mangrove forests are a vital component of coastal wetlands that serve many physical, chemical and ecological functions. Mangroves and other wetlands are valuable resources that support numerous forms of wildlife, particularly fish and birds, and should be protected and preserved. An important service provided by mangroves is the provision as a nutrient and carbon sink. Denitrification in the anaerobic environment and nitrogen fixation by certain bacteria and cyanobacteria associated with mangrove mud and with aboveground root systems can improve water quality by removing wastewater pollutants (Ewel *et al.*, 1998). Mangroves are effective in carbon sequestration and in provision of many ecological services, particularly that of coastal protection against coastal disturbances.

10.5.2 *Protection against tsunami*

Mangroves play an important role in protecting coastal communities from coastal disturbances such as storm surges, tsunamis and floods that can damage property and cause deaths and severe injury. The protection provided by mangroves has been observed and measured in laboratory experiments and in numerical simulations and field studies. The magnitude of energy reduction offered by the presence of mangroves strongly depends on three major factors, namely, mangrove structures, topography-bathymetric features and wave characteristics. Understanding this dependency is crucial to improving the ecosystem services provided by mangroves, to reducing mangrove vulnerability and to enhancing their resilience to these coastal disturbances.

10.5.3 *Vulnerability and resilience*

Around one quarter of the world's mangroves have been lost due to anthropogenic activities, particularly through conversion to aquaculture, agriculture and urban land uses (Friess and Webb, 2014). In the future, mangroves may suffer more loss and further degradation due to GCC and SLR. Of the 15 million ha of mangroves worldwide, over two-third are distributed in eighteen countries, most of which are in developing economies, such as Indonesia, Brazil,

Malaysia, the Philippines, Thailand, Vietnam, Bangladesh and India (Giri *et al.*, 2011). The underdevelopment of developing countries exposes mangroves to unmitigated anthropogenic and natural disturbances because of inadequate resources allocated for mangrove protection. The global disappearance of mangroves may therefore continue, which will have a major impact on the vulnerability of coastal communities in these developing countries, from damaging and life-threatening storm surges and coastal floods (Barbier and Lee, 2014). The vulnerability to GCC and SLR of rural populations living in the low-elevation coastal zones of developing countries is a concern (Barbier, 2015), directly threatening their local water security and food security because of soil salinization. The impacts of GCC arise in two ways. First, changes in precipitation, temperature and hydrology in response to GCC may threaten coastal and near-shore eco-hydrology and ecosystems (Spalding *et al.*, 2014). Second, GCC and SLR may cause mangrove loss and degradation due to altered eco-hydrology. This mangrove loss will lead to less protection against coastal storm surges, SLR, saline intrusion, seawater inundation and coastal erosion. The impact is particularly damaging to low-lying atolls across the world. The ecosystem impairment from GCC, SLR, saline intrusion, seawater inundation and coastal erosion is gradual and long term. The damages due to storm surges and coastal floods are both acute and chronic. These impairments and their response require further research to enhance foreshore protection.

10.5.4   *Foreshore protection*

Mangrove ecosystems are located at the interface between three distinct but interconnected ecosystems: coast, land and watershed ecosystems. This unique land–sea interface provides a high degree of connectivity across these three systems, which could lead to the interconnected provision of ecosystem services. Hence, Alongi (2008) suggests that the extent to which mangroves offer protection against catastrophic natural disasters, such as tsunamis, may depend on the relevant features and conditions within the mangrove ecosystem, as well as those within the foreshore habitats. For mangroves, the important characteristics critical to wave attenuation include width of forest, slope of forest floor, forest density, tree diameter and height. The proportions of aboveground biomass in the roots, soil texture and forest location (open coast versus lagoon) are features that may determine the degree of protection provided by mangroves. The presence of foreshore habitats, such as coral reefs, seagrass beds and dunes, may provide additional protection. In the Caribbean, mangroves appear to protect shorelines from coastal storms, but may also enhance the recovery of coral reef fish populations from disturbances due to hurricanes and other violent storms (Mumby and Hastings, 2008). Modeling simulations for an interconnected reef–seagrass–mangrove seascape confirm that the storm protection service of the whole system is greater than for a single coastal habitat on its own. In addition, modeling of this connectivity in providing storm protection and other ecosystem benefits of a reef–seagrass–mangrove seascape may also determine the optimal spatial location of development activities in the mangrove portion of the seascape (Barbier and Lee, 2014). This approach depends on reliable and robust economic valuation of ecosystem services provided by mangroves.

10.5.5 *Ecosystem services valuation*

Mangroves provide protection against storms and coastal floods, mainly through their ability to attenuate waves or buffer winds (Marois and Mitsch, 2015). Past valuations of the storm protection benefit of mangroves have relied on the replacement cost method, by estimating this protective value with the cost of building man-made storm barriers as a replacement. More reliable methods instead model the production of the protection service of mangroves and estimate its value in terms of reducing the expected damages or deaths avoided by coastal communities. Understanding the economic value of mangroves in providing protection against storms, flood damages and other coastal hazards is important for the broader policy issue of evaluating the vulnerability of the rural poor due to the continual loss of mangroves. The economic valuation of the coastal protection benefit provided by mangroves is undergoing an important transformation in valuation methods, leading to continual progress in the valuation methods employed to estimate this benefit. However, beyond storm protection, mangrove ecosystems provide other important economic benefits including carbon sequestration, collected wood and non-wood products, and support for offshore fisheries (Huxham *et al.*, 2015). Uncertainty and complexity in valuation give rise to an ongoing debate over whether the cost of mangrove restoration is higher than the value of the coastal protection service provided by these ecosystems (Sandilyan and Kathiresan, 2015). Coastal reclamation in areas fringing mangroves has been rampant in many Asian coastal cities such as Singapore, Shanghai and Seoul. Coastal reclamation often results in intense disturbances to coastal physical environment and its ecosystems, leading to loss of habitats and biodiversity, and to depletion of fishery resources. These negative impacts call into question the sustainability of coastal reclamation.

10.5.6   *Sustainable coastal reclamation*

These adverse and significant impacts arising from coastal reclamation complicate the uncertainty in the ICZM debate that seeks to protect coastal resources such as mangroves and their associated ecosystems. The impacts upset the natural balance between upland hydrology, tidal wave systems and sediment transport, and set in motion a process that is detrimental to long-term sustainability of the coastal physical and other resources. Policymakers must be mindful about the trade-offs between the interest to cater to transient demands for coastal land and the permanent interest to ensure sustainable use of the marine and coastal ecosystems over the long term. We may be able to meet the demand for land for economic development and maintain the health and resilience of the coastal ecosystems, by evaluating and deciding on the optimal scales and best locations for reclamation.

Since 2001, the Chinese government issued a series of policies on coastal reclamation and sea-use laws to cope with changing conditions in domestic and international economic environments. The major goal is to maintain a balance between economic development and marine environmental protection. The analysis by remote sensing data from 2002 to 2018 indicated that the total area of reclamation reached an alarming 2,892 km^2. However, the low utilization ratio of reclaimed land reflects significant idleness of the land post reclamation. It mandates the necessity of formulating a policy to revitalize the existing stocks of reclaimed land instead of reclaiming new land. Special efforts should be devoted to reinforcing law enforcement, to adequate supervision, and to policy risk assessment and simulation. The sea-use policy system has contributed significantly to managing reclamation in China. However, the main sea-use policy direction changed to development oriented from constraint oriented at a time when China attempted to increase the GDP during the global financial crisis around 2008, triggered by the subprime mortgage crisis in the US. The sharp turn of policy orientation toward development had resulted in the surge of reclamation areas, leading to much of the reclaimed land being still idle. Hence, subsequently, the policy made another sudden turnaround to a constraint-oriented approach to tighten reclamation in order to reduce idle land and to improve utilization of reclaimed land. The concept of sustainable sea use and integrated marine ecosystem-based management should be advocated and enforced in China and indeed in other developing countries with a similar reclamation status (Li *et al.*, 2020). Moreover, the GDP-oriented government performance appraisal system practiced in China should be shifted to a sustainability-oriented appraisal system, or an integration of both appraisal systems. Sustainability in the context of climate change is therefore complex. Hence, effective ESD is a critical success factor.

10.6 Resolving Urban Flood and Poverty Trap

Urban floods have great adverse socio-economic impacts, by causing disruptions to city services such as transportation, sewerage, communication, electricity supply and commerce. Floods inflict grave damages to urban infrastructure and cost human life. An efficient urban drainage system (UDS) is, therefore, an essential part of city infrastructure required to permit an acceptable level of urban services each time it rains. Climate change (CC) and urbanization are the two most influential impacts that challenge urban stormwater management strategies for the current and future cities (Miller and Hutchins, 2017; Zhou *et al.*, 2019). Both can exert large impacts on water cycles, patterns of precipitation extremes and surface runoffs. They increase the frequency, duration and intensity of urban floods. Well-planned CC adaptation measures must incorporate efficient UDS that helps prevent or reduce the adverse impacts of floods in order to minimize economic losses, prevent damage to infrastructure and save lives. Robust adaptation decisions and measures for flood mitigation can be achieved by benefit and cost analysis, i.e., by critically evaluating the potential reduction of risks or losses versus the costs incurred by installing effective adaptation measures (Halsnæs *et al.*, 2015). Recent innovations in efficient UDS measures incorporate soft approaches such as nature-based low-impact development (LID) that can contribute to enhancing UDS resilience. LID refers to the technology that mimics the natural drainage system existing before development. LID consists of two broad categories: (a) distributed small-scale measures, such as retention pond, permeable pavement, bio-retention systems, vegetated swales, rain barrel and rain garden, and (b) centralized large-scale measures such as retention basins and wetlands. Both categories of LID seek to mimic natural hydrologic functions through retention, infiltration, evaporation and recycle of stormwater on-site (Ahiablame *et al.*, 2012). In the context of LID as an efficient approach to UDS for flood mitigation, the following observation regarding the total lack of UDS in George Town of Penang is relevant.

10.6.1   *Perennial floods in Penang*

The impacts of CC and urbanization on aggravating flooding in Sg Pinang Basin (51 km^2) on the Penang island of Malaysia are severe. To evaluate the effectiveness of LID measures for alleviating floods in the Sg Pinang Basin, simulations were performed using the popular Storm Water Management Model (SWMM) developed by the US Environmental Agency (Figure 10.2). Simulation results indicate that climate change and urbanization will continue to contribute to severe and frequent floods along the Sg Pinang basin located in George Town (Koh *et al.*, 2005). SWMM simulation suggested that the use of LID technology such as distributed flood retention ponds along the Sg Pinang river is an effective adaptation measure for flood mitigation under the scenario of CC (Teh and Koh, 2020). Further, these adaptation measures can be designed to add aesthetic and cultural values to the cities. However, unlike the Greater Everglades CERP, the government of Penang does not have the political wisdom or the courage to implement appropriate flood mitigation and adaptation measures to confront the combined impacts of CC and urbanization.

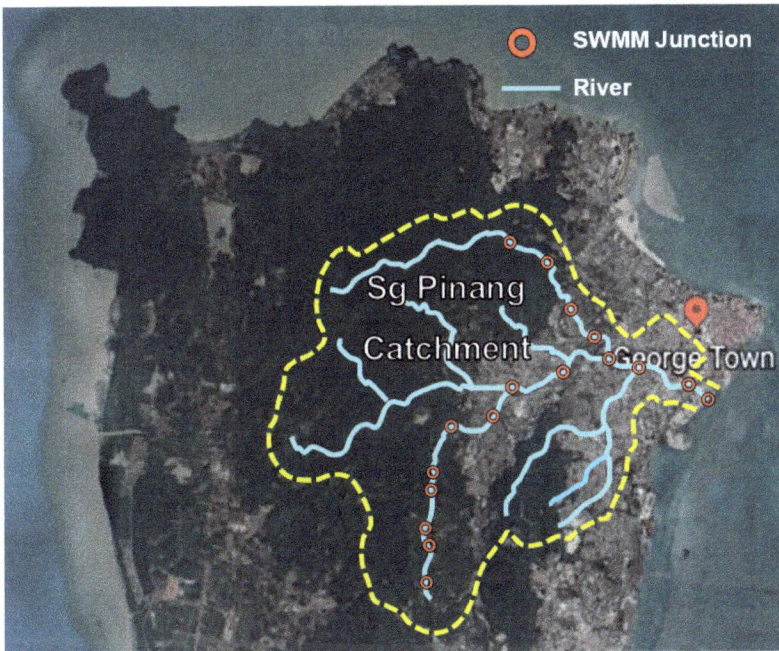

Figure 10.2: SWMM flow network for the Sg Pinang catchment area.

Literature review reveals that the vicious cycle of floods and poverty has the tendency to trap emerging economies, such as George Town and Penang, into perpetual loops of eternal poverty and floods (Koh and Teh, 2021). Innovative concepts and best practices in LID for flood prevention performed in other advanced economies, such as the Netherlands, are urgently needed for Penang to break out of this poverty trap. Without effective flood mitigation, Penang will be doomed, incapable of achieving the goals of SDG1 (no poverty), SDG3 (good health and well-being) and SDG6 (clean water and sanitation). Based upon the hydro-socio-economic scenarios envisaged for Penang, it is unlikely that George Town of Penang will break out from the low-income trap, given its weak political leadership, ineffective governance and absence of investments in flood mitigation measures. A sustainable future city in the context of urban poverty, population growth and climate change is therefore complex. Hence, effective ESD is a critical success factor.

10.7   Education for Sustainable Development

A contested concept, Sustainability hinges on several critical success factors, the most important one of which is ESD. Higher education institutes (HEI) play an important role in providing quality education (SDG4) and in cultivating generations of talented global citizens, committed to a shared vision of a sustainable future, and devoted to the missions of balancing responsibility and privileges across temporal and spatial divides. Toward these visions and missions, HEIs must help develop three Key Sustainability Competences (KSCs): Skills, Spirituality and Strategy. Skills and technical knowledge form the foundation for sustainability science, research and community outreach. Spirituality and social-cultural cohesion provide the fabric for binding diverse stakeholders together for the global common goods. Strategic visions and missions empower self-actualization, the highest level of human needs that ultimately sustain the keystones of sustainability at large (Koh and Teh, 2020b). This book revolves primarily around the visions and missions of ESD and UNSDGs by supporting the development of the three KSC throughout the entire community, of which the HEIs are an integral and important component. The goal is to help readers achieve excellence in sustainable development education and become global citizens toward the goals of the UNSDGs.

References

Ahiablame, L.M., Engel, B.A. and Chaubey, I. (2012). Effectiveness of low impact development practices: Literature review and suggestions for future research. *Water, Air, & Soil Pollution*, 223, 4253–4273. doi: 10.1007/s11270-012-1189-2.

ADB (2014). Regional state of the Coral Triangle — Coral Triangle marine resources: Their status, economies, and management. Asian Development Bank, p. 94.

Alcamo, J., Florke, M. and Marker, M. (2007). Future long-term changes in global water resources driven by socio-economic and climatic change. *Hydrological Sciences*, 52(2), 247–275.

Allen, G.R. (2008). Conservation hotspots of biodiversity and endemism for Indo–Pacific coral reef fishes. *Aquatic Conservation: Marine and Freshwater Ecosystems*, 18(5), 541–556.

Alongi, D.M. (2008). Mangrove forests: resilience, protection from tsunamis and responses to global climate change. *Estuarine, Coastal and Shelf Science*, 76, 1–13. doi: 10.1016/j.ecss.2007.08.024.

Australian Bureau of Statistics (2006). Water account Australia 2004–2005. Australian Government.

Barbier, E.B. (2014). A global strategy for protecting vulnerable coastal populations. *Science*, 345, 1250–1251.

Barbier, E.B. (2015). Climate change impacts on rural poverty in low-elevation coastal zones. *Estuarine, Coastal and Shelf Science*, 165, A1–A13.

Barbier, E.B. and Lee, K.D. (2014). Economics of the marine seascape. *International Review of Environmental and Resource Economics*, 7, 35–65.

Barbier, E.B., Koch, E.W., Silliman, B.R. *et al.* (2008). Coastal ecosystem-based management with nonlinear ecological functions and values. *Science*, 319, 321–323. doi: 10.1126/science.1150349.

Botts, L. and Muldoon, P. (2005). Evolution of the Great Lakes water quality agreement. Michigan State University Press.

Burke, L., Selig, E. and Spalding, M. (2002). Reefs at Risk in Southeast Asia. Washington, DC: World Resources Institute.

Burke, L., Reytar, K., Spalding, M. and Perry, A.L. (2011). Reefs at Risk Revisited. Washington, DC: World Resources Institute.

Burke, L., Reytar, K., Spalding, M. and Perry, A.L. (2012). Reefs at Risk Revisited in the Coral Triangle. Washington, DC: World Resources Institute.

Cabral, R.B. and Aliño, P.M. (2011). Transition from common to private coasts: Consequences of privatization of the coastal commons. *Ocean and Coastal Management*, 54(1), 66–74.

Earle, A., Cascão, A.E., Hansson, S., Jägerskog, A., Swain, A. and Öjendal, J. (2015). Transboundary water management and the climate change debate. Routledge.

Ellison, A.M. (2000). Mangrove restoration: Do we know enough? *Restoration Ecology*, 8(3), 219–229. doi:10.1046/j.1526-100x.2000.80033.x.

Ewel, K.C., Twilley, R.R. and Ong, J.E. (1998). Different kinds of mangrove forests provide different goods and services. *Global Ecology and Biogeography Letters*, 7, 83–94.

Falkenmark, M., Fox, P., Persson G., and Rockstrom, J. (2001). Water Harvesting for Upgrading of Rainfed Agriculture, SIWI Report 11, Stockholm International Water Institute, Sweden. ISBN: 91-974183-0-7.

FAO (2010). Fishery and Aquaculture Statistics Food Balance Sheets. Food and Agriculture Organization of the United Nations. http://www.

fao.org//fishery/publications/yearbooks/en (accessed 19 November 2011).

Friess, D.A. and Webb, E.L. (2014). Variability in mangrove change estimates and implications for the assessment of ecosystem service provision. *Global Ecology and Biogeography*, 23, 715–725.

Giri, C., Ochieng, E., Tieszen, L.L. *et al.* (2011). Status and distribution of mangrove forests of the world using earth observation satellite data. *Global Ecology and Biogeography*, 20, 154–159. doi: 10.1111/j.1466-8238.2010.00584.x.

Green, S.J., White, A.T., Flores, J.O., Carreon III, M.F. and Sia, A.E. (2003). Philippine Fisheries in Crisis: A Framework for Management. Coastal Resource Management Project of the Department of Environment and Natural Resources (DENR). Cebu City: DENR.

Green, A., Ramohia, P., Ginigele, M. and Leve, T. (2006). Fisheries Resources: Coral Reef Fishes. In A. Green, P. Lokani, W. Atu, P. Ramohia, P. Thomas, and J. Almany, eds. Solomon Islands Marine Assessment. Technical Report of Survey conducted from 13 May to 17 June 2004. TNC Pacific Island Countries Report No. 1/06.

Halsnæs, K., Kaspersen, P.S., Drews, M. (2015). Key drivers and economic consequences of high-end climate scenarios: Uncertainties and risks. *Climate Research*, 64, 85–98. doi: 10.3354/cr01308.

Huxham, M., Emerton, L., Kairo, J., Munyi, F., Abdirizak, H., Muriuki, T., Nunan, F. and Briers, R.A. (2015). Applying climate compatible development and economic valuation to coastal management: A case study of Kenya's mangrove forests. *Journal of Environmental Management*, 157, 168–181.

ITLOS (2003). International Tribunal for the Law of the Sea, Hamburg Germany. Public sitting held on Thursday, 25 September 2003. Case concerning Land Reclamation by Singapore in and around the Straits of Johor. Request for provisional measures, Malaysia v. Singapore. Verbatim Record.

ITLOS (2016). The Digest of Jurisprudence of the International Tribunal for the Law of the Sea, 1996–2016, p. 448.

Koh, H.L. and Teh, S.Y. (2020a). Sustainable and Resilient Cities: A Discourse on the Water Nexus. In: Leal Filho, W., Marisa Azul, A., Brandli, L., Gökçin Özuyar, P., Wall, T. (eds.) *Sustainable Cities and Communities*. Encyclopedia of the UN Sustainable Development Goals. Springer, Cham, pp. 694–705. doi: 10.1007/978-3-319-71061-7_111-1.

Koh, H.L. and Teh, S.Y. (2020b). University and community engagement: Toward transformational sustainability-focused problem solving. In: Leal Filho, W., Tortato, U., Frankenberger, F. (eds.) *Universities* and *Sustainable Communities: Meeting the Goals of the Agenda 2030*. World Sustainability Series. Springer, Cham, pp. 791–804. doi: 10.1007/978-3-030-30306-8_49.

Koh, H.L. and Teh, S.Y. (2021). *Urban Storm-water Management for Future Cities: Sustainable and Innovative Approaches*. World Sustainability Series. Springer Nature Switzerland AG, in press.

Koh, H.L., Syamsidik, Lee, H. and Zubir, D. (2004). Pulau Tekong and Changi Reclamation: Modeling and Institutional Perspectives. In: Shinichiro Ohgaki, Kensuke Fukushi, Hiroyuki Katayama and Satoshi Takizawa (eds.) *Proceedings Second International Symposium on Southeast Asian Water Environment*, University of Tokyo, pp. 271–278.

Koh, H.L., Lee, H.L., Saw, S.K., Teh, S.Y. and Izani, A.M.I. (2005). Flood Simulation for Sg. Pinang: Assessing the Potential Impacts of Climate Change. In: Dutta, D., Babel, M.S. (eds.) *Proceedings of the International Symposium on Floods in Coastal Cities under Climate Change Conditions*, 23–25 June 2005, Asian Institute of Technology, Bangkok, Thailand, pp. 45–50.

Li, F., Ding, D., Chen, Z., Chen, H., Shen, T., Wu, Q. and Zhang, C. (2020). Change of sea reclamation and the sea-use management policy system in China. *Marine Policy*, 115, 103861.

Lymer, D., Funge-Smith, S. and Miao, W. (2010). Status and Potential of Fisheries and Aquaculture in Asia and the Pacific 2010. RAP Publication 2010/17. FAO Regional Office for Asia and the Pacific. Bangkok.

Marois, D.E. and Mitsch, W.J. (2015). Coastal protection from tsunamis and cyclones provided by mangrove wetlands — a review. *International Journal of Biodiversity Science, Ecosystem Services and Management*, 11, 71–83.

Maruyama, H. (2016). Resilience, advanced sciences and technologies for security applications. Springer International Publishing, Switzerland.

McKee, K.L. and Rooth, J.E. (2008). Where temperate meets tropical: Multi-factorial effects of elevated CO_2, nitrogen enrichment, and competition on a mangrove-salt marsh community. *Global Change Biology*, 14, 971–984.

McLeod, E., Chmura, G., Bouillon, S., Salm, R., Bjork, M., Duarte, C.M., Lovelock, C.E., Schlesinger, W.H. and Silliman, B.R. (2011). A blueprint for blue carbon: Toward an improved understanding of the role of vegetated coastal habitats is sequestering CO_2. *Frontiers in Ecology and the Environment*, 9, 552–560.

Michel, D. and Pandya, A. (2009). Troubled waters. Climate change, hydropolitics, and transboundary resources. Washington, USA: Stimson Institute.

Miller, J.D. and Hutchins, M. (2017). The impacts of urbanisation and climate change on urban flooding and urban water quality: A review of the evidence concerning the United Kingdom. *Journal of Hydrology: Regional Studies*, 12, 342–362. doi: 10.1016/j.ejrh.2017.06.006.

Mumby, P.J. and Hastings, A. (2008). The impact of ecosystem connectivity on coral reef resilience. *Journal of Applied Ecology*, 45, 854–862.

Nañola, C.L. Jr, Alcala, A.C., Aliño, P.M., Arceo, H.O., Campos, W.L., Gomez, E.D., Licuanan, W.Y., Quibilan, M.C.C., Uychiaco, A.J. and White, A.T. (2006). Status report of the coral reefs in the Philippines. *Proceedings International Coral Reef Symposium*, 4–1, pp. 1055–1061.

Nañola, C.L., Aliño, P.M. and Carpenter, K.E. (2011). Exploitation-related reef fish species richness depletion in the epicenter of marine biodiversity. *Environmental Biology of Fishes*, 90, 405–420. doi: 10.1007/s10641-010-9750-6.

Ngigi, S.N. (2003). What is the limit of up-scaling rainwater harvesting in a river basin? *Physics and Chemistry of the Earth*, 28, 943–956.

Perillo, G., Wolanski, E., Cahoon, D. and Brinson, M. (2009). Coastal Wetlands: An Integrated Ecosystem Approach. Elsevier, Amsterdam.

Reef Check (2010). Reef Check Malaysia Annual Survey Report, 2010. Kuala Lumpur: Reef Check Malaysia. http://ftp01.economist.com.hk/oceans2011/reef_check_malaysia_2010.pdf.

Ren, X., Jia, Z., Chen, X., Han, Q. and Li, R. (2008). Effects of a rainwater-harvesting furrow-ridge system on spring corn productivity under simulated rainfalls. *Acta Ecologica Sinica*, 28(3), 1006–1015. Online English edition of the Chinese language journal.

Renner, R. (2018). Urban being – anatomy & identity of the city. Niggli, Salenstein, Switzerland.

Sandilyan, S. and Kathiresan, K. (2015). Mangroves as bioshields: An undisputed fact. *Ocean and Coastal Management*, 103, 94–96.

Snedaker, S.C. and Araujo, R.J. (1998). Stomatal conductance and gas exchange in four species of Caribbean mangroves exposed to ambient and increased CO_2. *Marine and Freshwater Research*, 49, 325–327.

Spalding, M.D., Ruffo, S., Lacambra, C., Meliane, I., Hale, L.Z., Shepard, C.C. and Beck, M.W. (2014). The role of ecosystems in coastal protection: Adapting to climate change and coastal hazards. *Ocean and Coastal Management*, 90, 50–57.

Stokstad, E. (2005). Taking the pulse of Earth's life-support systems. *Science*, 308, 41–43.

Syamsidik and Koh, H.L. (2003). Impact Assessment Modeling on Coastal Reclamation at Pulau Tekong, p. 8. In: *Proceedings of Regional*

Conference on Integrating Technology in the Mathematical Sciences, 2003, 14–15 April 2003, USM, Penang.

Talukder, B. and Hipel, K.W. (2020). Diagnosis of sustainability of transboundary water governance in the Great Lakes basin. *World Development*, 129, 104855.

Tam, V.W.Y., Tam, L. and Zeng, S.X. (2010). Cost effectiveness and trade-off on the use of rainwater tank: An empirical study in Australian residential decision-making. *Resources, Conservation and Recycling*, 54, 178–186.

Tian, Y., Li, F. and Liu, P. (2003). Economic analysis of rainwater harvesting and irrigation methods, with an example from China. *Agricultural Water Management*, 60, 217–226.

Teh, S.Y. and Koh, H.L. (2020). Education for sustainable development: The STEM Approach in Universiti Sains Malaysia. In: Leal Filho, W. *et al.* (eds.) World Sustainability Series. Springer, Cham, pp. 567–568. doi 10.1007/978-3-030-15604-6_35.

Uitto, J.I. and Duda, A.M. (2002). Management of transboundary water resources: Lessons from international cooperation for conflict prevention. *Geographical Journal*, 168(4), 365–378.

UN DESA (United Nations Department of Economic and Social Affairs) (2012). World urbanization prospects: The 2011 revision. United Nations, New York.

UNISDR (2005). Hyogo framework for action 2005–2015: Building the resilience of nations and communities to disasters. United Nations International Strategy for Disaster Reduction, Geneva, Switzerland. Available via UNISDR. https://www.unisdr.org/2005/wcdr/intergo ver/official-doc/L-docs/Hyogo-framework-for-action-english.pdf (accessed 2 January 2019).

Veron, J.E.N. (2000). Corals of the World. Townsville, Australia: Australian Institute of Marine Sciences.

Veron, J.E.N., Devantier, L.M., Turak, E., Green, A.L., Kininmonth, S., *et al.* (2009). Delineating the Coral Triangle. *Galaxea, Journal of Coral Reef Studies*, 11, 91–100.

Ward, R.D., Friess, D.A., Day, R.H. and MacKenzie, R.A. (2016). Impacts of climate change on mangrove ecosystems: A region by region overview. *Ecosystem Health and Sustainability*, 2(4), e01211.

White, A.T. and Cruz-Trinidad, A. (1998). The Values of Philippine Coastal Resources: Why Protection and Management Are Critical. Cebu City: Coastal Resource Management Project of the DENR.

Woodroffe, C.D. (1990). The impact of sea-level rise on mangrove shorelines. *Progress in Physical Geography*, 14, 483–520.

Zhang, Y., Huang, G., Wang, W., Chen, L. and Lin, G. (2012). Interactions between mangroves and exotic Spartina in an anthropogenically-disturbed estuary in southern China. *Ecology*, 93, 588–597.

Zhou, Q., Leng, G., Su, J., Ren, Y. (2019). Comparison of urbanization and climate change impacts on urban flood volumes: Importance of urban planning and drainage adaptation. *Science of the Total Environment*, 658, 24–33. doi: 10.1016/j.scitotenv.2018.12.184.

Chapter 11

Concluding Remarks

11.1 Introduction

We conclude this book with the following remarks. Global climate change (GCC) may alter local precipitation patterns, while sea level rise (SLR) has the propensity to increase coastal soil and groundwater salinity. In combination, SLR and GCC may cause salinization of coastal groundwater along low-elevation coasts, rendering the groundwater unsuitable for drinking, domestic, agricultural and industrial purposes. Diversion of water away from the Florida Greater Everglades region over the past century has caused a fragmentation of the Everglades ecosystem and has induced salinization of soil and groundwater. The Comprehensive Everglades Restoration Plan (CERP) was authorized by the US Congress in 2000 to rehabilitate the Greater Everglades. Salinization of groundwater due to SLR will pose a daunting challenge to water supply and crop cultivation and will undermine local water security (SDG6) and food security (SDG2) in Pantai Acheh of Penang (Kh'ng *et al.*, 2021). Yet, unlike the Everglades CERP, Penang has no plan at all to safeguard water and food security for the future. This failure on the part of the Penang government to address the impending disaster is criminal negligence. A positive way forward would depend on the vigilance of the communities and on the progress in education on sustainable development.

11.2   Education for Sustainable Development

The 1987 Brundtland Report (WCED, 1987) initiated a transformational process for developing integrated sustainability concepts and best practices to embrace the three pillars of sustainability (environment, society and economy). The goal was to promote an equitable distribution of benefits and responsibility between the current and future generations. This ongoing transformation stimulates an educational paradigm shift toward integrating socio-cultural and economic dimensions with ethical and esthetical values in education for sustainable development (ESD). ESD plays two important roles in achieving the United Nations Sustainable Development Goals (UN SDGs). First, ESD bridges the gaps and disconnect among the three sustainability pillars. Second, ESD trains future professionals and citizens to manage and adapt to complex sustainability challenges. The Decade of Education for Sustainable Development (DESD 2005–2014) highlighted the importance of developing four capacities in students: (1) an integrative capacity that allows for a holistic perspective; (2) a critical capacity that questions taken-for-granted patterns; (3) a transformative capacity that moves from simple awareness to transformative change; and (4) a contextual capacity that embeds the values of pluralism in learning (Wals, 2012). In reality, the complexity of the interconnected sustainability challenges is, however, viewed and interpreted within the context of vastly different values, divergent beliefs and conflicting cultures. It is not surprising that the progress of sustainability has been sluggish, uneven and negligible (Leal Filho *et al.*, 2015), three decades after the launch of the Brundtland report. Designed to ensure academic rigor within the discipline, the silo-mentality of individual discipline within the academic fields is part of the flaw. It did ensure that curriculum and learning outcomes met the needs of the respective professional organizations and the wider community that they served. However, silo-mentality hinders progress in transdisciplinary fields such as sustainability that are problem driven and solution oriented (Koh and Teh, 2020). Sustainability solutions rely on the integration of diverse knowledge and multi-stakeholder action across multiple disciplines to solve complex, real-world problems (Wiek *et al.*, 2011). To overcome this dilemma, sustainability-focused universities require innovations in curriculum and pedagogy to achieve sustainability learning outcomes.

11.3   Sustainability Curriculum and Pedagogy

Sustainability-focused universities should have the capacity and the mandate to develop curricula and pedagogy that are problem driven and solution oriented. These universities should devise curricula and pedagogy that are effective in developing holistic, integrated and use-inspired knowledge. Knowledge per se is of little use if it cannot address learning outcomes. The goal is to deliver sustainability learning outcomes (SLOs) tailored to serve the society and the environment (Koh and Teh, 2021). However, delivery of sustainability-focused curriculum to undergraduate students can be problematic due to the silo-mentality of curriculum in traditional disciplines. Few entry-level courses focusing on sustainability that are broad enough to accommodate all disciplines are available to students. In promoting and supporting sustainability curriculum development, it is essential to develop a university-wide program to connect different disciplines working and teaching in sustainability areas. Since 2015, the University of Vermont has adopted a model in which SLO is a university-wide requirement and in which sustainability is embedded throughout the university curriculum and pedagogy (Hill and Wang, 2018). Since 1997, the University of British Columbia (UBC) adopted a sustainable development policy that encouraged sustainable practices in all aspects of its actions and that mandated all students be educated about sustainability (Moore *et al.*, 2005). UBC offers an entry-level, interdisciplinary sustainability course (SUST 101), available to all students across the university, regardless of faculty or year level. SUST 101 strives for interdisciplinary content, broad-based student enrolment, use of innovative pedagogies that promote sustainability behavior and dialogue both in teaching and content development. The gap between cognitive goals (skills and knowledge) and affective goals (love, community esteem and self-actualization) can be wide in an education system that focuses primarily on the attainment of cognitive achievement of knowledge and skills to fulfill the economic demand of human resources. ESD must effectively address this disconnect among knowledge, attitude and behavior for SDGs to be achieved. This will require a transformational and innovative paradigm shift in curriculum and pedagogy for ESD to narrow the gaps and to bridge the disconnect among the three pillars of sustainability.

11.4   UN Sustainable Development Solutions Network

There are only ten more years left to achieve a sustainable world by 2030, as targeted by Agenda 2030. Hence, in the coming decade, coordinated actions at local, national, regional and global levels need to be intensified to achieve the 17 SDGs and 169 targets as enshrined in Agenda 2030. The UN Sustainable Development Solutions Network (SDSN) was established in 2012 to work closely with UN agencies, multilateral financing institutions, the private sector and civil society to support coordinated implementation of the SDGs at local, national, regional and global levels. Currently, SDSN has 40 national and regional networks comprising over 1,000 universities, research centers and other knowledge institutions in different countries and regions (UNSDSN, 2020), including Malaysia (Figure 11.1). The Jeffrey Sachs Center on Sustainable Development (JSC) was established in 2016 in Sunway University, a collaboration between the Jeffrey Cheah Foundation and SDSN. JSC hosts the SDSN Malaysia Chapter with 19 member institutions from academia, government, civil society and the private sector. JSC is a regional center of excellence that advances SDG achievement in Malaysia and Southeast Asia by creating world-class programs to train a new generation of students, practitioners and policy leaders (Woo *et al.*, 2020). JSC actively develops linkages with major universities in Malaysia and around the world for solving real-life problems related to the SDGs. The JSC-SDSN collaboration was further enhanced through a gift of another USD 10 million from the Jeffrey Cheah Foundation to SDSN. This enhanced collaboration creates three new thrusts: (a) The ASEAN Green Deal, (b) Deep Decarbonization and (c) Mission 4.7. A new SDSN-Asia office will be set up at Sunway University in 2021 to coordinate SDSN activities of members and to promote inter-regional collaboration between SDSN-Asia members and SDSN members in the Americas, Europe and Africa. The SDG-Academy, the educational arm of SDSN, will be relocated to Kuala Lumpur as well. The SDG-Academy has just launched a high-quality 12-month online program known as "Global Master's Program on Sustainable Development" in collaboration with the University College Dublin and JSC. For this purpose, JSC has created a 12-month

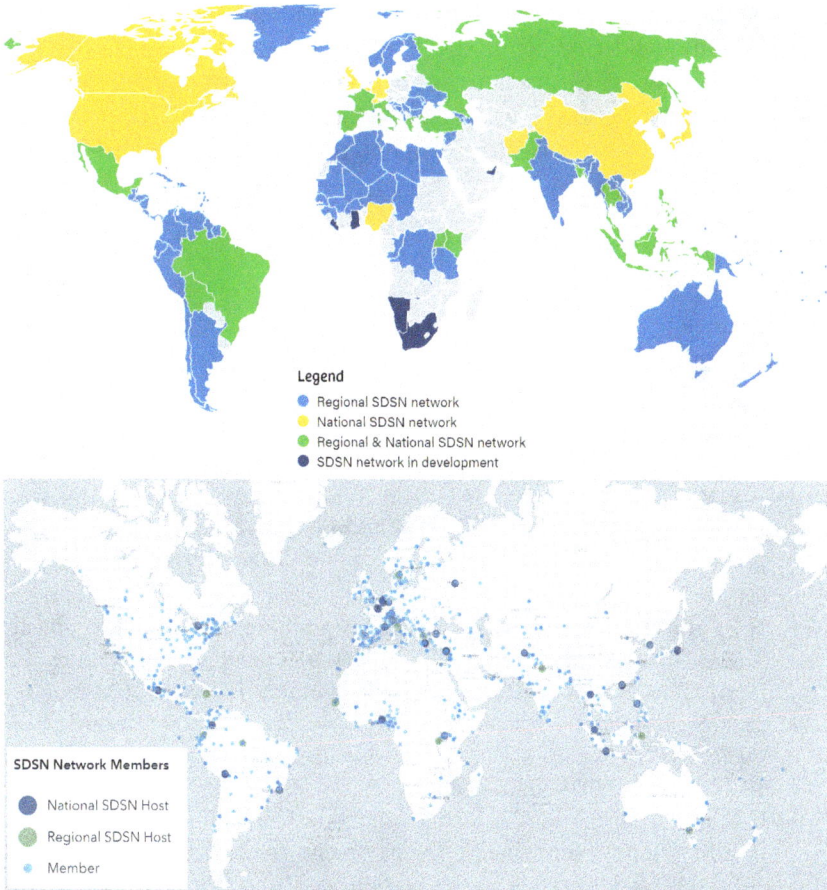

Figure 11.1: National and regional networks of SDSN (top) and member institutions denoted by red dots (bottom). Images adapted from UNSDSN (2020).

Open Distance Learning (ODL) version of its popular 18-month Masters in Sustainable Development Management (MSDM) program for the global community. The ODL version and the traditional version of MSDM now offer courses jointly taught by JSC faculty and experts from the SDG-Academy. The Mission 4.7 project of SDSN seeks to provide "Education for Sustainable Development and Global Citizenship". Mission 4.7 is a joint project among SDSN, UNESCO, the Ban Ki-moon Center for Global Citizens, Columbia University and Sunway University, with Tan Sri Jeffrey Cheah being the Co-Chair.

Mission 4.7 brings together national, regional and global leaders from government, academia, civil society and business to accelerate the implementation of Education for Sustainable Development around the world. This book serves to complement Mission 4.7 by inspiring education for sustainable development programs and action toward achieving UNSDGs.

References

Hill, L.M. and Wang, D. (2018). Integrating sustainability learning outcomes into a university curriculum: A case study of institutional dynamics. *International Journal of Sustainability in Higher Education*, 19(4), 699–720. doi: 10.1108/IJSHE-06-2017-0087.

Kh'ng, X.Y., Teh, S.Y., Koh, H.L. and Shuib, S. (2021). Sea Level Rise Undermines SDG2 and SDG6 in Pantai Acheh, Penang, Malaysia. *Journal of Coastal Conservation*, 25, 9. doi: 10.1007/s11852-021-00797-5.

Koh, H.L. and Teh, S.Y. (2020). University and community engagement: Toward transformational sustainability-focused problem solving. In: Leal Filho, W., Tortato, U., Frankenberger, F. (eds.) *Universities and Sustainable Communities: Meeting the Goals of the Agenda 2030*. World Sustainability Series. Springer, Cham, pp. 791–804. doi: 10.1007/978-3-030-30306-8_49.

Koh, H.L. and Teh, S.Y. (2021). Innovations in Curriculum and Pedagogy in Education for Sustainable Development. In: Leal Filho, W., Salvia, A.L., Frankenberger, F. (eds.) *Handbook on Teaching and Learning for Sustainable Development*. Edward Elgar Publishing (EEP), Cheltenham, UK, pp. 219–237.

Leal Filho, W., Manolas, E. and Pace, P. (2015). The future we want: Key issues on sustainable development in higher education after Rio and the UN decade of education for sustainable development. *International Journal of Sustainability in Higher Education*, 16(1), 112–129. doi: 10.1108/IJSHE-03-2014-0036.

Moore, J., Pagani, F., Quayle, M., Robinson, J., Sawada, B., Spiegelman, G. and VanWynsberghe, R. (2005). Recreating the university from within: Collaborative reflections on the University of British Columbia's engagement with sustainability. *International Journal of Sustainability in Higher Education*, 6(1), 65–80. doi 10.1108/14676370510573140.

Wals, A.E.J. (2012). Shaping the Education of Tomorrow: 2012 Full-Length Report on the UN Decade of Education for Sustainable Development. Paris: UNESCO.

WCED (1987). Our Common Future — The Bruntland Report. World Commission on Environment and Development, Oxford University Press, Oxford.

Wiek, A., Withycombe, L., Redman, C.L. and Mills, S.B. (2011). Moving forward on competencies in sustainability. *Environment: Science and Policy for Sustainable Development*, 53(2), 3–13. doi: 10.1080/00139157.2011.554496

Woo, W.T., Koh, H.L. and the, S.Y. (2020). Achieving Excellence in Sustainable Development Goals in Sunway University Malaysia. In: Leal Filho, W. *et al.* (eds.) Universities as Living Labs for Sustainable Development. World Sustainability Series. Springer, Cham, pp. 265–282. doi: 10.1007/978-3-030-15604-6_17.

UNSDSN (2020). Sustainable Development Solutions Network: A Global Initiative for the United Nations. Available at https://www.unsdsn.org/networks-overview (accessed on 10 December 2020).

Index